Robert A. Nelson

Mobile Communications
Engineering

Other McGraw-Hill Telecommunications Books of Interest

Mobile Communications Engineering

Theory and Applications

William C. Y. Lee
Vice President
Strategic Technology
Airtouch Communications, Inc.
Walnut Creek, California

Second Edition

McGraw-Hill

New York   San Francisco   Washington, D.C.   Auckland   Bogotá
Caracas   Lisbon   London   Madrid   Mexico City   Milan
Montreal   New Delhi   San Juan   Singapore
Sydney   Tokyo   Toronto

Library of Congress Cataloging-in-Publication Data

Lee, William C. Y.
    Mobile communications engineering / William C. Y. Lee. — 2nd ed.
      p.   cm.
    Includes index.
    ISBN 0-07-037103-2
    1. Mobile communication systems.   I. Title.
    TK6570.M6L44   1997
    621.3845—dc21   97-30668
                                        CIP

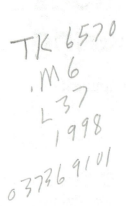
McGraw-Hill

A Division of The McGraw·Hill Companies

1 2 3 4 5 6 7 8 9 0 DOC/DOC 9 0 2 1 0 9 8 7

ISBN 0-07-037103-2

The sponsoring editor for this book was Steve Chapman, and the production supervisor was Pamela Pelton. It was set in Century Schoolbook by North Market Street Graphics.

Printed and bound by R. R. Donnelley & Sons Company.

McGraw-Hill books are available at special quantity discounts to use as premiums and sales promotions, or for use in corporate training programs. For more information, please write to the Director of Special Sales, McGraw-Hill, 11 West 19th Street, New York, NY 10011. Or contact your local bookstore.

 This book is printed on recycled, acid-free paper containing a minimum of 50% recycled de-inked fiber.

Contents

Preface

Students of Communications Engineering will find the material and analyses presented within this book helpful in extending their theoretical knowledge and understanding into the field of mobile-radio communication design. The theory supports the practical design and development of mobile-radio transmitters, receivers, and antenna systems operating in the VHF and UHF ranges and configured for high-capacity mobile telephone communication in both the commercial and military operational environments. The supervision, control, and switching elements of the mobile-radiotelephone network are only briefly discussed to clarify the reader's understanding of how mobile-radio equipment is affected by the operational constraints of the standard telecommunications network.

Portions of the material are extracted from published literature and from lecture notes developed in support of an in-house continuing education program (INCEP) for engineers first offered at Bell Telephone Laboratories in 1978.

Additional studies were conducted at the ITT Defense Communications Division to further document the requirements for mobile-radio design for tactical military applications. The introduction to this book will bridge the gap for students unfamiliar with mobile-radio communications and will serve as a comprehensive review for engineers entering the field of advanced mobile-radio communications. The design equations and associated design theory will be particularly useful to mobile-radio system designers.

Part I of the introduction chronicles the development of mobile-radio communication from the initial experiments of the German physicist Heinrich Rudolph Hertz in 1880 to present-day advances in mobile-radio design. Part II of the introduction presents an overview of the services, transmission criteria, and switching plan for the Advanced Mobile Phone Service (AMPS), an advanced concept in cellular network design. Cellular systems increase the traffic capacity within existing allocated regions of the radio spectrum through the reuse of assigned frequency channels. The assignment of channel frequencies is made on the basis of bandwidth, discretionary use in accordance with a special distribution matrix, and optimal use of diversity techniques for improved coverage of specified geographic areas (cells). In Part III of the introduction, the basic differences between mobile-radio systems designed essentially for military use and those intended for the commercial market are summarized.

Chapter 1 discusses the general characteristics of the mobile-to-fixed radio communication medium and its effects on mobile-radio performance, including propagation loss, multipath fading, and time-delay spread.

Chapter 2 presents tutorial material covering the general principles of statistical communications theory. These fundamental principles will be referenced in other chapters and will minimize the reader's dependence upon other reference sources.

Chapters 3, 4, and 5 deal with new concepts of mobile-radio propagation from the viewpoint of the long-term signal effects resulting from such factors as antenna height, topography, surface textures, the slope of path loss for hilly areas, and the type of antenna and RF design.

Chapters 6 and 7 describe the characteristics of the received signal envelope and phase, respectively. These are defined as short-term signal effects. Chapters 6 and 7 conclude the prerequisite studies for the functional design of a mobile-radio communication system and prepare the reader for the material contained in Chapters 8 through 14.

Chapters 8 through 11 discuss the functional design parameters for mobile-radio communication systems. Chapter 8 describes various types of modulation, including spread-spectrum systems; Chap. 9 analyzes the effects of different types of diversity schemes; Chap. 10 examines the different types of combining techniques; and Chap. 11 discusses signaling aspects and voice companding techniques.

Chapter 12 examines the effects of cochannel and adjacent channel interference, which in part determine the parameters for frequency reuse and adjacent-channel frequency assignments. Other types of interferences, such as near-end to far-end ratio and intermodulation and intersymbol interferences, are also described.

Chapters 13 and 14 describe various methods for analyzing signal error and voice quality in terms of the criteria prescribed for different systems.

Chapter 18 generally describes military mobile communications, which require strategies applicable in a jamming environment. The spread-spectrum, coding, and diversity schemes and adaptive antenna-nulling techniques are analyzed.

The Second Edition

This book is intended to be used as a textbook for engineering students in their senior year, and for those in their first year of graduate school. The reader will find a set of work problems at the end of each chapter that can be used to verify his or her understanding of the material covered in that chapter.

Written in 1982, this was my first published book. Since then, I have written and revised two other textbooks. I revised the other two books due to the rapid growth in the wireless communication field. Mr. Steve Chapman of The McGraw-Hill Companies and my students at George Washington University Continuous Education program have kept asking me to revise this book. This book was used in graduate schools as a textbook.

Since then, a lot of material needed to be added. It took me much longer to revise than I expected. I have added three chapters to enrich this book. Most of the material in these three chapters is from my own research.

In Chap. 15, multiple access schemes are addressed, as is the comparison of various multiple access schemes for wireless communications. Also, the multiple access schemes for virtual channels are described. Since the mobile satellite systems use different multiple access, the newly developed LEO (low earth orbit) satellite systems are included in this chapter.

In Chap. 16, the clarification of the concepts on many sensitive topics are for the readers to enjoy in thinking differently.

In Chap. 17, topics with new concepts are included. They are important to update our knowledge.

I also added materials in each chapter to update its content.

This book is a complement to my other two books on the wireless communication field. The other two books are *Mobile Cellular Telecommunications: Analog and Digital Systems,* 2d ed., McGraw-Hill, 1995, and *Mobile Communications Design Fundamentals,* 2d ed., Wiley & Sons, 1993. Among the three books, this book carries more theoretical and mathematic derivation and offers tools that are used more with the applications. Therefore, this book has been renamed *Mobile Communications Engineering: Theory and Applications* to be more suitable to the content. I hope this book can generate more new thoughts and ideas and provide the tools to solve more upcoming problems.

William C. Y. Lee

Acknowledgments

The first edition of this book was published in 1982. After 15 years, I did not realize the time it would take to do a revision. The revision process was very much like altering a dress. The effort in cutting and patching an old dress can become very time-consuming. I deleted some out-of-date material and added a great deal of new material, since the content and structure of the first edition was good. Fortunately, I had an excellent staff to assist me with the revision. I would like to extend my thanks to Mrs. Carla Scherbert, Ms. Atyeh Sohrabzadeh, Mr. Walter Strohmayer, and Ms. Maribeth Eisenmann. Also, for the full support from my colleagues, my students, my readers, and my family, I owe my deepest appreciation. I am pleased with this revised edition, and I think it will serve my readers better than an entirely new volume.

Acknowledgments for the First Edition

The author is deeply indebted to a number of individuals who helped make this book possible.

During the period when the book was being written, inspiration and encouragement were received constantly from Dr. F. H. Blecher and Dr. C. C. Culter of Bell Laboratories. A special thanks to Dr. Blecher who advised me to write this book based on my lecture notes.

Appreciation is owed to Mr. G. E. Huffman, Dr. M. R. Sambur, and Mr. P. Torrione of the ITT Defense Communications Division for their full support on this project, to Mr. R. A. Coleman for his thorough editing, and to Mrs. M. Erdos and Miss L. Rykowski for their help in preparation of the manuscript.

Credit must also be given to my former colleagues in the Bell Laboratories, and my present colleagues at ITT. Many of these individuals stimulated and encouraged the project.

Finally, special recognition is hereby expressed to my wife, Margaret, for her constant understanding and encouragement, to my children, Betty and Lily, for their consideration in giving up so many weekends and evenings of our family time, and to my parents and my in-laws for their commitment to education and guidance for many years that resulted in this book.

Mobile Communications
Engineering

Introduction

Part I Mobile Radio—The First 100 Years

By definition, the term "mobile-radio communications" describes any radio communication link between two terminals of which one or both are in motion or halted at unspecified locations and of which one may actually be a fixed terminal such as a base station. This definition applies to both mobile-to-mobile and mobile-to-fixed radio communication links. The mobile-to-mobile link could in fact consist of a mobile-to-fixed-to-mobile radio communication link. The term "mobile" applies to land vehicles, ships at sea, aircraft, and communications satellites. In tactical situations, mobile-radio systems may include any or all of these types of mobile terminals.

Mobile-radio systems are classified as radiophones, dispatching systems, radio paging systems, packet radios, or radiotelephones (also known as mobile phones), including train phones.

1. Radiophones (or walkie-talkies) are two-way radios, such as CB (citizens band) radios, which are allocated 40 channels for anyone to use whenever the channels are free. This system affords no privacy to the user.

2. Dispatching systems use a common channel. Any vehicle driver can hear the operator's messages to other drivers in the same fleet. The drivers can talk only to the control operator. In military applications, the users can also talk to each other on an open channel.

3. Radio paging customers carry personal receivers (portable radios). Each unit reacts only to signals addressed to it by an operator. A beep sounds to alert the bearer, who then must go to a nearby telephone to receive the message.

4. Packet radio requires a form of multiple-access control that permits many scattered devices to transmit on the same radio channel without interfering with each other's transmissions. Packet radios can be configured as either mobile or portable terminals. This system may become important in the future.* Each terminal is attached to a transmission control unit equipped with a radio transmitter and receiver. The data to be transmitted are formed into a "packet" within the transmission control unit. The packet contains the addresses of the receiving location and the originating terminal. A receiving device receives any packet addressed to it and transmits an acknowledgment if the packet appears to be free of error. The sending station waits a predetermined period for the acknowledgment. If it does not receive an acknowledgment, it transmits the packet again. For example, CDPD (cellular digital packet data) is a packet radio system.[†]

5. Radiotelephones include MTS (Mobile Telephone Service), IMTS (Improved Mobile Telephone Service), the Metroliner telephone, TACS (total access communication system),[†] and AMPS (Advanced Mobile Phone Service).[†] The Metroliner telephone is briefly discussed here, and the other types of radiotelephones are described, in greater detail, in subsequent paragraphs. The Metroliner telephone operates in the 400-MHz frequency range on the high-speed train between New York and Washington, D.C. The 225-mi railway distance is divided into nine zones. Each zone has a fixed radio transceiver located adjacent to the track right-of-way. As a train moves from one zone to another, calls that are in progress must be automatically switched from one fixed radio transceiver to the next without the customer's being aware of any changes or interference in communication.

6. Digital Cellular[†] and PSC (personal communication service)[†] are for high capacity and data transmission. Digital Cellular in Europe is called GSM (Global System Mobile), a standard system using TDMA (time division multiple access). PCS is a cellular-like system applied at 1.8–1.9 GHz instead of 800–900 MHz for cellular. Other than that, the system protocols are the same as cellular systems. Digital cellular in North America has two standards: IS-136 (TDMA) and IS-95 (CDMA). Digital cellular in Japan is called PDC (personal digital phone), a TDMA system.

7. TDD (Time Division Duplexing) systems[†] such as DECT (digital European cordless telephone), PHS (personal handy-phone system),

---

* S. Fralick and J. Garrett, "Technology for Packet Radio," *AFIPS Conf. Proc.,* vol. 44, 1975, AFIPS, Montvale, N.J.; and R. E. Kahn, S. A. Gronemeyer, J. Burchfield, and R. C. Kunzelman, "Advances in Packet Radio Technology," *Proc. IEEE,* vol. 66, no. 11, November 1978, pp. 1468–1496.

† W. C. Y. Lee, *Mobile Cellular Telecommunications, Analog and Digital Systems,* 2d ed. McGraw Hill Co., 1995.

and PACS (personal access communication system) use one frequency for both transmission and reception on a time-sharing basis. These systems are for low mobility and in-building communications.

8. Mobile Broadband Systems* will be the future public land mobile telecommunication system (FPLMTS). It will operate at a higher spectrum band (20–60 GHz), using ATM (asynchronous transfer mode) for broadband packet switching, and it will be compatible with the B-ISDN (broadband ISDN). It will be the future wireless information superhighway system.

Let's pause momentarily to review some of the historical highlights of mobile-radio communication. The first practical use of mobile-radio communication was demonstrated in 1897 by Marchese Guglielmo Marconi, who is credited with first successfully establishing radio transmission between a land-based station and a tugboat, over an 18-mi path. The following summary shows some of the important milestones in the history of mobile-radio communication:

1880: Hertz—Initial demonstration of practical radio communication

1897: Marconi—Radio transmission to a tugboat over an 18-mi path

1921: Detroit Police Department—Police car radio dispatch (2-MHz frequency band)

1932: New York Police Department—Police car radio dispatch (2-MHz frequency band)

1933: FCC—Authorized four channels in the 30- to 40-MHz range

1938: FCC—Ruled for regular service

1946: Bell Telephone Laboratories—152 MHz (simplex)

1956: FCC—450 MHz (simplex)

1959: Bell Telephone Laboratories—Suggested 32-MHz band for high-capacity mobile-radio communication

1964: FCC—152 MHz (full duplex)

1964: Bell Telephone Laboratories—Active research at 800 MHz

1969: FCC—450 MHz (full duplex)

1974: FCC—40-MHz bandwidth allocation in the 800- to 900-MHz range

1981: FCC—Release of cellular land mobile phone service in the 40-MHz bandwidth in the 800- to 900-MHz range for commercial operation

---

* W. C. Y. Lee, *Mobile Cellular Telecommunications, Analog and Digital Systems,* 2d ed. McGraw Hill Co., 1995.

1981: AT&T and RCC (Radio Common Carrier) reach an agreement to split 40-MHz spectrum into two 20-MHz bands. Band A belongs to nonwireline operators (RCC), and Band B belongs to wireline operators (telephone companies). Each market has two operators.

1982: AT&T is divested, and seven RBOCs (Regional Bell Operating Companies) are formed to manage the cellular operations.

1982: MFJ (modified final judgment) is issued by the government DOJ. All the operators were prohibited to (1) operate long-distance business, (2) provide information services, and (3) do manufacturing business.

1983: Ameritech system in operation in Chicago

1984: Most RBOC markets in operation

1986: FCC allocates 5 MHz in extended band

1987: FCC makes lottery on the small MSA and all RSA licenses

1988: TDMA voted as a digital cellular standard in North America

1992: GSM operable in Germany D2 system

1993: CDMA voted as another digital cellular standard in North America

1994: American TDMA operable in Seattle, Washington

1994: PDC operable in Tokyo, Japan

1994: Two of six broadband PCS license bands in auction

1995: CDMA operable in Hong Kong

1996: U.S. Congress passes Telecommunication Reform Act Bill. "Apparently anyone can get into anyone else's business."

1996: The auction money for six broadband PCS licensed bands (120 MHz) almost reaches 20 billion U.S. dollars.

1997: Broadband CDMA considered as one of the third-generation mobile communication technologies for UMTS (universal mobile telecommunication systems) during the UMTS workshop conference held in Korea.

In 1970, the FCC allocated the following frequencies for domestic public mobile-radio use on land:

Base transmit	Mobile transmit	Number of channels	Channel spacing	Total bandwidth	Name
35.26–35.66 MHz	43.26–43.66 MHz	10	40 kHz	0.8 MHz	MTS
152.51–152.81 MHz	157.77–158.07 MHz	11	30 kHz	0.6 MHz	IMTS (MJ)
454.375–454.65 MHz	459.375–459.65 MHz	12	25 kHz	0.55 MHz	IMTS (MK)

Although a total of 33 channels are provided for within these frequency allocations, the actual number of channels used in a specified area is much smaller on account of restrictions imposed by prevailing FCC regulations, which are explained later on.

In 1974, the FCC allocated a 40-MHz bandwidth in the 800- to 900-MHz frequency region for mobile telephone use. During the initial trial tests, it was utilized as follows:

Base transmit	Mobile transmit	Number of channels	Channel spacing	Total bandwidth
870–890 MHz	825–845 MHz	666	30 kHz	40 MHz

The Bell Telephone System used this narrow band of frequencies in trial tests of its new, high-capacity Advanced Mobile Phone Service (AMPS), which is designed for use in a cellular planned network. Because the system design is based on the reuse of allocated frequencies, the number of customers served is greatly increased; hence the term "high-capacity" system. Part II of this introduction describes a cellular system in greater detail.

By 1976, the Bell System served approximately 40,000 mobile-telephone customers within the United States. Of this number, 22,000 were able to dial directly, whereas 18,000 required operator assistance to place a call. The various systems that serve mobile radiotelephones are classified according to their assigned frequency range. For example, the MJ system operates in the 150-MHz range, whereas the MK system operates in the 450-MHz range. Each system can provide from 1 to as many as 12 channels, with FCC regulations requiring that 12 channels of an MK system serve an area of 50 miles in diameter. To illustrate how few channels are available and how overloaded they are, in 1976 the New York Telephone Company (NYTC) operated 6 channels of the MJ system serving 318 New York City mobile-telephone subscribers, approximately 53 customers per channel, and there were 2400 applicants wait-listed for MJ mobile-telephone service. NYTC also operated six MK channels serving 225 customers, approximately 38 customers per channel, and 1300 applicants were wait-listed for MK mobile-telephone service. New York City was limited to only six MK channels out of the maximum of 12 available because of the FCC regulation requiring that 12 channels serve an area of 50 miles in diameter.

In 1976, there were a total of 1327 mobile-telephone systems in operation across the United States. The Bell System operated 637 mobile-telephone systems within its coast-to-coast network, whereas 690 were operated by independent telephone companies. The market demand for mobile-telephone service is already much greater than the existing available supply and is increasing very rapidly because of the great, undisputed popularity of CB radio, which is very busy and congested, with only 40 assigned operating channels (26.96 to 27.41 MHz). When the new, cellular mobile-radio systems are fully operational across the United States, high-capacity direct-dialing service at reasonable cost will entice large numbers of CB radio users to subscribe to mobile radiotelephone. The obvious advantages of mobile radiotelephone over the heavily saturated CB radio channels are:

1. Direct-dialing features equivalent to those offered to fixed-telephone subscribers

2. Absolute privacy of communication, with greatly improved quality

3. An extended range of communication utilizing the total switching resources of the commercial telephone networks

4. A theoretically unlimited number of communication channels that can be provided

In this book, the theory and analyses are aimed at the mobile-to-fixed radio communication links that are designed to fit the cellular requirements of the VHF and UHF mobile-radiotelephone systems of the 1980s and that operate in the 30-MHz to 1-GHz mobile-radio frequency ranges. For systems operating above 1 GHz, atmospheric conditions such as moisture and climatic effects must be taken into consideration. These effects are minimal at operating frequencies below 1 GHz. Below 30 MHz, path loss and signal fading are not severe; but since there are few mobile-radio frequency allocations in this region of the radio spectrum, the primary emphasis of this book is on the design of 30-MHz to 1-GHz mobile radio.

Looking toward the future, the portable telephone and ultimately a pocket telephone are potential product designs that will share mobile-radiotelephone transmission facilities. Some of the major problems that must be solved before these designs are realized are the limitations of battery size, weight, and power capacity; radiation safety hazards to the user; and signal interference problems unique to the portable-telephone user's environment.

Part II   Cellular Network Planning

The future of mobile-radiotelephone communication is dependent upon techniques of network planning and mobile-radio equipment design that will enable efficient and economical use of the radio spectrum. One possible solution to the problem of meeting the steadily increasing customer demand for mobile-radiotelephone service, within the limitations of available FCC frequency allocations, is to develop a workable plan for reusing the assigned channels within each band of frequencies. To encourage the mobile-radiotelephone industry in its development of advanced high-capacity systems, the FCC in 1974 allocated a 40-MHz bandwidth in the 800- to 900-MHz frequency range for this purpose. Subsequent design research and trial tests conducted by the Bell Telephone Laboratories concluded that high-capacity systems based on the reuse of assigned channel frequencies in a cellular planned network are a practical solution. The system evolving from this work is known as the Advanced Mobile Phone Service (AMPS), and its functional capabilities are described in the following paragraphs.

AMPS service features*

In describing the service features of the AMPS, our primary area of interest is that of land mobile telephone service, which includes all of the features ascribed to normal telephone service to the extent that such services are compatible with the special characteristics of the mobile environment. This does not preclude the AMPS from providing other mobile-radio services, such as those services associated with direct dispatch, air-to-ground, and other types. Land mobile telephone service is offered as a subscriber service for privately owned vehicles, and as a public telephone service on commercial ground carriers such as buses and trains. We know from past experience that the special characteristics of the mobile-radio environment can have an adverse effect upon radio propagation, and consequently can affect the quality of the services provided. It is therefore essential to know the cause, extent, and methods for minimizing these effects in order to improve the quality and reliability of mobile-radiotelephone communication. The effects of the mobile environment on mobile-radio performance are further examined in later chapters covering the theory of functional design.

Radio enhancement techniques   As previously mentioned, the FCC has allocated a 40-MHz bandwidth in the 800- to 900-MHz frequency range for high-capacity mobile radiotelephone service. On the basis of the concept of cellular network planning, the 40-MHz bandwidth is separated into a 20-MHz base-station transmit band in the 870- to 890-MHz range and a mobile-radio 20-MHz transmit band in the 825- to 845-MHz range. The total 40-MHz bandwidth is further subdivided into 666 two-way channels, each channel consisting of two frequencies having chan-

---

* "Advanced Mobile Phone Service," special issue, *Bell System Technical Journal,* vol. 58, January 1979.

nel bandwidths of 30 kHz each. To enable frequency separation between channels within a given area, the 666 channels are arranged for two operators in the form of a distribution matrix, as illustrated in Fig. I.1. In Fig. I.1, the Block A and Block B operators have 333 channels each. Among 333 channels, 21 channels indicated in Fig. I.1 are the setup channels. The matrix can be considered to be 21 sets of channels. To simplify distribution, the 21 sets are arranged into 3 groups of 7 and assigned suffix letters A, B, and C, respectively. The distribution of channels and channel frequencies obtained by this arrangement ensures that assignments within one geographic cell area will not interfere with channels assigned in adjacent cell locations. Cells that are separated by a minimum distance determined by propagation variables can simultaneously use the same channels with no risk of interference. The sample cell structure shown in Fig. I.1 illustrates the method for assigning channels among contiguous cell locations.

A system operator serving a particular population center, such as a major city and its surrounding suburban communities, could provide mobile-radio coverage to large numbers of users based on cellular reuse of assigned channel frequencies. The basic cell structure is conceptually hexagonal in shape and can vary in size according to the number

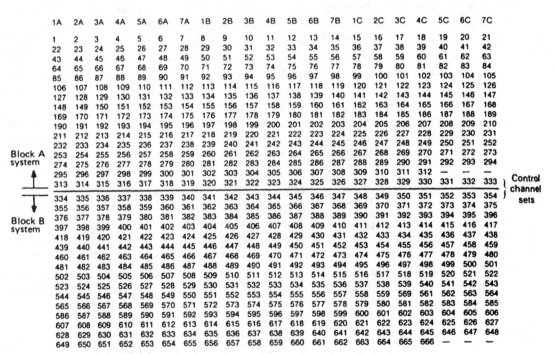

Figure I.1   Frequency-management chart.

of channels, traffic variables, and the effectiveness of propagation-enhancement techniques. For purposes of explanation, we will temporarily disregard cell size. A typical area divided into cells is shown in Fig. I.2. Each block of seven cells is repeated in such a manner that corresponding numbered cells in adjacent seven-cell blocks are located at a predetermined distance from the nearest cell having the same number. Correspondingly, the 20-MHz-bandwidth radio spectrum is divided into seven disjoint sets, with a different set allocated to each one of the seven cells in the basic block. With a total of 333 channels in 21 sets available, it is possible to assign as many as three sets to each of the seven cells constituting the basic block pattern.

For blanket coverage of cell areas, each cell site is installed at the center of the cell (the dotted-line cell) and covers the whole cell, as shown in Fig. I.3. There is another way of looking at the locations of the cell sites. The three cell sites are installed, one at each alternate corner of the cell and cover the whole cell, as shown in Fig. I.3. In both cases, although the boundary of a cell is defined differently, the cell sites do not need to be moved. For convenience, the cells illustrated in Fig. I.3 are pictured as hexagonal in shape. In actual practice, the cell boundaries are defined by the minimum required signal strength at distances determined by the reception threshold limits. In the AMPS, base stations are referred to as cell sites because they perform supervision and control in addition to the transmitting and receiving functions normally associated with the conventional base station. Mobile-telephone subscribers within a given cell are assigned to a particular cell site serving that cell simply by the

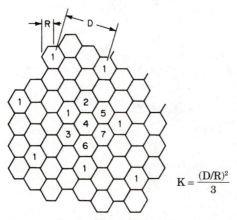

$$K = \frac{(D/R)^2}{3}$$

Figure I.2   Basic cell block: R = radius of each cell; D = distance between two adjacent frequency-reuse cells; K = number of cells in a basic cell block. $K = 7$ in this illustration, and $D/R = 4.6$.

assignment of an idle channel frequency under the control of the mobile-telephone switching office (MTSO). When a mobile unit crosses a cell boundary, as determined by the signal reception threshold limits, a new idle channel frequency is assigned by the new serving cell site. This automatic switching control function is referred to as a "handoff."

The problems of cochannel interference are avoided by ensuring a minimum distance between base stations using the same channel frequencies, and by enhancing signal level and reducing signal fading through the use of diversity schemes. These constraints limit any potential cochannel interference to levels low enough to be compatible with the transmission quality of landline networks.

Two forms of diversity are used to enhance radio propagation, thus improving AMPS cell coverage. These are defined as "macroscopic" and "microscopic" diversity. Macroscopic diversity compensates for large-scale variations in the received signal resulting from obstacles and large deviations in terrain profile between the cell site and the mobile-telephone subscriber. Macroscopic diversity is obtained by installing directional antennas, one for each sector of three sectors at the cell center, or installing at the alternate corners of cells, as shown in Fig. I.3, and transferring control to the antenna providing the strongest average signal from the mobile subscriber in any given time interval. For example, the three cell-site transmitters serving a particular cell area would not radiate simultaneously on an assigned channel frequency. On the basis of a computer analysis of the signals received from the mobile subscriber at each of the three sites, the one with the strongest

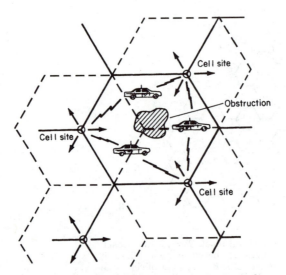

Figure I.3 Use of inward-directed antennas at alternate cell corners to achieve macroscopic diversity with respect to large obstructions.

signal would be selected for use as the serving cell site. Periodic analysis of channels in use would determine the necessity for handoff to a new sector in a cell or an alternate cell site within the cell area (intrahandoff), or handoff to a cell site in an adjacent cell area (interhandoff). All of these decisions would be made automatically without the knowledge or intervention of the user or the operator, and without interruption of the call in progress.

Microscopic diversity compensates for fast variations in the received signal resulting from multipath fading. Microscopic diversity is obtained by receiving dual inputs at both the mobile and cell-site receivers. These dual inputs can be two different frequencies, time slots, antennas, polarizations, etc. The diversity schemes associated with the combining techniques are described in subsequent chapters of this book.

Switch planning The cellular mobile-radiotelephone system can be expected to accommodate the growth of new subscribers in two ways. First, not all of the channels allocated to a cell are initially placed into service. As the numbers of mobile subscribers increase and the traffic intensity increases, transmission facilities for the additional channels are modularly expanded to keep pace with the demand. Second, as the number of channels per cell site approaches the maximum within the channel allocation plan, the area of individual cells can be reduced, thus permitting more cells to be created with less physical separation but with increased reuse of assigned channel frequencies. This reconfiguration of the cellular network permits the same number of assigned channels to adequately serve greater numbers of mobile units within a greater number of smaller cells. The ideal, customized cellular network would not be uniformly divided into cells of equal size but would contain cells of different sizes based on the density of mobile units within the various cell coverage areas. The concept of variable cell size is illustrated in Fig. I.4.

The interface between land mobile units and the commercial telephone landline network is illustrated in Fig. I.5. A call originating from or terminating at a mobile unit is serviced by a cell site connected via landlines to a mobile-telephone switching office (MTSO). The MTSO provides call supervision and control, and extends call access to a commercial telephone landline network via a local central-office (CO) telephone exchange, a toll office, and any number of tandem offices required to complete the call path. The terminating central office completes the connection to the called subscriber at the distant location. Two types of mobile-radio channels are used in setting up a call: paging channels and communication channels. The mobile unit is designed to automatically tune to the strongest paging channel in its local area for continuous monitoring, and to automatically switch to another paging channel when approaching the threshold transition level of reception.

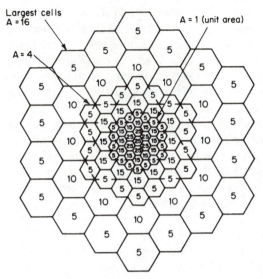

Figure I.4 Tailoring a cell plan to a severe density gradient. Maximum of 10 channels per cell in largest cells = 0.625 (= $^{10}\!/_{16}$) voicepaths per unit area. Maximum of 15 channels per cell in medium cells = 3.75 (= $^{15}\!/_{4}$) voicepaths per unit area. Maximum of 25 channels per cell in smallest cells = 25 (= $^{25}\!/_{1}$) voicepaths per unit area.

The following is a typical call sequence for processing a call from a fixed-telephone subscriber to a mobile-telephone subscriber. The call from the fixed-telephone subscriber is originated in the normal manner of direct dialing under control of the local telephone exchange. The telephone switching network translates the called number and routes the call to the MTSO in the cell area closest to the called mobile-telephone subscriber. The terminating MTSO determines whether the number called is busy or available. If it is busy, the MTSO causes a busy signal to be returned to the calling party. If the mobile subscriber's telephone is available, the called number is broadcast over all paging channels assigned to cell sites in that area. The called mobile unit automatically recognizes its number and responds by acknowledgment over the corresponding paging-channel frequency. On the basis of the paging-channel response, the MTSO will identify the serving cell site and automatically switch the mobile unit to an idle communication channel from among the channels allocated to that serving cell site. The MTSO, after selecting an idle communication channel, causes the mobile unit to tune to that channel by means of a data command over it. The incoming call is connected to the appropriate circuit serving the mobile unit, and a ringing signal is sent to the mobile unit.

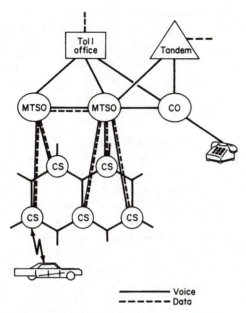

Figure I.5 System block diagram: MTSO = mobile-telephone switching office; CO = central office (telephone exchange); CS = cell site.

Each cell site has a locating receiver which monitors all active channels in discrete time intervals. After having served as the central office in completing the call setup, the MTSO continues to access the serving cell's locating receiver, thereby monitoring the mobile-radio transmissions at prescribed intervals. Should the received average signal strength drop below the prescribed level in any given time interval, the MTSO will automatically, and without interruption, switch the call to whatever idle channel at any cell site serving the mobile unit has the strongest received signal above the prescribed level. This is the handoff process, and it can be performed within the same cell or in a new cell.

In the processing of a call originating from a mobile unit, the mobile unit's telephone in going off the hook signals a request for service over the paging channel chosen by its receiver. After the MTSO identifies the location of the serving cell site, an idle communication channel is assigned to the mobile unit via that cell site and a dial tone is returned to the mobile user. The MTSO will perform the services of the central office in setting up the call and will continue to monitor the received signal strength of the mobile-radio transmissions from the locating receiver while the call is in progress, performing handoff whenever required.

In summary, the mobile-radiotelephone user is never involved in or even aware of the MTSO channel assignment and handoff processes, which are performed automatically under control of the MTSO processor. In addition to serving as the central office for all calls originating and terminating at mobile units, the MTSO controls and supervises the assignment of all communication channels and the quality of radio transmission over assigned channels at multiple numbers of cell sites. The MTSO also communicates with other MTSO facilities in much the same manner as one central office communicates with other central offices.

Mobile-telephone channels are uniquely different in many respects from conventional line circuits. Mobile-telephone channels are treated as trunks by the MTSO and thus appear on the trunk side of the switching matrix serving the various cell sites. The concept of a typical mobile-radio communications network is shown in Fig. I.5; however, in actual practice, there are many more cell sites under the control of each MTSO and many more MTSO facilities serving the mobile-telephone community. The existing commercial fixed-telephone network overlies the mobile-radiotelephone cellular network; the primary interface is via the MTSO trunking facilities.

Extended spectrum

FCC has allowed an additional 10 MHz for two operators. Each one has 5 MHz, i.e., 2.5 MHz one way. The total number of channels for each operator is 83. Therefore, the total voice channels increase to 395. The new channel-numbering scheme is shown in Fig. I.6. The new frequency management for band A (or block A) and band B (or block B) are shown in Tables I.1 and I.2, respectively.

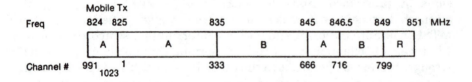

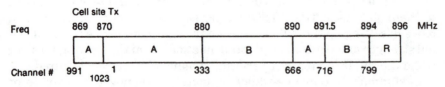

Figure I.6 New additional spectrum allocation.

TABLE I.1 New Frequency Management (Full Spectrum), Block A

1A	2A	3A	4A	5A	6A	7A	1B	2B	3B	4B	5B	6B	7B	1C	2C	3C	4C	5C	6C	7C
1	2	3	4	5	6	7	8	9	10	11	12	13	14	15	16	17	18	19	20	21
22	23	24	25	26	27	28	29	30	31	32	33	34	35	36	37	38	39	40	41	42
43	44	45	46	47	48	49	50	51	52	53	54	55	56	57	58	59	60	61	62	63
64	65	66	67	68	69	70	71	72	73	74	75	76	77	78	79	80	81	82	83	84
85	86	87	88	89	90												102	103	104	105
106	107	108	109	110	111												123	124	125	126
127	128	129	130	131	132												144	145	146	147
148	149	150	151	152	153												165	166	167	168
169	170	171	172	173	174												186	187	188	189
190	191	192	193	194	195												207	208	209	210
211	212	213	214	215	216												228	229	230	231
232	233	234	235	236	237												249	250	251	252
253	254	255	256	257	258												270	271	272	273
274	275	276	277	278	279												291	292	293	294
295	296	297	298	299	300												312	X	X	X
313*	**314**	**315**	**316**	**317**	**318**	**319**	**320**	**321**	**322**	**323**	**324**	**325**	**326**	**327**	**328**	**329**	**330**	**331**	**332**	**333**
667	668	669	670	671	672	673	674	675	676	677	678	679	680	681	682	683	684	685	686	687
688	689	690	691	692	693	694	695	696	697	698	699	700	701	702	703	704	705	706	707	708
709	710	711	712	713	714	715	716	X	991	992	993	994	995	996	997	998	999	1000	1001	1002
1003	1004	1005	1006	1007	1008	1009	1010	1011	1012	1013	1014	1015	1016	1017	1018	1019	1020	1021	1022	1023

* Boldface numbers indicate 21 control channels for Block A and Block B, respectively.

TABLE I.2 New Frequency Management (Full Spectrum), Block B

1A	2A	3A	4A	5A	6A	7A	1B	2B	3B	4B	5B	6B	7B	1C	2C	3C	4C	5C	6C	7C
334*	**335**	**336**	**337**	**338**	**339**	**340**	**341**	**342**	**343**	**344**	**345**	**346**	**347**	**348**	**349**	**350**	**351**	**352**	**353**	**354**
355	356	357	358	359	360	361														375
376	377	378	379	380	381	382														396
397	398	399	400	401	402	403														417
418	419	420	421	422	423	424														438
439	440	441	442	443	444	445														459
460	461	462	463	464	465	466														480
481	482	483	484	485	486	487														501
502	503	504	505	506	507	508														522
523	524	525	526	527	528	529														543
544	545	546	547	548	549	550														564
565	566	567	568	569	570	571														585
586	587	588	589	590	591	592	593	594	595	596	597	598	599	600	601	602	603	604	605	606
607	608	609	610	611	612	613	614	615	616	617	618	619	620	621	622	623	624	625	626	627
628	629	630	631	632	633	634	635	636	637	638	639	640	641	642	643	644	645	646	647	648
649	650	651	652	653	654	655	656	657	658	659	660	661	662	663	664	665	666	717	718	719
720	721	722	723	724	725	726	727	728	729	730	731	732	733	734	735	736	737	738	739	740
741	742	743	744	745	746	747	748	749	750	751	752	753	754	755	756	757	758	759	760	761
762	763	764	765	766	767	768	769	770	771	772	773	774	775	776	777	778	779	780	781	782
783	784	785	786	787	788	789	790	791	792	793	794	795	796	797	798	799				

(Cellular reuse diagram with cells labeled 3A/3B/3C, 4A/4B/4C, 5A/5B/5C, 6A/6B/6C, 7A/7B/7C.)

* Boldface numbers indicate 21 control channels for Block A and Block B, respectively.

16

Digital cellular

Digital cellular will share the same spectrum with the analog cellular. In digital cellular, there are two standards, North American TDMA (tune division multiple access) and CDMA (code division multiple access). These two systems will be discussed in Chap. 15.

PCS (Personal Communications Service)

Broadband PCS.   FCC has licensed six broadband PCS bands, totalling 120 MHz in 1.9 GHz. The frequency distribution is shown in Fig. I.7(a), three 30 MHz licenses and three 10 MHz licenses. Every market can have six system operators operating on six licenses. Besides, two unlicensed bands—one for voice (10 MHz) and one for data (10 MHz)—are also shown in Fig. I.7(a).

There are six possible standard systems:

1. *DCS (Digital Cellular System)-1900.*   A GSM-version system

2. *CDMA-1900.*   A cellular CDMA-version system

3. *NA-TDMA-1900.*   A cellular NA-TDMA (IS-136)-version system

4. *Omnipoint.*   A hybrid system with CDMA, TDMA, and FDMA

5. *B-CDMA.*   A broadband CDMA (5-MHz or 10-MHz) system

6. *PACS-1900.*   Personal Access Communications System, a Bellcore-developed system

The protocol in each PCS system is in general adapting its corresponding cellular system.

Narrowband PCS.   FCC has also licensed three narrowed band PCS for two-way paging as shown in Fig. I.7(b). There are three kinds of two-way paging channels:

1. Five 50-kHz channels paired with 50-kHz channels

2. Three 50-kHz channels paired with 12.5-kHz channels

3. Three 50-kHz unpaired channels

In a two-way paging channel, the reverse link (pager to base) has power limitations and needs an arrangement different than the cellular system.

Part III   Summary—Commercial versus Military

Mobile communication has been, and continues to be, an essential tactical requirement for many types of military operations involving mobile forces. Specific mobile-radiotelephone requirements vary widely because of different needs and different traditional methods of satisfying such needs within the several branches of the United States Armed Forces. Traditional methods were based on military radio-frequency allocations

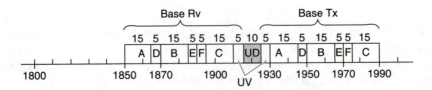

UV-unlicensed voice
UD-unlicensed date

Name of Band	Spectrum Bandwidth	Base Rv (MHz)	Base Tx (MHz)
Band A	30 MHz	1850–1865	1930–1945
Band D	10	1865–1870	1945–1950
Band B	30	1870–1885	1950–1965
Band E	10	1885–1890	1965–1970
Band F	10	1890–1895	1970–1975
Band C	30	1895–1910	1975–1990
Unlicensed voice	10	1910–1915	1925–1930
Unlicensed data	10	1915–1925	

(a)

Five 50-kHz channels paired with 50-kHz channels

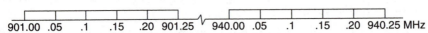

Three 50-kHz channels paired with 12.5-kHz channels

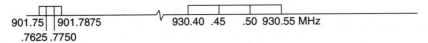

Three 50-kHz unpaired channels

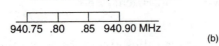

(b)

Figure I.7 (*a*) Wideband PCS for cellular-like systems; (*b*) narrowband PCS for two-way paging systems.

in all bands from VLF all the way through SHF, depending upon the particular transmission medium and the operational environment of the mobile user. Often, the equipment and methods used were inefficient, cumbersome to set up and move, wasteful of the assigned radio spectrum, and restricted by limitations imposed by design application and regulatory authority.

As new systems and concepts in military communication evolved from advances in solid-state digital technology, the role of mobile radiotelephone as an integrated extension of the global military switching networks became technologically significant. Before long, the advanced military systems were using digital modulation techniques in place of the traditional amplitude- and frequency-modulation techniques for voice and data transmission. In recent years, there has been considerable activity in the design and development of spread spectrum systems for military security applications. The spread-spectrum design concept has also been applied to military mobile radio. Some of the basic differences between commercial and military systems that affect the mobile radio system design approach are summarized in Table I.3.

TABLE I.3   Mobile Radio: Commercial vs. Military Design Considerations

Commercial	Military
1. Voice quality must approach that of normal speech.	1. Communications security and integrity of error-free transmissions must be ensured.
2. Signaling performance must be equal to landline telephones.	2. Signaling must comply with military signaling specifications.
3. Designed to minimize interference from predictable, unintentional sources.	3. In addition to interference from unintentional sources, interference from intentional sources must be anticipated and counteracted.
4. Compatible with frequencies allocated by the FCC, and often dependent upon FCC approval of a commercial proposal to use a particular frequency band.	4. Compatible with frequencies allocated by the IRAC (Interdepartment Radio Advisory Committee).
5. Base-station site is carefully selected and subject to approval by local government authority.	5. Base-station site is limited to military options and tactical operational requirements.
6. Base station is a fixed plant installation.	6. Base station is often transportable and designed for over-the-road hauling.
7. Base-station antenna is designed for optimal planned coverage of the area it serves.	7. Base-station antenna is designed for use under potentially hostile wartime conditions and for minimum visibility and quick setup and teardown.
8. Base station often serves densely populated urban areas.	8. Base station more often serves sparsely populated field deployment areas.
9. Noise levels based on the worst case for a normal environment are predictable.	9. Noise levels based on tactical wartime situations require special design consideration.
10. User expects a full range of customer service features and options.	10. Ruggedized construction capable of surviving in unfavorable environments.
11. Attractive appearance and styling.	11. Lightweight, simple to operate, and easy to remove and replace.
12. Designed for theftproof installation.	12. Design must use military-standard parts meeting military specifications.
13. Mobile-to-base or mobile-to-base-to-mobile (for billing purposes).	13. Mobile-to-base or mobile-to-mobile.
14. Cost consciousness.	14. High technology initiative.

The Mobile-Radio Signal Environment

1.1 The Mobile-Radio Communication Medium

Radio signals transmitted from a mobile-radio base station are not only subject to the same significant propagation-path losses that are encountered in other types of atmospheric propagations, but are also subject to the path-loss effects of terrestrial propagation. Terrestrial losses are greatly affected by the general topography of the terrain. The low mobile-antenna height, usually very close to ground level, contributes to this additional propagation-path loss. In general, the texture and roughness of the terrain tend to dissipate propagated energy, reducing the received signal strength at the mobile unit and also at the base station. Losses of this type, combined with free-space losses, collectively make up the propagation-path loss.

Mobile-radio signals are also affected by various types of scattering and multipath phenomena—which can cause severe signal fading—attributable to the mobile-radio communications medium. Mobile-radio signal fading compounds the effects of long-term fading and short-term fading, which can be separated statistically and are described in Chap. 3. Long-term fading is typically caused by relatively small-scale variations in topography along the propagation path. Short-term fading is typically caused by the reflectivity of various types of signal scatterers, both stationary and moving. Fading of this kind is referred to as "multipath" fading.

Propagation between a mobile unit and a base station is most susceptible to the effects of multipath fading phenomena, because all communication is essentially at ground level. The effects of multipath phenomena are not significant in air-to-ground and satellite-to-earth-station communications, because the angle of propagation precludes most types of interference caused by surrounding natural land features and man-made structures. The major concern in air-to-ground communication is the Doppler effect resulting from the relatively high flying speed of the communicating aircraft. The important considerations in satellite-to-earth-station communication are direct-path attenuation in space, which severely reduces the level of received signals; the signal delay time resulting from the long-distance up-down transit; and the requirement for highly directional earth-station antennas capable of tracking the satellite beacon.

Generally speaking, the signal strength of a signal transmitted from a base station decreases with distance when measured at various points along a radial path leading away from the base station. Ideally, signal-strength measurements would be made by monitoring and recording the signal received by a mobile unit as it moves away from the base station along a radial route at a constant rate of speed. This measurement technique would be repeated over many different radial routes in order to obtain a significant number of readings that would enable statistical analysis of the overall zone of mobile-radio coverage for a particular base station. In actual practice, it is difficult to achieve the ideal conditions for signal-strength measurements, since existing roads must be used and traffic conditions determine the actual rate of travel, necessitating occasional stops along the way. For optimal radio-signal reception in the mobile-radio use area, both the base-station antenna and the mobile-unit antenna should be located at the highest available point along the propagation path. However, even under the most optimal siting conditions, there are often hills, trees, and various man-made structures and vehicles that can adversely affect the propagation of mobile-radio signals.

A typical graphic plot of the instantaneous signal strength of a received signal as a function of time, or of $s(t)$, is shown in Fig. 1.1(b). The starting time t corresponds to the starting point x_1 for route x, as shown in Fig. 1.1(a). The route x is called the mobile path. If it is possible for the speed V of the mobile unit to remain constant during the recording period, then $s(t)$ recorded on a time scale can be used for $s(x)$ by simply changing the time scale into a distance scale, where $x = Vt$. However, if the speed of the mobile unit varies during the recording period, then $s(t)$ must be velocity-weighted in order to obtain a true representation, as illustrated by $s(x)$ in Fig. 1.1(c). The graph of the instantaneous signal strength of a received signal as a function of distance, $s(x)$, from the base station along a particular route is used to calculate the path loss of that route, even though the graph of $s(t)$ is the actual raw data obtained from the field.

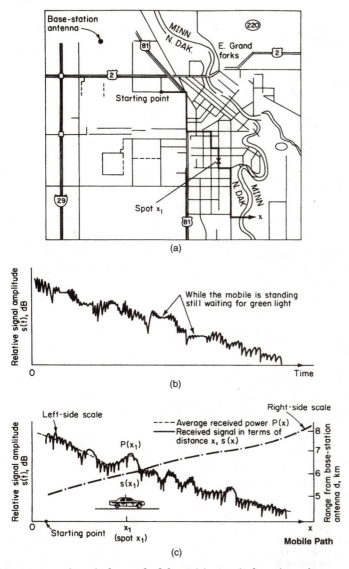

Figure 1.1 A typical record of data: (*a*) a typical navigator's map; (*b*) $s(t)$ expressed on time scale; (*c*) $s(t)$ expressed on distance scale.

Example 1.1 The following example illustrates the procedure for plotting $s(t)$ from the signal data recorded as the mobile unit is in motion with respect to the base-station antenna. During the time when signal data are being recorded, a "wheel-tick" device is used to record the actual speed of the mobile unit on a corresponding time scale, as shown in Fig. E1.1.1, where distance is plotted on the y axis and time is plotted on the x axis.

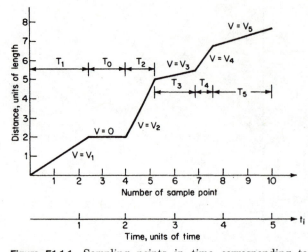

Figure E1.1.1 Sampling points in time corresponding to unequally spaced samples in length due to the variation of vehicular speed.

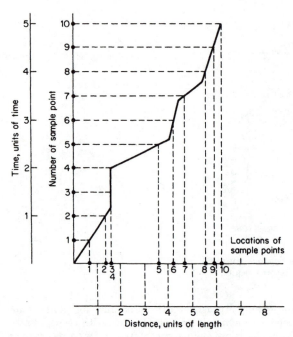

Figure E1.1.2 Converting sampling points from time frame to distance frame.

In the example shown in the figure, the mobile unit has recorded five different speeds and one complete stop. The y coordinate shows the distance of the mobile unit from the base-station antenna at any given time along the x coordinate. The correlation between sampling points in the time frame and those in the distance frame shows that the speed of the mobile unit was not constant throughout the period when measurements were being taken.

A different but related perspective is obtained from Fig. E1.1.2, where distance is plotted along the x axis and time along the y axis. Since the average field strength at each sampling point is the average of the field-strength measurements within the distance interval, the resultant plot shows that the distribution along the distance scale is not evenly distributed. It is therefore required that the engineer conducting the study determine that the distance intervals at which field-strength samples are recorded are consistent with the speed and motion characteristics of the mobile unit.

The measurements here were recorded at the mobile receiver as the mobile unit traveled from the starting point along a route x, as indicated on the map of Fig. 1.1(a). The dotted line in Fig. 1.1(c) represents the average power of the signal received at the mobile unit as a function of distance, or $P(x)$, for that particular path. In practice, the average received power at a distance x_1, $P(x_1)$, can be obtained directly, by averaging the instantaneous signal-strength measurements recorded within an interval of a certain length at a specified distance from the base station. The methods used to calculate the average received power and to determine the propagation-path loss at various radial distances along a path are described in greater detail in Chap. 3.

In the mobile-radio communications environment there are times when the mobile unit will be in motion, and other times when the mobile unit will be stationary. When the mobile unit is moving, it moves at various rates of speed and travels in various directions. As the mobile unit proceeds along its route, it passes many types of local scatterers, including numerous other vehicles in motion.

The presence of scatterers along the path constitutes a constantly changing environment that introduces many variables that can scatter, reflect, and dissipate the propagated mobile-radio signal energy. These effects often result in multiple signal paths that arrive at the receiving antenna displaced with respect to each other in time and space. When this happens, it has the effect of lengthening the time allotted to a discrete portion of the signal information and can cause signal smearing. This phenomenon is referred to as "delay spread." Also, the arrival of two closely spaced frequencies with different time-delay spreads can cause the statistical properties of the two multipath signals to be weakly correlated. The maximum frequency difference between frequencies having a strong potential for correlation is referred to as the "coherence bandwidth" of the mobile-radio transmission path. Coherence correlation then can be avoided by discretionary assignment of channel frequencies on the basis of frequency distribution and geographic separation. Time-delay spread and coherence bandwidth are discussed in Secs. 1.5 and 1.6, respectively.

Example 1.2   Under data-sampling conditions in which the speed of the mobile unit $V(t)$ is continuously varying, the data-sampling points may appear either stretched out or compressed in converting from the time frame to the space frame. In this case, it is first necessary to find the mean velocity of the mobile vehicle, by the equation

$$\overline{V} = \frac{\sum\limits_{i=1}^{N} D_i}{\sum\limits_{i=1}^{N} T_i} \tag{E1.2.1}$$

where $\sum_{i=1}^{N} D_i$ is the total distance and $\sum_{i=1}^{N} T_i$ is the total time of N intervals. The individual vehicular velocity V_i is then expressed as follows:

$$V_i = \frac{D_i}{T_i} = \frac{d_i - d_{i-1}}{t_i - t_{i-1}}$$

The ratio δ can be defined as follows:

$$\delta = \frac{\overline{V}}{V_i}$$

If the ratio of $\delta = \overline{V}/V_i > 1$, then the data-sampling rate is stretched. If the ratio of $\delta = \overline{V}/V_i < 1$, then the data-sampling rate is compressed. Figure E1.2.1 plots the ratio δ for vehicular speeds of $\overline{V}$ and V_i.

1.2   Propagation-Path Loss

Mobile-radio propagation-path loss is primarily due to terrain effects and the presence of radio-wave scatterers along the path within the

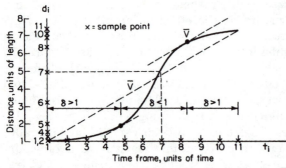

Figure E1.2.1   Plot of ratio δ for vehicular velocities of $\overline{V}$ and V_i.

mobile-radio environment. Variations in the contour and roughness of the terrain, including any scatterers that are present, cause changes in propagation as the result of specular reflection, diffuse reflection, and diffraction.

Specular reflection occurs when radio waves encounter a smooth interface between two dissimilar media and the linear dimensions of the interface are large in comparison with the wavelength of the radiated signal. This type of reflection is analogous to the imaging properties of mirrors as defined by Snell's law. The principles of specular reflection as they apply to images are illustrated in Fig. 1.2. The reflected wave, reflected at point Q, is essentially a reflection of the incident wave from antenna T. However, it can be considered to have been radiated by the fictitious image antenna T' and to have passed through the surface without refraction.

In some instances, the height of the antenna and the elevation of the terrain are significantly shorter than the link path between the transmitting antenna and the receiving antenna. For this reason, different scales are often used for the vertical and horizontal axes in plotting the contour of the terrain on graph paper. The vertical axis is usually scaled in feet or meters, whereas the horizontal axis is usually scaled in miles or kilometers. Regardless of the scaling used, as long as the reflection plane is horizontal, the incident and reflected angles will be equal in accordance with Snell's law. However, when different scales *are* used and the terrain contour is sloping and the reflection plane is not horizontal, then the incident and reflected angles will be unequal on the graph paper. Therefore, when this relationship is plotted on graph paper with different scales used to represent the two axes, a special method may be needed to find the reflection point. Example 1.3 illustrates three methods that can be used to find the reflection point on a sloping plane.

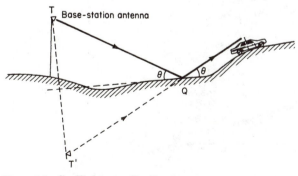

Figure 1.2 Snell's law application.

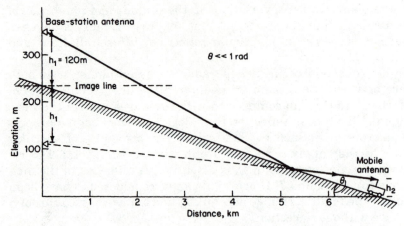

Figure E1.3.1 Image method for approximating the reflection point on a sloping plane.

Example 1.3 Figure E1.3.1 shows a typical sloping plane and the parameters needed to find the reflection point with the image method. This method is used when the link path is much greater than the height of the antenna. In plotting a graph, the vertical-to-horizontal scale ratio is usually smaller than 0.1, and the reflection point can be readily approximated by using the information shown in Fig. E1.3.1.

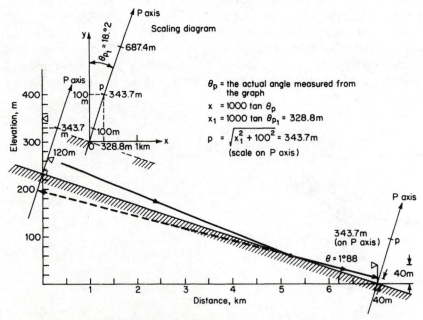

Figure E1.3.2 Scaling method for precise determination of the reflection point on a sloping plane.

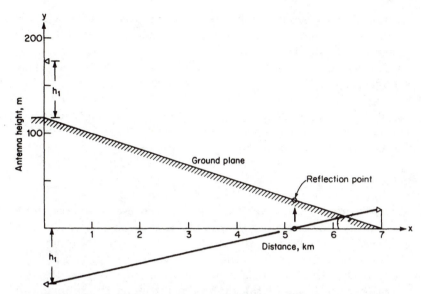

Figure E1.3.3   Simple method for finding the reflection point on a sloping plane.

Figure E1.3.2 shows a typical sloping plane and the parameters needed to find the reflection point by using the scaling method. This method is used when the vertical-to-horizontal scale ratio is greater than 0.1 and it is necessary to use a more precise means for finding the reflection point. This method uses the technique of converting the antenna height scale to a p scaling vector and then applying the image method of Fig. E1.3.1.

Figure E1.3.3 shows the simple method for finding the reflection point on a sloping plane when the vertical-to-horizontal scale ratio is much smaller than 0.1, by using the following simple two-step procedure.

Step 1—Locate a point along the negative y axis that is equal to the height of the base-station antenna above the terrain (h_1). Draw a line connecting that point and the mobile-unit antenna and mark the point where the connecting line intercepts the horizontal (x) axis.

Step 2—Draw a second line perpendicular to the x axis that intercepts the ground plane and the marked crossover point on the horizontal plane. Mark the intercept point where the vertical perpendicular line crosses the sloping ground plane. This is the reflection point.

Diffuse reflection occurs when radio waves encounter a rough-textured surface of which the roughness is compatible with the wavelength of the radiated signal. Unlike specular reflection, which follows Snell's law, diffuse reflection scatters the energy and causes the reflected radio waves to follow a divergent path. Huygen's principle can be used to explain the properties of diffuse reflection. In general, the intensity of diffusely reflected radio waves is less than that of specularly reflected radio waves, because of the scattering of energy along the path over the rough surface. In considering the mobile radio environment, it is necessary to appraise the reflective properties of the surrounding terrain between the mobile unit and the base station. The criteria for analyzing the reflectivity characteristics of the mobile-radio environment are discussed in Chap. 3.

In mobile-radio communication, the line-of-sight condition is considered to have been satisfied when only specular and diffuse reflection are present, the latter due to scatterers along the propagation path. However, if diffraction of radio waves is present because of variations in terrain contour obscuring the propagation path, the line-of-sight condition is no longer valid. In this respect, the definition of the line of sight is different for mobile radio from what it is for tropospherical propagation.

Diffraction of radio waves occurs when the propagation path is obstructed by features of the intervening terrain between the transmitting antenna and the receiving antenna. The severity of signal attenuation depends on whether the obstruction extends through the propagation path, protrudes into the line-of-sight propagation path, or merely approaches the line-of-sight propagation path. In practice, it is not always possible to select the highest point along the propagation path as the ideal location for the base station. In hilly areas, even with good siting of the base station, there will frequently be occasions when the mobile unit is out of the line of sight of the propagation path, as shown in Fig. 1.3. The calculation for determining the propagation-path loss, based on a knife-edge diffraction model, is described in Chap. 4.

The various situations and representations for predicting propagation-path losses are helpful in understanding the effects of multipath phenomena on mobile-radio signals; however, the summing of all these effects would introduce so many variables that a mathematical solution would be too complex for practical application. A recommended alternate approach to reaching a solution is to combine the techniques of analytic and statistical analysis. The first requirement is to obtain measured data from which statistical results can be obtained. When the statistical results are analyzed against the parameters and criteria that are unique for the particular situation, it will be possible to draw certain analytic conclusions based on electromagnetic theory. This method for predicting propagation-path loss is a powerful tool that can produce results closer to the actual path losses than is possible from using either analytic or statistical approaches alone. Data measured and recorded at several points along selected paths are used to evaluate propagation-path loss within a given area.

The dotted line in Fig. 1.1(c) represents a plot of the average received power as a function of distance from the base station along a particular route. These data are used in calculating the propagation-path loss with respect to the radial distance. To analyze the propagation-path loss slope with respect to radial distance for a general area of coverage, data would be recorded for many routes in that area.

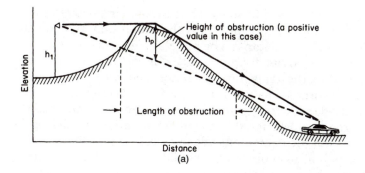

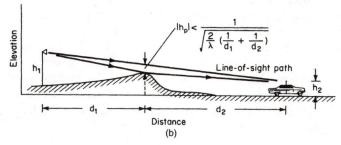

Figure 1.3 Propagation path: (*a*) out of sight; (*b*) line of sight.

The propagation-path loss slope in the mobile-radio environment is always greater than the $1/d^2$ free-space path loss. The path loss is also greater in hilly areas than in flat areas. The techniques and procedures for analyzing propagation-path loss are described at greater length in Chaps. 3, 4, and 5.

1.3 Multipath Fading Due to Scattering Factors

Multipath fading is a common occurrence in the mobile-radio environment and is therefore of major concern to the mobile-radio system designer.

To quantitatively evaluate the mobile-radio environment, it is first necessary to examine the properties of the carrier signal $s_0(t)$, which can be expressed in any one of the following complex forms:

$$s_0(t) = a_0 \exp\left[j(\omega_0 t + \phi_0)\right] \tag{1.1}$$

$$s_0(t) = \text{Re}\{a_0 \exp\left[j(\omega_0 t + \phi_0)\right]\} \tag{1.2}$$

$$s_0(t) = a_0 \cos\left(\omega_0 t + \phi_0\right) \tag{1.3}$$

These three equations are used to analyze the carrier signal $s_0(t)$ transmitted from a base station; a_0 represents the amplitude and ϕ_0 the phase. Both are treated as constants. The angular frequency ω_0 is equal to $2\pi f_0$, where f_0 is the carrier frequency. It is important to recognize that the power within the transmitted carrier frequency is the real part of $s_0(t)$; this is expressed as Re $[s_0(t)]$. Since we are only interested in the received baseband signal, which is derived from the amplitude and phase of the received RF signal, the complex notation is much easier to manipulate than the equivalent trigonometric expression.

The multipath fading phenomena discussed in the following paragraphs occur primarily in the following three situations: (1) where the mobile unit and nearby scatterers are all standing still; (2) where the mobile unit is standing still and nearby scatterers are moving; and (3) where the mobile unit and nearby scatterers are all moving.

In the first, static multipath situation, where the mobile unit and nearby scatterers are all standing still, the various signal paths from all reflecting scatterers reaching the mobile unit can be individually identified, in theory. The model for this situation is illustrated in Fig. 1.4. The received signal $s(t)$ at the stationary mobile unit, coming from N signal paths, can be expressed by the following equation:

$$s(t) = \sum_{i=1}^{N} a_i s_0(t - \tau_i) \tag{1.4}$$

The total propagation time for the ith path can be expressed by the equation

$$\tau_i = \bar{\tau} + \Delta\tau_i \tag{1.5}$$

where $\Delta\tau_i$ is the additional relative delay on the ith path, expressed as either a positive or a negative value with respect to the mean, and the average value of τ_i, $\bar{\tau}$, is defined as follows:

$$\bar{\tau} = \frac{1}{N} \sum_{i=1}^{N} \tau_i \tag{1.6}$$

In Eq. (1.4), a_i is the ith path transmission-attenuation factor, which can be a complex value. If we substitute the value of $s_0(t - \tau_i)$ from Eq. (1.4) in place of $s_0(t)$ in Eq. (1.1), then

$$s(t) = x(t - \bar{\tau}) \exp\left[j2\pi f_0(t - \bar{\tau}) + j\phi_0\right] \tag{1.7}$$

where the envelope $x(t)$ of the received signal $s(t)$ is expressed:

$$x(t) = a_0 \left[\sum_{i=1}^{N} a_i \exp\left(-j2\pi f_0 \Delta\tau_i\right)\right] \tag{1.8}$$

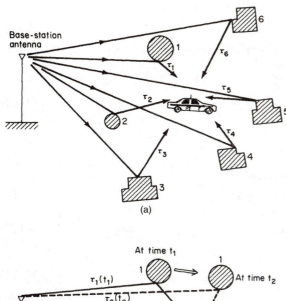

(a)

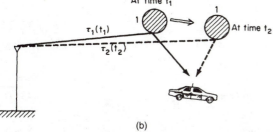

(b)

Figure 1.4 Multipath phenomenon: (*a*) mobile reception at one instant; (*b*) mobile reception at different instants.

with a_0 being constant, as previously stated, and the right side of Eq. (1.8) representing the multipath phenomenon. Since Eq. (1.8) is independent of time t, then $x(t)$ is also constant with respect to time. The received signal envelope of $s(t)$ therefore remains unchanged, so long as the mobile unit and the nearby scatterers are all standing still.

In the second situation, where the mobile unit is standing still and nearby scatterers (passing cars and trucks) are moving, the time delay τ_i and the attenuation factor a_i are uniquely different at any instant of time along the ith path. Under these conditions, the received signal $s(t)$ in Eq. (1.4) must be changed to

$$s(t) = x(t) \exp(j\phi_0) \exp(j2\pi f_0 t) \tag{1.9}$$

where

$$x(t) = \sum_{i=1}^{N} a_0 a_i(t) \exp[-j2\pi f_0 \tau_i(t)] \tag{1.10}$$

Let

$$R = \sum_{i=1}^{N} a_i(t) \cos \left[2\pi f_0 \tau_i(t)\right] \qquad (1.11)$$

$$S = \sum_{i=1}^{N} a_i(t) \sin \left[2\pi f_0 \tau_i(t)\right] \qquad (1.12)$$

Then

$$\begin{aligned} x(t) &= a_0 \{R - jS\} \\ &= A(t) \exp \left[-j\psi(t)\right] \end{aligned} \qquad (1.13)$$

where the amplitude and phase of $x(t)$ are expressed as time-dependent variables:

$$A(t) = a_0 \sqrt{R^2 + S^2} \qquad (1.14)$$

and

$$\psi(t) = \tan^{-1} \frac{S}{R} \qquad (1.15)$$

Since it is virtually impossible to isolate and identify each path of a reflected wave while the scatterers are in motion, it is thus necessary to perform a statistical analysis of the time-dependent variables [amplitude, Eq. (1.14), and phase, Eq. (1.15)]. Since the characteristics of amplitude and phase are similar to those encountered in narrowband thermal noise, they are discussed in Sec. 1.4.

It can be assumed that while the mobile unit is moving there are three extremes that must be considered: (1) the absence of scatterers, (2) the presence of a single scatterer, and (3) the presence of many scatterers in the vicinity of the mobile unit. We will further assume that the mobile unit is traveling in the direction of the positive x axis at a velocity V and is receiving a signal at an angle with respect to the plane of the x axis. Figure 1.5(a) illustrates the parameters of concern for a mobile unit in motion; the received signal is expressed as:

$$s(t) = a_0 \exp \left[j(\omega_0 t + \phi_0 - \beta V t \cos \theta)\right] \qquad (1.16)$$

where $\beta = 2\pi/\lambda$, λ being the wavelength. An additional frequency is contributed as a result of the motion of the mobile unit and is due to the Doppler effect. This additional Doppler frequency is expressed as follows:

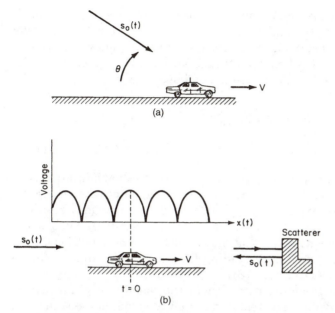

Figure 1.5   Signal reception while the mobile unit is in motion: (*a*) Doppler effect; (*b*) fading due to standing wave pattern.

$$f_d = f_m \cos \theta \tag{1.17}$$

where $f_m = V/\lambda$ is the maximum Doppler frequency. The Doppler frequency f_d can be either positive or negative, depending upon the arrival angle θ. In air-to-ground radio communication the Doppler effect on frequency modulation can be very significant because of the relatively high flying speed of the aircraft.

To properly understand the effects of multipath phenomena, it is necessary to understand the concept of standing waves as applied to radio signals. If a radio signal arrives from one direction and is reflected by a perfect reflecting scatterer in an opposite direction, as shown in Fig. 1.5(*b*), then the resultant signal received by the mobile unit moving at a speed V is as expressed in Eq. (1.18). For simplicity, we can assume that the arrival angle $\theta = 0$.

$$s(t) = a_0 \exp\left[j(\omega_0 t + \phi_0 - \beta V t)\right] - a_0 \exp\left[j(\omega_0 t + \phi_0 + \beta V t - \omega_0 \tau)\right]$$

$$= -j 2 a_0 \sin\left(\beta V t - \frac{\omega_0 \tau}{2}\right) \exp\left[j\left(\omega_0 t + \phi_0 - \frac{\omega_0 \tau}{2}\right)\right] \tag{1.18}$$

where τ is the time that it takes for the wave to travel to the scatterer and return to the $t = 0$ line. The envelope of Eq. (1.18) is the resultant standing wave pattern, and this can also be interpreted as a simple fading phenomenon. When $\beta Vt = n\pi + \omega_0\tau/2$, a fade with zero amplitude (or $-\infty$ dB) will be observed. In the case where square-law detection is applied after Eq. (1.18) has been obtained, then the envelope of Eq. (1.18) will be squared and expressed as follows:

$$x^2(t) = 4a_0^2 \sin^2 \left(\beta Vt - \frac{\omega_0\tau}{2} \right)$$

$$= 2a_0^2[1 - \cos(2\beta Vt - \omega_0\tau)] \tag{1.19}$$

Now the fading frequency is visualized as $2V/\lambda$, as shown in Eq. (1.19). Hence it can be shown that the same Doppler frequency received by the mobile antenna and passed through different detectors may result in different fading frequencies at the two detector outputs. In such cases, the fading frequency measured by square-law detection methods, as in Eq. (1.19), is always double the fading frequency measured by linear detection methods, as in Eq. (1.18).

In the third and final situation, where the mobile unit and nearby scatterers are all moving, the resultant received signal is the sum of all scattered waves from different angles θ_i, depending upon the momentary attitude of the various scatterers and whether or not the direct signal-transmission path is blocked. This complex situation is expressed in the following equations:

$$s(t) = \sum_{i=1}^{N} a_0 a_i \exp\left[j(\omega_0 t + \phi_0 - \beta Vt \cos\theta_i + \phi_i)\right]$$

$$= A_t \exp(j\psi_t) \exp\left[j(\omega_0 t + \phi_0)\right] \tag{1.20}$$

where

$$A_t = \left[\left(a_0 \sum_{i=1}^{N} a_i \cos\psi_i\right)^2 + \left(a_0 \sum_{i=1}^{N} a_i \sin\psi_i\right)^2\right]^{1/2} \tag{1.21}$$

$$\psi_t = \tan^{-1} \frac{\displaystyle\sum_{i=1}^{N} a_i \sin\psi_i}{\displaystyle\sum_{i=1}^{N} a_i \cos\psi_i} \tag{1.22}$$

and

$$\psi_i = \phi_i - \beta Vt \cos\theta_i$$

The characteristics of A_t and ψ_t are similar to those of narrowband thermal noise, illustrated in Fig. 1.8 and defined by Eqs. (1.45) and (1.46), respectively, and discussed in Sec. 1.4. It should be noted that

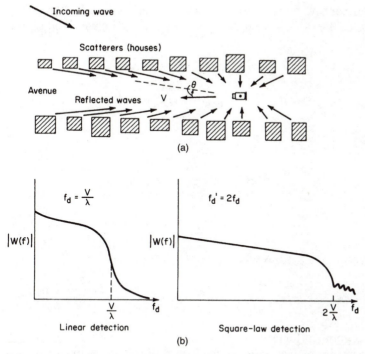

Figure 1.6 Amplitude spectrum of a fading signal: (*a*) mechanism of forming a fading signal; (*b*) continuous spectra of a fading signal after different detections.

Eq. (1.9), for moving scatterers, and Eq. (1.20), for the moving mobile unit, are similar in form.

As a mobile unit proceeds along a street, it is passing through a virtual avenue of scatterers, as shown in Fig. 1.6(*a*). The main components of the received signal are reflections from nearby scatterers. The highest Doppler frequency f_d can be calculated from Eq. (1.17), which is incorporated into Eq. (1.23):

$$f_m = \max \left(|f_d| \right) = \max \left(\frac{V}{\lambda} \left| \cos \theta \right| \right) = \frac{V}{\lambda} \qquad (1.23)$$

The frequency modulation resulting from the highest Doppler frequency for a given velocity V is the most probable cause of a Doppler shift in a mobile receiver. This phenomenon has been verified by statistical analysis of experimental data [1, 2].

Two typical experimental curves of the continuous spectrum, $W_0(f)$, are obtained from the outputs of two detections, linear detection and square-law detection, as shown in Fig. 1.6(*b*).

When two incoming waves have the same amplitude A but different incoming angles $\theta = \theta_1$ and $\theta = \theta_2$, respectively, then the received signal $S(t)$ becomes

$$S(t) = A \, e^{j2\pi ft} \, (e^{-j\beta x \cos \theta_1} + e^{-j\beta x \cos \theta_2})$$

$$= 2A \, e^{j2\pi ft} \, e^{-j\beta x (\cos \theta_1 + \cos \theta_2)/2} \cdot \cos \left(\frac{\beta x}{2} \cos \theta_1 - \frac{\beta x}{2} \cos \theta_2 \right) \qquad (1.24)$$

where $x = Vt$ and

$$\frac{\beta x}{2} \cos \theta_1 - \frac{\beta x}{2} \cos \theta_2 = 2\pi \cdot \frac{Vt}{2\lambda} (\cos \theta_1 - \cos \theta_2)$$

$$= 2\pi \left[\frac{V}{2\lambda} (\cos \theta_1 - \cos \theta_2) \right] t \qquad (1.25)$$

The fading frequency can be found from Eq. (1.25) as

$$f_d = \frac{V}{2\lambda} (\cos \theta_1 - \cos \theta_2) \qquad (1.26)$$

Eq. (1.26) is a general formula.

The phenomenon of frequency-selective fading also occurs in multipath situations, which cause a_0 in Eq. (1.1) to be treated as a time-varying signal $a_0(t)$ with its continuous spectrum expressed as $W_0(f)$:

$$W_0(f) = \int_{-\infty}^{\infty} a_0(t) \exp (-j2\pi ft) \, dt \qquad (1.27)$$

where $|W_0(f)|$ is the amplitude spectrum and the argument of $W_0(f)$ is the phase spectrum. The bandwidth B of $a_0(t)$ is defined later, in Eq. (1.42), as a characteristic of narrowband noise; however, in selective fading, $B \ll f_0$. When the mobile unit and nearby scatterers are both standing still, the received signal $s(t)$ remains the same as expressed previously in Eq. (1.7). However, the $x(t)$ of Eq. (1.7) used in the present case differs from that expressed in Eq. (1.8) and should be redefined accordingly as follows:

$$x(t) = \sum_{i=1}^{N} a_i a_0(t - \Delta\tau_i) \exp (-j2\pi f_0 \Delta\tau_i) \qquad (1.28)$$

where

$$a_0(t - \Delta\tau_i) = \int_{-\infty}^{\infty} W_0(f) \exp [j2\pi f(t - \Delta\tau_i)] \, df \qquad (1.29)$$

By substitution of Eq. (1.29) for the equivalent notation in Eq. (1.28), the following equation is obtained:

$$x(t) = \int_{-\infty}^{\infty} W_0(f) \left\{ \sum_{i=1}^{N} a_i \exp\left[-j2\pi(f+f_0)\,\Delta\tau_i\right] \right\} \exp\left(j2\pi ft\right) df \quad (1.30)$$

where $x(t)$ is the envelope of the received signal $s(t)$ and its continuous spectrum $W(f)$ is defined as follows:

$$W(f) = W_0(f)H(f) \quad (1.31)$$

$H(f)$ is the equivalent low-pass transfer function of a communication channel operating in the mobile-radio medium. By definition, a channel is a communications link between transmitting and receiving terminals. Equation (1.31) shows the relationship $H(f)$ between two terminals. $W(f)$, as we know, is the continuous spectrum of $x(t)$ expressed as follows:

$$x(t) = \int_{-\infty}^{\infty} W(f) \exp\left(j2\pi ft\right) df \quad (1.32)$$

Comparing Eq. (1.32) with Eq. (1.30) defines $H(f)$ as follows:

$$H(f) = \sum_{i=1}^{N} a_i \exp\left[-j2\pi(f+f_0)\,\Delta\tau_i\right] \quad (1.33)$$

For the mobile-radio environment, it is reasonable to assume that:

$$f\,\Delta\tau_i \ll 1 \quad (1.34)$$

With the assumption of Eq. (1.34), Eq. (1.33) can then be expanded by a Taylor series, retaining elements of the two leading terms, as follows:

$$H(f) = H_1 + jH_2(f) \quad (1.35)$$

where

$$H_1 = \sum_{i=1}^{N} a_i \exp\left(-j2\pi f_0\,\Delta\tau_i\right) \quad (1.36)$$

$$H_2(f) = -2\pi f \sum_{i=1}^{N} a_i\Delta\tau_i \exp\left(-j2\pi f_0\,\Delta\tau_i\right) \quad (1.37)$$

It should be noted that H_1 is frequency-independent and represents the distortionless portion of the transfer function $H(f)$ for the mobile-radio medium. H_1 can be a complex value. H_2 is frequency-dependent and represents the distortion elements of the transfer function $H(f)$ for the mobile-radio medium. Because H_2 is frequency-dependent, it can cause the signal strength of the received signal to vary as a function of the frequency of the mobile-radio channel. H_2 can also cause frequency-selective fading. In this case, as shown in Eq. (1.37), H_2 varies linearly with frequency across the mobile-radio band. In general, the following condition holds true:

$$|H_1| \gg |H_2(f)| \qquad (1.38)$$

In this situation, the selective-fading phenomenon is unnoticeable. However, there can be instantaneous situations in which a certain combination of phasors can cause $H_2(f)$ to become the dominant factor in Eq. (1.35), as the following becomes true:

$$|H_2(f)| > |H_1| \qquad (1.39)$$

When this occurs, the frequency-selective characteristic is obtained even though the situation is basically nonselective. This analysis also applies occasionally to the more complex situations in which the mobile unit and/or the nearby scatterers may be moving.

1.4   Thermal Noise and Human-Made Noise Characteristics

Thermal noise characteristics

The narrowband thermal noise characteristics of the received signal are very similar to the fading effects of the multipath signaling medium. Thermal noise is characterized as white noise, which is uniformly distributed throughout the mobile-radio-communications frequency band. White noise n_x is analogous to white light, having a constant power spectrum $S_x(f)$, expressed

$$S_x(f) = \frac{\eta}{2} \qquad (1.40)$$

and shown in Fig. 1.7(a), where η is the positive-frequency power density. By comparison, the spectrum of thermal noise is equivalent to white noise up to a frequency of approximately 10^{13} Hz (the infrared region); there is a rapid exponential decrease as the frequency increases above that region. In this context, the assumption that thermal noise has a constant power spectrum is both valid and justified so long as the thermal noise is confined to a frequency region lower than 10^{13} Hz. When thermal noise is passed through a narrowband filter having a frequency band Δf that is much less than the center frequency f_0 of the band (that is, when $f_0 \gg \Delta f$), the output consists of

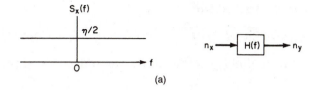

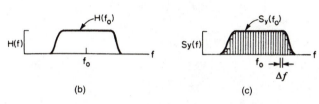

Figure 1.7 Spectrum of narrowband thermal noise: (*a*) spectrum of n_x; (*b*) filter transfer function; (*c*) spectrum after filtering.

narrowband thermal noise. The characteristics of the envelope and phase of the narrowband thermal noise output of the filter closely resemble the characteristics of a radio carrier that has passed through a mobile-radio multipath fading medium. The amplitude and phase of the output noise are very much like those of a continuous-wave (CW) radio signal passing through a multipath medium.

Since the analysis of the nature of thermal noise has been thoroughly developed, we may apply a similar analysis to mobile-radio signals. The following equation defines the power spectrum of thermal noise before and after filtering [3]:

$$S_y(f) = |H(f)|^2 S_x(f) \tag{1.41}$$

The filter transfer function $H(f)$ is illustrated in Fig. 1.7(*b*). The bandwidth B of $H(f)$ can be expressed as follows:

$$B = \frac{\int_{-\infty}^{\infty} |H(f)|^2 \, df}{|H(0)|^2} \tag{1.42}$$

The spectrum is partitioned into very narrow frequency bands having a width Δf and centered about a frequency $f_0 \pm k \, \Delta f$, as shown in Fig. 1.7(*c*). If there are $M = B/\Delta f$ bands and each band represents a sinusoid of frequency $f_0 \pm k \, \Delta f$, where $k = 1, M/2$, then the amplitude can be expressed through its relationship to average power, as follows:

$$\frac{A_k^2}{2} = 2S_y(f_0 + k \, \Delta f) \, \Delta f = \eta \, |H(f_0 + k \, \Delta f)|^2 \, \Delta f \tag{1.43}$$

Thus, the amplitude can be derived from Eq. (1.43):

$$A_k = \sqrt{2\eta \, | H(f_0 + k \, \Delta f) |^2 \, \Delta f} \qquad (1.44)$$

The total narrowband thermal noise is the sum of the individual sinusoid waves:

$$
\begin{aligned}
n(t) &= \sum_{k=-M/2}^{M/2} A_k \exp \left[j2\pi(f_0 + k \, \Delta f)t + j\theta_k \right] \\
&= [n_c(t) + jn_s(t)] \exp (j2\pi f_0 t) \\
&= A_t(t) \exp [j\psi_t(t)] \exp (j2\pi f_0 t) \qquad (1.45)
\end{aligned}
$$

where

$$n_c(t) = \sum_{k=-M/2}^{M/2} A_k \cos (2\pi k \, \Delta f \, t + \theta_k) \qquad (1.46)$$

$$n_s(t) = \sum_{k=-M/2}^{M/2} A_k \sin (2\pi k \, \Delta f \, t + \theta_k) \qquad (1.47)$$

$$A_t(t) = [n_c^2(t) + n_s^2(t)]^{1/2} \qquad (1.48)$$

and

$$\psi_t(t) = \tan^{-1} \frac{n_s(t)}{n_c(t)} \qquad (1.49)$$

Equation (1.45) has the same form as Eqs. (1.9) and (1.20), and the statistical characteristics of A_t and ψ_t in all three equations are the same [3, 4].

The amplitude and phase are illustrated in Fig. 1.8, and the phasor diagram for narrowband noise with $A_t(t)$ and $\psi_t(t)$ are analyzed in greater depth in Chaps. 6 and 7.

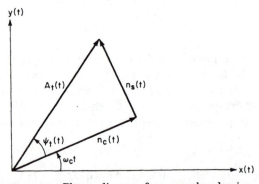

Figure 1.8 Phasor diagram for narrowband noise.

Example 1.4   The power of white noise becomes finite after filtering and can be expressed as follows:

$$N = \int_{-\infty}^{\infty} |H(f)|^2 \frac{\eta}{2} \, df = \eta \int_0^{\infty} |H(f)|^2 \, df \tag{E1.4.1}$$

Noise power after filtering is dependent upon the filter transfer function, which is related to noise bandwidth B_N:

$$B_N = \frac{1}{|H(f)|_{\text{max}}^2} \int_0^{\infty} |H(f)|^2 \, df \tag{E1.4.2}$$

where $|H(f)|_{\text{max}}$ is the center-frequency amplitude response (voltage); denoted by H_m. Therefore

$$N = H_m^2 \eta B_N \tag{E1.4.3}$$

For RC low-pass filters,

$$H(f) = \frac{1/j\omega C}{R + 1/j\omega C} \quad \text{and} \quad |H(f)| = \frac{1}{\sqrt{1 + (f/f_r)^2}} \tag{E1.4.4}$$

where $f_r = 1/2\pi RC$. Therefore

$$B_N = \int_0^{\infty} \left(\frac{1}{\sqrt{1 + (f/f_r)^2}}\right)^2 \, df = \int_0^{\infty} \frac{df}{1 + (f/f_r)^2} = \frac{\pi f_r}{2} = 1.57 f_r \tag{E1.4.5}$$

and B_N is approximately 50 percent greater than f_r.

Human-made noise characteristics

Human-made noise within the mobile-radio environment is primarily from unintentional sources, such as vehicular ignition, radiated noise from power lines, and industrial equipment. Two reference documents have been published that contain much information on human-made noise. The first is ITT's *Reference Data for Radio Engineers* [5], from which Fig. 1.9 has been obtained. In this figure, noise has been classified into the following six categories:

1. Atmospheric noise

2. Urban human-made noise

3. Suburban human-made noise

4. Galactic noise

5. Solar (quiet-sun) noise

6. Typical internal receiver noise, which is not considered an environmental noise

The noise figure F_a, in dB above kT_0B, appears on the y axis of Fig. 1.9, where:

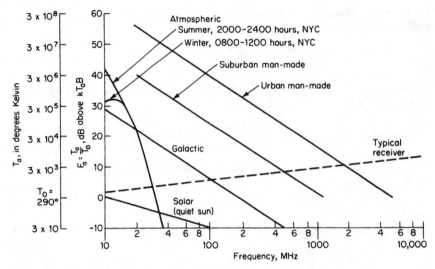

Figure 1.9   Human-made noise. *(From Ref. 5.)*

$$k = \text{Boltzmann's constant}$$
$$B = \text{the effective receiver noise bandwidth}$$

and

$$T_0 = \text{the reference temperature (290 K)} = 17°\text{C (room temper-ature)}$$
$$kT_0B = -174 \text{ dBm/Hz (at } 17°\text{C)}$$

Among the six curves plotted in Fig. 1.9, the first five decrease as frequency increases. The sixth curve, typical internal receiver noise, is not considered an environmental noise. As modern technology advances, it is expected that future mobile-radio receivers will show a reduction in internal receiver noise in the higher-frequency region. Hence, the dotted line in Fig. 1.9 is used to indicate the typical receiver internal-noise curve currently used.

The second reference document for human-made noise is one published by the National Bureau of Standards [6, 7]. In these references, NBS has conducted extensive studies on human-made noise, which it classifies into two categories:

1. The mean and the standard deviation of human-made noise

2. Average automotive traffic noise

Mean and standard deviation   Three types of areas should be considered in determining the values of the mean noise figure F_a as shown in Fig.

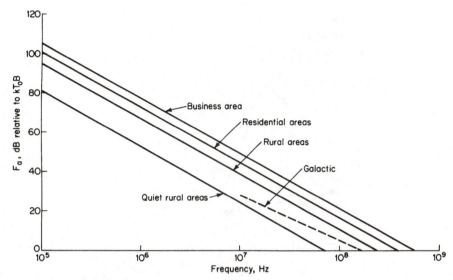

Figure 1.10   Mean values of F_a for three types of areas. *(From Ref. 6.)*

1.10, business, residential, and rural areas. Note that in Fig. 1.10 all three areas have similar slopes of approximately 28 dB per decade. Figure 1.10 shows that the greatest amount of noise occurs in the business area: the level is 6 dB higher than the noise level in the residential area and 12 dB higher than that in the rural area. In a comparison of the results of Figs. 1.9 and 1.10, Fig. 1.10 shows that at frequencies below 30 MHz the predicted human-made noise in business areas is *slightly greater* and the slope is somewhat steeper than that of suburban areas shown in Fig. 1.9. However, the human-made noise of business areas shown in Fig. 1.10 is *much less* than that of urban areas shown in Fig. 1.9. The comparison shows that the human-made noise measurement depends greatly on when the sample is taken and on how we define the types of areas.

Figure 1.11 shows the standard deviation σ of human-made noise F_a for the business, residential, and rural areas, respectively. Note that the value of σ is highest for the business area and fluctuates greatly as the frequency increases. By comparison, in the residential and rural areas, the value of σ decreases as the frequency increases.

Average automotive traffic noise   The average automotive traffic noise power [7] is shown in Fig. 1.12, for traffic densities of 1000 vehicles per hour and 100 vehicles per hour. Note that the average automotive traffic noise decreases as the frequency increases. By comparing Fig. 1.10 with Fig. 1.12, it can be seen that the predominant human-made noise

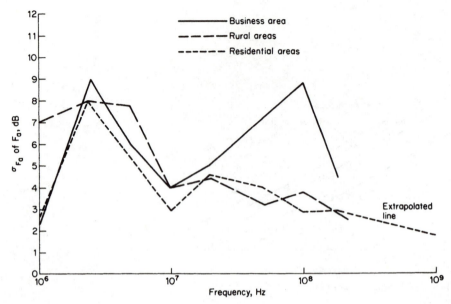

Figure 1.11   Standard deviation of F_a noise figure, σ_{F_a}. *(From Ref. 6.)*

is automotive traffic noise. When the traffic density is greater than 1000 vehicles per hour, a noticeable increase can be observed as the frequency approaches 1 GHz.

Additional information on human-made noise can be found in Skomal [8].

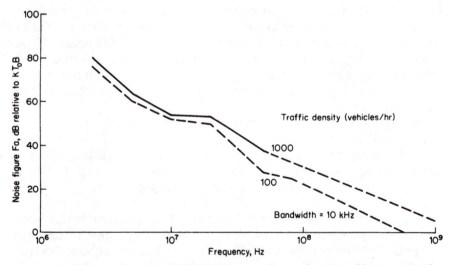

Figure 1.12   Average automotive traffic noise power F_a as a function of frequency with a bandwidth of 10 kHz. *(From Ref. 7.)*

1.5 Delay Spread

The phenomenon of delay spread occurs when the base station transmits an impulse signal $s_0(t) = a_0\,\delta(t)$ to the mobile unit and when because of multipath scattering the impulse signal becomes many impulse signals that arrive at the mobile unit at different times. The total delay spread time is significantly lengthened. The phenomenon is like the voice echoes received on the top of a mountain. The model for this situation is shown in Fig. 1.13. The received impulse signal at the mobile unit is

$$s(t) = a_0 \sum_{i=1}^{n} a_i\,\delta(t - \tau_i) \cdot e^{j\omega t}$$

$$= E(t) \cdot e^{j\omega t} \tag{1.50}$$

Equation (1.50) represents a train of discrete impulses at frequency ω arriving at the mobile receiver, as shown in Fig. 1.13(a). As the number of scatterers in the vicinity of the mobile unit increases, the received discrete impulses become a continuous signal pulse with a pulse length Δ, commonly referred to as the delay spread, as shown in Fig. 1.13(b). In general, an impulse traveling a short distance should arrive earlier with a strong power density, as illustrated in Fig. 1.14(a). However, in a

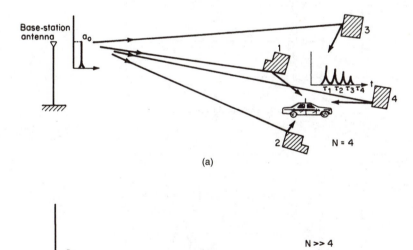

Figure 1.13 Illustration of delay spread: (a) four-scatterer case; (b) N-scatterer case.

mobile radio environment, the human-made structures use different materials. Some are more reflective. Some are more absorptive. Therefore, the first arriving impulse may not be the strongest one, as shown in Fig. 1.14(c). Depending on the location of the mobile unit in, for example, New York City, a tall metal and glass building can result in a strong reflected signal, even though it's at a far distance. The phenomenon can be seen in Fig. 1.14(d). This delay spread imposes a waiting period that determines when the next pulse can be transmitted by the base station. This requires that the signaling rate be slowed down to a period much less than $1/\Delta$, to prevent intersymbol interference in the Rayleigh-fading environment as described in Chaps. 6 and 7.

The use of a broadband pseudo-noise signal at 850 MHz to record the delay envelope in suburban and urban environments has been documented by Cox and Leck [9–13] and others [14–19]. This technique, used for obtaining the set of four discrete recordings shown in Fig. 1.14, is similar to the Rake technique [20]. These actual recordings were taken at various locations in the borough of Manhattan, New York City. Each recording shows the relative power density in decibels (on the y

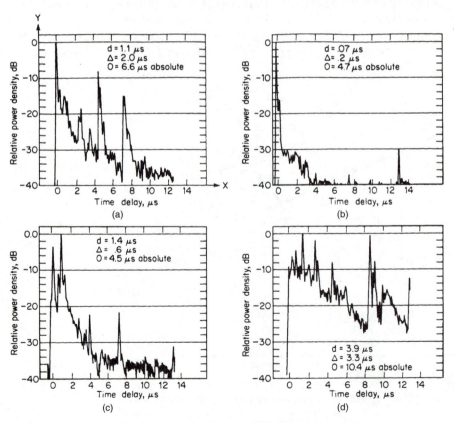

Figure 1.14   Delay envelopes in New York City. © 1975 IEEE. *(From Ref. 10.)*

axis) and the excess delay in microseconds (on the x axis). The initial peak corresponds to the direct or shortest-path signal and is not necessarily the highest peak in the waveform. The exponential characteristics of the envelope are typical of the general trend; however, significant peaks can occur at considerable delay times, and can usually be attributed to tall buildings. The parameters used are the same as those used by Cox and Leck [10], except that the names of the parameters have been changed to clarify their meaning. The parameters shown in Fig. 1.15 are the first moment or mean delay time d and the standard deviation or delay spread Δ, calculated as follows:

$$d = \int_0^\infty tE(t)\,dt \tag{1.51}$$

$$\Delta^2 = \int_0^\infty t^2 E(t)\,dt - d^2 \tag{1.52}$$

where $t = 0$ is designated the leading edge of the envelope, $E(t)$ of $s(t)$ shown in Eq. (1.50).

The typical ranges for these two parameters are summarized as follows:

Parameter	Urban	Suburban
Mean delay time d	1.5–2.5 µs	0.1–2.0 µs
Corresponding path length	450–750 m	30–600 m
Maximum delay time (–30 dB)	5.0–12.0 µs	0.3–7.0 µs
Corresponding path length	1.5–3.6 km	0.9–2.1 km
Range of delay spread Δ_i	1.0–3.0 µs	0.2–2.0 µs
Mean delay spread	1.3 µs*	0.5 µs
Maximum effective delay spread	3.5 µs	2.0 µs

   * 3.0 µs is normally used for calculation.

The urban values are representative of Manhattan, whereas the suburban values are representative of Keyport, Hazlet, and Middletown, all located in New Jersey. The delay spread Δ is longer in urban areas than it is in suburban areas. The maximum delay time, measured

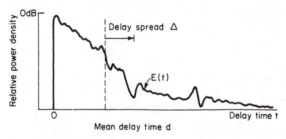

Figure 1.15   Typical delay envelope.

30 dB down from the highest level of the envelope, can be as much as 12 μs over a delay path length of 4 km in an urban area. The mean delay spread Δ for a suburban area is 0.5 μs and for an urban area is 1.3 μs. However, for studying the transmission data rate in an urban area, 3 μs is normally used. It is therefore probable that any signaling rate in excess of 2 MHz will cause intersymbol interference when no diversity schemes are used. In a mobile-radio environment, multipath fading causes performance degradation, and therefore the signaling rate must be less than 2 MHz. The rate drops as the required bit-error rate at the receiver decreases. In general, the delay spread can be considered independent of the transmitting frequency. The delay spread values are unchanged for any operating frequency above 150 MHz, because, above 150 MHz, the wavelengths are always much less than the sizes of human-made structures. Usually, the structure size greater than six wavelengths to ten wavelengths for sure is claimed as a scatterer, which can reflect the energy from any wave incident on its surface. For this reason, all human-made structures can be treated as reflectors for any operating frequency above 150 MHz. The number of scatterers counted will be the same independent of the operating frequencies as long as they are above 150 MHz. Thus, the number of reflected-wave paths occurring is the same regardless of the spectrum of the frequency. The delay spread is therefore the same.

A delay-spread model can be expressed as

$$p(T_i) = \frac{1}{\Delta} \exp\left(\frac{T_i}{\Delta}\right) \tag{1.50}$$

where T_i is the time delay of ith wave arrival. This model is assumed by N equal amplitude reflected waves, most of which arriving earlier are spaced closely. Fewer arriving later are spaced far apart, as shown in Fig. 1.16(a). The distribution of delay spread $p(T_i)$ is shown in Fig.

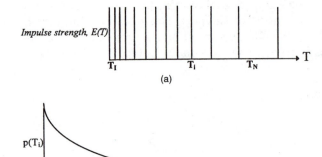

Figure 1.16   The distribution of delay spread: (a) most arrive closely near T_I; (b) delay spread $p(T_i)$.

1.16(*b*). Another model uses the exponential amplitudes at equal time intervals. These two models are equivalent, but the model of Eq. (1.50) is easier to use for mathematical manipulation.

Delay spread is caused by the multipath wave arrival after reflection from the human-made structures. The following mean delay spread data due to different human-made environments are listed as an aid to the reader.

Type of environment	Mean delay spread (Δ)
In building	<0.1 µs
Open area	<0.2 µs
Suburban area	0.5 µs
Urban area	3 µs

Delay spread in VHF

Previous studies of time-delay spread at 450 MHz [14] and 900 MHz [9, 15] have been thoroughly documented. However, only limited empirical data are available on time-delay spread in the VHF domain, and the data available show that the delay spread is independent of the operating frequency at frequencies above 30 MHz [16]. Two arguments can be used to explain these findings. The first argument states that since the path loss is less at lower frequencies, the area of scattering around the mobile unit increases, and therefore the delay spread may also increase as frequency decreases.

The second argument states that since the wavelength increases as the frequency decreases, the size of the local scatterers approaches the wavelength at 30 MHz, and therefore most of the radio-wave energy penetrates the scatterers and the delay spread decreases as the operating frequency decreases. The scattered area becomes smaller, less reflected energy is received by the mobile unit, and the delay spread decreases. These two arguments contradict each other and reinforce the conclusion that delay spread is independent of frequency above 30 MHz [16, 21].

Example 1.5 Figure E1.5.1 shows the relative configuration for a communications link and a signal-jamming source. In order to determine how quickly frequency hopping must be accomplished to prevent the jammer from following, it is necessary to calculate the time delay between signal arrival from the transmitter T_x and the arrival of the jammer signal J. Since the jammer must first receive the signal from the transmitter, detect it, and transmit the jamming signal, the time delay between $T_x \rightarrow R$ and $J \rightarrow R$ is

$$T = T_{T_x \rightarrow J} + T_{J \rightarrow R} - T_{T_x \rightarrow R} + \delta_T$$

Assuming that the jammer detection time δ_T is negligible, the frequency-hopping rate should be greater than $1/T$ in order to avoid jamming.

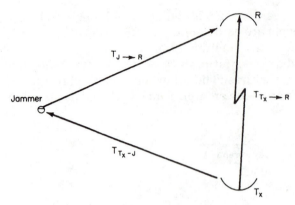

Figure E1.5.1 Typical communications link and jammer.

Note that the time delay associated with jamming is not to be confused with the time-delay spread, which is described in Example 1.6.

Example 1.6 In a certain case in which the mobile environment introduces a 5-μs delay spread, the signaling rate must be less than 20 kb/s. As shown in Fig. E1.5.1, if the time delay between $T_x \rightarrow R$ and $J \rightarrow R$ is 0.5 ms, then the frequency-hopping rate should be at least 2 kHz to ensure that 10 bits (b) or less are sent out during each hopping interval.

1.6   Coherence Bandwidth

The existence of different time delays in two fading signals that are closely spaced in frequency can cause the two signals to become correlated. The frequency spacing that allows this condition to prevail is dependent upon the delay spread Δ. This frequency interval is called the "coherence" or "correlation" bandwidth (B_c).

For purposes of discussion, the model used to illustrate the delay-spread envelope has an initial impulse corresponding to the received specular energy, followed by a decaying exponential corresponding to the received scattered energy. The initial impulse is assumed to have arrived undistorted via a nondispersive path, whereas the latter, decaying portion is assumed to have arrived by way of a large number of scattered paths, and therefore to have a finite correlation bandwidth. The normalized magnitude, using the Laplace transform of the delay-spread envelope, gives the correlation function $C(f)$ of the signal. Figure 1.17(a) shows an idealized representation of the impulse response model shown in Figs. 1.15 and 1.16. The corresponding magnitude of the bandwidth correlation function is shown in Fig. 1.17(b).

The precise definition of coherence bandwidth often differs from one reference to another (see Refs. 22 and 23) and tends to depend on the

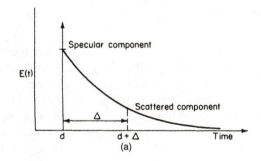

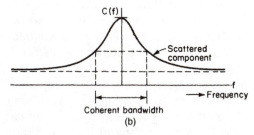

Figure 1.17 Channel impulse response model: (a) model presentation; (b) correction function.

extent of the correlation, determined subjectively, over which the two signals are correlated. A typical definition [23] is

$$B_c \approx \frac{1}{8\Delta} \tag{1.53}$$

The coherence bandwidths for path lengths of approximately 2 to 4 km are typically in the range of 100 kHz to 1 MHz, varying inversely with the number of multipaths present in the mobile-radio environment. In Chap. 6, it will be shown that

$$B_c = \frac{1}{2\pi\Delta} \qquad \text{for amplitude-modulated systems} \tag{1.54}$$

based on an amplitude correlation equal to 0.5, and that

$$B_c = \frac{1}{4\pi\Delta} \qquad \text{for frequency- or phase-modulated systems} \tag{1.55}$$

based on a phase correlation equal to 0.5. By averaging Eqs. (1.54) and (1.55), Eq. (1.53) can be obtained.* Equation (1.53) can be used in general if the modulation scheme is unknown.

---

* Equation (1.50) is obtained from a different approach in Ref. 23.

1.7   Multipath Fading Phenomenon in the 800- to 900-MHz Region

During the 1960s, Bell Telephone Laboratories chose the 800- to 900-MHz frequency region for preliminary study and planning of advanced mobile-telephone systems. One of the recognized advantages of operating mobile telephones in this higher frequency region was the ability to increase spectrum utilization. For example, 1 percent of the bandwidth at 35 MHz is only 35 kHz, whereas 1 percent of the bandwidth at 800 MHz approaches 1 MHz. However, there are practical limits to how much higher the frequency region can be extended for mobile-telephone applications. A prime consideration is that the severity of multipath fading greatly increases as the channel frequency increases. This effect is confirmed in Chaps. 3 to 7. At frequencies above 10 GHz, rainfall becomes a significant attenuation factor [24], in addition to the other causes of severe path loss. For this reason, frequencies above 10 GHz are not desirable for mobile-radio-telephone communications.

The severity of the multipath-fading phenomenon at 900 MHz is shown in Fig. 1.18, which presents a typical segment of a fading signal received at a mobile unit traveling at 24 km/h (15 mi/h). Variations as large as 40 dB in signal amplitude can occur as a result of fading, with nulls occurring about every half wavelength. Such severe fading will degrade the signal and produce poor voice quality. A complete understanding of the multipath-fading phenomenon is essential if we are to statistically analyze the fading characteristics and find suitable methods for counteracting fading effects.

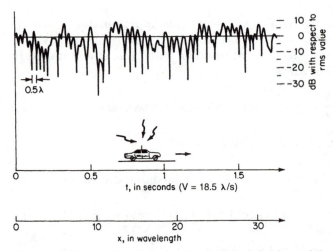

Figure 1.18 A typical fading signal received while the mobile unit is moving.

Problem Exercises

1.   A mobile unit traveling at a constant rate of speed experiences multipath fading while tuned to a mobile-radiotelephone channel operating at 850 MHz. Figure P1.1(*a*) shows the continuous spectrum after square-law detection on the mobile received signal. Figure P1.1(*b*) shows the multipath wave pairs directed toward the receiving antenna of the mobile unit.

    a.   What is the speed of the mobile unit?

    b.   What is the direction of wave arrival that contributes –5 dB to the amplitude of the signal at 0 Hz?

2.   A mobile unit is traveling in the vicinity of two nearby scatterers with a reflection coefficient of 0.5. Figure P1.2 shows three received-signal paths, A, B, and C.

    a.   What is the expression for the resultant signal from paths A, B, and C?

    b.   Plot the amplitude variation due to multipath fading and calculate the Doppler frequency.

3.   On the basis of the information plotted in Fig. P1.3, convert the signal $s(t)$ from the time frame to the distance frame while the speed of the mobile vehicle is varying.

4.   On the basis of the parameters illustrated in Fig. P1.4, find the reflection point on a sloping terrain between a base-station transmitting antenna and a mobile-unit receiving antenna.

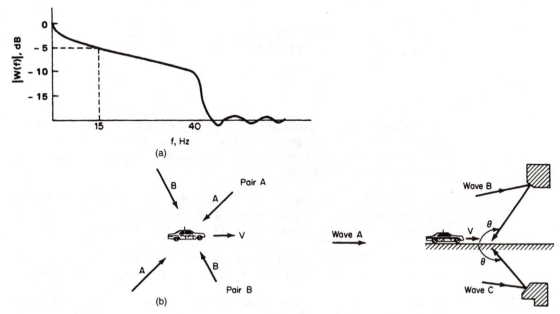

Figure P1.1   Problem 1: (*a*) continuous spectrum; (*b*) two pairs of incoming waves.

Figure P1.2   Problem 2.

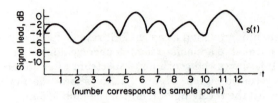

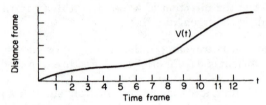

Figure P1.3   Problem 3.

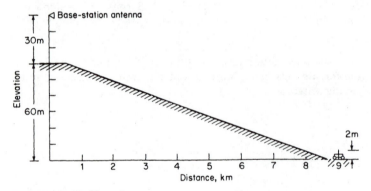

Figure P1.4   Problem 4.

5.   On the basis of an assumption that a delay spread of 1 μs has been found in the medium, how far apart in frequency separation should two frequencies carrying the same information be spaced to ensure fading characteristics that are independent of each other?

6.   Find the level-crossing rate N/T based on the number of times N that the signal shown in Fig. 1.17 drops below −10 dB during the total time interval T. Repeat the procedure for signal levels of 0, −15, and −20 dB, respectively.

7.   On the basis of the level-crossing rates for 0, −15, and −20 dB determined in Prob. 6, plot the level-crossing-rate curves versus signal level in decibels.

8.   If the signaling rate is increased and a bit-error rate of 0.5 is obtained, increasing the power does not decrease the error rate. What other means can be used to reduce the bit-error rate?

9. On the assumption that a low-pass filter is used, find the bandwidth B_N and the filtered white-noise power. The characteristics of the low-pass filter are as follows:

$$|H(f)| = \frac{1}{1 + 3(f/f_0)^2}$$

where f_0 is the center frequency.

References

1. W. C. Y. Lee, "Theoretical and Experimental Study of the Properties of the Signal from an Energy Density Mobile Radio Antenna," *IEEE Trans. Veh. Tech.,* vol. 16, October 1967, pp. 121–128.
2. R. H. Clarke, "A Statistical Theory of Mobile Radio Reception," *Bell System Technical Journal,* vol. 47, July 1968, pp. 957–1000.
3. S. O. Rice, "Mathematical Analysis of Random Noise," in N. Wax (ed.), *Selected Papers on Noise and Stochastic Process,* Dover, New York, 1954.
4. S. O. Rice, "Statistical Properties of a Sine Wave plus Random Noise," *Bell System Technical Journal,* vol. 24, January 1948, pp. 109–157.
5. *ITT Reference Data for Radio Engineers,* Sams, Indianapolis, 1968, p. 27-2.
6. A. D. Spaulding and R. T. Disney, *Man-Made Radio Noise, Part I: Estimate for Business, Residential, and Rural Areas,* U.S. Dept. of Commerce, OT Report 74-38, June 1974.
7. A. D. Spaulding, "The Determination of Received Noise Levels from Vehicular Traffic Statistics," *IEEE Nat. Telecomm. Conf. Rec.,* 19D-1-7, December 1972.
8. E. N. Skomal, *Man-Made Radio Noise,* Van Nostrand Reinhold, New York, 1978, chap. 2.
9. D. C. Cox and R. P. Leck, "Correlation Bandwidth and Delay-Spread Multipath Propagation Statistics for 910 MHz Urban Mobile Radio Channels," *IEEE Trans. Comm.,* vol. 23, November 1975, pp. 1271–1280.
10. D. C. Cox and R. P. Leck, "Distribution of Multipath Delay Spread and Average Excess Delay for 910 MHz Urban Mobile Radio Paths," *IEEE Trans. Anten. Prop.,* vol. 23, March 1975, pp. 206–213.
11. D. C. Cox, "910 MHz Urban Mobile Radio Propagation: Multipath Characteristics in New York City," *IEEE Trans. Comm.,* vol. 21, November 1973, pp. 1188–1194.
12. D. C. Cox, "A Measured Delay-Doppler Scattering Function for Multipath Propagation in an Urban Mobile Radio Environment," *Proc. IEEE,* vol. 61, April 1973, pp. 479–481 (letter).
13. D. C. Cox, "Delay-Doppler Characteristics of Multipath Propagation at 910 MHz in a Suburban Mobile Radio Environment," *IEEE Trans. Anten. Prop.,* vol. 20, September 1972, pp. 625–635.
14. H. Suzuki, "A Statistical Model for Urban Radio Propagation," *IEEE Trans. Comm.,* vol. 25, July 1977, pp. 673–680.
15. H. F. Schmid, "A Prediction Model for Multipath Propagation of Pulse Signals at VHF and UHF over Irregular Terrain," *IEEE Trans. Anten. Prop.,* vol. 18, March 1970, pp. 253–258.
16. J. J. Egli, "Radio Propagation above 40 Mc over Irregular Terrain," *Proc. IRE,* October 1957, pp. 1383–1391.
17. G. L. Turin et al., "A Statistical Model of Urban Multipath Propagation," *IEEE Trans. Veh. Tech.,* February 1972, pp. 1–8.
18. W. R. Young, Jr., and L. Y. Lacy, "Echoes in Transmission at 450 Megacycles from Land-to-Car Radio Units," *Proc. IRE,* vol. 38, March 1950, pp. 255–258.
19. D. L. Nielson, "Microwave Propagation Measurements for Mobile Digital Radio Application," *IEEE Trans. Veh. Tech.,* vol. 27, August 1978, pp. 117–132.
20. M. Schwartz, W. R. Barnett, and S. Stein, *Communication Systems and Techniques,* McGraw-Hill, New York, 1966, sec. 11-5.

21. D. W. Peterson, "Army Television Problems," *Final Report, Phase II, Tasks 1 and 2,* RCA, work performed under Signal Corps contract DA-36-039-SC-64438, January 1956.
22. W. C. Jakes, *Microwave Mobile Communications,* Wiley, New York, 1974, p. 51.
23. M. J. Gans, "A Power Spectral Theory of Propagation in the Mobile Radio Environment," *IEEE Trans. Veh. Tech.,* vol. 21, February 1972, pp. 27–38.
24. W. C. Y. Lee, "An Approximate Method for Obtaining Rain Rate Statistics for Use in Signal Attenuation Estimating," *IEEE Trans. Anten. Prop.,* vol. 27, May 1979, pp. 407–413. (A simple method first proposed in 1974 and widely used by Bell Labs.)

Statistical Communications Theory

2.1 The Statistical Approach

Statistical communications theory is a powerful tool for solving complex problems that are too complicated to solve by ordinary means. These complications arise when there are too many variations or too many variables and when the mobile-radio phenomena are too complicated to describe in simple terms. Statistical communications theory can be used to analyze various types of natural phenomena. For example, when the parameters for describing a medium are random in nature, there are always laws for defining them. The natural laws do not provide deterministic answers but render results that can be subjected to statistical evaluation. As an example, a resultant signal for N incoming waves received by a mobile unit in motion has already been expressed in Eq. (1.20) but can also be written in another way:

$$s(t) = (R + jS) \exp \left[j(2\pi f_0 t + \phi_0) \right] \tag{2.1}$$

where

$$R = a_0 \sum_{i=1}^{N} a_i \cos \psi_i \tag{2.2}$$

and

$$S = a_0 \sum_{i=1}^{N} a_i \sin \psi_i \tag{2.3}$$

as can be deduced from Eq. (1.21).

The relationships between the parameters of Eq. (1.20) and Eq. (2.1) are as follows:

$$A_t = (R^2 + S^2)^{1/2} \tag{2.4}$$

$$\psi_t = \tan^{-1} \frac{S}{R} \tag{2.5}$$

In general, the expressions for a resultant signal of N scattered waves, Eqs. (1.20) through (1.22), are used primarily to study the amplitude and phase of the baseband signal output, whereas Eqs. (2.1) through (2.3) are used primarily to study the natural characteristics of radio waves arriving at the RF input to the receiver.

The values R and S of Eqs. (2.2) and (2.3) separate the sum value of individual waves into real and imaginary parts, where a_i is the amplitude and ψ_i is the phase for the ith wave, as shown in Eqs. (1.21) and (1.22). It is practically impossible to measure a_i and ψ_i of the ith wave individually, and therefore it is impossible to know their exact values. For this reason, it is necessary to treat all a_i and ψ_i as random variables (or variates) within the bounds of any natural statistical laws which may apply.

A quantity consisting of two or more random variables forms a new random variable; hence R and S shown in Eqs. (2.2) and (2.3) respectively are also random variables that can be expressed as

$$R = \sum_{i=1}^{N} R_i \tag{2.6}$$

$$S = \sum_{i=1}^{N} S_i \tag{2.7}$$

If we assume that R_i is uniformly distributed within the $(0, 1)$ interval as shown in Fig. 2.1(a) and that R is merely the sum of three variables R_i, then the probability density of R is very close to a Gaussian- (normal-) distributed wave, as shown in Fig. 2.1(b). This is also true for S. This type of random variable (or variate) exhibits certain regularities that can best be described in terms of probability and statistical average. In mobile radio, there are always more than three direct and/or reflected waves arriving at the mobile receiver. Hence, the real and imaginary parts of a resultant mobile-radio wave are always Gaussian-distributed waves.

In addition to the Gaussian probability laws, there are numerous other probability laws and statistical techniques for analyzing both discrete and continuous variables. Only those laws which are applied in subsequent chapters are discussed in this chapter. Many excellent books [1–3] are recommended for further study.

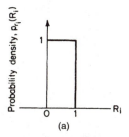
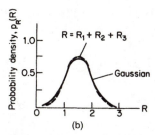

Figure 2.1 Distribution of the sum of three uniformly distributed variables: (*a*) uniformly distributed; (*b*) Gaussian distributed.

2.2 Averages

In determining average values for statistical analysis, there are four separate definitions, each with different notations and each yielding different statistical results. It is therefore essential to use the appropriate averaging technique if valid results are to be obtained. The different averages for a random variable are defined as follows:

Sample averages are those obtained by using conventional arithmetical solutions for any finite number of trials; they include the case where individual trials are repeated any number of times. The expression for conventional arithmetical averages of any finite number of trials N is

$$\bar{x} = \frac{\sum\limits_{i=1}^{N} x_i}{N} \tag{2.8}$$

where all x_i values are assumed to have equal probability. In the case where x_1 is observed n_1 times and x_2 is observed n_2 times, etc., the sum of the observed values can be expressed:

$$n_1 x_1 + n_2 x_2 + n_3 x_3 + \cdots + n_N x_N = \sum\limits_{i=1}^{N} n_i x_i \tag{2.9}$$

Then the sample average becomes:

$$\bar{x} = \sum\limits_{i=1}^{N} n_i x_i \Bigg/ \sum\limits_{i=1}^{N} n_i = \sum\limits_{i=1}^{N} \frac{n_i}{n} x_i \tag{2.10}$$

When all n_i's are equal, then Eq. (2.10) can be reduced to Eq. (2.8).

Statistical averages (expectation or ensemble averages) are those obtained for a very large number of trials, as applied to a conventional arithmetical solution such as that expressed in Eq. (2.10). In such cases, the average is based on the probability of occurrence and therefore becomes a statistical average. The expression for a statistical average is

$$E[x] = \lim_{n \to \infty} \sum_{i=1}^{N} \frac{n_i}{n} x_i = \lim_{n \to \infty} \sum_{i=1}^{N} \left(\frac{n_i}{n} \right) x_i$$

$$= \sum_{i=1}^{N} P(x_i) x_i \qquad (2.11)$$

where $E[x]$ represents the expectation of x and $P(x_i)$ is the probability of occurrence for x_i.

Biased time averages are those obtained when a continuous variate $x(t)$ occurs within a time domain. The mean value of $x(t)$ in a finite period T can be defined:

$$\tilde{x} = \frac{1}{T} \int_0^T x(t) \, dt \qquad (2.12)$$

Unbiased time averages are those obtained when a continuous variate $x(t)$ occurs over a period T that approaches infinity. This condition can be expressed:

$$\langle x(t) \rangle = \lim_{T \to \infty} \frac{1}{T} \int_0^T x(t) \, dt \qquad (2.13)$$

2.3   Ergodic Processes

Analysis of the statistical parameters for determining averages reveals that many of the random signals or variates present in the mobile-radio communication medium have statistically identical time and ensemble characteristics, such as autocorrelation and probability distribution. These identical characteristics result in certain equivalences between the expressions for statistical averages and unbiased time averages shown in Sec. 2.2; e.g., Eq. (2.11) is equivalent to Eq. (2.13):

$$E[x(t)] = \langle x(t) \rangle \qquad (2.14)$$

$$E[x^2(t)] = \langle x^2(t) \rangle \qquad (2.15)$$

$$E[x^n(t)] = \langle x^n(t) \rangle \qquad (2.15a)$$

The type of random process for which these equivalences are true is said to be "ergodic."

An ergodic process is also a "stationary" process, one in which statistics are not affected by shifting the time origin of the data from t to $t + \tau$, or in which the following equivalence is true:

$$\langle x(t) \rangle = \langle x(t + \tau) \rangle \qquad (2.16)$$

The reason the ergodic process is stationary is that the ensemble averages shown in Eq. (2.11) are independent of the time of observation, and the relationships of both Eq. (2.14) and Eq. (2.15) must hold. However, a stationary process does not necessarily imply ergodicity. Fortunately, in practice, random data representing stationary physical phenomena are generally ergodic, as are the data collected in the mobile-radio field. The ergodic random process is very important to the mobile-radio field, since all properties of ergodic random processes can be determined by performing time averaging over a single sample function, $x(t)$. In the analyses of mobile-radio random signals, it is necessary to deal almost exclusively with sample functions that are essentially ergodic processes and require the application of statistical theory in order to arrive at a satisfactory solution. Hence, in subsequent discussions the notations $E[\ ]$ and $\langle\ \rangle$ are interchangeable.

2.4 Cumulative Probability Distribution (CPD)

In the case where x is the random variable and X is a specified value of x, then the cumulative probability distribution (cpd) is defined as the probability that the random variable event x is equal to or less than the value X. The notation for this cpd function can be expressed as follows:

$$F_x(X) = \text{prob}\ (x \leq X) \tag{2.17}$$

$F_x(X)$ is called the "probability distribution function"; $F_x(X)$ of an event has the following general properties:

$$0 \leq F_x(X) \leq 1 \tag{2.18a}$$

$$F_x(-\infty) = 0 \quad \text{and} \quad F_x(\infty) = 1 \tag{2.18b}$$

$$F_x(X) \text{ is monotonic increasing with } X \tag{2.18c}$$

$$\text{prob}\ (X_1 < x \leq X_2) = F_x(X_2) - F_x(X_1) \tag{2.18d}$$

$$P(x \leq M) = 50\% \quad \text{when} \quad M = \text{the median value} \tag{2.18e}$$

In calculating the cpd of a mobile-radio signal, the sample functions $x(t)$ can be digitized, so that each sample of x_i can be correlated with the different signal levels of X. The number of samples which have a signal strength less than a predefined level can then be divided by the total number of samples to determine the cumulative probability at that level. The speed of the vehicle can be disregarded, as long as the vehicle is in motion at a constant speed, since the distribution is independent of time.

2.5   Probability Density Function (PDF)

Although the cpd process of Sec. 2.4 can be used to describe a complete probability model for a sample random variable, it is not the best choice for other types of calculations. For example, to find the cpd for a random variable $z = x_1 + x_2$ by using $F_{x_1}(X_1)$ and $F_{x_2}(X_2)$, it is easier and preferable to use the derivative of $F_{x_i}(X_i)$ (where $i = 1$ and 2) rather than $F_{x_i}(X_i)$ itself. The derivative is known as the "probability density function" (pdf).

PDF of a single variable

In determining the pdf of a single variable, the derivative of the cpd is the probability density function $p_x(X)$:

$$p_x(X) = \frac{d}{dX} F_x(X) \tag{2.19}$$

or

$$p_x(X)\, dX = \text{prob}\,(X \le x \le X + dX) \tag{2.20}$$

Its general properties may be summarized as follows:

$$p_x(x) \ge 0 \qquad -\infty < x < \infty \tag{2.21a}$$

$$\int_{-\infty}^{\infty} p_x(x)\, dx = 1 \tag{2.21b}$$

$$F_x(X) = \int_{-\infty}^{X} p_x(u)\, du \tag{2.21c}$$

$$\int_{X_1}^{X_2} p_x(x)\, dx = \text{prob}\,(X_1 \le x \le X_2) \tag{2.21d}$$

$$E(x) = \int_{-\infty}^{\infty} x p_x(x)\, dx \tag{2.21e}$$

$$E(x^n) = \int_{-\infty}^{\infty} x^n p_x(x)\, dx \tag{2.21f}$$

The notation $P(x)$ can be used in place of $P_x(x)$ when there are no complications. The pdf method also indicates where the greatest number of random variables is concentrated. The different types of pdf calculations are defined in the following examples and subsections.

Example 2.1   At times, it is desirable to bin the digitized samples x in incremental 1-dB bins instead of in linear-scale bins; i.e., samples between 5.5 and 6.5 dB will be in the 6-dB bin. In this instance, where N_1 is the number of samples x in the first bin, N_2 is the number of samples x in the second bin, and so on, then the pdf of this set of data is:

$$p(y \text{ in bin } j) = \frac{N_j}{\sum_{i=1}^{N} N_i} \tag{E2.1.1}$$

Where N is the total number of bins, the pdf curves can be predicted. It is more practical to use the decibel-bin collection method than the linear-scale-bin collection method. However, because statistical distributions are usually expressed in terms of linear-scale bins, it is advantageous to convert from linear to decibel scale. If the pdf of x in the linear scale is $p(x)$ and the pdf of y in the decibel scale is $p(y)$ and it is assumed that $y_1 = 20 \log x$ (where x is in volts), then the pdf of y_1 in the decibel scale is

$$p(y_1) = p(x) \frac{dx}{d(20 \log x)} = p(x) \frac{x}{20 \log e} \tag{E2.1.2}$$

The pdf expressed in terms of decibel-scale bins can thus be used in place of the more commonly used linear-scale statistical distribution.

Example 2.2   If we divide the total number of bins into half-dB increments and it is assumed that $y_2 = 40 \log x$, then the pdf of y_2 will be:

$$p(y_2) = p(x) \frac{x}{40 \log e} \tag{E2.2.1}$$

Example 2.3   For bins of different sizes, the relationships between $p_{y_1}(y_1)$ and $p_{y_2}(y_2)$ and between $p(y_1 \leq A)$ and $p(y_2 \leq A)$ as indicated in Examples 2.1 and 2.2 are as follows:

$$p_{y_2}(y_2) = \frac{1}{2} p_{y_1}(y_1) \tag{E2.3.1}$$

and

$$P(y_1 \leq A) = P(y_2 \leq A) \tag{E2.3.2}$$

Figure E2.3.1 shows the typical plots for the functions expressed in Eqs. (E2.1.2), (E2.3.1), and (E2.3.2), respectively.

Joint PDF of two variables

Let the joint cumulative probability distribution of the random variables x and y be defined by the equation

$$F_{xy}(X, Y) = P(x \leq X, y \leq Y) \tag{2.22}$$

If it is assumed that $F_{xy}(x, y)$ has partial derivatives of order 2, then this quantity can be expressed as the joint pdf of the random variables x and y, defined by the equation

$$p(x, y) = \frac{\partial^2 F_{xy}(X, Y)}{\partial X \, \partial Y} \bigg|_{x=X, y=Y} \tag{2.23}$$

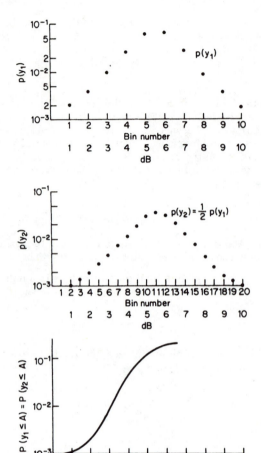

Figure E2.3.1   Plots of statistical distribution of functions.

where the following properties always hold true:

$$p(x, y) \geq 0 \tag{2.24}$$

$$\int_{-\infty}^{\infty} \int_{-\infty}^{\infty} p(x, y)\, dx\, dy = 1 \tag{2.25}$$

In situations in which it is desirable to find the pdf for random variable z as a function of x and y [i.e., where $z = f(x, y)$], the joint pdf, or $p(x, y)$, should be found first. The application of this technique is discussed in Chap. 6.

Marginal PDF

At times, it may be necessary to determine the relationship between the marginal and joint densities. The distribution $F_x(X)$ can be expressed in terms of the joint density as follows:

$$F_x(X) = F_{xy}(X, \infty) = \int_{-\infty}^{\infty} \int_{-\infty}^{X} p_{xy}(\xi, \eta) \, d\xi \, d\eta \tag{2.26}$$

By differentiating Eq. (2.26) with respect to X, the marginal pdf, or $p_x(X)$, can be defined as follows:

$$p_x(X) = \int_{-\infty}^{\infty} p_{xy}(X, y) \, dy \tag{2.27}$$

Similarly,

$$p_y(Y) = \int_{-\infty}^{\infty} p_{xy}(x, Y) \, dx \tag{2.28}$$

However, when two random variables x and y are mutually independent, then

$$p_{xy}(x, y) = p_x(x)p_y(y) \tag{2.29}$$

The notation $p(x, y)$ can be used in place of $p_{xy}(x, y)$ when there is no confusion.

Joint characteristic functions

There are instances when it is difficult to obtain $p(z)$ of the random variable z directly from $p(x, y)$, where $z = f(x, y)$. In these instances, the method of joint characteristic functions can be used as a bridge to obtain $p(z)$ from $p(x, y)$. The joint characteristic functions of two random variables x and y can be defined as follows:

$$\Phi_{xy}(v_1, v_2) = E\{\exp [j(v_1 x + v_2 y)]\}$$

$$= \int_{-\infty}^{\infty} \int_{-\infty}^{\infty} \exp [j(v_1 x + v_2 y)] \, p_{xy}(x, y) \, dx \, dy \tag{2.30}$$

Thus, $\Phi_{xy}(v_1, v_2)$ is the double Fourier transform of $p_{xy}(x, y)$, and it may easily be shown that

$$p_{xy}(x, y) = \int_{-\infty}^{\infty} \int_{-\infty}^{\infty} \exp [-j(v_1 x + v_2 y)] \, \Phi_{xy}(v_1, v_2) \, dv_1 \, dv_2 \tag{2.31}$$

The marginal characteristic functions $\Phi_x(v_1)$ and $\Phi_y(v_2)$ can be expressed in terms of $\Phi_{xy}(v_1, v_2)$:

$$\Phi_x(v_1) = E[e^{jv_1x}] = \Phi_x(v_1, 0) \tag{2.32}$$

Similarly,

$$\Phi_y(v_2) = E[e^{jv_2y}] = \Phi_y(0, v_2) \tag{2.33}$$

For example, if the values of $p_{x_i}(x_i)$ are known, $\Phi_{x_i}(v)$ can be found from Eq. (2.32).

In the situation where $z = \sum_{i=1}^{N} x_i$, it is easy to determine from Eq. (2.30) that

$$\Phi_z(v) = \prod_{i=1}^{N} \Phi_{x_i}(v) \tag{2.34}$$

and the pdf of z can be found as follows:

$$p_z(z) = \int_{-\infty}^{\infty} \exp{(-jvz)}\, \Phi_z(v)\, dv \tag{2.35}$$

Conditional PDF

There are times when it is useful to find the pdf of one variable with respect to another variable that remains fixed at a particular level. This comparison makes it possible to understand and evaluate the effects of the selected variable upon other variables. For example, the conditional pdf of y, assuming $x = X$, can be expressed:

$$p_y(y \mid x = X) = \frac{p_{xy}(x, y)}{p_x(X)} \tag{2.36}$$

Similarly,

$$p_x(x \mid y = Y) = \frac{p_{xy}(x, y)}{p_y(Y)} \tag{2.37}$$

Therefore, by combining Eqs. (2.36) and (2.37), the following expression is derived:

$$p_y(y \mid x = X) = \frac{p_x(x \mid y = Y)p_y(Y)}{p_x(X)} \tag{2.38}$$

The conditional pdf method is a powerful tool for calculating $p(x, y)$ from either Eq. (2.36) or Eq. (2.37). Equation (2.38) is a continuous version of Bayes' theorem.

Example 2.4   The following sequence is used to find the cpd for a sampling of mobile-radio signal data that has an amplitude variable x that is always less than a given value a.

$$P(x \le X \mid x \le a) = \frac{P(x \le X, x \le a)}{P(x \le a)}$$

(E2.4.1)

If $X \ge a$, then

$$P(x \le X \mid x \le a) = \frac{P(x \le a)}{P(x \le a)} = 1$$

(E2.4.2)

If $X < a$, then

$$P(x \le X \mid x \le a) = \frac{P(x \le X)}{P(x \le a)} = \frac{F_x(X)}{F_x(a)}$$

(E2.4.3)

Example 2.5   When the signal distribution is $F_x(X) = 1 - \exp(-X^2/2)$, then the relationship $P(x \le X \mid x \le a) = F_x(X \mid x \le a)$ can be found by using the method shown in Example 2.4. However, for $X \ge a$,

$$F_x(X \mid x \le a) = 1$$

(E2.5.1)

For $X < a$,

$$F_x(X \mid x \le a) = \frac{F_x(X)}{F_x(a)} = \frac{1 - e^{-X^2/2}}{1 - e^{-a^2/2}}$$

(E2.5.2)

Figure E2.5 shows the distribution response for Eqs. (E2.5.1) and (E2.5.2).

Example 2.6   The following method is used to evaluate the approximate mean of the function $g(x)$ where the pdf, $p_x(x)$, of the variable x is known.

Let

$$\eta = E(x) \qquad \text{the mean of } x$$

$$E[(x - \eta)^n] = \mu_n \qquad \text{the central moment of } x \text{ for each } n$$

Then

$$\mu_1 = 0$$

$$\mu_2 = \sigma^2 \qquad \text{the variance of } x$$

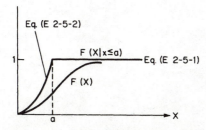

Figure E2.5   Signal distribution for $F(X)$.

The mean and mean square variance for the function $g(x)$ are expressed as follows:

$$E[g(x)] = \int_{-\infty}^{\infty} g(x)p_x(x)\, dx \text{ (mean)} \tag{E2.6.1}$$

$$E[g^2(x)] = \int_{-\infty}^{\infty} g^2(x)p_x(x)\, dx \text{ (mean square)} \tag{E2.6.2}$$

If $g(x)$ is a smooth function, then

$$E[g(x)] = g(\eta) \int_{-\infty}^{\infty} p_x(x)\, dx = g(\eta) \tag{E2.6.3}$$

Expansion of the function $g(x)$ into a series around η is expressed:

$$g(x) = g(\eta) + g'(\eta)(x - \eta) + g''(\eta)\frac{(x - \eta)^2}{2!} + \cdots + g^{(n)}(\eta)\frac{(x - \eta)^n}{n!} \tag{E2.6.4}$$

By substituting Eq. (E2.6.4) in Eq. (E2.6.1), the following general form is obtained:

$$E[g(x)] = g(\eta) + g''(\eta)\frac{\sigma^2}{2} + \cdots + g^{(n)}(\eta)\frac{\mu_n}{n!} \tag{E2.6.5}$$

If $g(\eta)$ is sufficiently smooth, as in Fig. E2.6, then the following term applies:

$$E[g(x)] = g(\eta) + g''(\eta)\frac{\sigma^2}{2} \tag{E2.6.6}$$

2.6   Useful Probability Density Functions

Uniform distribution

A variate x is said to be uniformly distributed when the probability exists that all random values chosen within a prescribed range (a, b) are statistically equal. A graph illustrating this type of distribution is shown in Fig. 2.2; its density function can be expressed as follows:

$$p_x(x) = \begin{cases} \dfrac{1}{b - a} & a \leq x \leq b \\ 0 & \text{otherwise} \end{cases} \tag{2.39}$$

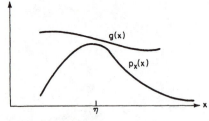

Figure E2.6   Plot of $g(x)$ and $p_x(x)$.

For example, the phase angle $p(\theta)$ of multipath signal fading with uniform distribution over $[-\pi, \pi]$ would be expressed as:

$$p_\theta(\theta) = \frac{1}{2\pi} \qquad -\pi \leq \theta \leq \pi \tag{2.40}$$

Gaussian distribution

The summing up of a large number of variates will normally produce a Gaussian distribution. Hence, a Gaussian distribution is generally referred to as a "normal" distribution. The density function for a Gaussian-distributed variate is illustrated in Fig. 2.3 and can be expressed as follows:

$$p_x(x) = \frac{1}{\sqrt{2\pi}\sigma} \exp\left[-\frac{(x-m)^2}{2\sigma^2}\right] \tag{2.41}$$

where the mean m is defined

$$m = \int_{-\infty}^{\infty} x p_x(x)\, dx = E(x) \tag{2.42}$$

and the mean square of x is $E(x^2)$, defined

$$E(x^2) = \int_{-\infty}^{\infty} x^2 p_x(x)\, dx \tag{2.43}$$

and the standard deviation σ is defined

$$\sigma = [E(x^2) - m^2]^{1/2} \tag{2.44}$$

and σ^2 is the variance.

For example, the probability distribution of a Gaussian variate x in the interval $(m, m + k\sigma)$ can be determined by the equation

$$P(m < x \leq m + k\sigma) = \frac{1}{\sqrt{2\pi}\sigma} \int_{m}^{m+k\sigma} \exp\left[-\frac{(x-m)^2}{2\sigma^2}\right] dx \tag{2.45}$$

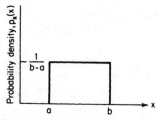

Figure 2.2 Uniform distribution in the range of $x(a, b)$.

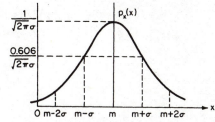

Figure 2.3 Gaussian distribution of the variate x.

Rayleigh distribution

The Rayleigh distribution is frequently used in mobile radio to represent the short-term amplitude of mobile-radio signals. The application of the Rayleigh distribution is explained in greater detail in Chap. 6.

To illustrate the concept of Rayleigh distribution, consider the case where two independent variates, x_1 and x_2, are Gaussian-distributed, both having zero mean and the same standard deviation σ. Under these conditions, the following equation applies:

$$p_{x_1 x_2}(x_1, x_2) = p(x_1)p(x_2) = \frac{1}{2\pi\sigma^2} \exp\left[-\frac{(x_1^2 + x_2^2)}{2\sigma^2} \right] \qquad (2.46)$$

If we let

$$r^2 = x_1^2 + x_2^2 \quad 0 \le r \le \infty \qquad (2.47)$$

and

$$\theta = \tan^{-1} \frac{x_2}{x_1} \quad 0 \le \theta \le 2\pi \qquad (2.48)$$

then a Jacobian transformation of the original variables x_1 and x_2 to new variables r and θ can be used:

$$J = \frac{\partial(x_1, x_2)}{\partial(r, \theta)} = \begin{vmatrix} \dfrac{\partial x_1}{\partial r} & \dfrac{\partial x_2}{\partial r} \\ \dfrac{\partial x_1}{\partial \theta} & \dfrac{\partial x_2}{\partial \theta} \end{vmatrix} \qquad (2.49)$$

Note that the old and the new joint probability density functions are related by the absolute value of J, $|J|$, as indicated in the equation

$$p_{r\theta}(r, \theta) = p_{x_1 x_2}(x_1, x_2)|J| \qquad (2.50)$$

In this case, $p(r, \theta)$ becomes:

$$p_{r\theta}(r, \theta) = \frac{r}{2\pi\sigma^2} \exp\left(-\frac{r^2}{2\sigma^2} \right) \qquad (2.51)$$

The probability density function of θ can be obtained by integrating Eq. (2.51) over r, which results in a uniform distribution:

$$p_\theta(\theta) = \int_0^\infty p_{r\theta}(r, \theta)\, dr = \begin{cases} \dfrac{1}{2\pi} & 0 \le \theta \le 2\pi \\ 0 & \text{otherwise} \end{cases} \qquad (2.52)$$

Otherwise, the probability density function of amplitude r can be obtained by integrating Eq. (2.51) over θ.

The Rayleigh distribution function $p(r)$ is plotted in Fig. 2.4 and expressed as follows:

$$p_r(r) = \int_0^{2\pi} p_{r\theta}(r,\,\theta)\,d\theta = \frac{r}{\sigma^2}\exp\left(-\frac{r^2}{2\sigma^2}\right)\text{ for } r \geq 0 \qquad (2.53)$$

The expressions for the mean $m = E(r)$, mean square $E(r_2^2)$, and standard deviation σ of r are:

$$m = E(r) = \int_0^\infty rp(r)\,dr = \sqrt{\frac{\pi}{2}}\,\sigma \qquad (2.54)$$

$$E(r^2) = \int_0^\infty r^2 p(r)\,dr = 2\sigma^2 \qquad (2.55)$$

and

$$\sigma_r = \sqrt{E(r^2) - E^2(r)} = \sqrt{2 - \frac{\pi}{2}}\,\sigma \qquad (2.56)$$

The results of the above three equations are a function of the standard deviation σ only where σ is as expressed in Eq. (2.44). The relationships among the three can be expressed as follows:

$$E(r^2) = \frac{4}{\pi}E^2(r) = \frac{4}{4-\pi}\sigma_r^2 \qquad (2.57)$$

Example 2.7 If a variable x is Rayleigh-distributed, and if x is to be displayed in the decibel scale, then a new variable, $y = 20\log x$, is introduced. The pdf of y, $p(y)$, can be obtained if the size of the decibel bins containing the samples is known. Assuming 1-dB bins are used, $p(y)$ can be found by using Eq. (E2.1.2), expressed:

$$p(y = 20\log x) = p(x = 10^{y/20})\frac{10^{y/20}}{20\log e} \qquad (E2.7.1)$$

If the value of x is normalized to its rms value $\sqrt{2}\,\sigma$, then $x' = x/\sqrt{2}\,\sigma$ and Eq. (2.53) becomes:

$$p(x') = p(x)\frac{dx}{dx'} = \sqrt{2}\,p(x)\sigma = 2x'\exp(-x'^2) \qquad (E2.7.2)$$

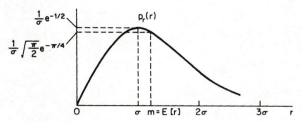

Figure 2.4 Rayleigh distribution of the variate r.

By changing x' to x in Eq. (E2.7.2) and substituting Eq. (E2.7.2) into Eq. (E2.7.1), the following is obtained:

$$p(y) = 2 \exp(-10^{y/10}) \frac{10^{y/10}}{(20 \log e)} \qquad \text{(E2.7.3)}$$

Fig. E2.7 shows the comparison between the functions of Eqs. (E2.7.2) and (E2.7.3). It should be noted that the maximum value of $p(y)$ obtained by Eq. (E2.7.3) is at the $y = 0$ dB point.

Lognormal distribution

The lognormal distribution is usually used to represent the long-term characteristics of the mobile-radio signal. The application of this distribution is described in greater detail in Chap. 3.

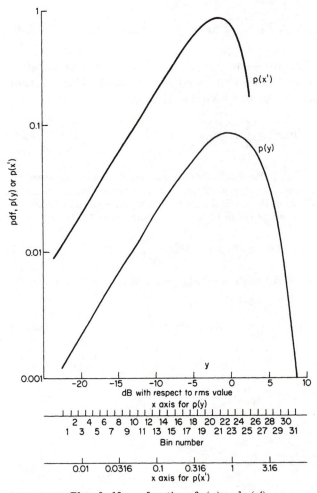

Figure E2.7 Plot of pdf as a function of $p(y)$ and $p(x')$.

If the variate x is defined as a long-term signal fade, the probability density of x can be expressed:

$$p(x) = \frac{1}{\sqrt{2\pi}D_1 x} \exp\left[-\frac{(w - \mu_1)^2}{2D_1^2}\right] \qquad x \geq 0$$

$$= 0 \qquad\qquad\qquad\qquad\qquad\qquad x < 0 \qquad (2.58)$$

where $w = \ln x$ and is assumed to be a random Gaussian process; μ_1 is the mean value of w; and

$$D_1^2 = \langle w^2 \rangle - \mu_1^2 \qquad (2.59)$$

The mean value x and the mean square value x^2 can also be found with the help of Eq. (2.58); they are expressed in decibels as follows:

$$20 \log \langle x \rangle = 2a\mu_1 + aD_1^2 \qquad \text{dB} \qquad (2.60)$$

$$10 \log \langle x^2 \rangle = 2a\mu_1 + 2aD_1^2 \qquad \text{dB} \qquad (2.61)$$

where $a = 10 \log e$. This means that the recorded decibel amplitude of the signal $x(t)$ is already converted into $w(t)$. Usually, all data are recorded in the form of $w(t)$, which makes it easy to find the mean μ_1 and the variance D_1^2 of $w(t)$. Since $\langle x(t) \rangle$ and $\langle x^2(t) \rangle$ are of greatest interest, they can be obtained from μ_1 and D_1, as shown in Eqs. (2.60) and (2.61).

Binomial distribution

The binomial distribution is frequently used to represent the probability of an event. The signaling-error rate can be found from the binomial distribution. Consider an experiment where there are two alternatives so that the possible outcomes are 0 and 1. The probabilities can be expressed as $P(0) = p$ and $P(1) = q = 1 - p$. If the experiment is repeated m times and 0 occurs n times, then the binomial distribution is

$$P_m(n) = \binom{m}{n} p^n q^{m-n} \qquad (2.62)$$

When Eq. (2.62) is inserted into Eq. (2.11), the mean of n is

$$E(n) = mp \qquad (2.63)$$

and the variance σ_n^2 is

$$\sigma_n^2 = mpq = E(n)(1 - p) \qquad (2.64)$$

If 0 represents the error bit and 1 the correct bit, then Eq. (2.62) can be interpreted as the probability that errors will occur n times.

Example 2.8   Assume that five jammers are operating at five different frequencies and that the angle of wave arrival is between 0 and 180° with reference to the base station. The probability that any one jammer signal will arrive within 0 to 20° is

$$p = \frac{20}{180} \qquad \text{(E2.8.1)}$$

and within 20 to 180° it is:

$$q = \frac{160}{180} \qquad \text{(E2.8.2)}$$

Therefore, if there are five jammer signals, the probability that three or more jammer signals will arrive within 0 to 20° can be expressed:

$$C = p^5 + \binom{5}{4} p^4 q^1 + \binom{5}{3} p^3 q^2$$

$$= 0.01153 \qquad \text{(E2.8.3)}$$

Poisson distribution

The Poisson distribution is used to represent the natural distribution of unexpected short pulses or spikes that are present in the incoming signal.

In the binomial-distribution experiment of Eq. (2.62), large values of m and very small values of p make the equation unsolvable. However, if the product of m and p remains finite, then the results of Eq. (2.62) can be approximated by the Poisson distribution, expressed

$$p(n) = e^{-\overline{n}} \frac{(\overline{n})^n}{n!} \qquad \text{(2.65)}$$

where, as before,

$$\overline{n} = E(n) = mp \qquad \sigma^2 = \overline{n}q \approx \overline{n} \qquad \text{(2.66)}$$

For example, if there are random N points in the interval $(0, T)$, then the probability of having n points in a ΔT within the time interval $(0, T)$ is:

$$p(n) = e^{-\mu \, \Delta T} \frac{(\mu \, \Delta T)^n}{n!} \qquad \text{(2.67)}$$

where:

$$\mu = \frac{N}{T} \qquad \text{(2.68)}$$

2.7 Level-Crossing Rate (LCR) [4]

The LCR formula for mobile radio was first derived by Lee [4] in 1967. The level-crossing rate lcr is regarded as a second-order statistic, since it is time-dependent and is also affected by the speed of the mobile-radio vehicle. The lcr of the mobile-radio signal provides additional information which when combined with other statistical data enables the radio designer to make intelligent decisions. To derive the expected number of level crossings $n(\Psi)$ of a given signal level Ψ, the random function ψ is assumed to be statistically stationary in time. The joint probability density function of ψ and its slope $\dot{\psi}$ are represented as $p(\psi, \dot{\psi})$.

Any given slope $\dot{\psi}$ can be obtained by the equation

$$\dot{\psi} = \frac{d\psi}{\tau} \tag{2.69}$$

where τ is the time required for a change of ordinate $d\psi$. The parameters for determining the slope $\dot{\psi}$ are illustrated in Fig. 2.5.

The expected number of crossings for a random function ψ within an interval $(\Psi, \Psi - d\psi)$ for a given slope $\dot{\psi}$ within a specified time dt is:

$$\left. \frac{\text{Expected amount of time spent in the interval } d\psi \text{ for a given } \dot{\psi} \text{ in time } dt}{\text{Time required to cross once for a given } \dot{\psi} \text{ in the interval } d\psi} \right|_{\psi = \Psi} = \frac{E[t]}{\tau}$$

$$= \left. \frac{p(\psi, \dot{\psi}) \, d\psi \, d\dot{\psi} \, dt}{d\psi / \dot{\psi}} \right|_{\psi = \Psi} = \dot{\psi} p(\psi, \dot{\psi}) \, d\dot{\psi} \, dt \tag{2.70}$$

The expected number of crossings for a given slope $\dot{\psi}$ in time T can be expressed:

$$\int_0^T \dot{\psi} p(\Psi, \dot{\psi}) \, d\dot{\psi} \, dt = \dot{\psi} p(\Psi, \dot{\psi}) \, d\dot{\psi} \, T \tag{2.71}$$

The total expected number of upward crossings in time T can be expressed:

$$N(\psi = \Psi) = T \int_0^{\infty} \dot{\psi} p(\Psi, \dot{\psi}) \, d\dot{\psi} \tag{2.72}$$

And finally, the total number of expected crossings per second—the level-crossing rate (lcr)—can be defined:

$$n(\psi = \Psi) = \frac{N(\psi = \Psi)}{T} = \int_0^{\infty} \dot{\psi} p(\Psi, \dot{\psi}) \, d\dot{\psi} \tag{2.73}$$

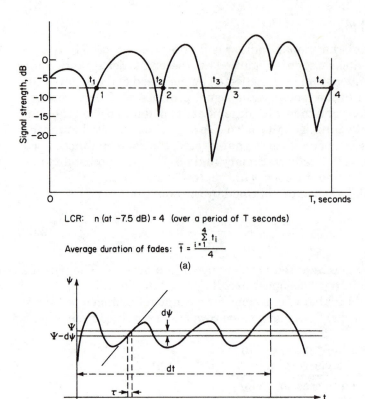

Figure 2.5 The notation used for level-crossing rates: (*a*) illustration of lcr and average duration of fades; (*b*) graphic illustration of Eq. (2.70).

Once the joint probability density function $p(\Psi,\dot{\psi})$ is found, the lcr, $n(\psi = \Psi)$, can be obtained. The application of Eq. (2.73) to mobile radio is discussed in Chap. 6.

Example 2.9 It is possible to find the statistical values of a long-term-fading signal directly from its decibel values by using its lcr curve [5].

Figure E2.9 illustrates how a measured data sample is used to find the number of crossings at five different decibel levels. First, five lines are drawn at different decibel levels across the data sample under study. Either the positive-slope or negative-slope crossings are counted, but not both. The level-crossing rate (lcr) is then plotted and normalized as shown in Fig. E2.9. Once this is done, the mean and variance of the data can easily be obtained.

Since it is assumed that the long-term-fading signal x is lognormal-distributed, then the measured signal strength, $w = \ln x$ in decibels, is normal-distributed, and its time derivative is also a normal process with zero mean, expressed as follows:

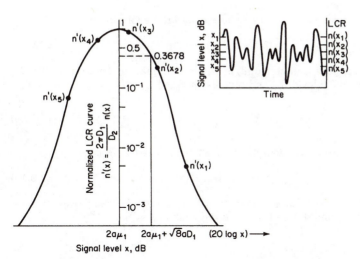

Figure E2.9 Finding the parameters μ_1, D_1, and D_2 from a measured lcr curve.

$$p(w, \dot{w}) = \frac{1}{2\pi D_1 D_2} \exp\left[-\frac{(w - \mu_1)^2}{2D_1^2} - \frac{(\dot{w} - \mu_2)^2}{2D_2^2}\right] \qquad \text{(E2.9.1)}$$

where D_1 and D_2 are variances of w and $\dot{w}$, respectively. Since $\mu_1 = \langle w \rangle$ and $\mu_2 = \langle \dot{w} \rangle = 0$, then

$$D_1^2 = \langle w^2 \rangle - \mu_1^2 \qquad \text{(E2.9.2)}$$

$$D_2^2 = \langle \dot{w}^2 \rangle - \mu_2^2 = \langle \dot{w}^2 \rangle \qquad \text{(E2.9.3)}$$

From Eq. (2.73), the average lcr becomes:

$$n(x) = \int_0^\infty \dot{x} p(x, \dot{x})\, d\dot{x} = \frac{D_2}{2\pi D_1} \exp\left[\frac{-(20 \log x - 2a\mu_1)^2}{2(2aD_1)^2}\right] \qquad \text{(E2.9.4)}$$

Now we redefine w as $w = 20 \log x$. It can also be shown that

$$n(w) = \int_0^\infty \dot{w} p(w, \dot{w})\, d\dot{w} = \frac{D_2}{2\pi D_1} \exp\left[-\frac{(w - 2a\mu_1)^2}{2(2aD_1)^2}\right] \qquad \text{(E2.9.5)}$$

The normalized lcr at level w, in dB, becomes:

$$n'(w) = \frac{2\pi D_1}{D_2} n(w) = \exp\left[-\frac{(w - 2a\mu_1)^2}{2(2aD_1)^2}\right] \qquad \text{(E2.9.6)}$$

where $a = 10 \log e$.

The maximum of $n(w)$ occurs at $2a\mu_1$, as shown in Eq. (E2.9.5) and as illustrated in Fig. E2.9. Therefore, the maximum lcr corresponds to the level $W_0 = 2a\mu_1$, and μ_1 can be determined by the equation $\mu_1 = W_0/2a$. Let

$$w = W_1 = \sqrt{8}\, aD_1 + 2a\mu_1 \qquad \text{(E2.9.7)}$$

and substitute this term into Eq. (E2.9.6) to obtain the following:

$$n'(W_1) = \exp(-1) = 0.3678 \qquad \text{(E2.9.8)}$$

Hence, the value of D_1 can be found, based on a normalized lcr value of 0.3678, which corresponds to level W_1, and can be expressed:

$$D_1 = \frac{W_1 - 2a\mu_1}{\sqrt{8a}} \qquad \text{(E2.9.9)}$$

The value of D_2 can then be found by Eq. (E2.9.6), by letting $W_0 = 2a\mu_1$:

$$D_2 = 2\pi D_1 n(W_0) \qquad \text{(E2.9.10)}$$

By using all these procedures together with Fig. E2.9, all three parameters, μ_1, D_1, and D_2, can be found from the measured lcr curve.

2.8   Duration of Fades

The duration of fades formula for mobile radio was first derived by Lee [4]. The duration of a signal fade determines the probable number of signaling bits that will be lost during the fade. The duration of the signal fade is affected by several factors, among which a primary factor is the travel speed of the mobile-radio vehicle. The relation between the average duration of the signal fade and the expected number of crossings at a particular level Ψ per second can be stated:

$$n(\psi = \Psi) = \frac{\text{expected amount of time where the function } \psi \text{ is below level } \Psi \text{ within 1 s}}{\text{average duration of fades below level } \Psi \text{ in 1 s}}$$

or expressed by notation:

$$n(\psi = \Psi) = \frac{P(\psi \leq \Psi)}{\bar{t}(\psi = \Psi)} \qquad \text{(2.74)}$$

From Eq. (2.74), the average duration of fades below the level Ψ can be calculated:

$$\bar{t}(\psi = \Psi) = \frac{P(\psi \leq \Psi)}{n(\psi = \Psi)} \qquad \text{(2.75)}$$

Equation (2.75) shows that the cumulative probability distribution, the lcr, and the average duration of fade are interrelated. Since two of them are known, the third one can be determined.

Example 2.10   If the variable r is Rayleigh-distributed, as defined in Eq. (2.53), and its time derivative $\dot{r}$ is Gaussian-distributed, according to the following definition:

$$p(\dot{r}) = \frac{1}{\sqrt{2\pi\sigma_2}} \exp\left(-\frac{\dot{r}^2}{2\sigma_2^2}\right) \tag{E2.10.1}$$

then both r and $\dot{r}$ are uncorrelated and the lcr at level R is expressed:

$$n(R) = \int_0^\infty \dot{r}\, p(R,\dot{r})\, d\dot{r}$$

$$= p(R) \int_0^\infty \dot{r}\, p(\dot{r})\, d\dot{r} = \frac{\sigma_2}{\sigma_1^2}\, \frac{R}{\sqrt{2\pi}} \exp\left(-\frac{R^2}{2\sigma_1^2}\right) \tag{E2.10.2}$$

The average duration of fades can then be obtained from Eq. (2.75), as follows:

$$\overline{t}(R) = \frac{P(r \le R)}{n(R)} = \frac{\int_0^R p(r)\, dr}{n(R)} = \frac{1 - \exp\left(-\dfrac{R^2}{2\sigma_1^2}\right)}{n(R)} \tag{E2.10.3}$$

$$= \frac{\sigma_1^2\sqrt{2\pi}}{\sigma_2 R}\left[\exp\left(+\frac{R^2}{2\sigma_1^2}\right) - 1\right] \tag{E2.10.4}$$

2.9   Correlation Functions

Correlation functions are useful in analyzing the statistical parameters of linear systems. The correlation technique can determine the expected value of the product of two random variables, or of two random processes. The terms "random variable" and "random process" are defined as follows:

Random variable—A real-valued function x defined on a sample space of points s is termed a "random variable" x_s. For example, a sample space representing the outcome for the flipping of a coin would result in $s = 2$ (heads or tails). Correspondingly, the sample space representing the outcome for the rolling of a single die would result in $s = 6$ (any one of six sides of the die). Among s points, point k (1 of s) is identified with an event, of flipping the coin or rolling the die, as the case may be, and is expressed as a function: $x_s(k) = k$. In the case where a die is rolled and $k = 3$, then $x_s(k) = 3$ is the expression for the event of rolling the die and obtaining a 3. The function $x_s(k)$ is therefore a random variable. The use of the term "random variable" to describe this type of function is in accord with tradition.

Random process—An extension of the concept of random variables wherein a collection of time functions constitutes an ensemble within the constraints of a suitable probability description. For example, the joint probability density function $p(x(t_1), x(t_2))$, where

$x(t_1)$ and $x(t_2)$ are two random variables occurring at two instants of time, is the result of a random process $x(t)$.

The correlation methodology that can be used will be either autocorrelation, if the two random variables are obtained from the same random process, or cross-correlation, if they are obtained from different random processes.

Autocorrelation

If two random variables x_1 and x_2 are obtained from the same random process—e.g., if

$$x_1 = x_1(t) \tag{2.76}$$

and
$$x_2 = x_2(t) \tag{2.77}$$

then the autocorrelation obtained from the ensemble average of the product $x_1 x_2$ can be defined as:

$$R_x(t_1, t_2) = E[x_1 x_2] = \int_{-\infty}^{\infty} dx_1 \int_{-\infty}^{\infty} x_1 x_2 p(x_1, x_2) \, dx_2 \tag{2.78}$$

The autocorrelation function for a stationary process, where the time t_1 can be any value and the result depends only on τ, can be expressed:

$$R_x(t_1, \tau) = E[x(t_1)x(t_1 + \tau)]$$
$$= R_x(\tau) \tag{2.79}$$

where

$$\tau = t_1 - t_2$$

An autocorrelation function obtained from the time average of $x(t)x(t + \tau)$ can be defined

$$\mathcal{R}_x(\tau) = \lim_{T \to \infty} \frac{1}{2T} \int_{-T}^{T} x(t)x(t + \tau) \, d\tau = \langle x(t)x(t + \tau) \rangle \tag{2.80}$$

Additional properties of autocorrelation that may be worthy of consideration are:

$$\mathcal{R}_x(\tau) = R_x(\tau) \qquad \text{for an ergodic process} \tag{2.81a}$$

$$R_x(0) = E(x^2) \qquad \text{average power} \tag{2.81b}$$

$$R_x(\tau) = R_x(-\tau) \qquad \text{even function (symmetry with respect to } \tau) \tag{2.81c}$$

$$R_x(0) \geq |R_x(\tau)| \qquad \text{maximal value when } \tau = 0 \tag{2.81d}$$

Cross-correlation

If two random variables x and y are obtained from different random processes—e.g., if

$$x_1 = x(t_1) \tag{2.82}$$

and

$$y_2 = y(t_1 + \tau) \tag{2.83}$$

and a stationary process is assumed, the cross-correlation obtained from the ensemble average of x_1 and y_2, can be defined:

$$R_{xy}(\tau) = E[x_1 y_2] = \int_{-\infty}^{\infty} dx_1 \int_{-\infty}^{\infty} x_1 y_2 p(x_1 y_2)\, dy_2 \tag{2.84}$$

Also, the cross-correlation obtained from the time average of $x(t)y(t + \tau)$ can be defined:

$$\mathscr{R}_{xy}(\tau) = \lim_{T \to \infty} \frac{1}{2T} \int_{-T}^{T} x(t)y(t + \tau)\, dt \tag{2.85}$$

Additional properties of cross-correlation that may be worthy of consideration are:

$$\left. \begin{aligned} \mathscr{R}_{xy}(\tau) &= R_{xy}(\tau) \\[4pt] \mathscr{R}_{yx}(\tau) &= R_{yx}(\tau) \end{aligned} \right\} \text{ for jointly ergodic processes} \tag{2.86a}$$

$$\tag{2.86b}$$

$$R_{xy}(0) = R_{yx}(0) \tag{2.86c}$$

$$R_{yx}(\tau) = R_{xy}(-\tau) \tag{2.86d}$$

$$|R_{xy}(\tau)| \le [R_x(0)R_y(0)]^{1/2} \tag{2.86e}$$

When the two random processes are uncorrelated, then the following expression applies:

$$R_{xy}(\tau) = E[x_1 y_2] = E[x_1]E[y_2]$$

$$= R_{yx}(\tau) \tag{2.87}$$

The correlation function is not necessarily a function of time; it can also be a function of antenna spacing, or of both time and spacing. The extended correlation function for the two random variables time and antenna spacing for two mobile-radio signals can be expressed [6]:

$$R_x(t_1, t_2; d_1, d_2) = E[x(t_1, d_1)x(t_2, d_2)] \tag{2.88}$$

where t_i and d_i are time and antenna spacing, respectively, and $i = 1, 2$. The cross-correlation coefficient $\rho_{xy}(\tau)$ is defined:

$$\rho_{xy}(\tau) = \frac{R_{xy}(\tau) - m_x m_y}{\sqrt{E[x^2] - m_x^2} \sqrt{E[y^2] - m_y^2}} \tag{2.89}$$

The autocorrelation coefficient $\rho_x(\tau)$ can be deduced from Eq. (2.89):

$$\rho_x(\tau) = \frac{R_x(\tau) - m_x^2}{E[x^2] - m_x^2} \tag{2.90}$$

The range of $\rho_{xy}(\tau)$ or of $\rho_x(\tau)$ is:

$$|\rho(\tau)| \leq 1 \tag{2.91}$$

The correlation coefficient is often used because it is a normalized value by which to compare different signals within different systems.

2.10   Power Spectral Density and Continuous Spectral Density

The power spectral density is commonly referred to as the "power spectrum." The power spectrum differs from the continuous spectrum as follows. When a signal $x(t)$ has a finite total energy

$$E_T = \int_{-\infty}^{\infty} x^2(t) \, dt < \infty \tag{2.92}$$

then the power averaged over all time is zero and the distribution of signal energy across the frequency band is described by the continuous spectrum $W(f)$, $|W(f)|^2$. The continuous spectrum is expressed in Eq. (1.24). When a signal $x(t)$ whose total energy is infinite has a finite average power

$$P = \lim_{T \to \infty} \frac{1}{T} \int_0^T x^2(t) \, dt \tag{2.93}$$

then the distribution of signal power across the frequency band is described by the power spectrum $S(f)$.

Therefore, the preferred method for calculating the power spectrum of a random signal is to use power equations. The power P is defined:

$$P = \lim_{T \to \infty} \frac{E_T}{T} = \int_{-\infty}^{\infty} S(f) \, df \tag{2.94}$$

where E_T is defined:

$$E_T = \int_{-T/2}^{T/2} x^2(t) \, dt = \int_{-\infty}^{\infty} |X_T(f)|^2 \, df \tag{2.95}$$

By substituting the terms of Eq. (2.95) into Eq. (2.94), the following is obtained:

$$P = \lim_{T \to \infty} \frac{1}{T} \int_{-\infty}^{\infty} |X_T(f)|^2 \, df = \int_{-\infty}^{\infty} \lim_{T \to \infty} \frac{|X_T(f)|^2}{T} \, df \qquad (2.96)$$

From Eq. (2.96), the power spectrum $S(f)$ can be calculated as follows:

$$S(f) = \lim_{T \to \infty} \frac{1}{T} |X_T(f)|^2 \qquad (2.97)$$

The changing of the order that takes place in Eq. (2.96) (i.e., from limiting and then integrating to integrating and then limiting) to result in Eq. (2.97) is valid only for a special class of signal, i.e., the sum of sinusoids. However, since mobile-radio signals belong to this class, as mentioned in Chap. 1 [see Eq. (1.20)], it is proper and valid to use Eqs. (2.94) through (2.97) to calculate the power spectrum of mobile-radio signals.

The power spectrum derived from Eq. (2.97) can also be used to find the correlation function:

$$S(f) = \int_{-\infty}^{\infty} R(\tau) e^{-j\omega\tau} \, d\tau \qquad (2.98)$$

Then, on the basis of Fourier integral theory, the following is true:

$$R(\tau) = \int_{-\infty}^{\infty} S(f) e^{j\omega\tau} \, df \qquad (2.99)$$

Similarly, the Fourier transform method can be used to calculate the cross-correlation function on the basis of cross-spectral densities:

$$S_{xy}(\omega) = \int_{-\infty}^{\infty} R_{xy}(\tau) e^{-j\omega\tau} \, d\tau \qquad (2.100)$$

$$R_{xy}(\tau) = \int_{-\infty}^{\infty} S_{xy}(f) e^{j\omega\tau} \, df \qquad (2.101)$$

where $S_{xy}(\omega)$ and $R_{xy}(\tau)$ can be complex functions and $S_{xy}(\omega) = S_{yx}^*(\omega)$. The continuous spectrum and power spectrum of mobile-radio signals are described further in Chap. 6.

2.11 Sampling Distributions

If a random variable x is characterized by the probability distribution function $F_x(X)$ and we assume that $x_1, x_2, x_3, \ldots, x_N$ are observed sample values of x, then it follows that any quantity derived from the sample values x_i will also be a random variable. Therefore, the mean value of x_i, using the form of Eq. (2.6), can be expressed:

$$\bar{x} = \sum_{i=1}^{N} \frac{x_i}{N} \tag{2.102}$$

When a random variable has the probability distribution function $F_x(\overline{X})$, it can be referred to as the "sampling distribution" of $\bar{x}$ and can be found whether the variable x is Gaussian (normal) or nonnormal.

In cases where x is Gaussian-distributed, the mean value m and the variance σ^2 of x are defined by Eqs. (2.42) and (2.44), respectively. Consequently, the sum of Gaussian variables also becomes a Gaussian variable. The mean value $\hat{m}$ and the variance $\hat{\sigma}^2$ of $\bar{x}$ can be obtained as follows:

$$\hat{m} = E(\bar{x}) = E\left[\frac{1}{N} \sum_{i=1}^{N} x_i \right]$$

$$= \frac{1}{N} E\left[\sum_{i=1}^{N} x_i \right] = E(x_i) = m \tag{2.103}$$

$$\hat{\sigma}^2 = E[(\bar{x} - \hat{m})^2] = E[(\bar{x} - m)^2]$$

$$= E\left[\left(\frac{1}{N} \sum_{i=1}^{N} (x_i - m) \right)^2 \right]$$

$$= \frac{1}{N^2} E\left[\left(\sum_{i=1}^{N} (x_i - m) \right)^2 \right]$$

$$= \frac{1}{N^2} \{ NE[(x_i - m)^2] + 2N(N-1)E[(x_i - m)(x_j - m)] \} \tag{2.104}$$

Since the sample observations x_i are independent, then

$$E[(x_i - m)(x_j - m)] = E[x_i - m]E[x_j - m] = 0 \tag{2.105}$$

and the following is obtained:

$$\hat{\sigma}^2 = \frac{1}{N} E[(x_i - m)^2] = \frac{\sigma^2}{N} \tag{2.106}$$

The standardized variable z for a Gaussian variable x is defined:

$$z = \frac{x - m}{\sigma} \tag{2.107}$$

and the probability that z is less than Z becomes:

$$P(z \le Z) = \int_{-\infty}^{Z} p(z)\, dz = \int_{-\infty}^{Z} \frac{1}{\sqrt{2\pi}} \exp\left(-\frac{z^2}{2}\right) dz$$

$$= C_Z + 0.5 \tag{2.108}$$

where C_Z is the percentage value obtained either from a table or by calculation. The new standard variable $\bar{z}$ of $\bar{x}$ is also a Gaussian value and can be expressed:

$$\bar{z} = \frac{\bar{x} - \hat{m}}{\hat{\sigma}} = \frac{\bar{x} - m}{\sigma/\sqrt{N}} \tag{2.109}$$

As a result of substituting $\bar{z}$ for z in Eq. (2.108), the following is obtained:

$$P(\bar{z} \le Z) = P\left(\frac{\bar{x} - m}{\sigma/\sqrt{N}} \le Z\right)$$

$$= P\left(\bar{x} \le \frac{Z\sigma}{\sqrt{N}} + m\right) = C_Z + 0.5 \tag{2.110}$$

In the case where x is distributed by nonnormal means, Eq. (2.110) still holds for $\bar{x}$, since the central limit theorem confirms that when the sample size N becomes large ($N > 10$), the sampling distribution of the mean of sample $\bar{x}$ approaches a normal distribution regardless of the distribution of the original variable x.

Example 2.11   Assume that x is a random variable having a Gaussian distribution with a mean of 0.5 and a variance of 1.0. The limited samples of x are:

$$x_1 = 0.5 \text{ V}$$

$$x_2 = 0.8 \text{ V}$$

$$x_3 = 0.3 \text{ V}$$

$$x_4 = 0.01 \text{ V}$$

$$x_5 = 0.95 \text{ V}$$

1. What is the mean value of x_i, that is, the $\bar{x}$ of Eq. (2.102)?
2. What is the mean value of $\bar{x}$?
3. What is the variance of $\bar{x}$?
4. Find the 90 percent confidence interval of true mean $E(\bar{x})$ from the data points. Note that $P(-1.65 \le z \le 1.65) = 90\%$ can be obtained from published mathematical tables. The confidence interval is described in Sec. 2.12.
5. How tight is the 90 percent confidence interval?
   a. Very tight
   b. Moderately tight
   c. Not very tight at all

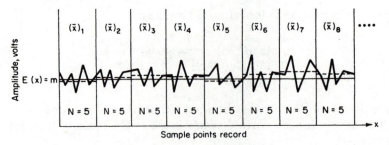

Figure E2.11   Reference data points for Example 2.11.

solution   (see Fig. E2.11)

Given $E(x) = E(x_i) = 0.5$, $\sigma_x = [E(x^2) - E^2(x)]^{1/2} = 1$, and $E(x^2) = 1.25$.

1. The sample average (or the mean value of the limited samples) is:

$$\overline{x} = \frac{\sum\limits_{i=1}^{N} x_i}{N} = 0.512 \tag{E2.11.1}$$

2. The statistical average (or the mean value) of $\overline{x}$ is:

$$E[\overline{x}] = E\left(\frac{\sum\limits_{i=1}^{N} x_i}{N}\right) = \frac{\sum\limits_{i=1}^{N}}{N} E(x_i) = 0.5 \tag{E2.11.2}$$

An alternate approach:

$$E[\overline{x}] = \int \overline{x} p(\overline{x})\, d\overline{x} = \int \left(\frac{\sum\limits_{i=1}^{N} x_i}{N}\right) p(x_i)\, dx_i$$

$$= \frac{1}{N} \sum_{i=1}^{N} \int x_i p(x_i)\, dx_i$$

$$= \frac{1}{N} \sum_{i=1}^{N} E(x_i) = E(x) \tag{E2.11.3}$$

3. The standard deviation $\hat{\sigma}_x$ is obtained from Eq. (2.106):

$$\hat{\sigma}_x^2 = E[(\overline{x} - \hat{m})^2] = E[(\overline{x} - m)^2] = \frac{\sigma_x^2}{N} \tag{E2.11.4}$$

and

$$\hat{\sigma}_x = \frac{1}{\sqrt{5}} = 0.4472$$

4. The 90 percent confidence interval can be found as follows:

$$P(0 \leq z \leq 1.65) = 0.45 \tag{E2.11.5}$$

$$P(-1.65 \leq z \leq 1.65) = 0.9 = 90\% \tag{E2.11.6}$$

Then, Eq. (2.109) provides:

$$\overline{z} = \frac{\overline{x} - \hat{m}}{\hat{\sigma}} \tag{E2.11.7}$$

Substituting Eq. (E2.11.7) into Eq. (E2.11.6) we obtain:

$$P\left(-1.65 \leq \frac{\overline{x} - \hat{m}}{\hat{\sigma}} \leq 1.65\right) = P\left(-1.65 \leq \frac{\overline{x} - m}{0.4472} \leq 1.65\right)$$

$$= P(-0.738 \leq \overline{x} - m \leq 0.738)$$

$$= P(\overline{x} - 0.738 \leq m \leq \overline{x} + 0.738)$$

$$= P(-0.226 \leq m \leq 1.25) = 90\% \qquad \text{(E2.11.8)}$$

5. The 90 percent confidence interval based on five samples is not tight at all, even though $\overline{x}$ is very close to m. This means that the probability that the value $\overline{x} = 0.512$, based on five samples, is very small.

2.12   Confidence Intervals

The range of a statistical interval within which a degree of certainty can be determined between the true mean m and the sample mean $\overline{x}$ is called a "confidence interval." This interval can be found by first confirming that

$$P(-Z_1 \leq z \leq Z_1) = \int_{-Z_1}^{Z_1} p(z)\, dz = 2C_{Z_1} \qquad (2.111)$$

Knowing that this relationship exists, one can then extend Eq. (2.110) to Eq. (2.111) to obtain:

$$P\left[\overline{x} - \frac{Z_1 \sigma}{\sqrt{N}} \leq m \leq \overline{x} + \frac{Z_1 \sigma}{\sqrt{N}}\right] = 2C_{Z_1} \qquad (2.112)$$

In Eq. (2.112), if $2C_{Z_1} = 90\%$, then the confidence interval has a degree of certainty of 90 percent. The value of Z_1 can be found from Eq. (2.111) by knowing C_{Z_1} and the value of $\overline{x}$ from Eq. (2.102). N is the number of samples, and σ can be either a known value or the value of σ_x, calculated as follows:

$$\sigma_x^2 = \frac{1}{N-1} \sum_{i=1}^{N} (x_i - \overline{x})^2 \qquad (2.113)$$

Therefore, the range of the statistical average for the true mean ($\hat{m} = m$) can be obtained from Eq. (2.112):

$$\overline{x} - \frac{Z_1 \sigma}{\sqrt{N}} \leq m \leq \overline{x} + \frac{Z_1 \sigma}{\sqrt{N}} \qquad (2.114)$$

Example 2.12   A Gaussian variable x has a mean value of 3 V and a variance σ^2 of 1 V. What is the probability that x is within 2σ of its mean value?

solution   (see Fig. E2.12)

Given $\sigma^2 = 1$ V and $m = 3$ V, the probability density function is:

$$p(x) = \frac{1}{\sqrt{2\pi}\sigma} \exp\left[-\frac{(x-m)^2}{2\sigma^2}\right]$$

$$= \frac{1}{\sqrt{2\pi}\sigma} \exp\left[-\frac{(x-3)^2}{2}\right] \tag{E2.12.1}$$

From the normal error curve, the value of

$$P(x_1 \leq z \leq x_2) = \int_{x_1}^{x_2} \frac{1}{\sqrt{2\pi}\sigma} \exp\left(-\frac{z^2}{2}\right) dz \tag{E2.12.2}$$

which can be found in published mathematical tables, can be expressed:

$$P(0 \leq z \leq 2) = 0.4733 \tag{E2.12.3}$$

Hence,

$$P(m - 2\sigma \leq x \leq m + 2\sigma) = P(1 \leq x \leq 5) = 2 \times 0.4733$$

$$= 94.66\% \tag{E2.12.4}$$

2.13   Error Probability

The probability density functions for different random variables have been previously discussed in Sec. 2.6, where it was shown that the binomial distribution calculation is frequently used to calculate the signaling-error rate in transmitting repetitive bits, or the signaling-error rate in transmitting repetitive words. By analysis, the error probability for transmitting 1 bit is derived.

To analyze the probability of bit error, consider the case where two random variables y_1 and y_0 are involved, as illustrated in Fig. 2.6. The

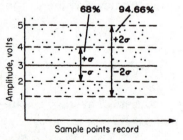

Figure E2.12   Reference data for Example 2.12.

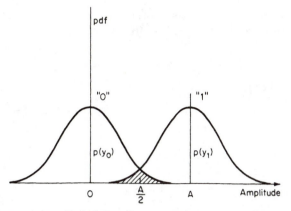

Figure 2.6 Probability density distribution of signal plus noise.

random variables correspond to their respective intended digits 1 and 0 as follows:

$$y_1 = \frac{A}{2} + x > 0 \qquad \text{digit 1} \qquad (2.115)$$

$$y_0 = \frac{-A}{2} + x < 0 \qquad \text{digit 0} \qquad (2.116)$$

where x is the random noise and y_1 and y_0 form a bipolar pulse. The probability density functions $p(y_1)$ and $p(y_0)$ are related to $p(x)$ of the random variable x, as shown in Fig. 2.6 and discussed in Sec. 2.6.

In this case, the probability for conversion error is:

$$P_{e_1} = P(y_1 < 0) = P\left(\frac{A}{2} + x < 0\right) \qquad (2.117)$$

$$P_{e_0} = P(y_0 > 0) = P\left(-\frac{A}{2} + x > 0\right) \qquad (2.118)$$

and the net error probability becomes:

$$P_e = P_1 P_{e_1} + P_2 P_{e_0} \qquad (2.119)$$

where P_1 and P_2 are the digit probabilities at the transmission source, and where they are usually, but not necessarily, equal. In either case, the relationship $P_1 + P_2 = 1$ is held to be true, since one or the other must be transmitted.

Equation (2.119) is also referred to as the "bit-error rate" and its application to mobile-radio digital communication is described in Chap. 8.

2.14   Impulse Response Measurements

The measurement of the impulse response $h(t)$ of a given channel can be carried out by two approaches.

1. *Transmit a strong impulse and then measure the impulse response at the receiving end.*
   The impulse $S(t)$ in time is

$$S\,(t - t_0) = \begin{cases} \infty & \text{when } t = t_0 \\ 0 & \text{when } t \neq t_0 \end{cases} \qquad (2.120)$$

The frequency response of an impulse $A \cdot S\,(t - t_0)$ is

$$\int_{-\infty}^{\infty} A \cdot S(t - t_0)\, e^{-j\omega t}\, dt = A \qquad (2.121)$$

Eq. (2.121) shows that the impulse energy will be spread across a very wide frequency band with a constant spectrum amplitude A. It is sometimes called *black modulation,* which will interfere with other frequency bands and cannot be tolerated by the industry.

2. *Transmit a pseudonoise code.*
   The pseudonoise code $x(t)$ is like white noise. The impulse response $h(t)$ from the medium (channel) or system can be obtained by sending a pseudonoise code that only needs low power, but the bandwidth needed is broader. In the receiver, there is a correlator, which consists of two parts, multiplier and averager, as shown in Fig. 2.7.

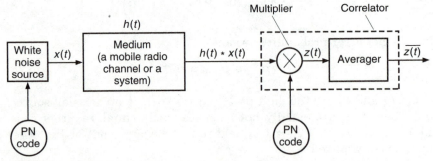

Figure 2.7   Impulse response measurement using white noise.

At the multiplier:

$$z(t) = [x(t - t_d)] \ [h(t) * x(t)]$$

$$= x(t - t_d) \int_{-\infty}^{\infty} h(\lambda) \, x(t - \lambda) \, d\lambda \tag{2.122}$$

At the averager:

$$\overline{z(t)} = \int_{-\infty}^{\infty} h(\lambda) \cdot E[x(t - t_d) \, x(t - \lambda)] \, d\lambda \tag{2.123}$$

where

$$E[x(t - t_d) \, x(t - \lambda)] = R_x \, (\lambda - t_d) \tag{2.124}$$

Since $x(t)$ is the pseudonoise code

$$R_x \, (\lambda - t_d) = (\eta/2) \, S(\lambda - t_d) \tag{2.125}$$

where $\eta/2$ is the power spectrum of $x(t)$.
   Therefore, $\overline{z(t)}$ from Eq. (2.123) becomes

$$\overline{z(t)} = \frac{\eta}{2} \int_{-\infty}^{\infty} h(\lambda) \, S(\lambda - t_d) \, d\lambda = \frac{\eta}{2} \, h(t_d) \tag{2.126}$$

Now the impulse response $h(t)$ of the medium (suburban or urban or open area) is obtained from Eq. (2.126). Also, the time t_d is found as the time delay spread of the impulse response at the reception.

2.15   Equalizers

The time delay spread causes intersymbol interference (ISI). An *equalizer* is a device that can remove all time delayed waves. Thus, the ISI can be reduced.
   Let the desired symbol sequence be $\{x_k\}$. When received, the $\{x_k\}$ can become an errored one $\{z_k\}$. Therefore, $\{z_k\}$ is the unequalized input to the equalizer, as shown in Fig. 2.8, in order to keep the sequence $\{x_k\}$ undistorted. The equalizer is used with the $2N + 1$ T-second taps, where N is the number of taps at one side from the center tap. The $(2N + 1)$ tap coefficients are $c_{-N}, c_{-N+1}, \ldots c_{-1}, c_0, c_1, \ldots c_{N-1}, c_N$. We may denote them by $\{c_n\}$. The output symbols $\hat{x}_k$ of the equalizer are in terms of the input $\{z_k\}$

$$\hat{x}_k = \sum_{n=-N}^{N} c_n z_{k-n} \qquad k = -2N \cdots 2N \tag{2.127}$$

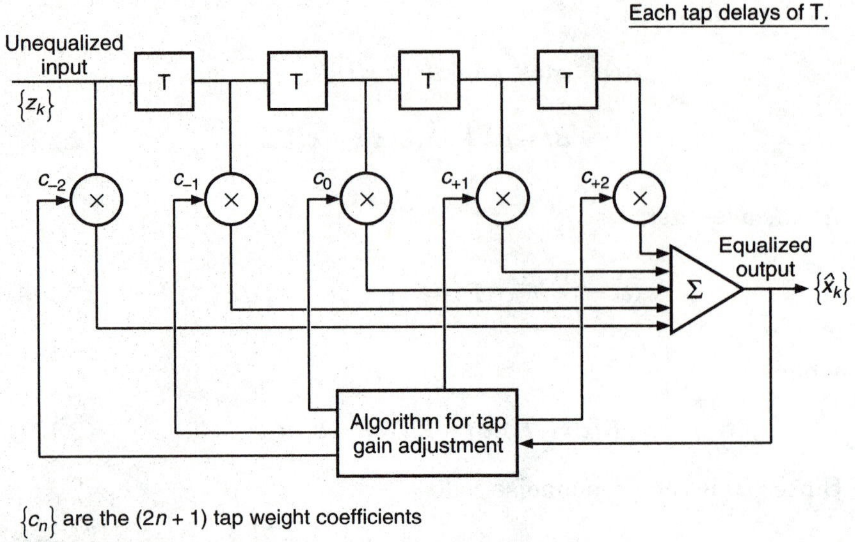

$\{c_n\}$ are the $(2n + 1)$ tap weight coefficients

Figure 2.8 Linear equalization.

The matrix representation of $\{\hat{x}_k\}$, $\{z_k\}$, and $\{c_n\}$ can be expressed as

$$\hat{X} = \begin{bmatrix} \hat{x}_{-2N} \\ \vdots \\ \hat{x}_0 \\ \vdots \\ \hat{x}_{2N} \end{bmatrix} \qquad C = \begin{bmatrix} c_{-N} \\ \vdots \\ c_0 \\ \vdots \\ c_N \end{bmatrix} \qquad (2.128)$$

and

$$Z = \begin{bmatrix} z_{-N} & 0 & 0 & \cdots & 0 & 0 \\ z_{-N+1} & z_{-N} & & & & \\ \vdots & & & & & \\ \vdots & & & & & \\ z_N & z_{N-1} & z_{N-2} & \cdots & z_{-N+1} & z_{-N} \\ \vdots & & & & & \\ 0 & 0 & 0 & \cdots & z_N & z_{N-1} \\ 0 & 0 & 0 & \cdots & 0 & z_N \end{bmatrix} \qquad (2.129)$$

We may use the vector representation

$$\hat{X} = ZC \qquad (2.130)$$

The criterion for selecting the proper c_n is either based on the peak distortion or mean square distortion.

**Minimizing the peak distortion
(the zero-forcing equalizer) [7]**

Calculate the tap coefficients c_n so that

$$\hat{x}_1 = 1 \qquad k = 0$$

$$\hat{x}_k = 0 \qquad k \neq 0$$

There are $2N + 1$ simultaneous equations from Eq. (2.130) to solve for obtaining all the proper tap coefficients c_n. The block diagram of channel with zero-forcing equalizer is shown in Fig. 2.9. $C(z)$ is the z-transform of the $\{c_n\}$

$$C(z) = \sum_{n=-\infty}^{\infty} c_n z^{-n}$$

and $F(z)$ is the z-transform of the linear filter model corresponding to the channel response $h(t)$.

Let the transfer function $C(z)$ in Fig. 2.9 be the inverse filter to the channel response $F(z)$.

$$C(z) = \frac{1}{F(z)} \tag{2.131}$$

Then the output $\{\hat{x}_k\}$ is distorted by the AWGN only after it is received. Thus, the output $\{\hat{x}_k\}$ is only slightly distorted from the input $\{x_k\}$. The AWGN (additive white Gaussian noise) is described in Chap. 16.

**Minimizing mean-square error
(MSE algorithm)**

The tap coefficients $\{c_n\}$ of the equalizer are adjusted to minimize the mean-square error

$$\varepsilon_k = x_k - \hat{x}_k \tag{2.132}$$

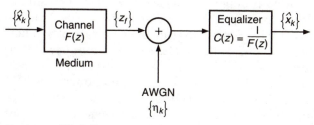

Figure 2.9   Block diagram of channel with zero-forcing equalizer.

The performance index for the MSE criterion is expressed by J, as

$$J = E\{|\varepsilon_k|^2\} \tag{2.133}$$

There are $2N + 1$ simultaneous equations in Eq. (2.133) to solve for obtaining proper $\{c_n\}$.

Making an automatic equalizer

Pretraining equalizer. Transmit a training sequence (known at the receiver) and compare with the locally generated known sequence. The difference between the two sequences are used to adjust the tap coefficients $\{c_n\}$. For example, x_k is the known training sequence, and $\hat{x}_k$ is the received training sequence at the output of equalizer shown in Eq. (2.130).

Adaptive equalizer. The coefficients are continually and automatically adjusted as a decision feedback scheme. The adaptive equalizer can work well if the adaptive can reduce the errors. If the adaptive cannot allow the channel errors to converge, the poor performance occurs. The disadvantage of having a training sequence is the constant repetition at a given break in transmission. In a time-varying channel such as the mobile unit, the field $\{c_n\}$ kept over a time break can generate ISI.

A common solution is to have a training sequence at the initial stage to provide a good channel error reduction. Then let the adaptive algorithm follow—when the adaptive algorithm is not working, hopefully the next train sequence will have already arrived.

Discussion of the equalizer

The equalizer has to be evaluated based on the number of taps. Each tap is spaced at the symbol interval T. For the design of the equalizer, the odd number of taps is chosen.

From the specification of GSM (global systems mobile, a European digital cellular standard) the equalizer is designed to reduce ISI errors within 16 μs. The transmission rate is 270 kbps, then the symbol interval $T = 3.7$ μs. The required number of taps is 5 (i.e., $N = 2$). From the specification of the North America TDMA, the equalizer is designed to reduce ISI errors within 60 μs. Since the transmission rate of symbols is 24 ksps, the symbol interval $T = 40$ μs. The required number of taps is 3 (i.e., $N = 1$). As number of taps $2N + 1$ increases, the equalizer operates as a discrete-transversal filter that spans a time interval of $2NT$ seconds. When the mobile is at a standstill, the larger number of taps can be beneficial, since the delay spread pulses between two adjacent symbols are alike. It is a time-invariant channel. The long delay due to the large number of taps does improve the solution. But when the mobile is moving, the environment is a time variant channel. Then the large number of taps causes a long time span in the equalizer, and the solution of finding the tap coefficients $\{c_n\}$ may not be convergent due to this long span.

Problem Exercises

1. In determining the binomial distribution, what method is used to obtain the mean of n, Eq. (2.65), and the variance of σ_n^2, Eq. (2.66)?

2. Show how the mean of $g(x)$ in Example 2.6 is obtained:

$$E[g(x)] = g(\eta) + g''(\eta)\,\frac{\sigma^2}{2} \tag{P2.2.1}$$

and how the standard deviation $\sigma_{g(x)}$ of $g(x)$ is obtained:

$$\sigma_{g(x)} = \left[g'^2(\eta)\sigma^2 - \frac{g''^2(\eta)\sigma^4}{4} \right]^{1/2} \tag{P2.2.2}$$

3. In the case of bipolar pulses $(-A/2, A/2)$, both y_0 and y_1 in Eqs. (2.117) and (2.118) are Gaussian, and either symbol (0 or 1) has an equal probability of being transmitted. Show that the error probability with noise power $\langle x^2 \rangle = N$ is:

$$P_e = \frac{1}{2}\,\text{erfc}\left(\frac{\sqrt{A^2/2N}}{2} \right) \tag{P2.3.1}$$

where erfc represents the complementary error function. If the bipolar pulse is replaced by an on-off pulse $(0, A)$, what effect will this have on error probability?

4. Figure P2.4 illustrates a typical long-term-fading signal with Gaussian distribution characteristics. Find the mean and variance for a sampling of signal data w, using the method described in Example 2.9.

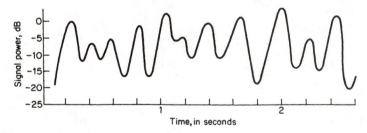

Figure P2.4 Reference data for Prob. 4.

5. Figure P2.5 illustrates a portion of the noise spectrum that is flat over a frequency bandwidth of $-B$ to B, with a two-sided spectral density of $\eta/2$. Prove that its correlation is:

$$R(\tau) = \eta B \, \frac{\sin\,(2\pi B\tau)}{2\pi B\tau} \qquad\qquad (P2.5.1)$$

6. If r_1 and r_2 are two independent Rayleigh variables with the ratio of r_1 to r_2 denoted by $R = r_1/r_2$ and the ratio of $\langle r_1^2 \rangle$ to $\langle r_2^2 \rangle$ denoted by $\Gamma = \langle r_1^2 \rangle/\langle r_2^2 \rangle$, prove that the probability density function $p_{R^2}(x)$ is

$$p_{R^2}(x) = \frac{\Gamma}{(\Gamma + x)^2}$$

If $R' = r_2/r_1$, what will $p_{R'^2}(x)$ be?

7. The random variables $x_1, \ldots, x_n$ are Gaussian and independent, with mean η and variance σ^2. The sample variance can be expressed as

$$\overline{v} = \frac{(x_1 - \overline{x})^2 + (x_1 - \overline{x})^2 + \cdots + (x_n - \overline{x})^2}{n}$$

Show that

$$E[\overline{v}] = \frac{n-1}{n}\,\sigma^2$$

$$E[\overline{v^2}] = \frac{n^2-1}{n^2}\,\sigma^4$$

8. The total number of mobile-telephone-call noncompletions in a day is a Poisson-distributed random variable with parameter a. The probability that an incomplete call occurs equals p. With n representing the number of incomplete calls in a day, show that

$$E[n] = pa$$

$$P(n = m) = \sum_{k=m}^{\infty} e^{-a}\,\frac{a^k}{k!}\binom{k}{m}p^m q^{k-m}$$

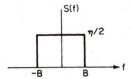

Figure P2.5 Reference data for Prob. 5.

9.   Let x be a Gaussian random variable with a mean of 0.5 and a variance of 1.0. The limited samples are

$$x_1 = 0.3, x_2 = 0.75, x_3 = 0.2, x_4 = 0.05, x_5 = 0.85$$

For these data, answer the questions asked in Example 2.11.

References

1. A. Papoulis, *Probability, Random Variables, and Stochastic Processes,* McGraw-Hill, New York, 1965.
2. W. B. Davenport, Jr. and W. L. Root, *An Introduction to the Theory of Random Signals and Noise,* McGraw-Hill, New York, 1958.
3. J. S. Bendat and A. G. Piersol, *Random Data, Analysis and Measurement Procedures,* Wiley/Interscience, New York, 1971.
4. W. C. Y. Lee, "Statistical Analysis of the Level Crossings and Duration of Fades of the Signal from an Energy Density Mobile Radio Antenna," *Bell System Technical Journal,* vol. 46, February 1967, pp. 412–448.
5. W. C. Y. Lee, "Finding the Statistical Properties of the Median Values of a Fading Signal Directly from Its Decibel Values," *Proc. IEEE,* vol. 58, February 1970, pp. 287–288.
6. W. C. Y. Lee, "An Extended Correlation Function of Two Random Variables Applied to Mobile Radio Transmission," *Bell System Technical Journal,* vol. 48, no. 10, December 1969, pp. 3423–3440.
7. John G. Proakis, *Digital Communications,* 2d ed., McGraw-Hill, 1989, p. 556.

Path Loss over Flat Terrain

3.1 Predicting Path Loss
by the Model-Analysis Method

In the analysis of radio-wave propagation for mobile communication, one of the major parameters of interest is propagation-path loss. A measure of propagation-path loss is the difference between the effective power transmitted and the average field strength of the received signal. The field strength of the received signal is an indication of the relative signal amplitude at the transmitted carrier frequency. As previously discussed in Chap. 1, the field strength recorded at the location of a moving mobile receiver fluctuates as a result of the effects of various multipath phenomena.

The average field strength can be calculated by averaging the random trial samples of instantaneous field-strength measurements recorded over a certain path length or distance. The variations in the average field strength are referred to as "long-term" fading; the short-term fading effects of multipath phenomena, which may be readily observed in the trial samples, are practically unnoticeable in the average-field-strength calculations.

General features and roughness of the terrain contour and the occasional scatterers along the distant portions of the propagation path tend to defocus the energy reaching the mobile receiver, and this also contributes to the overall path loss. The more substantial differences in terrain and the unique properties of individual scatterers can cause the path losses to be somewhat different in each of the respective areas studied.

The following factors should always be considered in attempting to predict the propagation-path loss for a particular mobile-radio environment.

Radio horizon

The typical maximum range for a mobile-radio propagation path is around 15 statute miles (24.1 km), depending upon the height of the transmitting and receiving antennas and the characteristics of the intervening terrain. A base-station antenna is usually installed 100 ft (30.5 m) or more above ground level. The mobile antenna is nominally 10 ft (3 m) above ground level. When these typical antenna heights are present, the maximum true distance d_t from the base-station transmitting antenna to the radio-path horizon can be calculated. The relationship between antenna height and maximum true distance is illustrated in Fig. 3.1.

The true distance d_t and the actual distance d_a are expressed:

$$d_t = \left(\frac{3h}{2}\right)^{1/2} \qquad \text{mi} \qquad\qquad (3.1a)$$

$$d_t' = 2.9\left(\frac{3h'}{2}\right)^{1/2} \qquad \text{km} \qquad\qquad (3.1b)$$

$$d_a = (2h)^{1/2} \qquad \text{mi} \qquad\qquad (3.2a)$$

$$d_a' = 2.9(2h')^{1/2} \qquad \text{km} \qquad\qquad (3.2b)$$

where h is in feet, d_t and d_a are in miles, h' is in meters, and d_t' and d_a' are in kilometers.

The actual distance d_a is of primary concern, since it represents the actual distance measured along the surface of the terrain. If we assume that the base-station antenna height h_1 is 100 ft (30.5 m), then the actual distance d_{a_1} is 14 mi (22.5 km). If we further assume that the mobile antenna height is 10 ft (3 m), then the actual distance d_{a_2} is 4.5 mi (7.24 km). In this example, the total actual distance d_a of the propagation path will be:

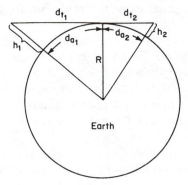

Figure 3.1   Antenna height vs. transmission distance.

$$d_a = d_{a_1} + d_{a_2}$$

$$14.0 + 4.5 = 18.5 \text{ mi} \qquad (29.77 \text{ km})$$

The total actual distance (d_a = 18.5 mi or 29.77 km) can be considered to be the radio-path horizon, and a propagation path that is shorter than 18.5 mi will be above the horizon and therefore will not experience path loss attributable to the curvature of the earth.

Propagation-path distances that extend beyond the radio horizon suffer additional path loss due to signal blockage by the earth's surface. However, propagation loss as a function of distance beyond the limits imposed by the radio horizon makes it possible to reject unwanted signals, thus reducing the potential for cochannel interference.

Sky reflections

Practically no mobile-radio propagation loss is attributable to sky reflections, because of the short ranges that are involved. Therefore, the major parameters of propagation loss are determined by earth or ground reflections and the scattering effects of buildings, structures, vehicles, and trees.

Signal averaging

Considering the relatively short range of mobile-radio transmissions, it is important to determine the average field strength of the received signal within a small angular sector at about 0.25-mi intervals over a distance of from 1 to 18 mi. This will provide a sufficient number of samples for a fine-scaled, detailed analysis of propagation-path loss characteristics along a given route.

Terminal in motion

In mobile-radio communication, the base station is normally at a fixed location while the mobile subscriber's terminal is free to travel. Therefore, a significant amount of the propagation-path loss is directly related to the general area in which the mobile unit is traveling. This establishes a set of variables for propagation-path loss that are different from those associated with radio communication between two fixed terminal locations, where the elements that cause signal fading are more predictable. Consequently, in mobile-radio situations, a more dynamic range of field-strength variations is expected at the mobile receiving location. The propagation-path loss characteristics that are attributable to the traveling mobile unit can be classified according to the different general areas in which they occur.

Mobile-unit antenna height

The typical height of the mobile-unit antenna is about 10 ft above the ground plane. Because of this relatively low height, the antenna is usually surrounded by a variety of different types of nearby scatterers, both moving and stationary. The resulting effects on propagation-path loss are significantly higher than those that occur in free-space propagation.

Effect of surface waves

Imperfectly conductive ground surfaces can produce surface waves in the region extending a few wavelengths above the ground plane at low frequency. For antenna heights many wavelengths above ground or frequencies 5 MHz and greater, the effects of surface waves are negligible and can be disregarded when calculating propagation-path loss in a mobile-radio environment [1].

Mobile-radio propagation-path loss is uniquely different from the kinds of path losses experienced in other communication media. The challenge to the designer is to develop a system that will operate satisfactorily within the constraints of recognized and predictable path-loss considerations, while providing the desired levels of mobile-telephone communication quality. It is possible to construct theoretical models based on analytical studies of measured data obtained in the mobile-radio environment that are representative of a predefined, fixed set of data parameters for prescribed situations. By comparing random samples of measured data against the appropriate model, predictions of propagation-path loss can be made with confidence and with a greater probability of success. This technique is growing in popular acceptance, since it saves considerable time and labor and provides predictable results within a range of acceptable limits.

3.2 Propagation Loss—Over Smooth Terrain

When energy is propagated over a smooth surface, specular reflection can be expected to occur and usually results in one reflected wave arriving at the mobile receiver, as shown in Fig. 3.2(a). Irregularities in the contour of an otherwise smooth, flat terrain will produce scattered reflected waves varying in number according to the number and location of such irregularities, as shown in Fig. 3.2(b) and (c). The properties of propagation loss over smooth terrain are discussed in greater detail in the following paragraphs.

The specular reflections from a smooth surface conform to Snell's law, which states that the product of the refraction index N_1 and the

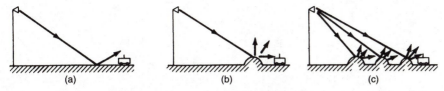

Figure 3.2 Contour irregularities that affect radio-wave scattering.

cosine of the grazing angle (cos θ) is constant along the path of a given ray of energy. This relationship is illustrated in Fig. 3.3. Since N_1 is always the same for both the incident and reflected waves, it follows that the incident and reflected wave angles are also the same.

Reflection coefficient

The ratio of the incident wave to its associated reflected wave is called the "reflection coefficient." The generalized Fresnel formulas for horizontally and vertically polarized radio waves [2] can be used to determine the power and boundary relationships of transmitted, incident, and reflected radio waves conforming to Snell's law. These relationships are shown in Fig. 3.4. By disregarding that part of the transmitted wave that is transmitted into the ground, the formulas can be applied, in terms of the incident and reflected waves, to calculate the reflection coefficient:

$$a_h e^{-j\phi_h} = \frac{\sin \theta_1 - (\varepsilon_c - \cos^2 \theta_1)^{1/2}}{\sin \theta_1 + (\varepsilon_c - \cos^2 \theta_1)^{1/2}} \quad \text{(horizontal)} \quad (3.3)$$

$$a_v e^{-j\phi_v} = \frac{\varepsilon_c \sin \theta_1 - (\varepsilon_c - \cos^2 \theta_1)^{1/2}}{\varepsilon_c \sin \theta_1 + (\varepsilon_c - \cos^2 \theta_1)^{1/2}} \quad \text{(vertical)} \quad (3.4)$$

where

$$\varepsilon_c = \varepsilon_r - j60\sigma\lambda \quad (3.5)$$

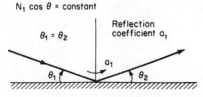

Figure 3.3 Smooth-surface reflections.

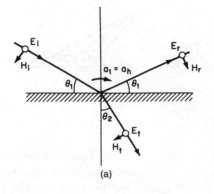

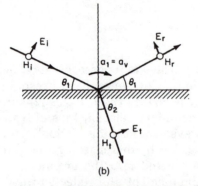

Figure 3.4 Wave relationships for horizontally and vertically polarized radio waves: (*a*) perpendicular (horizontal) polarization (may result from building reflection); (*b*) parallel (vertical) polarization (may result from ground reflection).

and where the permittivity ε_r is the principal component of the dielectric constant ε_c and σ is the conductivity of the dielectric medium in siemens per meter and λ is the wavelength.

Some typical values of permittivity and conductivity for various common types of media are shown in the following table:

Medium	Permittivity ε_r	Conductivity σ, S/m
Copper	1	5.8×10^7
Seawater	80	4
Rural ground (Ohio)	14	10^{-2}
Urban ground	3	10^{-4}
Fresh water	80	10^{-3}
Turf with short, dry grass	3	5×10^{-2}
Turf with short, wet grass	6	1×10^{-1}
Bare, dry, sandy loam	2	3×10^{-2}
Bare, sandy loam saturated with water	24	6×10^{-1}

For ground reflections in a mobile-radio environment, the permittivity ε_r of the dielectric constant ε_c is always large, and the angle θ_1 is always much less than 1 rad for the incident and reflected waves. When the incident wave is vertically polarized, then Eq. (3.4) can be applied, as illustrated in Fig. 3.4(b), with the result that

$$a_v \approx -1 \qquad\qquad (3.6)$$

When the incident wave E_i is horizontally polarized, then Eq. (3.3) can be applied, as illustrated in Fig. 3.4(a), with the result that

$$a_h \approx -1 \qquad\qquad (3.7)$$

In both of these cases, the actual value of the dielectric constant ε_c does not have a significant effect on the resultant values of a_v and a_h, which are symbolically the same as a_i in Eq. (1.20) of Chap. 1. It should be noted that Eqs. (3.6) and (3.7) are valid only for smooth terrain conditions, where $\theta_1 \ll 1$.

In the analysis of reflections from highly conductive surfaces of buildings and structures, the imaginary part of the dielectric constant ε_c is very large, and the following are true:

$$a_v \approx -1 \qquad \theta \text{ very small}$$

$$a_v \approx 1 \qquad \theta \text{ close to } (2n + 1)\pi/2 \qquad (3.8)$$

and $\qquad\qquad a_h \approx -1 \qquad$ either θ very small or ε_c very large $\qquad (3.9)$

Again, as in Eqs. (3.6) and (3.7), the actual value of the dielectric constant ε_c does not affect the resultant values of a_v and a_h in Eqs. (3.8) and (3.9), the values which are denoted a_i in Eq. (1.4) of Chap. 1.

When the incident wave E_i is vertically polarized (but horizontally with respect to the building walls) and the reflected wave is from a highly conductive surface, as in Fig. 3.4(a), then the reflection coefficient a_h is always approximately equal to -1, as shown in Eq. (3.9).

Figure 3.5 illustrates a hypothetical situation that is typical of an ideal mobile-radio environment. The surface of the terrain is smooth, and only one highly conductive building is reflecting the transmitted wave.

The received signal at the mobile unit can be calculated from the stationary properties of the energy rays by using Fermat's principle and by applying Snell's law:

$$E_r = E_i \left(1 - e^{-j\beta \, \Delta d_1} + a_h e^{-j\beta \, \Delta d_2}\right) \qquad (3.10)$$

where Δd_1 and Δd_2 are the propagation-path differences, defined:

$$\Delta d_1 = d_1 - d$$

$$\Delta d_2 = d_2 - d$$

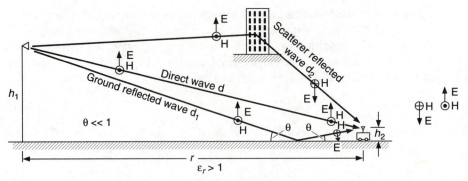

Figure 3.5   Reflections typically found in an ideal mobile-radio environment.

Also the incident angle θ is very small for the ground reflected wave. It is also proper to rewrite Eq. (3.10) in terms of time delay:

$$E_r = E_i(1 - e^{-j\omega\,\Delta\tau_1} + a_h e^{-j\omega\,\Delta\tau_2})\tag{3.11}$$

where

$$\Delta\tau_i = \frac{\Delta d_i}{v_c}$$

and where v_c is a constant equal to the speed of light.

3.3   Propagation Loss—Over Rough Terrain

To be able to predict the behavior of radio signals propagated over rough terrain, it is first necessary to define the criteria for determining the various degrees of roughness. The generally accepted criterion for determining roughness is the "Rayleigh criterion." A classic example of the effects of surface roughness on two rays of energy is illustrated in Fig. 3.6.

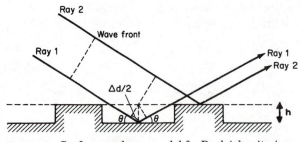

Figure 3.6   Surface roughness model for Rayleigh criterion.

Consider the two rays impinging on a rough surface characterized by ruts and ridges of varying height. The angle of the incident and reflected rays is called the grazing angle (θ), and the path difference Δd between the two rays is:

$$\Delta d = 2H \sin \theta \qquad (3.12)$$

The phase difference between the two rays is expressed:

$$\Delta \psi = \beta \, \Delta d = \frac{2\pi}{\lambda} \, \Delta d = \frac{4\pi H}{\lambda} \sin \theta \qquad (3.13)$$

When the phase difference $\Delta \psi$ is very small, the implication is that the difference in the height of the reflecting surfaces is also very small, and effectively, the surface is considered to be smooth. For the opposite condition, extreme roughness, the phase difference $\Delta \psi$ is equal to the value π, and the resultant signal from the two reflected rays with the same reflection coefficient a_i can be expressed:

$$a_i e^{j\psi} + a_i e^{j(\psi + \Delta \psi)} = a_i - a_i = 0 \qquad (3.14)$$

In practice, the extreme case for roughness, as shown in Eq. (3.14), is used infrequently, and a more arbitrary choice is the value $\pi/2$, which distinguishes a rough surface from a smooth surface. Rayleigh's criterion for roughness:

$$\Delta \psi > \frac{\pi}{2} \qquad (3.15)$$

Substituting $\Delta \psi = \pi/2$ into Eq. (3.13) and replacing h with h_R yields the following:

$$H_R = \frac{\lambda}{8 \sin \theta} \qquad (3.16)$$

The criterion for roughness then becomes:

$$H > H_R$$

The calculation of the roughness of a surface is primarily affected by the height of the irregularities, since the grazing angle for mobile-radio signals is relatively small. However, other criteria are sometimes used based on $\Delta \psi = \pi/4$ or $\Delta \psi = \pi/8$, which are described in other literature on the subject [3, 4]. Because the grazing angle θ for mobile-radio propagation is very small. Eq. (3.15) can be replaced by:

$$H_R = \frac{\lambda}{8\theta} \qquad (3.17)$$

The range d_1 out from the vicinity of the mobile unit for averaging the actual terrain height can be obtained by

$$d_1 = k_1 \cdot \frac{H_R}{\tan \theta} \tag{3.18}$$

where k_1 is usually one. But sometimes k_1 is less than one. The range d_1 is shown in Fig. 3.6. For example, if $\theta = 0.5°$ and $\lambda = 0.353$ m (for a frequency of 850 MHz), then $H_R = 5.06$ m.

In reality, ground surfaces are not smooth, but usually contain very gentle undulations. In the mobile-radio environment, the grazing angle is typically very small and the vector that is normal for the mean plane of a rough surface is also normal for the vectors associated with small irregularities or humps in the surface contour.

Propagation over an undulating surface is illustrated in Fig. 3.7. The scattered field E_s arriving at point P, the wave front at the mobile receiving antenna, can be calculated by using the Helmholtz integral method. The scattered field, for purposes of simplification, can be expressed in terms of two dimensions, x and z, assuming there are no ground variations in the y axis. The distance d_1 extends from the scattering point Q along the x and z axes to the point of observation P. The normal vector to the mean plane is n, and the incident wave is E. As determined by the Helmholtz integral method, the scattered field received at point P over the rough surface expressed by $z = S(x)$ is

$$E_S(P) = \frac{1}{4\pi} \int_S \left(E \frac{\partial \psi}{\partial n} - \psi \frac{\partial E}{\partial n} \right) dS$$

$$= \frac{y_1 - y_2}{4\pi} \int_{L_1}^{L_2} \left(E \frac{\partial \psi}{\partial n} - \psi \frac{\partial E}{\partial n} \right) dx \tag{3.19}$$

where

$$\psi = \frac{e^{-j\beta d_1}}{d_1}$$

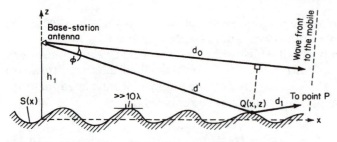

Figure 3.7   Propagation effects over an undulating surface.

For scattered plane waves, a number of which are usually assumed to be incident at the mobile receiving antenna, $d_1 \gg \lambda$, and the following relationship holds true:

$$\beta d_1 = \beta d_0 - \beta d' \cos \phi \qquad (3.20)$$

where d_0 is the direct-path distance between the base-station antenna and the mobile receiving antenna and d' is the distance from the base-station antenna to the scattering point Q. E is the incident wave, and $\partial E/\partial n$ is the normal derivative of S. Point P can be assumed as the position of the mobile unit.

When the Helmholtz integral method of Eq. (3.19) is used to calculate the scattered field over a rough surface at the point of reception, it is necessary to first determine the value of the rough surface $S(x)$. The value of $S(x)$ can be determined in two ways—by approximation, or by normal distribution. The approximation method assumes that the characteristics of the rough surface $S(x)$ are random and that the radius of the curvature of the surface irregularities is usually large in relation to the wavelength of the incident wave. Under this condition, the E_s field can be approximated:

$$E_s = (1 + a \, e^{j\beta(d' + d_1 - d_0)})E \qquad (3.21)$$

where a is the reflection coefficient of a smooth plane [5]. By applying Maxwell's equation, the magnetic field H_s can be approximated:

$$H_s = (1 + a \, e^{j\beta(d' + d_1 - d_0)})H \qquad (3.22)$$

Normally distributed surface characteristics

The $S(x)$ value for a normally distributed surface can be expressed:

$$E[S(x)] = 0 \qquad (3.23)$$

The summing up of a large number of variates produces a Gaussian distribution with a density probability as shown in Eq. (2.41) of Chap. 2. Similarly, if $z = S(x)$ and the standard deviation of z is σ_z, then the probability density function for a normally distributed surface can be expressed:

$$p_z(z) = \frac{1}{\sigma_z \sqrt{2\pi}} \exp\left(-\frac{z^2}{2\sigma_z^2}\right) \qquad (3.24)$$

The expected value of the scattered field expressed in Eq. (3.19) can be found from Eq. (2.42), as follows:

$$\mu_1 = E[E_s(P)] = \int_{-\infty}^{\infty} E_s(P) \, p(z) \, dz \qquad (3.25)$$

and the expected value for the scattered-field power can be expressed:

$$\mu_2 = E[E_s(P)E_s^*(P)] = \int_{-\infty}^{\infty} |E_s(P)|^2 p(z)\, dz \qquad (3.26)$$

Therefore, the standard deviation of the scattered field $E_s(P)$ is

$$\sigma_{E_s} = (\mu_2 - \mu_1^2)^{1/2} \qquad (3.27)$$

The solution to Eq. (3.19) is used to obtain the solutions of Eqs. (3.25) and (3.26), which are not easily solved, even when the incident field E is known. Fortunately, the approximation obtained by Eq. (3.21) is close enough for meaningful analysis of the scattered field in a mobile-radio situation.

Different deviations corresponding to different levels of surface roughness σ_z are illustrated in Fig. 3.8. $\sigma_z = 0$ indicates a smooth surface and $\sigma_z \gg \lambda$ indicates a rough surface. The effects of the different levels of roughness on the scattering pattern of Eq. (3.26) are quantitatively shown in Fig. 3.8.

Example 3.1 The accuracy of path-loss predictions is affected by surface irregularities and by the wavelength of the transmitted signal. A typical situation is illustrated in Fig. E3.1. The following questions are based on the assumption that the maximum height of the irregular surface is 25 ft, as shown in Fig. E3.1, and the wavelength of the transmitted signal is 850 MHz ($\lambda = 1.16$ ft). Under these given conditions:

1. What is the grazing angle θ of the incident wave shown in Fig. E3.1?
2. Based on the grazing angle of the incident wave, what is the minimum spacing S that will result in two reflected waves?
3. If the actual spacing S is less than the minimum spacing of question 2, will the surface be considered rough or smooth?

solution

1. The grazing angle of the incident wave relative to a smooth surface is obtained by using the Rayleigh criterion of Eq. (3.17), as follows:

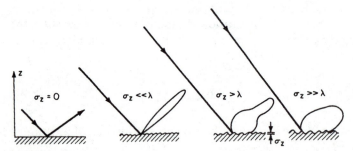

Figure 3.8 Effects of different types of surface roughness on radio-wave propagation.

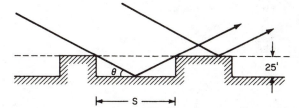

Figure E3.1   Reference data for Example 3.1.

$$\theta < \theta_R = \frac{\lambda}{8h_R} = \frac{1.16 \text{ ft}}{8 \times 25 \text{ ft}} = 0.0058 \text{ rad} = 0.332° \qquad \text{(E3.1.1)}$$

2. Based on the grazing angle of $\theta < 0.332°$, the minimum spacing S, shown in Fig. E3.1, can be calculated as follows:

$$S = 2\,\frac{h}{\tan \theta} = 2\,\frac{25}{0.0058} = 8620 \text{ ft} \qquad \text{(E3.1.2)}$$

3. If the actual spacing S is less than the minimum calculated value in question 2 (8620 ft), then one of the two reflected waves will be obstructed by the hump, and the surface can be considered a smooth one.

Surface roughness is a function of distance

Let the antenna height at the base station be h_i, the distance between the base station and the mobile unit d, and the incident angle θ as shown in Fig. 3.5.

Then

$$\theta = \frac{h_1 + h_2}{d} \qquad \text{(3.28)}$$

Substituting Eq. (3.28) into Eq. (3.17) yields

$$H_R = \frac{\lambda d}{8(h_1 + h_2)} = k \cdot d \qquad \text{(3.29)}$$

where

$$k = \frac{\lambda}{8(h_1 + h_2)} \qquad \text{(3.30)}$$

Let $\lambda = 0.333$ m (at 900 MHz), $h_1 = 30$ m, and $h_2 = 2$ m; then $k = 0.0013$. Equation (3.29) is plotted in Fig. 3.9, which indicates the region for the flat terrain criterion. It shows that, further out in distance, the flat-terrain-criterion region can tolerate higher terrain height. Therefore, in considering the rough terrain at a distance of 10 km, the terrain

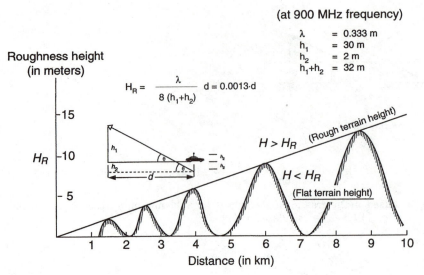

Figure 3.9   Flat terrain criterion.

height has to be more than 13 meters in meeting the rough terrain criterion. Also, H_R is lower when $h_1 + h_2$ is higher, as seen from Eq. (3.29).

The resolution interval of terrain data

The resolution interval of terrain data is always in our consideration for predicting the signal strengths. If the interval is very small, too many data points need to be stored in the database, which takes longer time to calculate the prediction of signal strength at each data point. Here is a method to use for justifying the resolution interval. Let the resolution interval be Δd and the corresponding increasing in roughness height be ΔH_R. Then:

$$\frac{H_R}{d} = \frac{\Delta H_R}{\Delta d} = \frac{\lambda}{8(h_1 + h_2)} = k \qquad (3.31)$$

In Fig. 3.9, $\Delta H_R = 0.0013\Delta d$. For $\Delta d = 100$ m, then $\Delta H_R = 0.13$ m. For $\Delta d = 50$ m, then $\Delta H_R = 0.075$ m.

Assume that, at the increment $\Delta d = 100$ m, the real terrain height increment is $\Delta H = 0.11$ m.

$$\Delta H > \Delta H_R$$

We have to use the data based with the smaller increment $\Delta d = 50$ m, from which we obtain $\Delta H_R = 0.075$ m. Under this condition, ΔH is smaller than ΔH_R, and the resolution interval is justified.

Analysis of the active scattering region

Careful analysis of the first Fresnel zone can confirm, with high probability, any region suspected of active scattering. The waves reflected within the first Fresnel zone typically have a phase difference of less than $\pi/2$ from the direct ray, as illustrated in Fig. 3.10 and by the following equation:

$$(d' + d_1) - d_0 = \frac{\lambda}{4} \tag{3.32}$$

The elliptic zone in the ground plane xy is determined by the intersection of that plane with an ellipsoid of revolution having as its foci terminals a and b. Equation (3.28) is the expression for an ellipsoid of revolution. Finding the elliptic zone is a relatively straightforward but tedious procedure. The solution for determining the center of the ellipse is:

$$x_0 = \frac{r}{2}\left[1 - \frac{\tan\theta}{\left(\dfrac{\lambda/4}{r} + \sec\theta\right)^2 - 1}\right] \tag{3.33}$$

and

$$\theta = \tan^{-1}\frac{h_1 - h_2}{r}$$

The solution for determining the area of the elliptic zone is

$$A = \frac{\pi r\sqrt{\lambda r}}{4}\frac{3/2}{[1 + (h_1 + h_2)^2/\lambda r]^{3/2}} \tag{3.34}$$

The family of curves that are obtained from Eq. (3.30) is shown in Fig. 3.11. As an example, if $h_1 = 100$ ft (the height of the base-station antenna), $h_2 = 10$ ft (the height of the mobile antenna), and $r = 4$ mi, then the active scattering region has a radius $\sqrt{A/\pi}$ of approximately 989 ft.

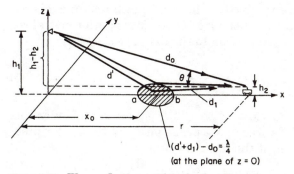

Figure 3.10 Wave reflections within the first Fresnel zone.

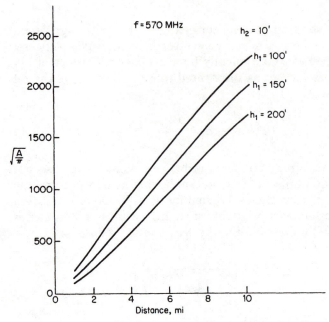

Figure 3.11 Graphical analysis of scattering within the first Fresnel zone.

3.4 Two-Wave Model [6]—Explaining Mobile Radio Path Loss and Antenna Height Effects

The mobile radio path loss and the effects of antenna height on the scattered field E_s can be evaluated by using the characteristics of the scattered field, as previously shown in Eq. (3.21):

$$E_s = (1 + a\ e^{j\beta(d' + d_1 - d_0)})E \qquad (3.21)$$

where E is the direct wave received at the mobile antenna and a is the reflection coefficient. Because of the free-space propagation-path loss, E is linearly attenuated in proportion to the distance d:

$$|E|^2 \propto \frac{1}{(d/\lambda)^2} \qquad (3.35)$$

If there are no reflected waves, a in Eq. (3.21) equals zero. Under this condition, according to Friis' free-space propagation-path-loss formula, the average received power at the mobile would be:

$$P_r = \frac{|E|^2}{2\eta_0} = P_0\left(\frac{1}{4\pi d/\lambda}\right)^2 \qquad (3.36)$$

where $P_0 = P_t G_t G_m$
$\eta_0 =$ intrinsic impedance of free space
$P_t =$ transmitted power
$G_t =$ gain of the transmitting antenna
$G_m =$ gain of the receiving antenna

Assuming that a vertically polarized wave is transmitted and two waves (direct and reflected) are observed along the propagation path, then the small difference $\Delta\Phi$ caused by the difference between the two ray paths Δd can be expressed:

$$\Delta\Phi = \beta \, \Delta d \tag{3.37}$$

From Eqs. (3.4) and (3.6), a vertically polarized transmitted wave has a reflection coefficient of $a \approx -1$ for a typical mobile-radio environment. The received power P_r can be expressed:

$$P_r = \frac{|E_s|^2}{2\eta_0} = P_0\left(\frac{1}{4\pi d/\lambda}\right)^2 |1 + a_v \exp{(j\Delta\Phi)}|^2 \tag{3.38}$$

The expression Δd can be obtained in a straightforward manner (see Fig. 3.5):

$$\Delta d = \sqrt{(h_1 + h_2)^2 + d^2} - \sqrt{(h_1 - h_2)^2 + d^2} \tag{3.39}$$

where h_1 is the height of the base-station antenna, h_2 is the height of the mobile antenna, and d is the distance between the two antennas. The relationship $h_1 + h_2 \ll d$ is predominantly true and Δd is typically a very small value, expressed:

$$\Delta d \approx d\left[1 + \frac{(h_1 + h_2)^2}{2d^2} - 1 - \frac{(h_1 - h_2)^2}{2d^2}\right] = \frac{1}{d} 2h_1 h_2 \tag{3.40}$$

Substituting Eq. (3.40) into Eq. (3.37), and recognizing that $\Delta\Phi$ is also a phase difference expressed in radians, yields the following:

$$\sin \Delta\Phi \approx \Delta\Phi = \frac{4\pi h_1 h_2}{\lambda d} \tag{3.41}$$

$$\cos \Delta\Phi \approx 1 \tag{3.42}$$

Then, Eq. (3.38) becomes:

$$P_r = \frac{|E_s|^2}{2\eta_0} = P_0\left(\frac{1}{4\pi d/\lambda}\right)^2 |1 - \cos \Delta\Phi - j \sin \Delta\Phi|^2$$

$$\approx P_0\left(\frac{1}{4\pi d/\lambda}\right)^2 (\Delta\Phi)^2$$

$$= P_0 \left(\frac{1}{4\pi d/\lambda} \right)^2 \left(\frac{4\pi}{\lambda d} \, h_1 h_2 \right)^2$$

$$= P_0 \left(\frac{h_1 h_2}{d^2} \right)^2 \tag{3.43}$$

Note Eq. (3.43) is not a precisely accurate formula for predicting path loss, since it does not include the path loss variable based on wavelength. However, Eq. (3.39) is useful in explaining the 40 dB/dec path loss in the field and also making relative comparisons based on the effects of antenna height.

$$L = 40 \log (d'/d) \qquad \text{in dB} \tag{3.44}$$

$$\Delta G = 20 \log (h_1'/h_1) \qquad \text{in dB} \tag{3.45}$$

where L is the path loss when d' is greater than d. ΔG is the antenna-height gain when h_1' is greater than h_1. When $d' < d$ and/or $h_1' < h_1$, both L and ΔG are negative in dB.

Measurement studies published in reference literature [6–9] and based on the mean value of many data points predict gains of up to 6 dB per octave for the received gain versus base-station antenna height. The difference in gain, which can be obtained by doubling the height of the base-station antenna, is:

$$\frac{P_{r_2}}{P_{r_1}} = \left(\frac{2h_1}{h_1} \right)^2 = 4 \qquad \approx 6 \text{ dB per octave} \tag{3.46}$$

Hence, the 6 dB per octave rule is derived from Eq. (3.45) and verified by the measured data. It is important to note that changing the height of the mobile antenna produces gains of only 3 dB per octave, as shown by actual measurements [6, 10], because of the effects of nearby scatterers. Because of practical limitations, the usual height of the mobile antenna is approximately 10 ft above ground level. This should not be considered in using Eq. (3.43) to study the effects of mobile-antenna height on propagation-path loss.

From the measured and recorded data, analysis shows that the propagation-path loss is affected by the frequency, as shown in Fig. 3.12, and as noted in Eq. (3.36). It is not noted, however, in Eq. (3.43). The loss increases after 20 km due to the radio below the earth horizon.

The distance criterion for calculating mobile radio path loss

Equation (3.44) is obtained from Eq. (3.43), which is an approximation formula from Eq. (3.38) by assuming that the incident angle θ is very small; that is, $d \gg h_1 + h_2$. When d is small, we have to find the range

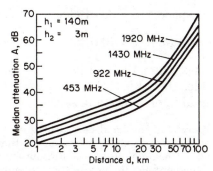

Figure 3.12 Path loss as a function of wavelength. *(From Ref. 6.)*

such that Eq. (3.44) still holds. It is related to the Fresnel zone distance. Let Δd in Eq. (3.40) be $\lambda/2$ as the first Fresnel zone.

$$\Delta d = \frac{\lambda}{2} = \frac{1}{D_f} \, 2h_1h_2 \tag{3.47}$$

Then distance criterion D_f is obtained from Eq. (3.47) as the Freznel zone distance:

$$D_f = \frac{4h_1h_2}{\lambda} \tag{3.48}$$

Eq. (3.48) is plotted in Fig. 3.13 for both frequencies 850 MHz and 1.9 GHz. The path loss curve is

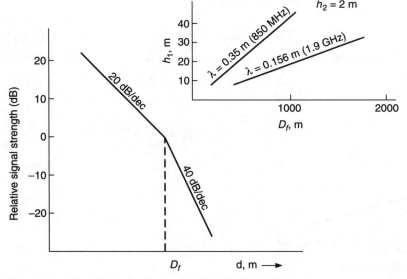

Figure 3.13 The distance criterion for plotting the path loss curves.

$$\text{Free space loss} = 20 \text{ dB/dec} \qquad d < D_f$$

$$\text{Mobile radio loss} = 40 \text{ dB/dec} \qquad d > D_f$$

When the distance is less than 1 km and if the information of building layout is not available, then Fig. 3.13 can be used to estimate the closed-in received signal in an averaging sense.

Example 3.2 An improvement in signal gain can be obtained by raising the height of the base-station antenna. On the assumptions that the received radio signal is −110 dBm and the height of the base-station antenna is 100 ft, as shown in Fig. E3.2, how much higher must the base-station antenna be raised to obtain an increase from −110 to −100 dBm (a gain of 10 dB) in the received signal?

solution To obtain a 10-dB increase in the received signal, the base-station antenna must be raised 216 ft above its original height of 100 ft, for a total height of 316 ft. The following relationships are valid:

$$\frac{P_r'}{P_r} = \frac{P_0\left(\dfrac{h_1'h_2}{d^2}\right)^2}{P_0\left(\dfrac{h_1 h_2}{d^2}\right)^2} = \frac{h_1'^2}{h_1^2} \tag{E3.2.1}$$

and

$$10 \log\left(\frac{P_r'}{P_r}\right) = 20 \log\left(\frac{h_1'}{h_1}\right) \tag{E3.2.2}$$

where h_1' is the new antenna height, $10 \log (P_r'/P_r) = 10$ dB, and $h_1 = 100$ ft. Then Eq. (E3.2.1) becomes:

$$10 = 20 \log \frac{h_1'}{100} \tag{E3.2.3}$$

Then

$$h_1 = 100 \text{ ft} \times 10^{0.5} = 316 \text{ ft} \tag{E3.2.4}$$

The expression for raising the base-station antenna is:

$$\Delta h = 316 \text{ ft} - 100 \text{ ft} = 216 \text{ ft}$$

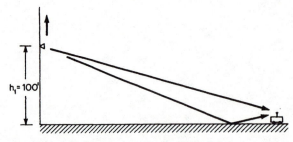

$h_1 = 100'$

Figure E3.2 Reference data for Example 3.2.

3.5   Rules for Calculating Average Signal Strength (Local Mean)

One of the first rules for statistical analysis of propagation-path loss is to determine the proper distance intervals that will enable long-term or local mean values to be obtained from received-field-strength data. There are two preferred methods for calculating the local mean: (1) the median approach and (2) the mean approach.

Median approach

Young [11] obtained sample measurements of the field strength taken at 200-ft intervals over a distance of approximately 1000 ft. The median value was then calculated from the samples obtained within this relatively small area along the propagation path. Okumura et al. [6] obtained sample measurements of the field strength taken at 20-m intervals over a distance of from 1 to 1.5 km. Again, the median was calculated from the samples obtained.

Mean approach [12]

In the mean approach, the value of the local mean can be estimated from the Rayleigh fading effects $r(y) = m(y)r_0(y)$ as follows (see Fig. 6.1):

$$\hat{m}(x) = \frac{1}{2L} \int_{x-L}^{x+L} r(y)\, dy = \frac{1}{2L} \int_{x-L}^{x+L} m(y)r_0(y)\, dy \qquad (3.49)$$

Under the assumption that $m(y)$ is the true local mean and $r_0(y)$ is a stationary Rayleigh process with unit mean, then

$$m(y) = \text{constant} \qquad x - L < y < x + L \qquad (3.50)$$

The estimator then becomes:

$$\hat{m}(x) = m(x) \frac{1}{2L} \int_{x-L}^{x+L} r_0(y)\, dy = m(x)\overline{r_0(y)} \qquad (3.51)$$

where $\overline{r_0(y)}$ should approach 1 when $\hat{m}(x)$ approaches $m(x)$.

Determining the value of L

The mean value and the correlation of a Rayleigh-fading signal are discussed in Eq. (2.54) and Chap. 6; they are expressed:

$$\langle r(y) \rangle = \sqrt{\frac{\pi}{2}}\, \sigma$$

$$R_r(y) \approx \frac{\pi}{2} \, \sigma^2 \left[1 + \frac{J_0^2(\beta y)}{4} \right] \tag{3.52}$$

where $2\sigma^2$ is the average power of the Rayleigh-fading signal, J_0 is the zero-order Bessel function of the first order, $\beta = 2\pi/\lambda$, and λ is the wavelength. The mean of the estimator expressed in Eq. (3.51) then becomes:

$$\langle \hat{m}(x) \rangle = \frac{1}{2L} \int_{x-L}^{x+L} \langle r(y) \rangle \, dy = \langle r(y) \rangle = \sqrt{\frac{\pi}{2}} \, \sigma \tag{3.53}$$

where $\langle r(y) \rangle = \langle m(y) r_0(y) \rangle = m(x) \langle r_0(y) \rangle = m(x)$. Hence $\langle \hat{m}(x) \rangle = m(x)$, and the variance of the estimator is:

$$\sigma_{\hat{m}}^2 = \langle \hat{m}^2(x) \rangle - \langle \hat{m}(x) \rangle^2$$

$$= \frac{1}{4L^2} \int_{x-L}^{x+L} \int_{x-L}^{x+L} E[r(y_1)r(y_2)] \, dy_1 \, dy_2 - \frac{\pi}{2} \, \sigma^2 \tag{3.54}$$

A more simplified expression [12] can be found:

$$\sigma_{\hat{m}}^2 = \frac{1}{L} \int_0^{2L} \left(1 - \frac{y}{2L} \right) \left[R_r(y) - \frac{\pi}{2} \, \sigma^2 \right] dy \tag{3.55}$$

Normalizing $\hat{m}(x)$ with respect to its mean value provides the expression

$$\langle \hat{m}(x) \rangle = 1 \tag{3.56}$$

and
$$\sigma_{\hat{m}}^2 = \frac{1}{4L} \int_0^{2L} \left(1 - \frac{y}{2L} \right) J_0^2 \, (\beta y) \, dy \tag{3.57}$$

Therefore, $\hat{m}(x)$ is within the spread of $1\sigma_{\hat{m}}$ or $2\sigma_{\hat{m}}$ and can be expressed in decibels as a function of $2L$. Since in Eq. (3.57) $\sigma_{\hat{m}}^2$ is a function of $2L$, then $20 \log (1 \pm \sigma_{\hat{m}})$ or $20 \log (1 \pm 2\sigma_{\hat{m}})$ is also a function of $2L$.

The following table shows a typical range of standard deviation values and spreads as a function of $2L$.

$2L$	$\sigma_{\hat{m}}$	$1\sigma_{\hat{m}}$ spread, dB	$2\sigma_{\hat{m}}$ spread, dB
5λ	0.165	3	6
10λ	0.122	2.1	4.2
20λ	0.09	1.56	3.14
40λ	0.06	1	2.12

From the table, a $2L$ value of 40λ is required to obtain a $1\sigma_{\hat{m}}$ spread of 1 dB. Other values are shown for reference purposes.

Any length over 40λ can be used in obtaining the local mean [12] since $1\sigma_{\hat{m}}$ spread of $\hat{m}(x)$ is less than 1 dB. The local scattering area in the vicinity of a mobile unit in a suburban area is also shown roughly 110 ft in diameter [13]. This implies that an interval of 100 to 200 ft for 800 MHz is a reasonable guideline for obtaining the local mean.

For 450 MHz or lower the wavelength becomes longer. In this case, we may use 20λ as the length of averaging window for obtaining the local mean values.

3.6 Models for Predicting Propagation-Path Loss

The local means obtained from field-strength measurements taken at predetermined sampling intervals can vary over a large range of decibels. It is therefore advantageous to classify the characteristics of the various sampling environments from which the data for local means are collected. Rules can then be formulated based on predictions of the value of local means associated with classified types of sampling environments. The two main factors that will determine the broad parameters for classification are (1) the characteristics of the terrain surface and contour and (2) the presence or absence of buildings, structures, and other human-made objects.

The degree of surface undulation present in the contour of the terrain enables a classification to be assigned within an "interdecile range" (Δh) for a given sampling interval, e.g., 10 km. The interdecile range typically classifies surface irregularities falling within the 10 to 90 percent vertical boundaries of the overall terrain-contour profile. These parameters for interdecile range are illustrated in Fig. 3.14.

The terrain contour is not the sole contributor to propagation-path loss in the mobile-radio environment, according to Okumura et al. [4]. Buildings and structures along the propagation path are also main contributors to path loss and must be classified in conjunction with the contour features of the surrounding terrain. The following general classifications are quite obvious:

1. **Open land**—Undeveloped or partially developed farmland with conventional small dwellings and barns, and sparsely populated.

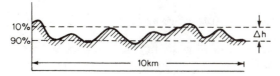

Figure 3.14 Interdecile range parameters.

2. **Industrialized open land**—Developed areas where large-scale farming activities are combined with limited industrial use. Such areas would contain large silos, high-voltage power-distribution lines, and occasional industrial facilities.

3. **Suburban areas**—Mixed residential and clean industrial uses; i.e., warehouses, shopping malls, light manufacturing, and moderate vehicular traffic conditions.

4. **Small to medium-sized city**—Densely populated residential and commercial areas with well-defined business districts characterized by numbers of high buildings and few, if any, skyscrapers. Traffic density is usually moderate to heavy, depending upon flow patterns and time of day.

5. **Large-sized city**—Heavily commercial and industrial with many high-rise residential structures. Business districts are characterized by many skyscrapers. Traffic density is heavy virtually at all times. Cities like New York, Boston, Philadelphia, Los Angeles, and Newark, New Jersey, are typical large cities.

Prediction model for UHF

Since suburban areas represent the mean and have relatively similar path-loss characteristics, the statistical average of the path losses for all suburban areas can also be used as a comparative reference for evaluating the path-loss characteristics for other types of areas, such as urban or open rural areas. A typical case would be a suburb in the Philadelphia-Camden area, as reported in Kelly [14]. The local mean data for that area were based on statistically measured data at a frequency of 800 MHz. A necessary parametric conversion was made, resulting in the following given conditions:

1. The base-station transmitted power was 10 W (40 dBm). Antenna gain was 6 dB with respect to a half-wave dipole. Antenna height was 100 ft above ground. The ERP (effective radiation power) is 40 dBm + 6 dB = 46 dBm.

2. The mobile antenna was a quarter-wavelength whip with a height of 10 ft above ground. The antenna gain was 0 dB with respect to the gain of a dipole antenna.

The measured results were as follows:

1. The local means were lognormal-distributed with a standard deviation σ of 8 dB.

2. The power at the 1-mi point of interception P_{r_1} was –61.7 dBm.

3. The slope γ was 38.4 dB per decade (or $\gamma \log 2 = 11.55$ dB per octave), or $\gamma = 3.84$ in linear scale.

A graphic representation of the propagation-path loss for this suburban area is shown in Fig. 3.13.

The field strength of the received signal (P_r) for any given system can be expressed as:

$$P_r = P_{r_0}\alpha_0$$
$$= P_{r_1}r^{-\gamma}\alpha_0 \tag{3.58}$$

where P_{r_1} is the strength at 1 mi, r is in miles, and α_0 is the adjustment factor.

The standard conditions are:

$$h_1 = 30.48 \text{ m (100 ft)}$$

$$P_t = 10 \text{ watts}$$

$$G_1 = 6 \text{ dB (at the base)} \tag{3.59}$$

The adjustment factor is derived as follows:

$$\alpha_0 = (\text{Adj})_1(\text{Adj})_2(\text{Adj})_3$$

$$\text{where} \quad (\text{Adj})_1 = \left(\frac{\text{new base-station heights (ft)}}{100 \text{ ft}}\right)^2$$

$$= \left(\frac{\text{new base-station heights (m)}}{30.48}\right)^2$$

$$(\text{Adj})_2 = \frac{\text{new transmitter power (W)}}{10 \text{ W}}$$

$$(\text{Adj})_3 = \frac{\begin{array}{c}\text{new base-station antenna}\\ \text{gain with respect to } \lambda/2 \text{ dipole}\end{array}}{4}$$

The field strength of the received signal (P_r) can also be expressed in decibels, as follows:

$$P_r = P_{r_0} + \alpha_0 = -61.7 - 38.4 \log r + \alpha_0 \tag{3.60}$$

where r is in miles. It is customary to use the received-power data collected at the 1-mi intercept, rather than the data collected in the area within a 1-mi radius of the transmitter. The area around the transmitter is often difficult to measure, since there are often few roads and therefore not enough sampling data to calculate a true median value.

In working in the metric system, the following equation can be used in place of Eq. (3.60); it converts the values shown in Fig. 3.15 to their metric equivalent.

$$P_r = P_{r_0} + \alpha_0 = -54 \text{ dB} - 38.4 \log r + \alpha_0 \qquad (3.61)$$

where −54 dB is the strength at 1 km and r is in kilometers.

For different environments and/or areas, the slopes and intercept points will probably differ. In some urban areas where the slope is flatter, its measurement at the 1-mi intercept point is lower [14]. The same criteria used for the UHF model can also be applied to VHF, as long as the 1-mi-intercept power and slope at the VHF frequency are known.

Example 3.3 A base station transmitting 20 W of power has an antenna gain of 6 dB and an antenna height of 200 ft. A mobile unit with an antenna gain of 0 dB and an antenna height of 10 ft receives a −79.5-dBm signal at 5 mi and a −91-dBm signal at 10 mi from the base station. Using the above data, plot the field strength of the received signal to measure the path loss, as previously illustrated in Fig. 3.15.

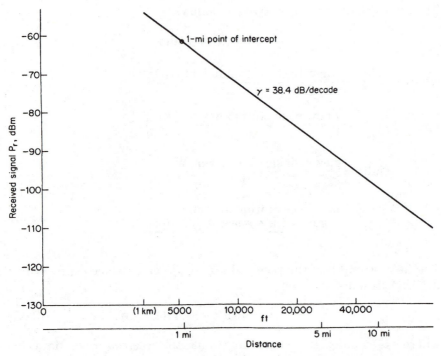

Figure 3.15 Path loss in a typical suburban area at 800–900 MHz.

The model of Okumura et al.

Of several models that are described in published studies, the Okumura et al. model [7] is the model of choice for analyzing mobile-radio propagation. The Rice-Longley-Norton-Barsis model [15] was found to be useful in predicting transmission loss for long-range radio propagation but was limited in its application to mobile-radio propagation. Longley subsequently published a modified model [16] that recognizes the differences between stationary-microwave-link propagation and mobile-radio-link propagation. However, this modified approach is not as simple as Okumura's approach [7]. Bullington has published a simplified method that describes radio propagation for vehicular communications [17] but makes no distinction between urban and suburban areas.

The model of Okumura et al. is based on empirical data, compiled into charts, that can be applied to VHF and UHF mobile-radio propagation. Path-loss predictions for open rural areas and suburban and urban areas can easily be distinguished by using the following procedure. First, a basic prediction of median attenuation over quasi-smooth terrain* in an urban area is obtained from Fig. 3.16. Second, the correction factor for either an open area or a suburban area must be subtracted from the median attenuation value obtained from Fig. 3.16. The two correction factors are shown in Fig. 3.17. Third, the correction factors for antenna heights must be applied, as follows:

6 dB per octave for base-station antenna height

3 dB per octave for mobile-antenna height $3 \text{ m} < h_2 < 5 \text{ m}$

$2h_2 \log (h_2/3 \text{ m})$ for mobile-antenna height $5 \text{ m} < h_2 < 10 \text{ m}$

Additional correction factors are described in the Okumura model, such as those for rolling, hilly terrain and for street orientation. As these additional correction factors are added to the basic median attenuation value of Fig. 3.16, the final prediction will more closely approach the statistical mean value. Chapter 4 describes the use of correction factors in greater detail and expands on those that are discussed in Okumura et al.

Example 3.4 Use the Okumura method to calculate the path-loss prediction for mobile-radio transmission over a quasi-smooth terrain in a suburban area for the following three cases:

Case 1—The mobile antenna height h_1 and the base-station antenna height h_2 are both 2 m. The distance between the antennas along the transmission path is 10 km.

---

* Interdecile range approximately equals 20 m [6].

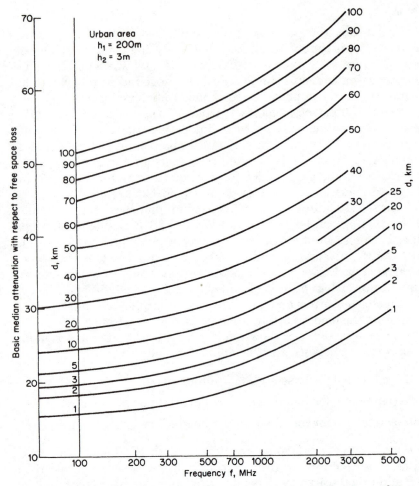

Figure 3.16 Prediction curve for basic median attenuation over quasi-smooth terrain in an urban area. *(From Ref. 6.)*

Case 2—The mobile antenna height h_2 is 2 m and the base-station antenna height h_1 is 10 m. The distance between antennas is 16 km.

Case 3—Both antenna heights, h_1 and h_2, are 10 m. The distance between antennas is 25 km.

The results of cases 1 through 3 are plotted in Fig. E3.4.

A general formula of path loss passing different environments

Usually a propagation path may run over more than one kind of terrain contour. Assume that the propagation-path loss slope γ_1 is predicted in area A and the slope γ_2 in area B. A path runs over at a

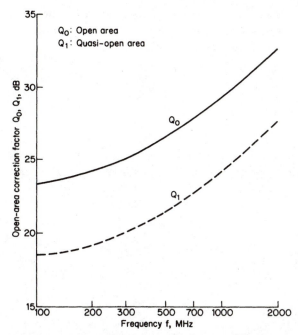

Figure 3.17 Correction factors for open and quasi-open (suburban) areas as a function of frequency. *(From Ref. 6.)*

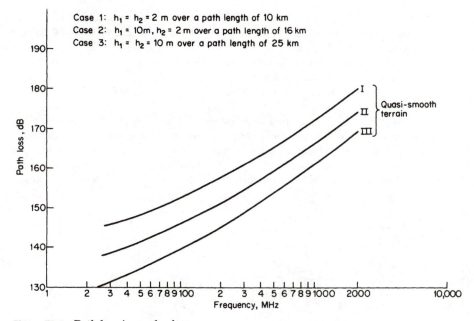

Figure E3.4 Path loss in a suburban area.

distance r where $r < r_1$ is in area A and $r_1 < r < r_2$ is in area B. Then the expected power P_r received in area B at a distance r from the transmitter P_0 which is in area A can be obtained as follows:

$$P_r = \frac{P_0}{\left(\dfrac{4\pi r_1}{\lambda}\right)^{\gamma_1} \left(\dfrac{r}{r_1}\right)^{\gamma_2}} \qquad r_1 \le r \le r_2 \qquad (3.62)$$

where γ is in linear scale.

The general formula of path loss L for a path length r which runs over N different areas with N different path loss slopes γ_i, $i = 1, N$, can be easily proved as

$$
\begin{aligned}
L &= \frac{P_r}{P_0} \\
&= \left(\frac{4\pi r_1}{\lambda}\right)^{-\gamma_1} \left(\frac{r_2}{r_1}\right)^{-\gamma_2} \left(\frac{r_3}{r_2}\right)^{-\gamma_3} \cdots \left(\frac{r}{r_{N-1}}\right)^{-\gamma_N} \qquad r_{N-1} \le r \le r_N \quad (3.63)
\end{aligned}
$$

Problem Exercises

1. The accuracy of path-loss predictions is affected by surface irregularities. From Fig. P3.1, determine the effects of ground-surface roughness, where $h_2 \gg h_1$ and $h_2 \gg h_3$.

2. Using Fig. P3.2, find the time-delay spread at antenna h_2 due to the active scattering region. Antenna height h_1 is 100 ft, antenna height h_2 is 10 ft, and the link distance is 6 mi. Find
    a. The delay spread between link a and link b_1
    b. The delay spread between link a and link b_2
    c. The range of maximum and minimum delay spread

3. Consider a link where reception over a distance of 10 km is satisfactory with a base-station antenna height h_1 of 50 m and a mobile antenna height h_2 of 2 m. If the base-station antenna height is lowered to 10 m, what will be the effect on reception in terms of distance? If the distance is the same (10 km), how high must the mobile antenna be raised to ensure satisfactory reception?

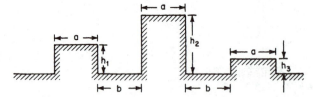

Figure P3.1 Surface roughness data for Prob. 1.

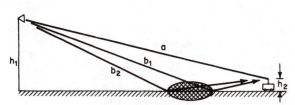

Figure P3.2   Delay-spread model for Prob. 2.

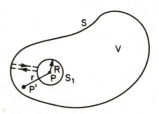

Figure   P3.6   Scattered-field data for Prob. 6.

4.   Apply the accurate prediction model for UHF (see Fig. 3.13) to find the path loss in a suburban area over a distance of 10 km at a frequency of 800 MHz. The following are given: Base-station transmitting power is 25 W and antenna gain is 8 dB referring to a dipole antenna. Base-station antenna height h_1 is 50 m. Mobile antenna gain is 0 dB using a dipole antenna, and the antenna height is 5 m.

5.   Using the same conditions given in Prob. 4, apply the Okumura model to find the path loss.

6.   The Helmholtz integral is used to calculate a scattered field over a rough surface, received at a point P. The parameters for deriving the Helmholtz integral are illustrated in Fig. P3.6. Equation (3.18) represents the scattered-field relationships:

$$E_s(P) = \frac{y_1 - y_2}{4\pi} \int_{L_1}^{L_2} \left(E \, \frac{\partial \psi}{\partial n} - \psi \, \frac{\partial E}{\partial n} \right) dx$$

Derive the Helmholtz integral of Eq. (3.18) from the stated assumptions.

References

1. K. A. Norton, "The Calculation of Ground Wave Field Intensity over a Finitely Conducting Spherical Earth," *Proc. IRE,* vol. 29, December 1941, p. 623.
2. E. C. Jordan, *Electromagnetic Waves and Radiating Systems,* Prentice-Hall, Englewood Cliffs, N.J., 1950, p. 141.
3. K. A. Norton and A. C. Omberg, "Maximum Range of a Radar Set," *Proc. IRE,* vol. 35, 1947, p. 4.
4. D. E. Kerr, *The Propagation of Short Radio Waves,* MIT Rad. Lab. Rep. no. 13, McGraw-Hill, New York, 1951.
5. P. Beckmann and A. Spizzichino, *The Scattering of Electromagnetic Waves from Rough Surfaces,* Macmillan, New York, 1963, p. 20.
6. H. Bremmer, "Terrestrial Radio Waves," Elsevier Publishing Company, Amesterdam, 1949; B. Van der Pol and H. Bremmer, phil. mag. LXXVI, November, 1937 (original paper).
7. Y. Okumura, E. Ohmori, T. Kawano, and K. Fukuda, "Field Strength and Its Variability in VHF and UHF Land Mobile Service," *Rev. Elec. Comm. Lab.,* vol. 16, September–October 1968, pp. 825–873.
8. W. C. Y. Lee, "Studies of Base-Station Antenna Height Effects on Mobile Radio," *IEEE Trans. Veh. Tech.,* vol. 29, May 1980, pp. 252–260.

9. W. C. Jakes, *Microwave Mobile Communications,* Wiley, New York, 1974, p. 98.

10. G. D. Ott and A. Plitkins, "Urban Path-Loss Characteristics at 820 MHz," *IEEE Trans. Veh. Tech.,* vol. 27, November 1978, pp. 189–197.

11. R. W. Young, "Comparison of Mobile Radio Transmission at 150, 450, 900 and 3700 Mc," *Bell System Technical Journal,* vol. 31, November 1952, pp. 1068–1085.

12. W. C. Y. Lee and Y. S. Yeh, "On the Estimation of the Second-Order Statistics of Log Normal Fading in Mobile Radio Environment," *IEEE Trans. Comm.,* vol. 22, June 1974, pp. 869–873.

13. W. C. Y. Lee, "Effects on Correlation between Two Mobile Radio Base-Station Antennas," *IEEE Trans. Comm.,* vol. 21, November 1973, pp. 1214–1224.

14. K. K. Kelly, II, "Flat Suburban Area Propagation at 820 MHz," *IEEE Trans. Veh. Tech.,* vol. 27, November 1978, pp. 198–204.

15. P. L. Rice, A. G. Longley, K. A. Norton, and A. P. Barsis, "Transmission Loss Prediction for Tropospheric Communication Circuits," *NBS Tech. Note no. 101,* 1967, vols. 1 and 2.

16. A. G. Longley, "Radio Propagation in Urban Areas," U.S. Dept. of Commerce OTR 78-144/April, 1978.

17. K. Bullington, "Radio Propagation for Vehicular Communications," *IEEE Trans. Veh. Tech.,* vol. 26, November 1977, pp. 295–308.

18. P. Beckmann and A. Spizzichino, *The Scattering of Electromagnetic Waves from Rough Surfaces,* The MacMillan Co., 1963, p. 11.

Path Loss over Hilly Terrain and General Methods of Prediction

4.1 Path-Loss Predictions Based on Model Analysis

As might be expected from the discussions in preceding chapters, the propagation-path loss for hilly areas is greater than that for nonhilly areas. It therefore follows that the methods used to predict the signal strength of received transmissions are more difficult in hilly areas than in nonhilly areas [1]. The various shapes encountered in the contour of hilly terrain are unpredictably irregular. Often, the direct line-of-sight path is obstructed by the tops of intervening hills. To properly analyze the path loss in hilly areas, it is necessary to classify hilly terrain into either one of two classic types that are described in terms of the physical characteristics of the propagation path [1]. Type 1 is the line-of-sight path over hilly terrain where surface irregularities do not obstruct the direct signal path. Type 2 is the out-of-sight path over hilly terrain where surface irregularities project into, and therefore obstruct, the direct signal path. In type 2, diffraction loss due to the obstruction of hilltops (the out-of-sight case) must be considered. A formula for approximating the diffraction-loss prediction, and other special considerations for predicting the path loss over hilly terrain, are discussed in subsequent paragraphs.

Figure 4.1 illustrates a simple model for predicting propagation-path loss over flat, unobstructed terrain. As was shown in Eq. (3.46), a base-station antenna-height gain factor of 6 dB per octave can be predicted for this model situation. Measured data recorded in flat suburban and urban areas support predictions based on this model. Similar models can be used to explain phenomena associated with hilly areas.

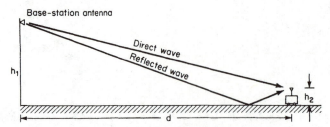

Figure 4.1   Model for propagation over flat terrain.

Figure 4.2 is the model for a base-station transmitter and a mobile receiver separated by a large distance r over terrain that combines level ground and a hill slope having an angle much less than 1 rad. Two sets of conditions can be considered. In type A the mobile antenna is located on the slope and the base-station antenna is located on level, flat terrain. Type B is the reverse situation, where the base-station antenna is located high on the slope and the mobile antenna is located on level, flat terrain.

As shown in Fig. 4.2, the key parameters are defined as follows:

R = distance over which terrain is flat
r = total distance between base-station antenna and mobile-unit antenna
h_1 = height of base-station antenna above ground level
h_2 = height of mobile antenna above ground level
H = difference in height between flat terrain and elevated terrain

In both situations, type A and type B, a wave is always reflected by the slope of the hill. Consistent with Snell's law, the spot where reflection occurs is always close to the top of the hill. Therefore, reflection would occur in the vicinity of the mobile unit in type A, or of the base-station antenna in type B. The existence of a second reflected wave depends upon the length of flat terrain and the prevailing antenna height, as described in the following subsection.

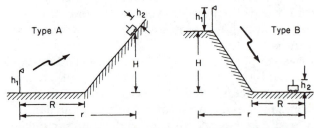

Figure 4.2   Model for propagation over terrain combining level and sloping contours.

For the purpose of illustrating the hilly contour, the scales on the x and y axis are not the same. x is in meters or feet, and y is in kilometers or miles. If x and y are used in the same scale in Fig. 4.2, which is in reality, the terrain contour will look like a flat ground. The variation in contour can hardly be seen. Thus we should remind the readers that the base-station antenna mast is always vertically drawn on the sloped hill.

Second-reflected-wave conditions present

Figure 4.3 is the model for conditions that will produce a second reflected wave where reflection occurs at the ground level. As in the previous model discussed, the combination of level and sloping terrain results in two sets of conditions, wherein the locations of the base-station antenna and the mobile-unit antenna alternate.

Snell's law applies to the conditions under which a second reflected wave can occur. In the models for type A and type B, the distance

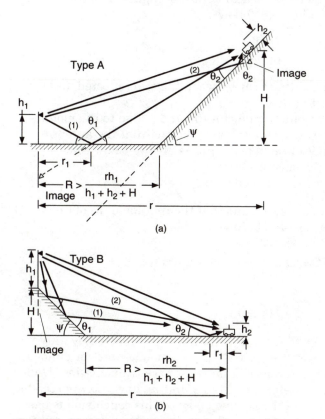

Figure 4.3   Model for analysis of second-reflected-wave conditions.

between the base-station antenna and the reflection point (distance r_1) is a key parameter, as indicated in the following equations:

$$\left.\begin{array}{c} \dfrac{h_1}{r_1} = \dfrac{h_2 + H}{r - r_1} \\[3mm] r_1 = \dfrac{h_1 r}{H + h_1 + h_2} \end{array}\right\} \quad \text{for type A} \qquad (4.1)$$

or

$$\left.\begin{array}{c} \dfrac{h_2}{r_1} = \dfrac{h_1 + H}{r - r_1} \\[3mm] r_1 = \dfrac{h_2 r}{H + h_1 + h_2} \end{array}\right\} \quad \text{for type B} \qquad (4.2)$$

or

Therefore, where the condition exists that the actual distance of level ground R, for either type A or type B, is $R > r_1$, there will always be a reflected wave. The method used to obtain a similar deviation in Eq. (3.38) can also be used here, as follows:

$$P_r = \alpha P_0 \frac{1}{(4\pi r/\lambda)^2} \, | \, 1 - e^{j\Delta\phi_1} - e^{j\Delta\phi_2} \, |^{\,2} \qquad (4.3)$$

where $\Delta\phi_1, \Delta\phi_2$ = phase differences between the two reflected and the direct path

α = the attenuation factor ($0 \le \alpha \le 1$) due to the mobile-radio environment; e.g., the fact that the antenna height of the mobile unit is always below that of nearby surroundings

$P_0 = P_t g_t g_m$ [as defined in Eq. (3.36)]

Assuming $\Delta\phi_1$ and $\Delta\phi_2$ are small and that the relationships of Eq. (3.41) and (3.42) hold, then Eq. (4.3) becomes:

$$P_r \approx \alpha P_0 \frac{1}{(4\pi r/\lambda)^2} \, (1 + 2 \, \Delta\phi_1 \, \Delta\phi_2)$$

$$\approx \alpha P_0 \frac{1}{(4\pi r/\lambda)^2} \qquad (4.4)$$

Equation (4.4) shows that when two waves are reflected by hilly terrain, the combined power received at the mobile-unit antenna approaches or equals that received from free-space transmission. Under such conditions, the antenna heights h_1 and h_2 have no appreciable effect on the magnitude of the received power. This conclusion is based on the mathematical approach. In the real mobile-radio communications environment the true free-space transmission path found in the space-communications environment does not exist. Hence, Eq. (4.4) only shows that under certain circumstances the received signal can be much stronger at the receiving antenna when two reflected waves are present. If it is assumed that the attenuation factor α, shown in Eq. (4.3), is different for each of the three waves, one direct wave and two reflected waves, then Eq. (4.3) becomes:

$$P_r = P_0 \frac{1}{(4\pi r/\lambda)^2} \, | \, \alpha_0 - \alpha_1 e^{j\Delta\phi_1} - \alpha_2 e^{j\Delta\phi_2} \, |^2 \qquad (4.5)$$

where α_0, α_1, and α_2 are related to the obstructions along the three respective individual paths.

As might be expected, the energy from the one wave whose reflection point is closer to the mobile antenna may contribute the greater amount of reflected power to the resultant signal received by the mobile-unit antenna. This type of reflection is known as "specular reflection" [2]. The other reflected wave, whose reflection point is farther away from the mobile-unit antenna, will disperse more of its energy along the path before reaching the mobile-unit antenna. This type of reflection is known as "diffusely scattered reflection" [2].

Hence, the value of the attenuation factor α_2 associated with the wave whose reflection point is closer to the mobile unit's antenna is much greater (less attenuated) than the value of the other wave's attenuation factor (α_1):

$$\alpha_2 \gg \alpha_1$$

Also, it is reasonable to assume that attenuation factors α_0 and α_1 are approximately equal, since the wave reflected from the point closest to the mobile unit, having a small grazing angle, disperses less energy and can be compared favorably with the energy contained in the direct wave. Then Eq. (4.5) becomes:

$$P_r = P_0 \frac{\alpha_0}{(4\pi r/\lambda)^2} \, | \, 1 - e^{j\Delta\phi_2} \, |^2 \qquad (4.6)$$

which is essentially the same as Eq. (3.38). Equation (3.43) can also be derived from Eq. (4.6). The main difference between Eq. (3.38) and Eq. (4.6) is that Eq. (3.38) is based on one reflected wave and one direct wave, whereas Eq. (4.6) is based on one effective reflected wave selected from two reflected waves and one direct wave. Figure 4.4 illustrates the model for Eq. (4.6).

Second-reflected-wave conditions absent

No second reflected wave can occur when the value of R in Fig. 4.3 is less than the value of r_1. This is true for either type A or type B, as shown by Eqs. (4.1) and (4.2). This condition is expressed:

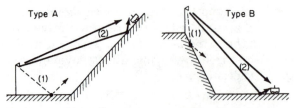

Figure 4.4 Model for predicting combined effects of reflected and direct waves.

$$R < r_1$$

Under this condition, Eq. (3.43) can be used to determine the received power at the mobile-unit antenna, as follows:

$$P_r = \alpha P_0 \left(\frac{h_1' h_2'}{r^2} \right)^2 \tag{4.7}$$

where

$$\left. \begin{aligned} h_1' &= h_1 \\ h_2' &= H + h_2 \end{aligned} \right\} \quad \text{for type A} \tag{4.8}$$

and

$$\left. \begin{aligned} h_1' &= h_1 \\ h_2' &\approx h_2 \end{aligned} \right\} \quad \text{for type B} \tag{4.9}$$

First-reflected-wave conditions absent

In Eq. (4.7), h_1' and h_2' are the effective antenna heights measured from the ground level at which reflections occur. The use of effective antenna heights instead of actual antenna heights is the main difference between the equations for propagation effects over hilly terrain and those for propagation effects over level, nonhilly terrain, as described in Chap. 3, Sec. 3.4.

The parameter h_1' is the effective antenna height and is of primary importance, as expressed in the following equations:

$$h_1' = h_1 + \frac{H r_a}{r - r_a} \quad \text{for type A terrain} \tag{4.10}$$

and $\quad\quad h_1' = h_1 + H \quad$ for type B terrain $\tag{4.11}$

The conclusion, as shown in Fig. 4.5, is that when there is only one reflected wave and one direct wave and the angle ψ is small, it is nec-

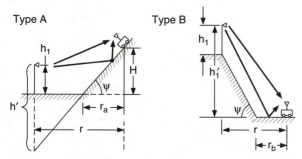

Figure 4.5   Model for one reflected wave and one direct wave when the angle of reflection is small.

essary to modify the base-station antenna height for type A terrain. The effective antenna height h_1' for type B is simply $h_1 + H$.

The typical-case example at 800 MHz, given in Sec. 3.5 of Chap. 3 for a suburb in the Philadelphia-Camden area, resulted in a 38.4-dB-per-decade slope of path loss at the 1-mi point of interception for a base-station antenna height of 100 ft and 40 dBm (10 W) of transmitted power. In this example, the height of the mobile-unit antenna was given as 10 ft. This typical case can be used as a reference to calculate the path loss for other, different cases.

Example 4.1   Given the terrain configuration of Fig. E4.1(a) and a base-station antenna height of 150 ft with 40 dBm of transmitted power, find the path loss along the path. Assume that the transmitting base-station antenna has a gain of 6 dB above that of a dipole, and that the mobile-unit antenna is a dipole with a height of 10 ft above ground.

solution   Since the height of the base-station antenna is 150 ft, then the 1-mi point of interception is $-61.7 + 20 \log 150/100$. The slope of path loss is 38.4 dB per decade, which is the same as for the 100-ft antenna height.

The minimum distance $R_{\min}$ in Fig. E4.1(b), which can produce a second reflected wave, can be determined from Eq. (4.2) as follows:

$$R_{\min} \simeq h_2 \times \frac{(3 \times 5280 \text{ ft})}{h_1 + H} = \frac{10 \times 3 \times 5280 \text{ ft}}{350 \text{ ft}} = 452.5 \text{ ft}$$

In theory, after traveling the minimum distance $R_{\min}$, the signal at the mobile-unit receiving antenna increases because of the second reflected wave produced by the flat terrain:

$$\text{Additional gain} = 20 \log \frac{350}{150} = 7.35 \text{ dB}$$

It is important to note that this additional gain introduces a transit region that occurs around the 3-mi area, as shown in Fig. E4.1(c), wherein the path loss suddenly decreases as the energy of reflected wave 1 carries over into reflected wave 2, as shown in Fig. E4.1(a).

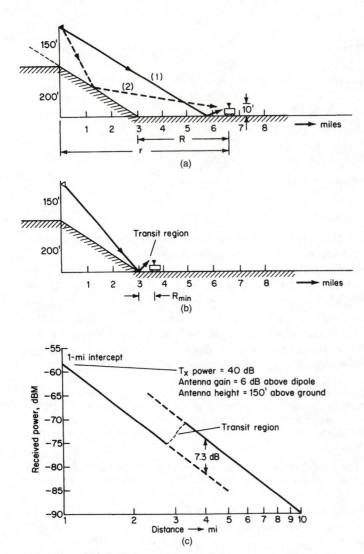

Figure E4.1   Model for determining path loss over type B terrain.

4.2   Diffraction Loss

Propagation over hilly terrain is often adversely affected by obstructions such as hilltops. Path loss resulting from such obstructions is termed "diffraction loss." Kirchhoff's theory on diffraction has been found useful for predicting path loss along a transmission path containing mountain ridges and similar obstructions [3, 4, 5, 6, 7, 8, 9]. Diffraction loss from the round hilltop that has a small radius of curvature will be described in Sec. 4.3.

In classic electromagnetic theory applications, the field strength of a diffracted radio wave associated with a knife edge can be expressed:

$$\frac{E}{E_0} = Fe^{j\Delta\phi} \tag{4.12}$$

where E_0 is the free-space electromagnetic field with no knife-edge diffraction present, F is the diffraction coefficient, and $\Delta\phi$ is the phase difference with respect to the path of the direct wave. The loss due to diffraction is [10, 11]

$$\mathscr{L}_r = 20 \log F \tag{4.13}$$

where

$$F = \frac{S + 0.5}{\sqrt{2} \sin (\Delta\phi + \pi/4)} \tag{4.14}$$

In the definition of F,

$$\Delta\phi = \tan^{-1}\left(\frac{S + 0.5}{C + 0.5}\right) - \frac{\pi}{4}$$

and C and S are the Fresnel integrals, expressed

$$C = \int_0^v \cos\left(\frac{\pi}{2} x^2\right) dx \tag{4.15}$$

$$S = \int_0^v \sin\left(\frac{\pi}{2} x^2\right) dx \tag{4.16}$$

where v is a dimensionless parameter, defined:

$$v = -h_p \sqrt{\frac{2}{\lambda}\left(\frac{1}{r_1} + \frac{1}{r_2}\right)} \tag{4.17}$$

In Eq. (4.17), r_1 and r_2 are the separation distances, and h_p is the height of the knife-edge shown in Fig. 4.6(a).

In predicting the effects of knife-edge diffraction, it is necessary to consider the two possible situations: first, where the wave is not obstructed; second, where the wave is diffracted by the knife-edge obstruction. Figure 4.6(b) illustrates the first condition, where the wave is not obstructed by the knife edge and therefore h_p is a negative value and v becomes a positive value, as shown in Eq. (4.17). The range of F in Eq. (4.14), when v is a positive value, can be expressed:

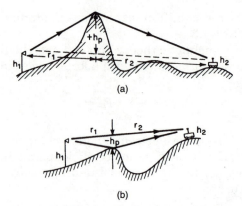

Figure 4.6 Effects of knife-edge obstructions on transmitted radio waves: (*a*) knife-edge diffraction; (*b*) nonobstructed knife-edge propagation effects.

$$0.5 \leq F \tag{4.18}$$

Figure 4.6(*a*) illustrates the second condition, where the wave is diffracted by the knife-edge obstruction and therefore h_p is a positive value and v becomes a negative value. In this situation, the range of F in Eq. (4.14), when v is negative, can be expressed:

$$0 \leq F \leq 0.5 \tag{4.19}$$

The exact solution for Eq. (4.14) is shown graphically in Fig. 4.7.

Since Eq. (4.14) includes terms associated with Fresnel integrals, it is difficult to find simple expression. An approximate solution can be obtained as follows:

$$_0L_r = 0 \text{ dB} \qquad 1 \leq v \tag{4.20}$$

$$_1L_r = 20 \log (0.5 + 0.62v) \qquad 0 \leq v \leq 1 \tag{4.21}$$

$$_2L_r = 20 \log (0.5e^{0.95v}) \qquad -1 \leq v \leq 0 \tag{4.22}$$

$$_3L_r = 20 \log \left(0.4 - \sqrt{0.1184 - (0.1v + 0.38)^2} \right) \qquad -2.4 \leq v < -1 \tag{4.23}$$

$$_4L_r = 20 \log \left(-\frac{0.225}{v} \right) \qquad v < -2.4 \tag{4.24}$$

Equations (4.20) through (4.24) greatly simplify the computation process.

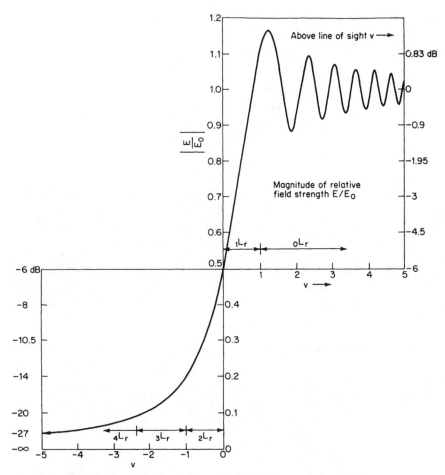

Figure 4.7 Magnitude of relative field strength E/E_0 due to diffraction loss.

Double knife-edge diffraction

In many instances, there is more than one knife-edge-diffraction source along a given propagation path, as shown in Fig. 4.8. The excess path loss (that loss that exceeds the free-space loss) resulting from double knife-edge diffraction can also be used to describe triple or more complex knife-edge diffraction that may occur along a given propagation path. The following discussion deals primarily with the phenomenon of double knife-edge diffraction.

There are many models that could be used to calculate path loss; however, there are basically three models that will enable calculation of excess path loss with results that are approximately equal to a theoretical solution. There are no exact solutions for double knife-edge diffraction.

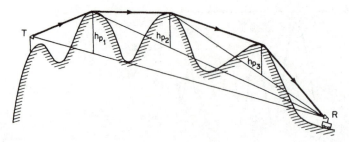

Figure 4.8 Multiple-knife-edge diffraction model.

In Bullington's model [12], two tangential lines are extended, one over each knife-edge obstruction, and the effective height of the knife-edge obstructions is measured as shown in Fig. 4.9(a). This model can be difficult to apply, since the same effective height may be associated with various obstruction heights and separations, as shown in Fig. 4.9(b). For example, the same effective height could be used to describe pair a or pair b. This model affords reasonable accuracy when two knife-edge obstructions are relatively close to each other.

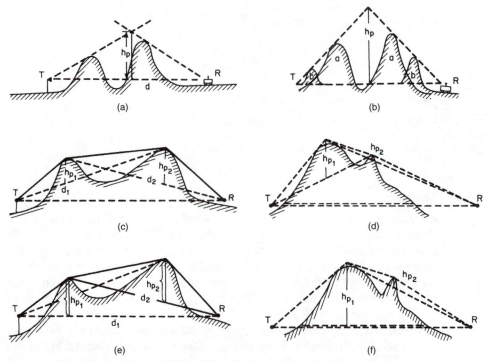

Figure 4.9 Double knife-edge diffraction: (a) and (b) Bullington's models; (c) and (d) Epstein and Peterson's models; (e) and (f) Picquenard's models.

In Epstein and Peterson's model [13], the heights h_{p_1} and h_{p_2} of the two effective knife-edge obstructions are obtained as shown in Fig. 4.9(c). The excess path loss is the loss due to h_{p_1} plus the loss due to h_{p_2}. This model is difficult to apply when the two obstructions are relatively close together, as shown in Fig. 4.9(d). This model provides the highest degree of accuracy when the two knife-edge obstructions are a large distance apart.

In Picquenard's model [14], the limitations that were apparent in Bullington's and Epstein and Peterson's models are not present. Picquenard's technique is illustrated in Fig. 4.9(e). The height of one obstruction, h_{p_1}, is obtained first without regard for the second obstruction, as though it did not exist. The height of the second obstruction, h_{p_2}, is then measured by drawing a line from the top of the first obstruction to the receiver. The total path loss is then calculated from two loss terms. One term is derived from h_{p_1} and d_1, and the other from h_{p_2} and d_2. Since this method provides accurate results without the limitations previously noted [see Fig. 4.9(f)], it is recommended for general use.

Example 4.2   Figure E4.2 is the model for calculating diffraction loss due to a knife-edge obstruction along a 4-mi transmission path. The base-station antenna is 100 ft above ground, and the mobile-unit antenna is 10 ft above ground. Based on the parameters shown in Fig. E4.2, calculate the diffraction loss at a transmitted frequency of 850 MHz.

solution   By analysis of Fig. E4.2, it becomes apparent that $h_p = 430$ ft. Then, from Eq. (4.17), the following information is derived:

$$v = -430 \sqrt{\frac{2}{1.157} \times 2 \left(\frac{1}{10,560}\right)} = -7.78 \tag{E4.2.1}$$

Finally, from Eq. (4.24), the diffraction loss is calculated:

$$\mathscr{L}_r = 20 \log \left(-\frac{0.225}{-7.78}\right) = -30.77 \text{ dB} \tag{E4.2.2}$$

Example 4.3   Figure E4.3 illustrates a situation where the base-station antenna is 200 ft in height and the mobile-unit antenna, 8 mi away, is 10 ft in height. An

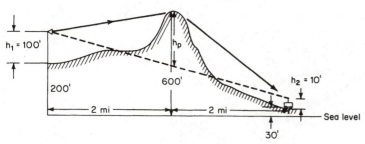

Figure E4.2   Model for calculating diffraction loss due to a knife-edge obstruction.

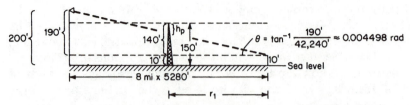

Figure E4.3   Model for calculating distance based on a 15-dB diffraction loss.

obstruction 150 ft in height is located somewhere along the propagation path and constitutes a 15-dB diffraction loss. The transmitted frequency is 850 MHz. How far is the obstruction located from the base-station antenna?

solution   According to Fig. 4.7, a 15-dB loss correlates to $v = -1.15$. Then,

$$v = h_p \sqrt{\frac{2}{\lambda}\left(\frac{1}{r_1} + \frac{1}{r_2}\right)}$$

$$= -h_p \sqrt{\frac{2}{\lambda}\left(\frac{1}{r_1} + \frac{1}{42{,}240 - r_1}\right)} \tag{E4.3.1}$$

The value of r_1 can then be obtained as follows:

$$\frac{140 - h_p}{r_1} \approx 0.004498 \tag{E4.3.2}$$

$$1.15 = h_p \sqrt{\frac{2}{1.157}\left(\frac{1}{r_1} + \frac{1}{42{,}240 - r_1}\right)} \tag{E4.3.3}$$

As a result, the following values are derived:

$$h_p = 82.559 \text{ ft}$$

$$r_1 = 12{,}770 \text{ ft}$$

Example 4.4   Based on the parameters given in Fig. E4.4, calculate the excess loss due to double knife-edge diffraction at 850 MHz, using Picquenard's methods.

solution   The loss due to the first obstruction can be obtained from parameter v of obstruction 1, expressed:

$$v = -h_{p_1} \sqrt{\frac{2}{\lambda}\left(\frac{1}{r_1} + \frac{1}{r_2}\right)} = -2.56 \tag{E4.4.1}$$

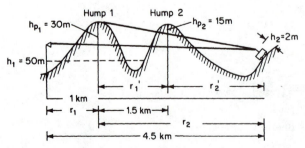

Figure E4.4   Model for Example 4.4.

As shown in Fig. 4.7, the loss due to the diffraction by obstruction 1 is:

$$\mathscr{L}_1 = 21.1 \text{ dB} \qquad (\text{E4.4.2})$$

The loss due to the second obstruction can be obtained from parameter v of obstruction 2, expressed:

$$v' = -h_{p_2} \sqrt{\frac{2}{\lambda} \left(\frac{1}{r_1'} + \frac{1}{r_2'} \right)} = -1.219 \qquad (\text{E4.4.3})$$

From the curves of Fig. 4.7, the additional loss due to the diffraction by obstruction 2 is:

$$\mathscr{L}_2 = 14.7 \text{ dB}$$

Therefore, the total loss due to diffraction by obstructions 1 and 2 is:

$$\mathscr{L}_1 + \mathscr{L}_2 = 35.9 \text{ dB}$$

4.3 Diffraction Loss over Rounded Hills

The measurement [25, 26] has shown that the knife-edge prediction value is always either above or below the measurement data values depending on the polarization of waves when the signal is diffracted over a rounded hill.

1. The formula of Eq. (4.14) for a knife-edge hill comes from the following integration.

$$F(v) = \frac{1-j}{2} \int_v^\infty \exp\left(\frac{j\pi z^2}{2}\right) dz \qquad (4.25)$$

The parameter v is shown in Eq. (4.17). It can be expressed another way, as

$$v = \left(\frac{2r}{\lambda}\right)^{1/2} \cdot \theta \qquad (4.26)$$

$$\frac{1}{r} = \frac{1}{r_1} + \frac{1}{r_2} \qquad (4.27)$$

θ is the scattering angle shown in Fig. 4.10 with h_p shown in Fig. 4.6.

$$\theta \approx -h_p \left(\frac{1}{r_1} + \frac{1}{r_2}\right) \qquad (4.28)$$

2. The calculation of the correction factors of using knife-edge formula due to a round hill top we apply the antenna shift [27] δ, which is also called the visual displacement in height. Assuming that $r_1 > r_2$, then

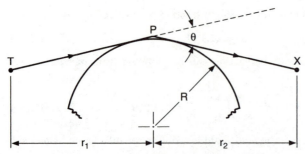

Figure 4.10   Geometry of obstacle arrangement for rounded hills.

$$\delta = +0.316 \left(1 + \frac{r_2}{r_1}\right) \cdot \lambda \cdot \left(\frac{R}{\lambda}\right)^{1/3} \qquad \text{vertical polarization} \qquad (4.29)$$

$$= -0.333 \left(1 + \frac{r_2}{r_1}\right) \cdot \lambda \cdot \left(\frac{R}{\lambda}\right)^{1/3} \qquad \text{horizontal polarization} \quad (4.30)$$

The difference σ in power between the knife-edge diffraction loss and the rounded-hill diffraction loss can be found as

$$\sigma = K \cdot \delta \qquad (4.31)$$

where

$$K = \frac{1}{2r_2} \cdot \frac{1}{2(r_1 + r_2)^2} \sqrt{\frac{2\beta r}{\pi}} \qquad \text{for } r_1 > r_2$$

The power difference σ is shown in Fig. 4.11. The received power for vertical polarization wave over the rounded hills is stronger

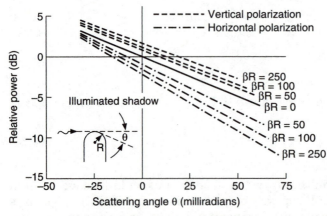

Figure 4.11   Variation of power behind a cylindrical mountain with scattering angle as a function of radius of curvature (far field).

than that over the knife-edge, but the received power for horizontal polarization wave over the rounded hills is weaker than that over the knife-edge. In other words, for vertical polarization, the region behind the hill is brighter than that behind a knife-edge; for horizontal polarization, the region behind the hill is darker.

4.4   Path-Clearance Criteria

Path clearance over a hilltop can be calculated by using Fresnel zone equations. Figure 4.12 illustrates the parameters for establishing the criteria for path-clearance calculations.

As a consequence of the intervening hilltop, it can be assumed that the direct-wave-path distance is d_0, whereas the reflected-wave-path distance is $d_1 + d_2$, as shown in Fig. 4.12. If the difference between the direct and reflected waves is more than $\lambda/2$ when received at the mobile unit as a result of the different path lengths, then it can be assumed that the reflected wave is not significantly affected by diffraction. This condition is expressed:

$$\beta(d_1 + d_2 - d_0) > \frac{\pi}{2} \tag{4.32}$$

The height H of the direct wave path above the top of the hill can be calculated from Eq. (4.32) and expressed in approximate value by:

$$H \geq \sqrt{\frac{\lambda d_1(d_0 - d_1)}{d_0}} \geq \sqrt{\frac{\lambda d_{0_1} d_{0_2}}{d_0}} \tag{4.33}$$

4.5   Lee's Macrocell Model

The macrocell model is used for the distance from the base station that is greater than 1 km. The microcell model is used for the distance that is less than 1 km. This is because the environment at a distance greater than 1 km is different from that at a distance less than 1 km, as shown

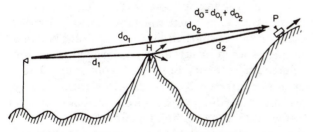

Figure 4.12   Model for calculation of path clearance over a hilltop.

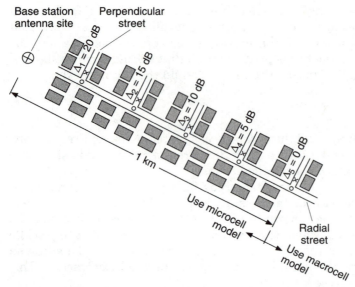

Figure 4.13 Power differences between that of the radial and perpendicular streets:

$P_\parallel$ = Power received from a radial street (o)
$P_\perp$ = Power received from a perpendicular street (x)
Δ_i = Power difference = $P_\parallel - P_\perp$ at ith location

in Fig. 4.13. Within 1 km, the power received from a perpendicular street as soon as the mobile unit is turning from a radial street, has drastically dropped. It is more pronounced at a close-in location as shown in Fig. 4.13. When the mobile unit is 1 km or further away from the base station, there is no power difference received at the mobile unit between two streets near a corner. The cause of this phenomenon is due to the layout of the building blocks.

It would be difficult for the signal to arrive at perpendicular streets when the streets are close to the base station, because there are only a few buildings surrounding the base station, and those buildings cause very few reflected waves that have less chance to reach the close-in corner. This is why the corner signal is weak, but there are many buildings in between the base station and the far-out corner, and those buildings cause many reflected waves that can reach the far-out corner with a great probability. Thus, the far-out-corner signal is no different between the radial street and the perpendicular street. Therefore, within 1 km, the received signal is affected by the city building layout map. Over 1 km, the building layout is not affected by the street orientation anymore. In this section, we are describing a macrocell prediction model (over 1 km) that is not affected by the building layout map.

The mobile radio signal arrival is caused by two factors. One factor is the human-made structures, and the other factor is the terrain contour.

Since the building structure or the layout in each human-made city is different, we have to find a way to characterize any given city.

The human-made effect—interpret path loss in a different way

Either we can collect enough measured data of signal strength to cover all possible terrain contours in one city, or we can collect enough measured data of signal strength from the selected high and low spots in the city. Although the standard deviation from all data points retains the terrain-contour variation, the generated path loss curve through the averaging process washes out the terrain contour variation. The path loss curve now represents the character of that given human-made city as if the city were situated on flat ground.

In Fig. 4.14, the path loss curve for each city is normalized to standard conditions (see Eq. 3.59) and thus can be plotted on the same

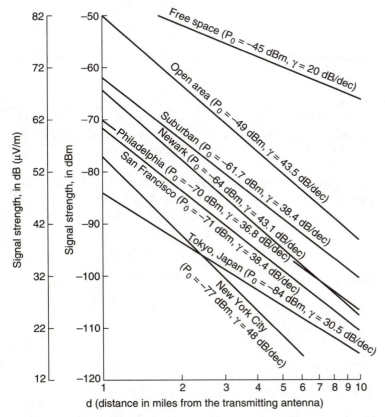

Figure 4.14 Propagation path loss in different areas. All data are normalized to the standard conditions (Eq. 3.59).

chart using Eq. (3.58). The one-mile intercept and the path loss slope for each city are indicated. The explanation of the different values for different cities is stated in Chap. 16.6.

The terrain-contour effect

The terrain-contour effect can be treated differently in three conditions: the nonobstructive condition, the obstructive condition, and the over-the-water condition.

The nonobstructive condition. We obtain the effective antenna height h_1' according to Eq. (4.10) for type A and Eq. (4.11) for type B to accommodate the terrain variation. Then, by Eq. (3.45) listed below,

$$\Delta G = 20 \log (h_1'/h_1) \qquad \text{in dB}$$

This equation is used to find a gain or loss as comparing h_1' with the actual antenna height h_1. If $h_1' > h_1$, ΔG is a gain. If $h_1' < h_1$, ΔG is a loss.

The obstructive condition. In the obstructive condition, the knife-edge diffraction loss calculation is used—this is described in Sec. 4.2. Also, the modification of the knife-edge diffraction due to rounded hilltops is described in Sec. 4.3.

Over-the-water condition. Since there are no human-made structures over the water, the reflected wave over the water (as shown in Fig. 4.15) is still considered a specular reflected wave, although the reflection point of this specular reflected wave is at a distant location away from the mobile unit.

We may use one direct wave and two reflected waves as shown in Eq. (4.3) to obtain the received signal. Equation (4.3) leads to the result

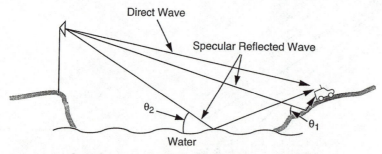

Figure 4.15 Over-the-water condition. If θ_1 and $\theta_2 \ll 1$ rad, then the three waves add → to create free space loss

$$|1 - e^{-j\phi_1} - e^{-j\phi_2}|^2 \doteq 1$$

shown in Eq. (4.4), which is the free-space formula. It is also verified by measuring the signal-strength-over-water condition from the San Diego base station to the Oceanside mobile unit. The received signals are very strong when the two specular reflected waves prevail at those locations. The received signal becomes weaker as soon as one specular reflected wave disappears due to the location changes.

Lee macrocell prediction model

The received signal can be predicted by the following:

Nonobstructive condition

$$P_r = P_{r_1} - \gamma \log \frac{r}{r_o} + \alpha_o + 20 \log \frac{h_1'}{h_1} - n \log \frac{f}{f_o} \tag{4.34}$$

$$\underbrace{\phantom{P_r = P_{r_1} - \gamma \log \frac{r}{r_o}}}_{\substack{\text{Effected by human-made} \\ \text{structure (Fig. 4.14)}}} \quad \underbrace{\phantom{20 \log \frac{h_1'}{h_1}}}_{\substack{\text{Effected by} \\ \text{terrain contour} \\ \text{(Eq. 3.45)}}} \quad \underbrace{}_{\text{(Fig. 3.12)}}$$

Obstructive condition

$$P_r = P_{r_1} - \gamma \log \frac{r}{r_o} + \alpha_o + L(v) - n \log \frac{f}{f_o} \tag{4.35}$$

$$\underbrace{\phantom{P_r = P_{r_1} - \gamma \log \frac{r}{r_o}}}_{\substack{\text{Effected by human-made} \\ \text{structure (Fig. 4.14)}}} \quad \underbrace{}_{\substack{\text{Diffraction} \\ \text{loss (Fig. 4.7} \\ \text{\& Fig. 4.11)}}}$$

Over-the-water condition—free-space loss [see Eq. (4.4)]

$$P_r = \alpha \cdot P_o \cdot \frac{1}{(4\pi r/\lambda)^2} \tag{4.36}$$

In all three equations, $f_o = 850$ MHz, and n is

$$n = \begin{cases} 20 & f < f_o \\ 30 & f > f_o \end{cases} \tag{4.37}$$

The early version of the Lee Model was first formed in 1976 at AT&T Bell Labs. It was used in the ACE tool for designing the AT&T cellular systems. Later, the name of the tool was changed to ADMS and PACE. Lee Model predicts the local means and uses the terrain database.

4.6   Lee's Microcell Model [28]

As mentioned in Sec. 4.5, the microcell model is used for a distance of up to 1 km. When the size of a cell is small—less than 1 km—the street orientation and individual blocks make a difference in signal reception.

The particular buildings directly affect the received signal strength level and are considered a part of the path loss. Although the strong received signal at the mobile unit is coming from the multipath reflected waves, not from the waves penetrating through the buildings, there is a correlation between the attenuation of the signal and the number of building blocks along the radio path. The larger the number of building blocks and the size of the blocks, the higher the signal attenuation. The propagation mechanics within the microcell are shown in Fig. 4.16. The microcell prediction formula is

$$P_r = P_t - L_{\text{los}}(d_A, h_1) - L_b \qquad \text{(dBm)} \qquad (4.38)$$

where P_t is the ERP in dBm and $L_{\text{los}}(d_A, h_1)$ is the line-of-sight path loss at distance d_A with the antenna height h_1. The $L_{\text{los}}(d_A, h_1)$ can be found in Fig. 4.17 as the curve a indicated at the antenna height $h_1 = 20$ feet.

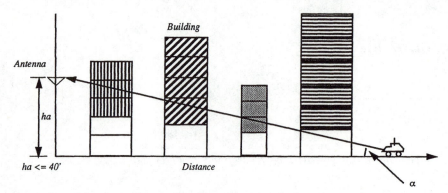

Figure 4.16 The propagation mechanics of low antenna height at the cell site.

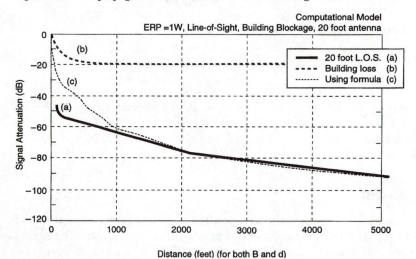

Figure 4.17 Microcell parameters.

In the microcell environment, the antenna height and gain relationship follows a 9-dB/dec rule [28] as

$$\Delta G = 30 \log \frac{h_1'}{h_1} \qquad (4.39)$$

L_B in Eq. (4.38) is the loss due to building blocks. The value of L_B is obtained by first calculating the sum of lengths of building blocks between antenna site and location A as shown in Fig. 4.18. The total block length is $B = a + b + c$. In Fig. 4.18, we can find L_b from the curve b. The L_B will remain unchanged when the total block length B exceeds 1000 feet. Curve a and curve b are empirical curves.

The theoretical $L_{los}(d_A, h)$ also can be obtained from Eq. (3.48) as

$$L_{los} = 20 \log \frac{4\pi d_A}{\lambda} \quad \text{(free space loss)} \; d_A < D_f$$

$$= 20 \log \frac{4\pi D_f}{\lambda} + \gamma \log \frac{d_A}{D_f} \qquad d_A > D_f \qquad (4.40)$$

where D_f is the Fresnel-zone distance, $D_f = 4h_1 h_2 / \lambda$. Equation (4.40) is plotted in Fig. 4.17 as curve c. The measurements are carried out along various routes, as shown in Fig. 4.19. The measurement on the zigzag route (R2) in San Francisco was shown in Fig. 4.20 with a frequency at 1.9 GHz, $h_1 = 13.3$ meters, and the power at ERP = 1 watt. The measured and the predicted curves agree fairly well with each other [29].

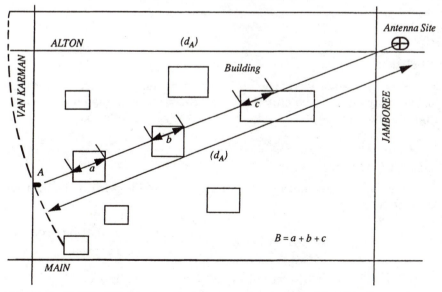

Figure 4.18 Building block occupancy between antenna site and location A.

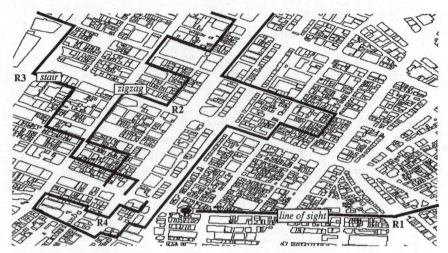

Figure 4.19   A sample of building layout with various combination of routes.

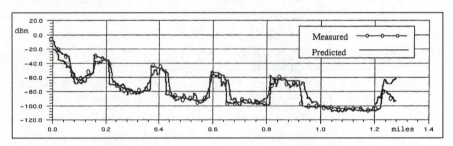

Figure 4.20   Measured and predicted vs. mobile distance from the start of the drive—Microcell model predictions compared to measurements vs. delta.

A simple method of prediction

Since the digitalization of the building layout map in a city is a tedious job, we may simplify by only digitizing the streets instead of the buildings, as shown in Fig. 4.21. The methodology is as follows:

1. Digitize the streets (see Fig. 4.21).

2. Identify the street blocks.

3. Calculate the percentage of building occupation within each street block (density).

$$P_i = \text{density of } i\text{th street block} = \frac{\text{total building occupation area}}{\text{street block area}}$$

Indicate P_i for each street block as shown in the figure.

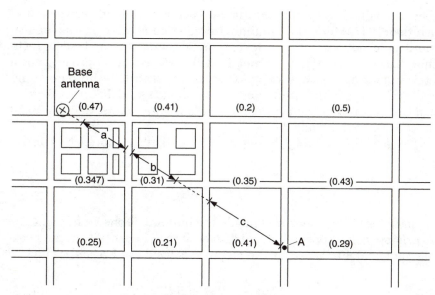

Figure 4.21 A simple method of estimating building blocks.

4. Take the line-of-sight path loss curve $L_{los}(d_A, h_1)$ from Fig. 4.17 curve a or c.

5. At point A, there are three street blocks—a, b, and c (see Fig. 4.21).

6. Calculate the total equivalent block length B_{eq} by taking care of the density of each street block shown in Fig. 4.21.

$$B_{eq} = a \cdot p_a + b \cdot p_b + c \cdot p_c$$
$$= a \cdot (0.347) + b \cdot (0.31) + c \cdot (0.41)$$

7. Find L_B from Fig. 4.17 curve b for the value of B_{eq}.

8. The predicted signal strength is

$$P_r(\text{at A}) = P_t - L_{los}(d_A, h_1) - L_{B_{eq}} \qquad (4.41)$$

Equation (4.41) is similar to Eq. (4.38). However, Eq. (4.41) is much easier to obtain.

4.7 Inbuilding Prediction Models [29]

The inbuilding prediction model is not as important as the macrocell prediction model (Sec. 4.5) and microcell prediction model (Sec. 4.6). The reason is that the structure of each building is different and the layout of each floor is different. To digitize each floor layout would be labor intensified. Why not just take a quick real-time measurement of each floor? Therefore, the inbuilding prediction is more or less an academic topic, not a practical application. Secondly, the ray tracing technique based on each floor layout is still a statistical approach, not a deterministic approach. It means that the ray tracing technique cannot predict the signal amplitude and phase at each spot. It can only get a local mean through the average of random data points. Section 16.13 has addressed this topic.

Here, a simple method based on the statistical approach is introduced.

Path loss slope

The path loss slope measured inside the building is shown in Fig. 4.22. Surprisingly, the slope follows 20 dB/dec for a distance within the Fresnel's zone and 40 dB/dec for the distance beyond the Fresnel's zone

$$L = 20 \log \frac{d'}{d} \qquad d \text{ and } d' < D_f \qquad (4.42)$$

$$L = 40 \log \frac{d'}{d} \qquad d \text{ and } d' > D_f \qquad (4.43)$$

where d and d' are the distance from the inbuilding base transmitter. Equation (4.43) shows that a two-wave model (Sec. 3.4) can be applied to the inbuilding propagation as well.

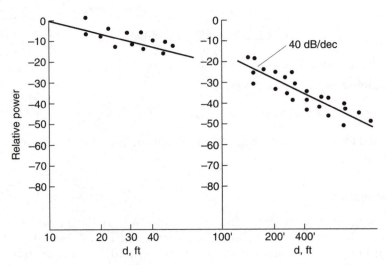

Figure 4.22 Inbuilding measured data.

The total received power

$$P_r = P_t + G_t + G_r - L \qquad (4.44)$$

where G_t and G_r are the gains of transmitter and receiver, respectively.

L is the path loss, which can be obtained from Eq. (4.42) or Eq. (4.43). When the signal penetrates through one of the special rooms, the walls of which are causing high penetration loss, then

$$L(d_2) = \gamma \log [(d_1 + d_2)/d_1]$$

where d_2 is the path length in the special room, and γ is the slope and is obtained by the measurement.

If d_1 is the distance passing through normal rooms and d_2 is the distance passing through one or more special rooms, then the losses $L(d_1)$ and $L(d_2)$ are obtained depending on whether d_1 or d_2 or $d_1 + d_2$ is beyond the Freznel's zone, as shown in Fig. 4.23. The loss L can be obtained as

$$L = \underbrace{20 \log \frac{4\pi D_f}{\lambda} + 40 \log \frac{d_1}{D_f}}_{L(d_1)} + \underbrace{\gamma \log \frac{d_1 + d_2}{d_1}}_{L(d_2)} \qquad \text{(for } D_f < d_1 + d_2\text{)} \quad (4.45)$$

The predicted curve obtained from Eq. (4.45) has been shown to agree fairly with the measured data [29].

4.8   Effects on Field-Strength Prediction

Various methods for predicting the field strength of radio waves arriving at the mobile receiving terminal have been discussed in Chap. 3 and in the preceding sections of Chap. 4. Since the reciprocity theorem always holds true, the predictions for field strength at the mobile

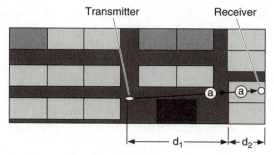

Figure 4.23   Top view, receiver not in Fresnel zone.

receiving terminal are the same as those predictions for the base-station location, under prevailing sets of circumstances. Such reciprocity means that the same findings and predictions for transmitted power and received field strength at either location are applicable to both the base-station and mobile-terminal locations.

In predicting the mean field strength of a signal received at a given distance r, the following equation is a valid assumption:

$$P_r = (P_{r_0}\alpha_0)\mathscr{L}_r\alpha_1\alpha_2 \cdots \alpha_n \tag{4.46}$$

where all α values are directly applicable and not caused by other factors. In Eq. (4.46), the value $P_{r_0}\alpha_0$ is the same as that shown in Eq. (3.58). $\mathscr{L}_r$ is the diffraction loss shown in Eqs. (4.21) through (4.24). α_i ($i = 1, 2, \ldots, n$) represents the additional path-loss factors due to many other causes. Equation (4.46) can be expressed in dB:

$$P_r = P_{r_0} + \mathscr{L}_r + \alpha_0 + \alpha_1 + \cdots + \alpha_n \quad \text{dBm} \tag{4.47}$$

All of the factors in Eq. (4.47) are median values based on the following types of causes:

Effects of street orientation

In urban areas, streets that run radially from the base station tend to enhance the received signal because of a phenomenon known as the "channeling effect." In New York City, the difference in signal strengths due to the different directions of streets may vary over a range of 10 to 20 dB at frequencies of 11 GHz. At frequencies of 1 GHz and lower, the channeling effect is appreciably less. In general, the area of signal-strength measurement should include an even mix of streets that are parallel and those that are perpendicular to the propagation path, so that the average values are not biased by street orientation.

Effects of foliage

Radio-wave propagation in forested environments has been investigated by many authors [15–22], and their studies of propagation conditions [16–18] reveal that the presence of vegetation produces a constant loss, independent of distance, between communication terminals that are spaced 1 km or more apart. Since the density of foliage and the heights of trees are not uniformly distributed in a forested environment, the results are not surprising. Tamir [19] measured the signal loss due to foliation at a distance of 1 km, with both the transmitting and receiving antennas at treetop heights. The signal loss versus frequency as a function of foliage effects is shown in Fig. 4.24. Tamir also

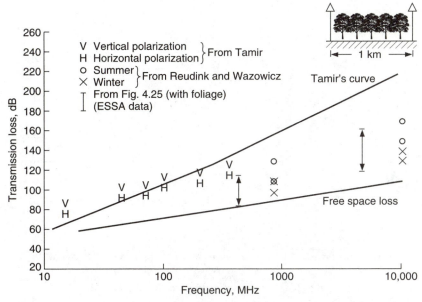

Figure 4.24 Signal loss due to the effects of foliage (ESSA measurements vs. Tamir measurements and Reudink and Wazowicz measurements).

made a theoretical prediction that the loss increases at the rate of f^4 as the frequency increases. His studies also indicate that the signal loss for vertically polarized waves is slightly greater than for horizontally polarized waves in areas with dense foliage. In Fig. 4.24, the measurement made by Reudink and Wazowicz [20], with one antenna clear of the treetops and the other antenna completely immersed in vegetation, shows that under these conditions losses were less than those measured by Tamir, except at one point.

The Environmental Sciences Services Administration (ESSA) [21] measured the signal loss during the summer when the trees were in full leaf, and again during the winter when the trees were bare. A 3-km path was chosen that was a line-of-sight path except for the existing intervening trees. The transmitting antenna was placed at a height of 6.6 m above the ground, whereas the receiving antenna was set at different heights. At heights up to 3 m above ground, the receiving antenna was below the average level of the tree foliage, with tree crowns extending upward to approximately 4 to 19 m above the ground. The curves shown in Fig. 4.25 were measured at a distance of 3 km and plotted for different antenna heights from 2 to 24 m. Since the signal loss due to foliage seems to be independent of the distance, the range of losses for different antenna heights and foliage conditions shown in Fig. 4.25 can be trans-

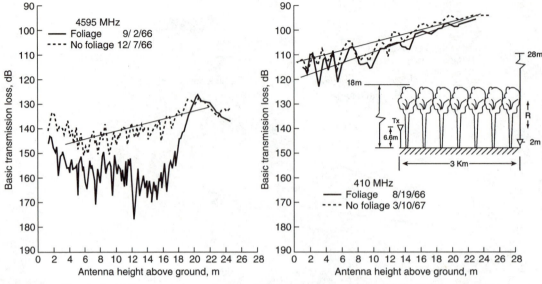

Figure 4.25   Transmission loss vs. antenna height along a 3-km path: effects of foliage. (*From Ref. 21.*)

lated into differences in path loss due to free-space loss from 3 km to 1 km, as plotted in Fig. 4.24.

ESSA's data show less loss than Tamir's, and among the three different sources, Tamir's data show the highest loss in worst-case dense-foliage areas at each different frequency. Based on signal-strength measurements along the same suburban streets in both summer and winter, variations of up to 10 dB have been observed. However, it is improbable that radio waves would propagate through heavily wooded areas that extended over an entire path.

Effects of attenuation within tunnels

Signal-attenuation measurements made within the Lincoln Tunnel, connecting New Jersey and New York, showed attenuations of 40 dB at 150 MHz, 14 dB at 300 MHz, 4 dB at 1 GHz, and negligible loss at 11.2 GHz when measured at a distance of 1000 ft inside the tunnel [23].

Measurements of signal attenuation in mines [24] show losses of 15 to 20 dB within the optimal frequency range of 500 to 1000 MHz at distances of 1000 ft from the mine entrance. The attenuation loss increases drastically at frequencies below 400 MHz, and losses gradually increase at frequencies above 1000 MHz.

Effects of buildings and structures

The P_{r_0} in Eq. (4.47) will be different in value for the different kinds of buildings and structures which are characterized by area and environmental classifications for both flat and hilly terrain. There are many human-made variables that contribute to path loss. To simplify the analyses and examples presented herein, the selected P_{r_0} values have been based on suburban areas rather than the more complex urban areas. A similar P_{r_0} can be found for urban areas too.

Relationship between path loss and local mean

In Eq. (4.47), assume that all additional path-loss factors are not mean values but are slow random variables received at a given distance r from the transmitting location, expressed as follows:

$$\mathscr{P}_r = [P_{r_0} + \mathscr{L}_r + \alpha_0 + \alpha_1 + \cdots + \alpha_n]_{\text{all are variables}}$$
$$= [\hat{m}(r)]^2 \tag{4.48}$$

where $\mathscr{P}_r$ is also a random variable and $\hat{m}(r)$ is the local mean shown in Eq. (3.49). $\mathscr{P}_r$ is called the signal strength of long-term fading and directly related to $\hat{m}(r)$. As previously discussed in Sec. 2.1, $\mathscr{P}_r$ becomes Gaussian-distributed for a sum of many variables. Since each element is in decibels, $\mathscr{P}_r$ will have a lognormal distribution at any given distance r. The equation for lognormal distribution given in Chap. 2—Eq. (2.41)—can also be expressed as follows:

$$p_{\mathscr{P}_r}(x) = \frac{1}{\sqrt{2\pi}\,\sigma_x} \exp\left[-\frac{(x - P_r)^2}{2\sigma_x^2}\right] \tag{4.49}$$

where σ_x is a standard deviation of 8 dB and P_r is the mean value in decibels obtained from Eq. (4.47). Averaging the long-term fading $\mathscr{P}_r$ becomes P_r:

$$P_r = \overline{\mathscr{P}_r} = \overline{[\hat{m}(r)]^2}$$

4.9   Signal-Threshold Prediction

There are two accepted methods for predicting signal coverage. One is based on signal-reception threshold values which is a function of time, and the other is based on signal field strength within the coverage area which is a function of space. As found in Eq. (4.49), the field-strength predictions are based on a lognormal distribution. The probability that the signal is above a predetermined threshold value A, expressed in decibels, is:

$$P_{p_r}(x \geq A) = \int_A^\infty \frac{1}{\sqrt{2\pi}\,\sigma_{P_r}} \exp\left[-\frac{(x-P_r)^2}{2\sigma_{P_r}^2}\right] dx$$

$$= \int_{(A-P_r)/\sigma_{P_r}}^\infty \frac{1}{\sqrt{2\pi}} \exp\left(-\frac{z^2}{2}\right) dz$$

$$= P\left(z \geq \frac{A-P_r}{\sigma_{P_r}}\right) \tag{4.50}$$

Example 4.6   Given that the mean field strength at 8 mi is −100 dBm and the standard deviation σ_P is 8 dB, then the percentage of signal field strength above −110 dBm can be obtained from Eq. (4.50):

$$P\left(z \geq \frac{-110+100}{8}\right) = P(z \geq -1.25)$$

$$= 89.44\% \tag{4.51}$$

4.10   Signal-Coverage-Area Prediction

The signal coverage within a given area is usually specified as a percentage of the signal field strength above a predefined threshold established for that area. The term R defines the radius of a cell within receiving range of a base-station antenna. For a signal received at $r = R$, 50 percent of the measured signal field strength would be greater than P_R and 50 percent would measure less than P_R. Signal-measurement techniques were discussed in Sec. 3.5 of Chap. 3. There are two approaches, stated as follows:

Approach A

Find the percentage of signal strength above a new threshold with a new cell radius. In this case, the new threshold level can be increased or decreased corresponding to the distance between the mobile unit and the base station. Equation (4.50) provides the probability that the signal is above a predetermined threshold value A for field strength $P_r = x$, which is in dB, and

$$\sigma_x = \sigma_{P_r} = 8 \text{ dB} \tag{4.52}$$

is assumed. P_r is obtained from Eq. (3.58) as follows:

$$P_r = P_{r_1} - \gamma \log r \tag{4.53}$$

where r is expressed in miles. For $r = R$, the equation can be written:

$$P_R = P_{r_1} - \gamma \log R \tag{4.54}$$

There are two conditions that can be stated in approach A.

Condition 1. When a radius r_a is less than R, find a level P_{r_a} at $r = r_a$ having the same percentage K of the signal strength x above the level A at $r = R$.

$$P\,(x \ge A) = P\,(x \ge P_{r_a}) = \int_{P_{r_a}}^{\infty} P_{r_r}(x)\,dx = K\% \qquad (4.55)$$
$$\text{(at } r = R) \qquad \text{(at } r = r_a)$$

or

$$P\,(x \ge P_R) = P\,(x \ge P_{r_a}) = 50\% \qquad (4.56)$$
$$\text{(at } r = R) \qquad \text{(at } r = r_a)$$

The interpretation of Eq. (4.55) and Eq. (4.56) is shown in Fig. 4.26.

Condition 2. When $K\%$ of the signal strength x is above the level P_R at $r = R$, find the percentage K' of the signal strength x above the same level P_R at $r = r_a$.

If

$$P\,(x \ge A) = K \qquad \text{at } r = R \qquad (4.57)$$

find $K'\%$ such that

$$P\,(x \ge A) = K'\% \qquad \text{at } r = r_a \qquad (4.58)$$

If Level A $= P_R$, which is the average signal strength level, then $K = 50$ percent.

Figure 4.26 graphically illustrates the Gaussian-type distribution of the field-strength variables.

To find the solutions of these two conditions, we may start with the calculation of $P\,(x \ge A)$ as follows.

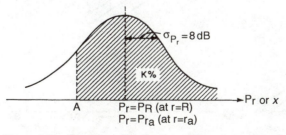

Figure 4.26 Gaussian distribution of long-term signal strength $\mathcal{P}_r$ for radius r.

Equation (4.55) can be expressed as follows:

$$P(x \geq A) = \int_A^\infty \frac{1}{\sqrt{2\pi}\ \sigma_{P_r}} \exp\left[-\frac{(x - P_r)^2}{2\sigma_{P_r}^2}\right]dx$$

$$= 0.5 - \int_0^{(A - P_r)/\sigma P_r} \frac{1}{\sqrt{2\pi}} \exp\left(-\frac{y^2}{2}\right)dy \qquad (4.59)$$

where

$$\frac{1}{\sqrt{2\pi}} \int_0^{(A - P_r)/\sigma_{P_r}} \exp\left(-\frac{y^2}{2}\right)dy = 0.5\ \mathrm{erf}\left(\frac{A - P_r}{\sqrt{2}\sigma_{pr}}\right) = -0.5\ \mathrm{erf}\left(\frac{P_r - A}{\sqrt{2}\sigma_{pr}}\right)$$

$$= 0.5 - K\% \qquad (4.60)$$

If K is given, and $P_r = P_R$ at $r = R$ then $r = r_a$ can be found from Eq. (4.60) for the same K at the threshold level P_R; and from Eqs. (4.53) and (4.54), the following is obtained:

$$A - P_R = \gamma \log \frac{R}{r} \qquad (4.61)$$

Also, if γ is known, then the ratio R/r can be found.

Figure 4.27 shows how the path-loss slopes for three values of γ can be plotted from the combined Eqs. (4.60) and (4.61). Where $\gamma = 38.4$ dB

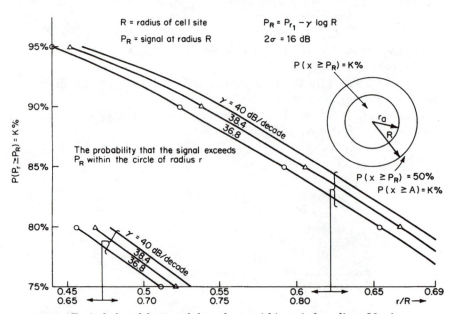

Figure 4.27   Typical plot of three path-loss slopes within a circle radius of 8 mi.

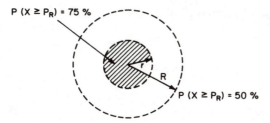

Figure E4.7 Model for calculating signal coverage as a percentage of signal greater than P_R.

per decade and $R = 8$ mi, then 90 percent of the signal is above P_R in a circle where the radius $r_a = 0.539\ R$, a radius of 4.31 mi. This approach shows that the probability of a signal strength above the level P_R is 90 percent at the circle of $r = r_a$.

Example 4.7 Figure E4.7 illustrates the signal-coverage parameters for an area having a circle radius P_R of 8 mi. Calculate the signal-coverage area in which 75 percent of the signal is above the P_R value.

solution With reference to Fig. E4.7, assume that:

$$\gamma = 38.4\ \text{dB per decade and } \sigma = 8\ \text{dB} \qquad (E4.7.1)$$

Solving for the distance along the radius r_a wherein 75 percent of the signal is above the P_R median value yields the following:

$$r_a = 0.72\ R = 5.76\ \text{mi} \qquad (E4.7.2)$$

From Eq. (E4.7.2), 75 percent of the signal strength above the P_R median value is contained in an area having a circle radius of 5.76 mi.

Approach B

Find the percentage of the area H that strength is the percentage of signal above a threshold.

The percentage of the area H can be calculated by integrating

$$H = \frac{\text{Area}}{\pi R^2} = \frac{1}{\pi R^2} \int P\ (x \geq A)\ da$$

$$= \frac{1}{\pi R^2} \int_0^R \left[0.5 - 0.5\ \text{erf}\left(\frac{A - P_r}{\sqrt{2}\,\sigma_{p_r}} \right) \right] d(\pi r^2)$$

$$= \frac{1}{\pi R^2} \left[\int_0^R 0.5\ (2\pi r)\ dr - \int_0^R 0.5\ \text{erf}\left(\frac{A - P_r}{\sqrt{2}\,\sigma_{p_r}} \right) \cdot (2\pi r)\ dr \right]$$

$$= 0.5 - \frac{1}{R^2} \int_0^R r \cdot \text{erf}\left(\frac{A - P_r}{\sqrt{2}\,\sigma_{p_r}} \right) dr \qquad (4.62)$$

where P_r is a function of distance

$$P_r = P_{r_1} - \gamma \log r \qquad (4.63)$$

and the mean value P_R at $r = R$ is

$$P_R = P_{r_1} - \gamma \log R \qquad (4.64)$$

If $A = P_R$ at $r = R$, then combining Eq. (4.63) and Eq. (4.64) yields

$$A - P_r = P_R - P_r = \gamma \log r - \gamma \log R = \gamma \log \frac{r}{R} \qquad (4.65)$$

Substituting Eq. (4.65) into Eq. (4.62), the results are as follows:

$$H = 0.5 + \frac{1}{2} \exp\left(\frac{\sqrt{2}\,\sigma_{p_r}}{0.434\lambda}\right)^2 \left[1 - \operatorname{erf}\left(\frac{\sqrt{2}\,\sigma_{p_r}}{0.434\lambda}\right)\right] \qquad (4.66)$$

If $A < P_R$, then $P_R - A = \delta$ in db at $r = R$:

$$A - P_R = P_R - \delta - P_r = \gamma \log \frac{r}{R} - \delta \qquad (4.67)$$

Substituting Eq. (4.67) into Eq. (4.62), the result is

$$H = 0.5 \left\{ 1 + \operatorname{erf}\left(\frac{\delta}{\sqrt{2}\sigma}\right) + \exp\left(\frac{2\delta}{0.434\gamma}\right) \right.$$
$$\left. - \left(\frac{2\sigma^2}{0.434\gamma}\right)\left[1 - \operatorname{erf}\left(\frac{\sqrt{2}}{0.434\lambda} - \frac{\delta}{\sqrt{2}\sigma}\right)\right]\right\} \qquad (4.68)$$

When $\delta = 0$, Eq. (4.68) equals Eq. (4.66).

Comparison of the two approaches

The difference between two approaches is addressed as follows. In approach A, we may find out $K\%$ of locations along the circumference that the signal $x \geq A$. We reduce the radius and keep the $K\%$ by increasing the threshold level.

In approach B, we may find out the $H\%$ of area in a cell that the signal $x \geq A$.

In designing a system, approach A is very important for getting a boundary of a cell at which the signal strength exceeds the threshold 90 percent of the time. We will take the following steps. First, calculate the threshold level A at which the signal is 50 percent above the level at the boundary of a cell. Then, if we want a new cell boundary that can provide the signal 50 percent above threshold B, we just simply mea-

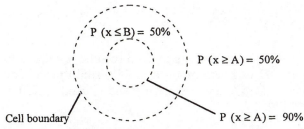

Figure 4.28 Illustration of $P\,(x \geq A) = 90$ percent within the cell.

sure the signal strength (average) at the level B. The interpretation is that the probability of signal x, $P\,(x \geq B) = 50$ percent, corresponds to $P\,(x \geq A) = 90$ percent if the signal strength $B = A + 10$ dB at the new cell boundary.

4.11   Wideband Signal Propagation

Path loss of a wideband signal

Suppose that the transmitting power is P_t and the antenna gain is G_t. The pointing vector (power density) U_t is

$$U_t = \frac{P_t G_t}{4\pi r^2} \tag{4.69}$$

At the receiving end, the signal is arrived after passing through the mobile radio environment. The received power of a narrowband signal can be expressed as

$$P_r = U_t \cdot C(d, f)\, A_e(f) \tag{4.70}$$

where $C(d, f)$ is the medium characteristic, which can be expressed as

$$C(d, f) = \frac{k}{r^2 f} \tag{4.71}$$

$A_e(f)$ is the effective operative of the receiving antenna at the receiving end, and it can be expressed as

$$A_e(f) = \frac{c^2 G_r}{4\pi f^2} \tag{4.72}$$

where c is the speed of light and G_r is the gain of receiving antenna. Substituting Eq. (4.71) and Eq. (4.72) into Eq. (4.70), the received power of a narrowband signal is

$$P_r = \frac{kc^z G_t G_r P_t}{(4\pi r^2)^2} \cdot \frac{1}{f^3} \tag{4.73}$$

Suppose that a transmit power P_t in watts is used to send a wideband signal with its bandwidth B in hertz along a mobile radio path. The power spectrum density $S_t(f)$ at the transmitting end is used to define the transmit power.

$$P_t = \int_{f_0 - B/2}^{f_0 + B/2} S_t(f)\, df \tag{4.74}$$

At the receiving end, the received power of a wideband signal after passing through a mobile radio environment can be obtained:

$$P_t = \frac{1}{4\pi r^2} \int_{f_0 - B/2}^{f_0 + B/2} S_t(f) \cdot C(d, f) \cdot A_e(f)\, df \tag{4.75}$$

which is different from receiving a narrowband signal, shown in Eq. (4.70). Substituting Eq. (4.71) and Eq. (4.72) into Eq. (4.75) yields

$$P_r = \frac{1}{4\pi r^2} \int_{f_0 - B/2}^{f_0 + B/2} S_t(f) \frac{K}{r^2 f} \frac{c^2 G}{4\pi r^2}\, df$$

$$= \frac{kc^2 G}{4\pi r^2} \int_{f_0 - B/2}^{f_0 + B/2} S_t(f) \frac{1}{f^3}\, df \tag{4.76}$$

Assume that the waveform of the sending pulse is

$$p(t) = A \sin c\,(t \cdot B) \qquad -\frac{1}{B} < t < \frac{1}{B} \tag{4.77}$$

where A is the amplitude of the pulse and B is the total bandwidth. The power spectrum density $S_t(f)$ of the pulse $p(t)$ is:

$$S_t(f) = \text{constant} = \frac{A}{B} \qquad f_0 - \frac{B}{2} \le f \le f + \frac{B}{2} \tag{4.78}$$

Substituting Eq. (4.78) into Eq. (4.76),

$$P_r = \frac{kc^2 G_t G_r A}{(4\pi r^2)^2 B} \int_{f_0 - B/2}^{f_0 + B/2} \frac{1}{f^3}\, df$$

$$= K \frac{1}{f_0^3 [1 - (B/(2f_0))^2]^2} \tag{4.79}$$

where

$$K = \frac{kcG_rG_rA}{(4\pi r^2)^2} \tag{4.80}$$

When the bandwidth B approaches zero, Eq. (4.79) becomes

$$P_r = K\frac{1}{f_0^3} \tag{4.81}$$

which agrees with Eq. (4.73) as the received power of a CW wave. From Eq. (4.79), we find that the received power of a wideband signal at a bandwidth $B = f_0/2$ is merely 0.28 dB higher than the received power of a CW wave. Equation (4.79) is plotted in Fig. 4.29. From the figure, we find that, when the bandwidth is extremely wide $(B = f_0)$, the received power is about 2.5 dB higher than that of a CW signal. Therefore, in any practical situation, the received power of a wideband signal is the same as that of a CW signal. The path loss rule of a CW signal can be used to estimate the path loss of a wideband.

Multipath fading characteristic on wideband

The wideband pulse waveform $p(t)$ has been shown in Fig. 4.29(a) and expressed in Eq. (4.77)

$$p(t) = A \sin c(t \cdot B) = A\frac{\sin(\pi Bt)}{\pi t} \tag{4.82}$$

where A is the amplitude. The received signal can be represented as

$$S(t) = \left(\frac{A}{B}\right)\sum_{m=-\infty}^{\infty} b_m(t)\frac{\sin \pi B(t - m/B)}{\pi(t - m/B)} \tag{4.83}$$

The pulse width of $1/B$ is the time interval of the pulse occupied. We count all b_m that do not vanish over a range of a finite number of m corresponding to a time delay spread Δ. Since the wideband signal has less fading than the narrowband signal, we may use an equivalent number of diversity branches M_{eq} to represent the less faded signal as

$$M_{eq} = \frac{\Delta + 1/B}{1/B} = B \cdot \Delta + 1 \tag{4.84}$$

The equivalent number of diversity branches M_{eq} varies from different human-made structure. Since the delay spread Δ is larger in the urban than in the suburban area, we may find the equivalent number of diver-

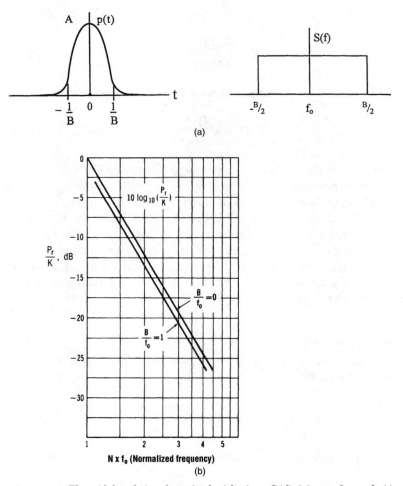

Figure 4.29   The wideband signal received with given $S_t(f)$: (a) waveform of $p(t)$ and band-limited $S_t(f)$; (b) the received signal due to the wide bandwidth.

sity branches M_{eq} of a fading signal with different bandwidths and different human-made environments as shown in the following table.

From Table 4.1, we find that the delay spread does not affect the narrowband fading signals but reduces the number of fades for wideband signals. For $B = 1.25$ MHz, the signal fading is reduced more in urban

TABLE 4.1   The Equivalent Diversity Branches M_{eq} for Fading Signal

Human-made environment	Δ (μs)	$B = 25$ kHz	$B = 30$ kHz	$B = 1.25$ MHz	$B = 5$ MHz
Enclosed area	0.1	1.0025	1.003	1.125	1.5
Open area	0.2	1.005	1.006	1.25	2
Suburban	0.5	1.0125	1.015	1.625	3.5
Urban	3	1.075	1.09	4.75	16

($M_{eq} = 4.75$) than suburban ($M_{eq} = 1.625$). The delay spread only reduces the short-term fading, not the long-term fading.

Problem Exercises

1. Based on the parameters given in Fig. P4.1, what is the number of reflected waves that may occur? Where are the reflection points located in Fig. P4.1?

2. Based on the parameters given in Fig. P4.2, calculate the increase in gain at the mobile receiver when the base-station height is raised from 30 m to 45 m.

3. Based on the parameters given in Fig. P4.3, assuming that the path loss is 40 dB per decade over the distance, calculate the distance D at which the received-signal power at the mobile unit is the same for the two base-station antenna transmissions.

4. How much higher must the antenna in Fig. P4.2 be raised in order to obtain the equivalent of free-space loss? The model for predicting propagation loss over flat terrain, described in Sec. 3.6, can be used for reference.

5. Based on the parameters given in Fig. P4.5, calculate the diffraction loss due to a single knife-edge obstruction.

6. If the height H_t of the obstruction in Fig. P4.5 remains constant but its location between the base-station and mobile antennas can be changed, at what location will H_t cause maximum diffraction loss?

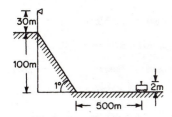

Figure P4.1 Data for Prob. 1.

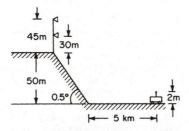

Figure P4.2 Data for Probs. 2 and 4.

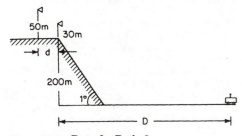

Figure P4.3 Data for Prob. 3.

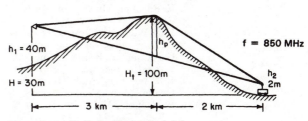

Figure P4.5 Data for Probs. 5 and 6.

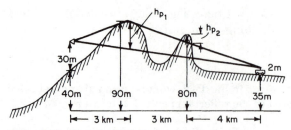

Figure P4.7   Data for Prob.7.

7.   Based on the double knife-edge diffraction parameters given in Fig. P4.7, use Picquenard's model [Fig. 4.9(*e*) and (*f*)] to calculate the total diffraction loss.

8.   Given a mean signal field strength of –90 dBm at 10 km and a standard deviation σ_{p_r} of 8 dB, what percentage of the field strength is above –100 dBm at a distance of 10 km from the base-station transmitting antenna?

9.   If the signal at the receiving location covers an area of 10 km what is the size of the coverage area in which 65 percent of the signal is above the $\acute{P}_r$ value?

Note   The P_r value is the mean signal field strength received at a given distance from the transmitting antenna.

References

1. W. C. Y. Lee, "Studies of Base Station Antenna Height Effects on Mobile Radio," *IEEE Trans. Veh. Tech.,* vol. 29, May 1980, pp. 252–260.
2. P. Beckmann and A. Spizzichino, *The Scattering of Electromagnetic Waves from Rough Surfaces,* Macmillan, New York, 1963, chaps. 14–17.
3. K. Bullington, "Radio Propagation at Frequencies above 30 Megacycles," *Proc. IRE,* vol. 35, no. 10, 1947, pp. 1122–1136.
4. J. Epstein and D. W. Peterson, "An Experimental Study of Wave Propagation at 850 Mc/s," *Proc. IRE,* vol. 41, no. 5, 1953, pp. 595–611.
5. P. L. Rice, A. G. Longley, K. A. Norton, and A. P. Barsis, "Transmission Loss Prediction for Tropospheric Communication Circuits," *NBS Tech. Note no. 101,* 1967, vols. 1 and 2.
6. F. H. Dickson, J. J. Egli, J. W. Herbstreit, and G. S. Wickizer, "Large Reduction of VHF Transmission Loss and Fading by the Presence of a Mountain Obstacle in Beyond Line-of-Sight Paths," *Proc. IRE,* vol. 41, August 1953, pp. 967–969.
7. K. Furutsu, "On the Theory of Radio Wave Propagation over Inhomogeneous Earth," *J. Res. NBS 67D (Radio Prop.),* no. 1, January 1963, pp. 39–62.
8. R. E. Wilkerson, "Approximation of the Double Knife-Edge Attenuation Coefficient," *Radio Science,* vol. 1, no. 12, December 1966, pp. 1439–1443.
9. G. Millington, R. Hewitt, and F. S. Immirzi, "Double Knife-Edge Diffraction in Field Strength Predictions," *IEEE Mon. 507E,* March 1962, pp. 419–429.
10. L. J. Anderson and L. G. Trolese, "Simplified Method for Computing Knife-Edge Diffraction in the Shadow Region," *IEEE Trans. Anten. Prop.,* July 1958, pp. 281–286.
11. W. C. Jakes, *Microwave Mobile Communications,* Wiley, New York, 1974, p. 87.

12. K. Bullington, "Radio Propagation at Frequencies above 30 Megacycles," *Proc. IRE,* vol. 35, no. 10, 1947, pp. 1122–1136.
13. J. Epstein and D. W. Peterson, "An Experimental Study of Wave Propagation at 850 Mc/s," *Proc. IRE,* vol. 41, 1953, pp. 595–611.
14. A. Picquenard, *Radio Wave Propagation,* Wiley, New York, 1974, p. 296.
15. A. Kinase, "Influences of Terrain Irregularities and Environmental Clutter Surroundings on the Propagation of Broadcasting Waves in the UHF and VHF Bands," *Japan Broadcasting Corp. Technical Research Laboratory Tech. Mon. no. 14,* March 1969.
16. J. W. Herbstreit and W. Q. Crichlow, "Measurement of the Attenuation of Radio Signals by Jungles," *J. Res. NBS (Radio Sci.),* vol. 68D, August 1964, pp. 903–906.
17. A. H. LaGrone and C. W. Chapman, "Some Propagation Characteristics of High UHF Signals in the Immediate Vicinity of Trees," *IEEE Trans. Anten. Prop.,* vol. 9, 1961, pp. 487–491.
18. G. H. Hagn, G. E. Barker, H. W. Parker, J. D. Hice, and W. A. Ray, "Preliminary Results of Full-Scale Pattern Measurements of a Simple VHF Antenna in a Eucalyptus Grove," *Special Tech. Rep. 19,* Stanford Research Institute, Menlo Park, Calif., January 1966.
19. T. Tamir, "On Radio-Wave Propagation in Forest Environments," *IEEE Trans. Anten. Prop.,* vol. 15, November 1967, pp. 806–817.
20. D. O. Reudink and M. F. Wazowicz, "Some Propagation Experiments Relating Foliage Loss and Diffraction Loss at X-Band and UHF Frequencies," *IEEE Trans. Comm.,* vol. 21, November 1973, pp. 1198–1206.
21. A. P. Barsis, M. E. Johnson, and M. J. Miles, "Analysis of Propagation Measurements over Irregular Terrain in the 96 to 9200 MHz Range," *ESSA Tech. Rep. ERL 114-ITS 82,* U.S. Dept. of Commerce, Boulder, Colo., March 1969.
22. Y. Okumura, E. Ohmuri, T. Kawano, and K. Fukuda, "Field Strength and Its Variability in VHF and UHF Land-Mobile Radio Service," *Rev. Elec. Comm. Lab.,* vol. 16, September–October 1968, pp. 825–873.
23. D. O. Reudink, "Mobile Radio Propagation in Tunnels," *IEEE Veh. Tech. Group Conf.,* December 2–4, 1968, San Francisco, Calif.
24. A. G. Emslie, R. L. Lagace, and P. F. Strong, "Theory of the Propagation of UHF Radio Waves in Coal Mine Tunnels," *IEEE Trans. Anten. Prop.,* vol. 23, March 1975, pp. 192–205.
25. H. E. J. Neugebauer and M. P. Bachynski, "Diffraction by Smooth Cylindrical Mountains," *Proc. IRE,* September 1958, pp. 1619–1627.
26. K. Hacking, "U.H.F. Propagation over Rounded Hills," *Proc. IEE,* March 1970, pp. 499–511.
27. K. Artmann, "Beugung polarisierten Lichtes an Blenden endlicher Dicke im Gebiet der Schattengrenze," *Z. für Phys.,* vol. 127, May 1950, pp. 468–494.
28. W. C. Y. Lee, *Mobile Communications Design Fundamentals* 2d ed., Wiley & Sons, 1993, p. 88.
29. W. C. Y. Lee and D. J. Y. Lee, "In-Building Prediction," *IEEE PIMRC '96,* Taipei, Taiwan, ROC, October 15–18, 1996, pp. 771–775.

Effects of System RF Design on Propagation

5.1 Effects of Antenna Design

In Chaps. 3 and 4, the effects of the natural environment on radio-wave propagation were described in some detail. Other factors associated with mobile-radio system RF design can also have a significant effect on radio-wave propagation. Such RF design factors include the type of antenna employed, the directivity characteristics of the antenna, and various field components of the transmitted energy, signal polarization, and gain factors.

To properly evaluate the effects of antenna design on mobile-radio propagation, it is first necessary to develop an expression for the amount of transmitted power received by the mobile unit. Since the typical environment of a mobile unit includes both natural and human-made obstructions—i.e., trees, hills, buildings, structures, and other vehicles—the expression for the amount of transmitted power received by the mobile unit is quite different from that used to describe conventional line-of-sight propagation parameters.

The power P_t radiated by a base-station antenna is equal to the surface integral of the radiation pattern distributed over the surface of a large sphere with the base-station antenna located at its center. The radiated power P_t can be expressed:

$$P_t = \int_S G_t(\theta, \phi) \, dS \qquad (5.1)$$

where S is the surface of a large sphere. Figure 5.1 illustrates the spherical-surface concept of a large sphere in the far field with a centrally located base-station transmitting antenna. With the coordinates illustrated in the figure, Eq. (5.1) may be written:

$$P_t = \int_S G_t(\theta, \phi) r_1^2 \sin \theta \, d\theta \, d\phi \tag{5.2}$$

where

$$G_t(\theta, \phi) = \frac{1}{2} \, \text{Re} \, (E_\theta H_\phi^*) = \frac{1}{2Z_0} \, |E_\theta|^2 \tag{5.3}$$

and

θ, ϕ = angles in different planes
E_θ = electric-field component
H_ϕ = magnetic-field component, H_ϕ^* indicates complex conjugate of H_ϕ
Z_0 = intrinsic impedance of free space (377 Ω)
r_1 = distance from transmitter to receiver

If the base-station antenna radiation pattern is the same in all directions (and thus is denoted by G_{iso}), then the radiated power P_t derived from Eq. (5.2), is:

$$P_t = G_{iso} \int_0^{2\pi} \int_0^{\pi} r_1^2 \sin \theta \, d\theta \, d\phi = G_{iso} r_1^2 4\pi \tag{5.4}$$

or

$$G_{iso} = \frac{P_t}{4\pi r_1^2} \tag{5.5}$$

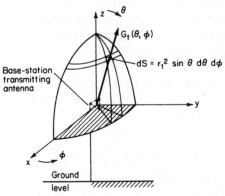

Figure 5.1 Spherical coordinates for radiated power P_t.

If the transmitting antenna is not omnidirectional but has a gain g_t, then:

$$G_{\text{gain}} = \frac{P_t g_t}{4\pi r_1^2} \tag{5.6}$$

The equation for expressing the received power at the mobile-unit receiver is

$$P_r = \alpha G_{\text{gain}} A_e \tag{5.7}$$

where α is the attenuation factor associated with the mobile-radio environment and A_e is the effective aperture. A_e can be expressed:

$$A_e = \frac{g_r \lambda^2}{4\pi} \tag{5.8}$$

where g_r is the gain factor for the receiving environment. Substituting Eqs. (5.6) and (5.8) into Eq. (5.7) gives the following:

$$P_r = \left[\frac{P_t g_t}{\left(\frac{4\pi r_1}{\lambda} \right)^2} \right] (\alpha g_r) \tag{5.9}$$

where on the right side of Eq. (5.9) the quantity within brackets is the free-space path-loss formula and α and g_r are the two factors affected by the local environment surrounding the mobile unit. Alpha is the loss factor depending on the angle of signal arrival, and g_r is the antenna gain factor as a function of the following two parameters at the receiving site: (1) the pattern of radiation received by the mobile unit's antenna, $G_r(\theta, \phi)$, and (2) a relatively new term associated with the probability density function (pdf) of angular wave arrival, $p_r(\theta, \phi)$, which represents the probability that radio waves arrive at the antenna from every angle in θ and ϕ, respectively, as shown in Fig. 5.2. On the basis of the parameters illustrated in Fig. 5.2, the g_r can be expressed:

$$g_r = \frac{\int_0^{2\pi} \int_0^{\pi} G_r(\theta, \phi) p_r(\theta, \phi) r_2^2 \sin \theta \, d\theta \, d\phi}{(1/k_1) \int_0^{2\pi} \int_0^{\pi} r_2^2 \sin \theta \, d\theta \, d\phi}$$

$$= \frac{k_1}{4\pi} \int_0^{2\pi} \int_0^{\pi} G_r(\theta, \phi) p_r(\theta, \phi) \sin \theta \, d\theta \, d\phi \tag{5.10}$$

where k_1 is a constant.

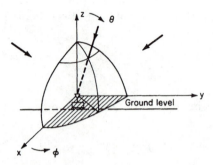

Figure 5.2 Coordinates of angular radio-wave arrival.

The term $p_r(\theta, \phi)$ can be obtained approximately by using either a highly directional antenna or a method introduced by Lee [1]. Equation (5.10) defines g_r more precisely than G_m was defined in Eq. (3.33) in Chap. 3. When $p_r(\theta, \phi)$ is uniformly distributed,

$$p_r(\theta, \phi) = \frac{1}{2\pi^2} \tag{5.11}$$

and the constant k_1 in Eq. (5.10) equals $2\pi^2$, then Eq. (5.10) becomes:

$$g_r = \frac{1}{4\pi} \int_0^{2\pi} \int_0^{\pi} G_r(\theta, \phi) \sin \theta \, d\theta \, d\phi = G_m \tag{5.12}$$

which is the antenna gain specified by the manufacturer. Equation (5.9) is primarily used to study effects that are caused by the mobile-radio environment.

Example 5.1 Figure E5.1(a) shows the configuration of a rectangular horn antenna. Assuming that at a frequency of 850 MHz the horn is 100 percent efficient, the following information can be obtained:

1. The gain of the antenna horn is

$$G = \frac{4\pi}{\lambda^2} \times \text{Aperture} = 113.1 \approx 20.5 \text{ dB} \tag{E5.1.1}$$

2. The directivity and gain are the same; therefore the beamwidth of the antenna pattern in the xz and yz planes can be expressed:

$$D = \frac{41,250}{\phi_1 \theta_1} = \frac{4\pi}{\lambda^2} (ab) \qquad \phi, \theta \text{ in degrees} \tag{E5.1.2}$$

$$\phi_1 \theta_1 = \left(\frac{57.3\lambda}{a}\right)\left(\frac{57.3\lambda}{b}\right) = (22°)(16°) \tag{E5.1.3}$$

as shown in Fig. E5.1(b).

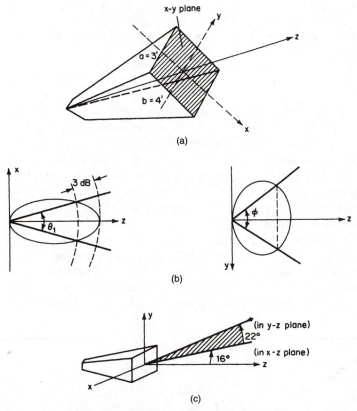

Figure E5.1 (*a*) Rectangular horn antenna; (*b*) beamwidth of antenna pattern; (*c*) angle of incident wave arrival.

3. Figure E5.1(*c*) shows that this type of rectangular horn antenna cannot cover 45° of incident wave arrival in the *xz* plane, since:

$$\theta_1 = 16° \tag{E5.1.4}$$

Example 5.2   Given an antenna beam pattern of $G_r(\theta, \phi) = \cos \theta$ and a probability density function (pdf) of angular wave arrival of $p_r(\theta, \phi) = \cos \theta / \pi$, then the gain g_r from Eq. (5.10) can be expressed:

$$g_r = \frac{k}{4\pi} \int_0^{2\pi} \int_0^\pi \frac{\cos^2 \theta \sin \theta}{\pi} \, d\theta \, d\phi \tag{E5.2.1}$$

Since the pattern of $\cos^2 \theta$ in Eq. (E5.2.1) is maximum at $\theta = 0$ and null at $\theta = 90°$, the range of θ is from 0 to $\pi/2$. Therefore, the gain can be expressed:

$$g_r = \frac{k}{4\pi} \int_0^{2\pi} \int_0^{\pi/2} \frac{\cos^2 \theta \sin \theta}{\pi} \, d\theta \, d\phi = \frac{k}{6\pi} \tag{E5.2.2}$$

In the mobile-radio environment, it is commonly assumed that the pdf of angular wave arrival $p_r(\theta, \phi)$ is uniformly distributed in both the azimuth angle ϕ and the elevation angle θ. If this were true, the gain factor g_r of Eq. (5.12) would be correct; however, the actual distribution of elevation angle θ is not exactly uniform and thus causes slight differences in the measured data. These effects are described in greater detail in Sec. 5.4.

5.2 Effects of Antenna Directivity

As might be expected, improving antenna directivity at the base station usually results in an increase in the power received at the mobile unit. Figure 5.3(a) shows how a directional antenna concentrates its radiated energy in the direction of the mobile unit. Figure 5.3(a) illustrates that even though the energy transmitted by the base-station antenna is directional, the mobile unit's received signal is the combined direct and reflected signals from scatterers located in all directions around the mobile unit.

The use of a directional antenna by the mobile unit [Fig. 5.3(b)] would be expected to add some additional gain to the received signal; however, a directional receiving antenna would limit reception to a certain angular sector of the field of propagation. Those radio waves outside the directional beamwidth would not contribute to the received

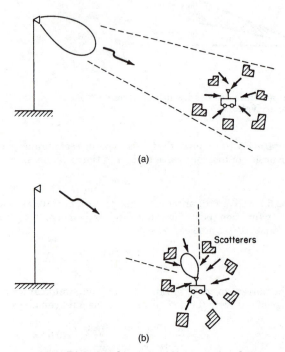

(a)

(b)

Figure 5.3   Directional antennas improve signal reception: (a) directional propagation increases signal reception; (b) effects of scatterers on directional propagation.

power. While it is true that a narrower beamwidth equates with higher gain, it is also true that fewer incoming radio waves are received, and therefore, the advantages of a directional antenna at the mobile unit are counterbalanced by the obvious disadvantages. Consequently, it is more or less to maintain a constant level of reception over an omnidirectional coverage area.

If Eq. (5.12) is applied for a directional antenna with a beamwidth of $\Delta\theta \, \Delta\phi$, where $\Delta\theta$ is the beamwidth of the elevation and $\Delta\phi$ is the beamwidth of the azimuth, then the following expressions can be derived:

$$g_r = \frac{1}{4\pi} \int_0^{2\pi} \int_0^{\pi} G(\theta, \phi) \sin \theta \, d\theta \, d\phi \approx \frac{k^2}{\Delta\theta \, \Delta\phi} \tag{5.13}$$

Equation (5.13) is a common equation found in many antenna textbooks [2]. The loss factor α in Eq. (5.9) due to an antenna beamwidth of $\Delta\theta \, \Delta\phi$ can be expressed:

$$\alpha = \int_0^{\Delta\theta} \int_0^{\Delta\phi} p_r(\theta, \phi) \, d\theta \, d\phi = \frac{\Delta\theta \, \Delta\phi}{2\pi^2} \tag{5.14}$$

since $p_r(\theta, \phi) = 1/2\pi^2$ is uniformly distributed. From the relationship between Eq. (5.13) and Eq. (5.14), it follows that

$$\alpha g_r = \frac{k_2}{2\pi^2} \tag{5.15}$$

Equation (5.15) indicates that the product of the loss factor α and the gain factor g_r will remain substantially constant within a mobile-radio environment, regardless of the beamwidth of the mobile unit's directional antenna. Substituting Eq. (5.15) into Eq. (5.9) gives the following expression:

$$p_r \approx \left[\frac{p_t g_t}{(4\pi r_1/\lambda)^2} \right] \left(\frac{k_2}{2\pi^2} \right) \tag{5.16}$$

The received power at the mobile unit, therefore, is also independent of the beamwidth of the mobile-unit directional antenna, as shown in Eq. (5.16). This result was confirmed in experiments performed by Lee [3] in the 836-MHz frequency range using linear monopole antenna arrays with half-power azimuth beamwidths of 45, 26, and 13.5°, respectively. The average received power of each linear monopole antenna array was about the same as that received by an omnidirectional monopole antenna independently of the beamwidth measured, in a plane 90° from the azimuth; i.e., the plane of θ was equal to 90 degrees, as illustrated in Fig. 5.2. The exact distribution of θ varies slightly from the assumed uniform distribution, as will be shown in Sec. 5.4; hence Eq. (5.16) is an approximate solution. However, Eq. (5.16) does explain the independence of antenna beamwidth and received power.

If there is any advantage in using a directional antenna on a mobile unit, it is that the relatively small number of incoming waves within the angular sector of the directional beamwidth will result in less susceptibility to severe fading. Multipath fading was described in Chap. 1, and the effects of short-term fading are further discussed in Chap. 6.

Example 5.3 The use of a corner-reflector antenna, as shown in Fig. E5.3, can sometimes be helpful in obtaining signal gain and coverage. Assuming that the frequency of the transmitted signal is 850 MHz and $\theta_1 = 78°$, and that the field intensity is expressed in volts per meter, what is the gain of the corner-reflector antenna in decibels? What is the coverage angle ϕ_1 of the corner reflector in degrees?

solution To find the gain and angle of coverage for the corner-reflector antenna, the following procedure is used. The four source points associated with the image planes are symmetrical. The pattern for the E field can be expressed:

$$E(\phi) \propto \cos (\beta S \cos \phi) - \cos (\beta S \sin \phi) \tag{E5.3.1}$$

The following parameters are given for Fig. E5.3:

$$S = \frac{\lambda}{2} \tag{E5.3.2}$$

$$E(\phi) \propto \cos (\pi \cos \phi) - \cos (\pi \sin \phi) \tag{E5.3.3}$$

$$E(\phi = 0) = -2 \tag{E5.3.4}$$

$$E(\phi_1 = 22°) = -1.414 \tag{E5.3.5}$$

Hence, the coverage angle is

$$\phi_1 = 44° \tag{E5.3.6}$$

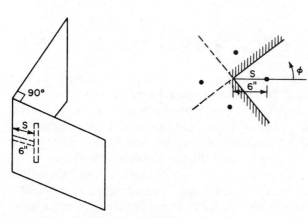

Figure E5.3 Corner-reflector antenna.

and therefore, the gain is:

$$G = \frac{29,000}{78 \times 44} = 8.45 \approx 9.3 \text{ db} \qquad \text{(E5.3.7)}$$

5.3   Antenna-Pattern Ripple Effects

When an antenna pattern is measured at the mobile receiver in the vicinity of a second, colocated antenna, small fluctuations or ripples can be observed along the fringes of the pattern due to the presence of the adjacent antenna. The effects of such ripples on an omnidirectional mobile antenna are not significant, since the patterns of the two antennas on the roof of the vehicle usually are not the same and the antenna pattern itself is dependent upon the antenna mounting site, and therefore the antenna pattern is somewhat deformed. On the other hand, the antenna pattern at the base-station multiple-antenna configuration conforms closely to the free-space pattern with regard to the antenna pattern ripples caused by the adjacent antenna.

As an example, suppose that a signal is transmitted from a mobile unit traveling at 15 mi/h and received by two identical base-station antennas separated by only 8 wavelengths. Under this set of conditions, one would intuitively suspect that the average power of the two received signals over the period of a short 10-s interval would be about the same. However, the results of studies prove that they are not the same, as shown in Fig. 5.4. In a 1-min time interval, the average signal level of one antenna with respect to the signal level from the other antenna can vary as much as 4 to 5 dB. In the next time interval, the reverse can be true. To eliminate the possibility that this phenomenon is the result of instability in the local oscillators of the two receivers, the data from a number of identical test runs conducted at various times were compared [4]. Analysis of the various test-run data proved that receiver oscillator instability was not the cause, since the data results were reproducible.

Another possible cause that was studied was the interaction of base-station antenna patterns where two colocated antennas are separated by very small distances. The following method was used to study the phenomenon of antenna-pattern ripples and their effects.

The first phase of the study was to obtain measurements of the free-space patterns of the received signal from two base-station antennas of a three-antenna multiple-array configuration, as shown in Fig. 5.5. This antenna multiple-array configuration consisted of three antennas mounted on a triangular structure at a height of 100 ft. One antenna of the multiple array was used for transmitting, while the other two antennas were used for diversity-combined reception of mobile transmissions. The three antennas of the multiple array were equally sepa-

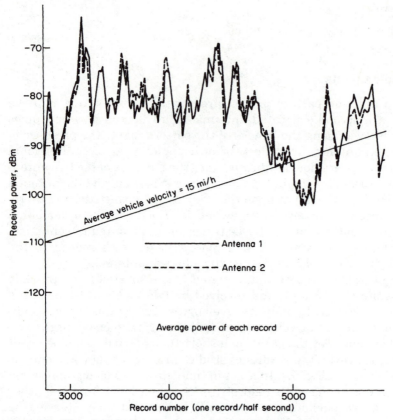

Figure 5.4   Plot of received power for two base-station antennas.

rated by about 8 wavelengths, based on an operational frequency of 850 MHz. The heights and spacing for multiple-array antennas operating at other particular frequencies are described in greater detail in Sec. 9.4 of Chap. 9. Analysis of the free-space patterns of the two receiving antennas revealed that the antenna-pattern ripple effects were caused by interaction of the adjacent, colocated antennas.

The second phase of the study consisted of road testing within a designated sector of approximately 15° between 330 and 345° with respect to the base-station orientation shown in Fig. 5.6(a). The two base stations are simultaneously receiving the signal from the mobile. The difference in power between the two received signals, $P_1 - P_2$, is plotted in Fig. 5.6(b). Note that the power difference is not dependent upon the path loss of each signal.

In the third phase of the study an analysis was made of the recorded power differences between the two base-station receiving antennas as the mobile unit traveled across a particular radial line within the test sector. For example, six radial lines (θ_1 to θ_6) were chosen, as shown in

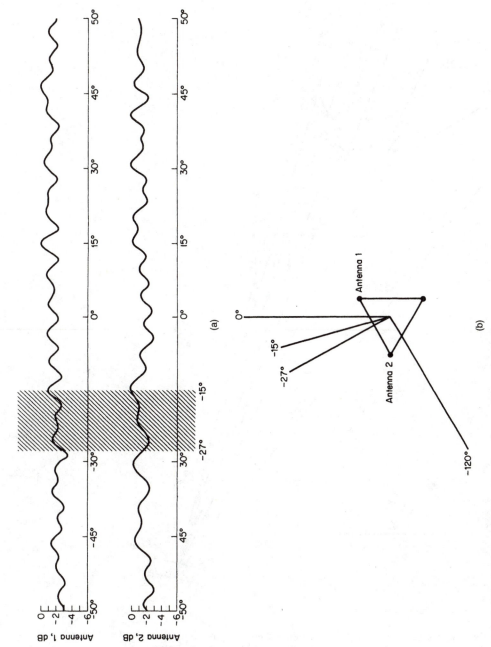

Figure 5.5 Free-space patterns for two receiving antennas in a multiple-array antenna configuration: (*a*) free-space pattern for two colocated receiving antennas; (*b*) physical orientation of antennas in a multiple-array configuration.

187

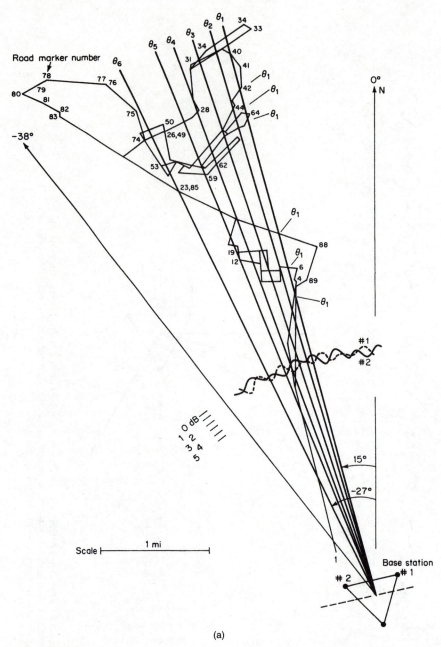

(a)

Figure 5.6 Demonstration of antenna-pattern ripple effects: (*a*) geographic sector and area of coverage during second phase of road testing of dual base-station antenna signal reception.

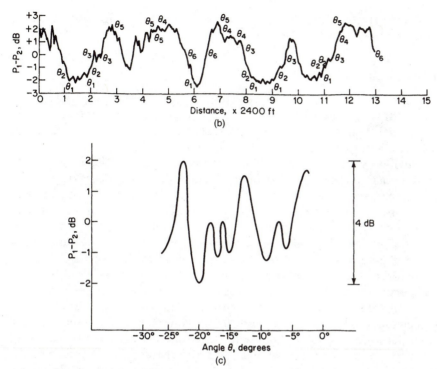

Figure 5.6 *(Continued)* (*b*) typical plot of received-power differences between antennas 1 and 2; (*c*) received-power differences of two free-space antenna patterns from colocated antennas.

Fig. 5.6(*a*). The power differences between the two antennas were recorded at the base station each time the mobile unit was traveling across each particular radial line. The power differences at six spots on the θ_1 radial line are shown in Fig. 5.6(*b*) at their corresponding spot locations. Note that a constant of –2 dB signal difference was observed for the mobile in passing each spot of the θ_1 radial line. The constants of power differences in other values are observed from other radial lines, as shown in the figure. It proves that only the constant difference between two base-station antenna patterns at any radial line due to the ripple effect can cause this result. By comparing the power differences of Fig. 5.6(*b*) with that from two free-space patterns plotted in Fig. 5.6(*c*), a distinct similarity can be observed, also.

The conclusion that can be drawn from this three-phase study is that the base-station antenna-pattern ripples are caused by the interaction between antennas in the multiple-array configuration. Consequently, the antenna-pattern ripples have a pronounced effect on signals received or transmitted at the base station, as was shown in Fig. 5.4. The range of ±2 dB in differences in power resulting from antenna-pattern ripple effects, as shown in Fig. 5.6(*b*), will degrade performance in a diversity-combining receiver. Other factors affecting diversity combining are discussed in Chap. 9.

5.4 Effects of High-Gain Antennas

High gain is obtained by suppressing the amount of high-elevation-angle radiation, which increases the amount of radiation propagated at low elevation angles. As previously discussed in Sec. 5.2, the antenna pattern should cover the angular sector—both in azimuth and elevation—wherein mobile radio waves are present. It has also been shown that increasing the directivity of an antenna does not necessarily result in an increase in the power of the received signal. These conclusions also apply, with certain limitations, to the effects of high-gain antennas.

To properly design a high-gain antenna, it is first necessary to determine the range of elevation angles within which mobile-radio signals are propagated for optimal mobile communication. The vertical-pattern characteristics of the high-gain antenna can then be designed to cover that particular area. The use of a properly designed high-gain antenna by the mobile unit will greatly improve signal reception. Studies have been conducted by Lee and Brandt [5] in which two types of mobile-unit antennas were evaluated. One had a gain of 6 dB with elevation coverage of 16°. This antenna's gain was 4 dB higher than the other antenna's, which was a $\lambda/4$ whip antenna with elevation coverage of 39°. Table 5.1 lists the results of field-strength data measurements in a typical suburban area of New Jersey.

In Table 5.1, the line-of-sight signal strength from the high-gain (vertical-directivity) antenna averaged 4 dB above that recorded for the whip antenna. The line-of-sight path for the greatest amount of incoming signal energy always lies near the horizontal plane, since the difference in antenna heights of the base-station and mobile antennas is less than 300 ft and the intervening distance is more than 1.5 mi.

For the out-of-sight condition, the signal strength from the same high-gain antenna was generally 2 to 3.5 dB above that of the whip antenna on virtually every location sampled. The results tabulated in Table 5.1 confirm that the major portion of signal arrival is within 16° of the horizontal plane. In a sampling of 14 locations on 11 streets, only one street (Maple Place, Keyport) showed no improvement in high-gain-antenna performance. This could mean that under certain conditions a major portion of the signal is above the 16° elevation coverage of the high-gain (vertical-directivity) antenna, but still within the 39° elevation coverage of the whip antenna. The conclusion is that the spread of the vertical incident angles from the horizontal can be somewhat larger than 16° but is usually smaller than 39°. It can also be concluded that the high-gain (vertical-directivity) antenna is the antenna of choice in suburban, residential, and small-town areas.

TABLE 5.1 Differences in Signal Strength between a Vertical-Directivity Antenna (6 dB) and a λ/4 Whip Antenna (2 dB)

Name of street	Town	Difference in dB average (out-of-sight)	Difference in dB (line-of-sight)
1. Main St. (S→N)	Keyport	2	3
Main St. (N→S)	Keyport	2	4
2. Maple Pl.	Keyport	1	3
Maple Pl.	Keyport	0	4
3. Broadway	Keyport	2	4
Broadway	Keyport	1	4
4. American Legion Dr.	Keyport	2	
5. First St.	Union Beach	2	
6. Stone Rd.	Union Beach	2	4
7. Poole Ave.	Raritan	3	4
8. Annapolis Dr.	Raritan	3.5	4
9. Idlebrook Rd.	Matawan	3	4
10. Idlestone Rd.	Matawan	3.5	4
11. Van Brackle Rd.	Matawan	2	4

5.5 Independence of Electric and Magnetic Fields in the Mobile-Radio Environment

When a vertically polarized electric wave is transmitted, the time-changing magnetic field generates a time-varying electric field, and vice versa. The resultant electromagnetic energy is propagated through time and space at the velocity of light. The coincidental electric and magnetic fields are considered interdependent.

When a vertically polarized electric wave is transmitted from an antenna through any radio environment, the received signal contains the information present in both the electric and magnetic fields of the propagated wave. The mathematical expression for the three field components of an incoming wave arriving at the mobile unit's receiving antenna can be written:

$$E_z = \tilde{a}_i \exp\left(-j\beta \mathbf{V} \cdot \mathbf{u}_i t\right) e^{j\omega_0 t} \qquad \text{V/m} \tag{5.17}$$

$$H_x = \eta_0 h_x = \tilde{a}_i \sin \phi_i \exp\left(-j\beta \mathbf{V} \cdot \mathbf{u}_i t\right) e^{j\omega_0 t} \tag{5.18}$$

$$H_y = \eta_0 h_y = -\tilde{a}_i \cos \phi_i \exp\left(-j\beta \mathbf{V} \cdot \mathbf{u}_i t\right) e^{j\omega_0 t} \tag{5.19}$$

where $\tilde{a}_i$ = complex term expressed in Eq. (1.4)
 $\mathbf{u}$ = direction vector of wave arrival
 ϕ_i = angle of wave arrival
 $\mathbf{V}$ = velocity vector of motion

The coordinates for the arrival of an incoming wave at the mobile unit are shown in Fig. 5.7.

To simplify the calculation of Eqs. (5.17), (5.18), and (5.19), where η_0 is the intrinsic impedance of 377 Ω, the time variation $e^{j\omega_0 t}$ can be dropped from the three equations. Also, all three field components E_z, H_x, and H_y can be expressed in volts per meter. The field component E_z is linearly polarized in the z direction only, whereas the field component H is linearly polarized in both the x and y directions. Equations (5.17) through (5.19) reasonably assume that electromagnetic waves normally propagate close to the ground plane in the mobile-radio environment.

When N vertically polarized electric waves coming from N directions are received by an isotropic antenna of a mobile-radio terminal, the three field components become:

$$E_z = \sum_{i=1}^{N} \tilde{a}_i \exp\left[-j\beta Vt \cos\left(\phi_i - \alpha\right)\right] \tag{5.20}$$

$$H_x = \sum_{i=1}^{N} \tilde{a}_i \sin\phi_i \exp\left[-j\beta Vt \cos\left(\phi_i - \alpha\right)\right] \tag{5.21}$$

$$H_y = -\sum_{i=1}^{N} \tilde{a}_i \cos\phi_i \exp\left[-j\beta Vt \cos\left(\phi_i - \alpha\right)\right] \tag{5.22}$$

Then, the three resultant field components at the mobile receiving antenna do not have an interdependent relationship but are consid-

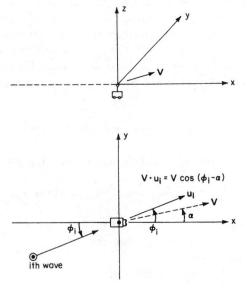

Figure 5.7 Coordinate system for an incoming wave.

ered mutually independent [6, 7]. Figure 5.8 illustrates the properties of an energy-density antenna which is capable of receiving the three field components E_z, H_x, and H_y separately [8]. The functional design of the energy-density antenna is based on the principle that the electric fields received along two edges of the loop and having the same polarity, simply add. This principle, applied to the E field component E_z, is illustrated in Fig. 5.8(b).

Figure 5.8(c) shows that the loop current induced by the H_y field component flows in opposite directions along the two edges; subtracting the outputs results in the H_y signal component. Similarly, the loop current induced by the H_x field component in Fig. 5.8(d) also is opposite and subtracts to produce the H_x signal component. Collectively, the functions of the energy-density antenna can be considered as field-component diversity [7], since each field component carries the same communication intelligence. Although a transmitted radio wave consists of both an E field and an H field, the receiving antenna normally uses the E field component to develop signal information. Since the receiving antenna does not use the H field component to develop signal information, this component is wasted.

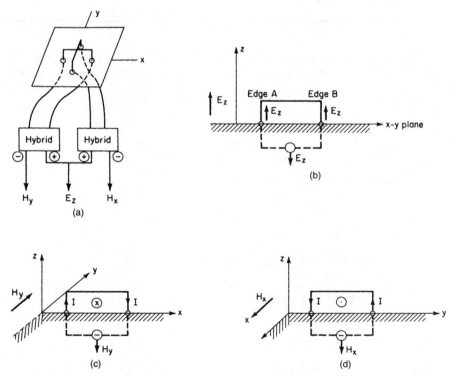

Figure 5.8 Structure and functions of an energy-density antenna.

One way to receive all three field components at the mobile receiver would be to have three omnidirectional antennas, one for each field component in the xy plane. There is no problem in having an omnidirectional antenna for the E field component as shown in Fig. 5.8(b); however, the semiloop antennas required for the H_y and H_x field components [Fig. 5.8(c) and (d)] are not omnidirectional antennas referenced to the xy plane. For example, Fig. 5.9 shows the antenna pattern for the H_x loop function in the xy plane shown in Fig. 5.8(d).

The antenna pattern $G_r(\theta, \phi)$ for a loop antenna can be expressed:

$$G_r(\theta, \phi) = \sin^2 \phi \tag{5.23}$$

Assuming that the angular probability density function (pdf) of wave arrival is uniform:

$$p_r(\theta, \phi) = \frac{1}{k_1} \tag{5.24}$$

then, by substituting Eqs. (5.23) and (5.24) into Eq. (5.10), the following equation for gain is obtained:

$$g_r = \frac{k_1}{4\pi} \int_0^{2\pi} \int_0^{\pi} \sin^2 \phi \, \frac{1}{k_1} \sin \theta \, d\theta \, d\phi = \frac{1}{2} \tag{5.25}$$

Equation (5.25) indicates that there is a 3-dB power loss due to the effects of the mobile-radio environment on the antenna pattern of a

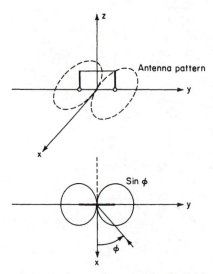

Figure 5.9 Antenna pattern for H_x field component.

loop antenna at the mobile unit. This is similar to the 3-dB power loss in receiving the H_y field components as shown in Fig. 5.8(c).

Example 5.4   When a radio wave is transmitted, the wave consists of an E field and an H field. The receiving antenna is normally used to extract either the E field or the H field, but not both.

It can be assumed that the E field and H field components are always present when a radio wave is propagated, and that the E field and H field components received at the same spot are always uncorrelated. Under these conditions, is it practical to use two antennas at the receiving terminal to extract components of both the E field and the H field?

solution   It is practical and desirable, in theory, to use two different receiving antennas to receive the E field and H field components. However, in certain circumstances it is desirable to use two identical receiving antennas to receive either two E fields or two H fields simultaneously. As noted in the assumptions, the E and H field components received at the same spot are always uncorrelated, whereas the two E field signals are only uncorrelated when the two E field antennas are placed at least $\lambda/2$ apart. At low transmission frequencies approaching the 10-m wavelength, $\lambda/2$ is 5 m, which is an impractical separation for mobile-radio antenna installations. It is therefore desirable to use E/H-type antennas for mobile installations. At higher frequencies, this physical constraint is no longer valid and the use of two identical E or H antennas becomes a practical solution. Besides, to use E/H-type antennas with the portable units is always a desirable solution to achieving diversity (see Chap. 9).

5.6   Effects of Radio-Wave Polarization

The effects of radio-wave polarization have been documented by Lee and Yeh [9], Rhee and Zysman [10], and Lee [11], which are the basis for the following discussion. When a vertically polarized wave E_v and a horizontally polarized wave E_h are transmitted simultaneously, a propagation effect known as "cross-coupling" can occur. Since the mobile-radio propagation medium is linear, the principles of reciprocity can be applied. The two differently polarized waves arriving at two colocated mobile-unit receiving antennas can be expressed:

$$E_v = \Gamma_{11} + \Gamma_{21} \tag{5.26}$$

$$E_h = \Gamma_{22} + \Gamma_{12} \tag{5.27}$$

When the principles of reciprocity are applied, the two differently polarized waves arriving at two colocated base-station receiving antennas are:

$$E'_v = \Gamma_{11} + \Gamma_{12} \tag{5.28}$$

$$E'_h = \Gamma_{22} + \Gamma_{21} \tag{5.29}$$

where Γ_{11} = transmit vertical, receive vertical

Γ_{12} = coupling between vertical at the base and horizontal at the mobile

Γ_{21} = coupling between horizontal at the base and vertical at the mobile

Γ_{22} = transmit horizontal, receive horizontal

Figure 5.10 illustrates the horizontal and vertical polarization relationships expressed in Eqs. (5.26) through (5.29). For purposes of discussion, assume that the two colocated mobile-unit antennas, one vertically polarized and the other horizontally polarized, are mounted in the same spot on top of the mobile unit, as shown in Fig. 5.11.

The following equation can be obtained from Eq. (1.20):

$$\Gamma_{11} = \sum_{i=1}^{N} a_i e^{j\psi_i} \tag{5.30}$$

where a_i and ψ_i are the amplitude and phase, respectively, of the ith incoming wave. If horizontal- and vertical-polarized antennas are colocated at both the base-station and mobile-unit terminals, then it can be assumed that the two transmitted waves will follow the same propaga-

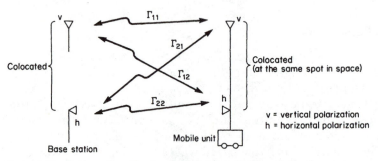

Figure 5.10 Reciprocity of horizontally and vertically polarized waves.

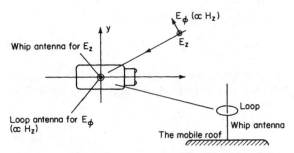

Figure 5.11 Physical arrangement of horizontal and vertical mobile-unit antennas.

tion path between the two terminals. It is important, however, to note that the reflection coefficients for the two waves are different, as expressed in Eqs. (3.6) through (3.9). The reflection coefficient for the vertically polarized wave is:

$$a_v \approx -1 \qquad \text{ground-reflected wave}$$

$$a_h \approx -1 \qquad \text{building-reflected wave}$$

The reflection coefficient for the horizontally polarized wave is:

$$a_h \approx -1 \qquad \text{ground-reflected wave}$$

$$-1 \leq a_v \leq 1 \qquad \text{building-reflected wave}$$

where

$$a_v \approx -1 \qquad \text{for incident angles to the building wall} \ll 10°$$

$$a_v \approx +1 \qquad \text{for incident angles to the building wall} \approx 90°$$

Therefore, the horizontally polarized wave Γ_{22} can be expressed:

$$\Gamma_{22} = \sum_{i=1}^{N} (a_i + \Delta a_i) e^{j(\psi_i + \Delta \psi_i)} \qquad (5.31)$$

where Δa_i and $\Delta \psi_i$ are the amplitude and phase difference, respectively, of each wave path as compared with Eq. (5.30). The expressions for the waves in Eqs. (5.30) and (5.31) are similar; however, because of the differences in amplitude and phase, Γ_{11} and Γ_{22} are proved statistically independent [9]. If the two waves, one horizontally polarized and the other vertically polarized, are transmitted simultaneously, then polarization cross-coupling occurs. The terms Γ_{12} and Γ_{21} in Eqs. (5.26) and (5.27) are also important and are discussed in subsequent paragraphs.

In analyzing the distribution of E_v and E_h, it was found that the local means or long-term-fading characteristics of the two signals Γ_{11} and Γ_{22} are lognormal-distributed. E_v and E_h also include the two signals Γ_{21} and Γ_{12}, as expressed in Eqs. (5.26) and (5.27). From the field studies described by Lee and Yeh [9], the local means of the two signals E_v and E_h exhibited the same shadowing effects. Both signals peaked up and down at approximately the same street locations among those where data were recorded. However, the local means of the two signals were by no means exactly equal. There are two main parameters for the statistical distribution of E_v and E_h: the cumulative probability distribution and the spread difference between the two local means.

The cumulative probability distribution of the local means of both E_v and E_h appeared to be lognormal-distributed, as were the distributions for Γ_{11} and Γ_{22}. The spread S, expressed in decibels, for the difference between the two local means of E_v and E_h was found to be lognormal-distributed, with a standard deviation of $\sigma = 2$ dB in suburban areas:

$$S = |E_v^2| - |E_h^2| \qquad \text{dB}$$

The 50 percent point is very close to 0 dB, which makes it simple to combine the two signals for diversity reception, as discussed in Chap. 9.

In comparing the local means of E_v and E_h in a suburban area, there are no gross differences between the average local mean levels of E_v and E_h. However, on the basis of measured data recorded within a 3-mi radius of the base station located in the Keyport-Strathmore area of New Jersey, it was noted that E_v was slightly higher than E_h [9].

In comparing the local means of Γ_{11} and Γ_{22} in urban areas, it was found that the local mean of Γ_{11} was always higher than that of Γ_{22}. Figure 5.12 plots the decibel difference between Γ_{11} and Γ_{22} based on measured data recorded in downtown Philadelphia in 1972 [10]. A 4 dB difference between signals Γ_{11} and Γ_{22} was found at distances of approximately 1 mi or more from the base-station antennas.

Published literature does not reveal very much information on polarization cross-coupling effects. It is therefore interesting to note the cross-coupling comparisons obtained from the same data measured in downtown Philadelphia in 1972, as shown in Figs. 5.13 and 5.14 [10, 11]. The conclusion that can be drawn from the polarization cross-coupling studies involving measured data is that the cross-coupling of energy from a transmitted vertically polarized wave into a horizontally polarized wave is appreciably less than the cross-coupling of energy from a transmitted horizontally polarized wave into a vertically polarized wave. The postulation for this phenomenon is that since most structures, buildings, electric poles, and similar reflectors are parallel to the vertically polarized wave, there is less chance of vertical-signal depolarization after each such reflection. Therefore the cross-coupling effects are noticeable in urban areas but not in suburban areas.

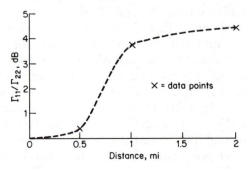

Figure 5.12   Plot of the difference between signals Γ_{11} and Γ_{22} (dB).

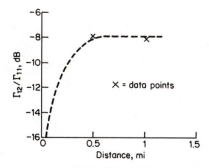

Figure 5.13 Plot of the difference between signals Γ_{11} and Γ_{12} (dB).

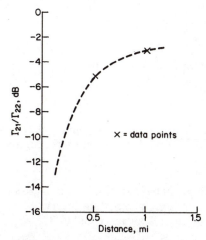

Figure 5.14 Plot of the difference between signals Γ_{21} and Γ_{22} (dB).

Base-station polarization antenna

Sometimes, the whip antenna's mounting on the mobile unit is not always erected vertically. Also, the antenna of portable phones used by people always tilt to an angle from the vertical position. Therefore, the cross-dipole polarization antenna is proposed for receiving signals at the base station. The transmitting signal still comes from the whip antenna of a mobile (or portable) terminal. The polarization antenna mounted at the base station can increase the signal gain and also provide the diversity gain at the base station. Thus, it tries to replace the space diversity antenna and save more space and cost. The antenna structure is shown in Fig. 5.15. The antenna structure is very favorable to mount on an antenna master. However, the diversity gain compared to the space diversity antenna is much weaker, as discussed in Chap. 9.

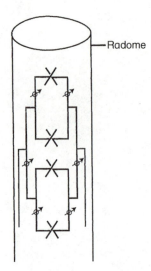

Figure 5.15   Sketch of cross-dipole polarization antenna.

Problem Exercises

1.   Explain in physical rather than mathematical terms why the average power received by a mobile directional antenna is roughly independent of the beam pattern.

2.   Explain why two identical base-station antennas receive different levels of average power within the same given time interval. Are the average power ratios of the two received signals always constant?

3.   Figure P5.3 shows the free-space propagation pattern for two colocated base-station antennas. Given the conditions that the mobile unit is traveling within a 5° sector (45° to 50°) of the antenna pattern at a distance of 2 km and that the two antennas are transmitting on different frequencies, how far would the mobile unit have to travel in a circular direction to cause the peak signal reception at the base station to change from one antenna to the other?

4.   A mobile unit is equipped with a 4-dB high-gain antenna (that is, 4 dB with reference to a dipole antenna); the field strength of the received signal is only 2 dB over that of a dipole antenna. Under this assumption, what is the elevation angle of wave arrival?

5.   Prove that the power-combining function of an energy-density antenna is less than the power-combining function of three E_z field antennas.

6.   Prove that the antenna configuration shown in Fig. P5.6 can be used as an energy-density antenna [12].

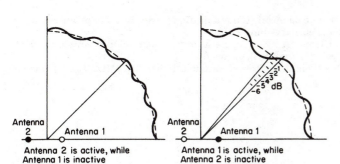

Antenna 2 is active, while
Antenna 1 is inactive

Antenna 1 is active, while
Antenna 2 is inactive

Figure P5.3 Free-space patterns for two colocated base-station antennas.

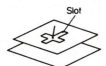

Figure P5.6 One kind of energy-density antenna.

7. Explain why a vertically polarized wave is preferred over a horizontally polarized wave for mobile-radio communications.

References

1. W. C. Y. Lee, "Finding the Approximate Angular Probability Density Function of Wave Arrival by Using a Directional Antenna," *IEEE Trans. Anten. Prop.,* vol. 21, no. 3, May 1973, pp. 328–334.
2. J. D. Kraus, *Antennas,* McGraw-Hill, New York, 1950, p. 25.
3. W. C. Y. Lee, "Preliminary Investigation of Mobile Radio Signal Fading Using Directional Antennas on the Mobile Unit," *IEEE Trans. Veh. Comm.,* vol. 15, no. 2, October 1966, pp. 8–15.
4. W. C. Y. Lee, "Antenna Pattern Ripple Effects," Bell Telephone Laboratories unpublished memorandum, June 11, 1976. This piece of work has been reported by R. J. Pillmeier, "Performance Evaluation of a Cellular Field Test System," Symposium on Microwave Mobile Communications, Boulder, Colo., Sept. 29–Oct. 1, 1976.
5. W. C. Y. Lee and R. H. Brandt, "The Elevation Angle of Mobile Radio Signal Arrival," *IEEE Trans. Comm.,* vol. 21, no. 11, November 1973, pp. 1194–1197.
6. E. N. Gilbert, "Energy Reception for Mobile Radio," *Bell System Technical Journal,* vol. 44, October 1965, pp. 1779–1803.
7. W. C. Y. Lee, "Statistical Analysis of the Level Crossings and Duration of Fades of the Signal from an Energy Density Mobile Radio Antenna," *Bell System Technical Journal,* vol. 46, no. 2, February 1967, pp. 417–448.
8. W. C. Y. Lee, "An Energy Density Antenna for Independent Measurement of the Electric and Magnetic Field," *Bell System Technical Journal,* vol. 46, no. 7, September 1967, pp. 1587–1599.
9. W. C. Y. Lee and Y. S. Yeh, "Polarization Diversity System for Mobile Radio," *IEEE Trans. Comm.,* vol. 20, no. 5, October 1972, pp. 912–923.
10. S. B. Rhee and G. I. Zysman, "Results of Suburban Base Station Spatial Diversity Measurements on the UHF Band," *IEEE Trans. Comm.,* vol. 22, October 1974, p. 1630.

11. W. C. Y. Lee, lecture note on mobile communication theory, Bell Telephone Laboratories in-house course, February–June 1978.

12. K. Itoh and D. K. Cheng, "A Slot-Unipole Energy-Density Mobile Antenna," *IEEE Trans. Veh. Tech.,* vol. 21, May 1972, pp. 59–62.

13. EMS Wireless, "Dual-Pole Base Station Antenna, 1850–1990 MHz," EMS, Norcross, Georgia.

Received-Signal Envelope Characteristics

6.1 Short-Term versus Long-Term Fading

When an instantaneously fading signal $s(t)$ is received at any time t in a mobile-radio environment, this signal can be expressed:

$$s(t) = r(t)e^{j\psi(t)} \qquad (6.1)$$

where $\psi(t)$ is the term for the phase of the signal $s(t)$ and $r(t)$ is the term for the envelope of the signal. Furthermore, $r(t)$ can be separated into two terms:

$$r(t) = m(t)r_0(t) \qquad (6.2)$$

where $m(t)$ represents long-term signal fading, as previously described in Sec. 3.5, and $r_0(t)$ represents short-term signal fading. If $m(y)$ represents any physical spot y corresponding to time t during the test runs, then $m(y)$ is the local mean. In this situation, Eq. (6.2) can be expressed:

$$r(y) = m(y)r_0(y) \qquad (6.3)$$

The instantaneous fading signal $s(t)$ received in the field contains both envelope information $r(t)$ and phase information $\psi(t)$. Although the phase information $\psi(t)$ is not used when calculating path loss, it directly affects signaling performance and voice quality, as is discussed later in Chap. 7.

In Chaps. 3 through 5, the model for explaining propagation-path loss and the methods for calculating the loss from the local means $m(y)$ are described. If no multipath fading is present, then the propagation-path loss is the only major factor that must be considered. However, if severe multipath fading is present in the mobile-radio environment, this means that $r_0(y)$ in Eq. (6.3) cannot be treated as a constant and in order to obtain $r_0(y)$ it is first necessary to obtain $m(y)$ by estimation of $\hat{m}(x = y)$, as shown in Eq. (3.43). The term $\hat{m}(y)$ is derived from an averaging process applied to the envelope $r(y)$ of an instantaneous fading signal $s(y)$ at any spot y. Thus, $\hat{m}(y)$ will be factored out from $r(y)$ in order to obtain $r_0(y)$. The envelope of a fading signal contains both long-term-fading and short-term-fading components. The long-term-fading components, which contribute only to propagation-path loss, must be removed, and the short-term-fading components, which are the result of the multipath phenomenon, must be retained.

As was previously shown by Eq. (3.43), the estimate of the local mean $\hat{m}(x)$ is different from the true local mean $m(x)$ and can be expressed [1]

$$\hat{m}(x) = m(x)\,\frac{1}{2L}\int_{x-L}^{x+L} r_0(y)\,dy \tag{6.4}$$

as illustrated in Fig. 6.1. The local means obtained from Eq. (6.4) are called the "running means"; i.e., the data points within a length L on both the left and right sides of point x are used to obtain an average for that point.

If the length L is not long enough, $m(x)$ itself retains partial short-term-fading information and therefore $\hat{m}(x)$ is different from $m(x)$. When L is too long, the details of the local means are wiped out from the averaging process of Eq. (6.4); again, $\hat{m}(x)$ is different from $m(x)$. If the length of L is chosen properly—more than 40λ but less than 200λ, as discussed in Sec. 3.5—then

$$\hat{m}(x) \to m(x)$$

The integral portion of Eq. (6.4) then becomes:

$$\frac{1}{2L}\int_{x-L}^{x+L} r_0(y)\,dy \to 1 \tag{6.5}$$

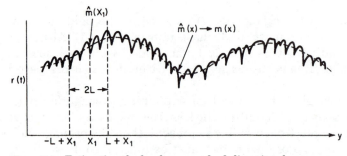

Figure 6.1   Estimating the local means of a fading signal.

The term $r_0(y)$ is an instantaneous signal free of local means and can be treated as though the $m(x)$ curve in Fig. 6.1 were straightened, as shown in Fig. 6.2.

In an arithmetical expression, $r_0(y)$ is obtained by dividing the received instantaneous signal $r(y)$ by the estimated local means $\hat{m}(x)$ as:

$$r_0(y) = \frac{r(y)}{\hat{m}(x = y)} \qquad (6.6)$$

and the logarithmic expression of $r_0(y)$ is the difference between $r(y)$ and $\hat{m}(x = y)$ in their respective decibel scales:

$$r_0(y)_{(dB)} = r(y)_{(dB)} - \hat{m}(y)_{(dB)} \qquad (6.7)$$

The method for determining how close $\hat{m}(y)$ approaches the value of $m(y)$ is described in greater detail in Sec. 3.5.

For flat-ground areas, $m(y)$ is a constant over the entire measured area. However, in hilly areas, $m(y)$ fluctuates with the terrain contour. m_y is the normalization factor in hilly areas and normalizes the instantaneous signal $r(y)$. Hence $r_0(y)$ is a normalized signal and therefore is not included in the local means. The short-term fading of $r_0(y)$ is also called "velocity-weighted fading," since it characterizes the mobile unit as if it were traveling at a constant velocity. However, if the mobile is traveling with changing velocity, then the envelope $r_0(t)$ of the received signal must be weighted by the changing velocity in order to obtain $r_0(y)$. In this context, $r_0(t)$ denotes the envelope of the received signal where the vehicle speed is constant.

6.2   Model Analysis of Short-Term Fading [2]

E. N. Gilbert was the first one to create the short-term fading model. The behavior of an E field signal can easily be described by using a theoretical model [2, 3]. Assume that the E field consists of the component E in the z direction only, as shown in Fig. 6.3. Then, E_z can be expressed from Eq. (5.20) as:

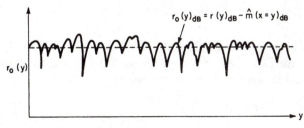

Figure 6.2   Normalized fading signal.

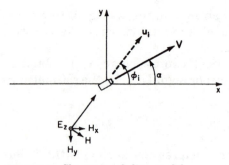

Figure 6.3 Short-term fading model.

$$E_z = \sum_{i=1}^{N} \tilde{a}_i \exp\left[-j\beta Vt \cos(\phi_i - \alpha)\right] \tag{6.8}$$

where ϕ_i is the angle between the positive x axis and the direction $\mathbf{u}_1$ and $0 \leq \phi_i \leq 2\pi$. α is the angle between the x axis and the velocity $\mathbf{V}$, and $0 \leq \alpha \leq 2\pi$. Both ϕ_i and α are shown in Fig. 6.3. $\tilde{a}_i$ is a complex variable that can be separated into real and imaginary parts:

$$\tilde{a}_i = R_i + jS_i \tag{6.9}$$

Hence, N incoming waves have N real values of R_i and S_i.

For the theoretical model analysis, suppose that all $2N$ real values are Gaussian independent variables with values of zero mean and a variance of 1. Under these conditions, the following expressions are true:

$$E[R_i] = E[S_i] = 0$$

and

$$E[R_i^2] = E[S_i^2] = 1$$

Also assume that all N waves have uniform angular distribution. For example, the kth wave u_k has an arrival angle ϕ_k, and the probability density function of ϕ_k is:

$$p(\phi_k) = \frac{1}{2\pi} \qquad 0 \leq \phi_k \leq 2\pi$$

Moreover, an infinite number of multiple reflected waves are assumed.

Inserting Eq. (6.9) into Eq. (6.8) and separating Eq. (6.8) into real and imaginary parts yields the following expression:

$$E_z = X_1 + jY_1 \tag{6.10}$$

The real part of E_z is

$$X_1 = \sum_{i=1}^{N} (R_i \cos \xi_i + S_i \sin \xi_i) \tag{6.11}$$

and the imaginary part of E_z is

$$Y_1 = \sum_{i=1}^{N} (S_i \cos \xi_i - R_i \sin \xi_i) \tag{6.12}$$

where

$$\xi_i = \beta V t \cos (\phi_i - \alpha) \tag{6.13}$$

Since all values N of R_i, S_i, and ϕ_i in Eqs. (6.11) and (6.12) are time-independent, from the central limit theorem it follows that X_1 and Y_1 are independent random variables which are normally distributed, as long as the value of N is greater than 10. The mean value of X_1 is obtained by taking a statistical average [see Eq. (2.11)] after time-averaging [see Eq. (2.13)] $\cos \xi_i$ and $\sin \xi_i$:

$$E[X_1] = \sum_{i=1}^{N} E[R_i \langle \cos \xi_i \rangle + S_i \langle \sin \xi_i \rangle]$$

$$= \sum_{i=1}^{N} \{E[R_i] \langle \cos \xi_i \rangle + E[S_i] \langle \sin \xi_i \rangle]\} = 0$$

This equation can be obtained either from

$$\langle \cos \xi_i \rangle = \langle \sin \xi_i \rangle = 0$$

or from given values of the described model:

$$E[R_i] = E[S_i] = 0$$

Similarly, the mean value of Y_1 is expressed:

$$E[Y_1] = 0$$

The mean square value of X_1 is obtained in the same manner as that described above:

$$E[X_1^2] = \sum_{i=1}^{N} \{E[R_i^2] \langle \cos^2 \xi_i \rangle + E[S_i^2] \langle \sin \xi_i \rangle\}$$

$$= \sum_{i=1}^{N} (\tfrac{1}{2} + \tfrac{1}{2}) = N \tag{6.14}$$

for $E[R_iS_i] = E[R_iS_j] = 0$. Similarly,

$$E[Y_1^2] = \sum_{i=1}^{N} \{E[S_i^2]\langle\cos^2 \xi_i\rangle + E[R_i^2]\langle\sin^2 \xi_i\rangle\}$$

$$= N$$

Let $N = \sigma^2$; then

$$E[X_1^2] = E[Y_1^2] = \sigma^2 \qquad (6.15)$$

Also,

$$E[X_1Y_1] = \sum_{i=1}^{N} \sum_{j=1}^{N} \{E[R_iS_j]\langle\cos \xi_i \cos \xi_j\rangle - E[S_iR_j]\langle\sin \xi_i \sin \xi_j\rangle$$

$$- E[R_iR_j]\langle\cos \xi_i \sin \xi_j\rangle + E[S_iS_j] \langle\sin \xi_i \sin \xi_j\rangle\}$$

$$= 0 \qquad (6.16)$$

where $E[R_iS_j] = E[S_iR_j] = 0$ and $\langle\cos \xi_i \sin \xi_j\rangle = \langle\sin \xi_i \cos \xi_j\rangle = 0$. The Gaussian functions X_1 and Y_1 from Eq. (6.16), then, are mutually independent.

6.3 Cumulative Probability Distribution (CPD)

All statistical characteristics that are not functions of time are called "first-order statistics." For example, the average power, mean value, standard deviation, probability density function (pdf), and cumulative probability distribution (cpd) are all first-order statistics.

The cpd of $r(t)$ or its normalized signal $r_0(t)$ shown in Eq. (6.3) is a first-order statistic. In the following paragraphs the pdf of different signal envelopes $r_0(t)$ will be obtained first. Then, by integrating those pdf's the cpd's of different signal envelopes $r_0(t)$ will be obtained, as previously discussed in Sec. 2.5.

The CPD of $r_e(t)$ from an E field signal

Since X_1 and Y_1 of Eq. (6.10) are Gaussian, as shown in Sec. 6.2, then the envelope of E_z, $r_e(t)$, is

$$r_e = (X_1^2 + Y_1^2)^{1/2} \qquad (6.17)$$

Equation (2.53) shows that the pdf of $r_e(t)$ is

$$p(r_e) = \frac{r_e}{\sigma^2} \exp\left(-\frac{r_e^2}{2\sigma^2}\right) \tag{6.18}$$

and that the cpd of $r_e(t)$, the probability that r_e is less than level A, is

$$P(r_e \le A) = \int_0^A \frac{r_e}{\sigma^2} \exp\left(-\frac{r_e^2}{2\sigma^2}\right) dr_e = 1 - \exp\left(-\frac{A^2}{2\sigma^2}\right) \tag{6.19}$$

where the mean square of r_e is

$$E[r_e^2] = E[X_1^2] + E[Y_1^2] = 2\sigma^2 \tag{6.20}$$

The function of Eq. (6.19) is plotted on Rayleigh paper as shown in Fig. 6.4. The mean value of r_e is

$$E[r_e] = \int_0^\infty r_e p(r_e) \, dr_e = \int_0^\infty \frac{r_e^2}{\sigma^2} \exp\left(-\frac{r_e^2}{2\sigma^2}\right) dr_e = \sqrt{\frac{\pi}{2}}\, \sigma \tag{6.21}$$

Hence the variable r_0, defined in Eq. (6.3), can be expressed:

$$r_0 = \frac{1}{E[r_e]}\, r_e = \frac{\sqrt{2\sigma^2}}{E[r_e]} \frac{r_e}{\sqrt{2\sigma^2}} = \sqrt{\frac{2}{\pi}} \frac{r_e}{\sigma} \tag{6.22}$$

Analysis shows that r_e is different from r_0 by a constant. The cpd of r_0 can be found from Eq. (6.19):

$$P(r_0 \le R_0) = 1 - \exp\left(-\frac{\pi}{4} R_0^2\right) \tag{6.23}$$

The variance of r_e is:

$$\sigma_{r_e}^2 = E[r_e^2] - (E[r_e])^2 = \left(1 - \frac{\pi}{4}\right)(2\sigma^2) \tag{6.24}$$

The median of r_e is

$$P(r_e \le M_{r_e}) = \int_0^{M_{r_e}} p(r_e) \, dr_e = 50\% \tag{6.25}$$

From Eq. (6.18), the following is obtained:

$$M_{r_e} = 0.832 \sqrt{2\sigma^2} \tag{6.26}$$

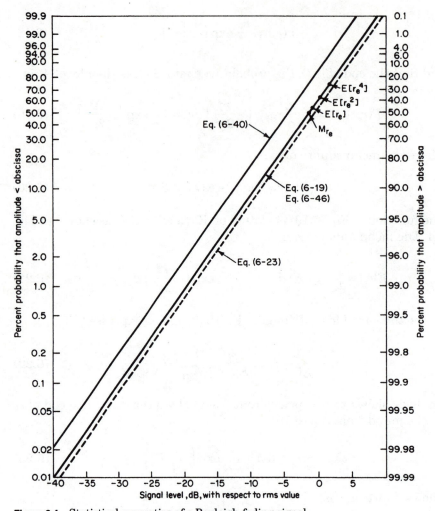

Figure 6.4 Statistical properties of a Rayleigh-fading signal.

Also, the mean square of r_e^2 can be expressed:

$$E[r_e^4] = \int_0^\infty \frac{r_e^5}{\sigma^2} \exp\left(-\frac{r_e^2}{2\sigma^2}\right) dr_e$$

$$= 2(2\sigma^2)^2 \tag{6.27}$$

and the variance of r_e^2 can be expressed:

$$\sigma_{r_e^2}^2 = E[r_e^4] - (E[r_e^2])^2$$

$$= 2(2\sigma^2)^2 - (2\sigma^2)^2 = 4\sigma^4 \tag{6.28}$$

Another application of r_e can be found in the following expression:

$$\frac{\sigma_{r_e^2}}{E[r_e^2]} = 1 \tag{6.29}$$

The properties of the Rayleigh-fading signal r_e are given in Table 6.1.

The CPD of H-field component signals

The cpd of $r_h(t)$ from H_x and H_y field signals can be derived from Eqs. (5.21) and (5.22), respectively, and Fig. 6.3, as expressed in the following equations:

$$H_x = \sum_{i=1}^{N} \tilde{a}_i \sin \phi_i \exp [-j\beta Vt \cos (\phi_i - \alpha)] \tag{6.30}$$

and

$$H_y = -\sum_{i=1}^{N} \tilde{a}_i \cos \phi_i \exp [-j\beta Vt \cos (\phi_i - \alpha)] \tag{6.31}$$

The quantities ϕ_i, α, and $\tilde{a}_i$ have been previously described in Eqs. (6.8) and (6.9). If the description of the theoretical model in Sec. 6.2 is applied to Eqs. (6.30) and (6.31), the following relationships are valid:

$$H_x = X_2 + jY_2 \tag{6.32}$$

and

$$H_y = X_3 + jY_3 \tag{6.33}$$

where

$$\left.\begin{array}{c} X_2 \\ X_3 \end{array}\right\} = \sum_{i=1}^{N} \left\{\begin{array}{c} \sin \phi_i \\ -\cos \phi_i \end{array}\right\} (R_i \cos \xi_i + S_i \sin \xi_i) \tag{6.34}$$

TABLE 6.1   Rayleigh-Fading-Signal r_e Properties

Parameters, α_i	Symbols	Values	dB over $2\sigma^2$	Percentage $p(r_e \leq \alpha_i)$	Eq.
Average power	$E[r_e^2]$	$2\sigma^2$	0	63	(6.20)
Mean value	$E[r_e]$	$\frac{\sqrt{\pi}}{2}\sqrt{2\sigma^2}$	−1.05	53	(6.21)
Median value	M_{r_e}	$0.832\sqrt{2\sigma^2}$	−1.59	50	(6.26)
Standard deviation of r_e	σ_{r_e}	$\frac{\sqrt{4-\pi}}{2}\sqrt{2\sigma^2}$	−6.68		(6.24)
Standard deviation of r_e^2	$\sigma_{r_e^2}$	$2\sigma^2$	0		(6.28)
Expectation of r_e^4	$E[r_e^4]$	$2(2\sigma^2)^2$	1.5	74	(6.27)

$$\left.\begin{array}{l} Y_2 \\ Y_3 \end{array}\right\} = \sum_{i=1}^{N} \left\{\begin{array}{c} \sin \phi_i \\ -\cos \phi_i \end{array}\right\} (S_i \cos \xi_i - R_i \sin \xi_i) \tag{6.35}$$

The term ξ_i is shown in Eq. (6.13). R_i and S_i are time-independent Gaussian variables, and ϕ_i is uniformly distributed. According to the central limit theorem, $X_2, X_3, Y_2,$ and Y_3 are Gaussian values. Following a procedure similar to that used in Eqs. (6.17) to (6.21) will give the mean and mean square values of these four Gaussian functions:

$$E[X_2] = E[X_3] = E[Y_2] = E[Y_3] = 0 \tag{6.36}$$

and

$$E[X_2^2] = E[X_3^2] = E[Y_2^2] = E[Y_3^2] = \frac{N}{2} = \frac{\sigma^2}{2} \tag{6.37}$$

Hence, the mean square values of envelopes H_x and H_y, and of r_{h_x} and r_{h_y}, respectively, are

$$E[r_{h_x}^2] = E[X_2^2] + E[Y_2^2] = \sigma^2 \tag{6.38}$$

and

$$E[r_{h_y}^2] = E[X_3^2] + E[Y_3^2] = \sigma^2 \tag{6.39}$$

The cpd of r_{h_x} or r_{h_y} has the same distribution as that expressed in Eq. (6.19); however, Eqs. (6.38) and (6.39) are both different from Eq. (6.20). The cpd expressed in the following equation is plotted in Fig. 6.4:

$$P\left(\begin{array}{c} r_{h_x} \\ r_{h_y} \end{array} \le A\right) = 1 - \exp\left(-\frac{A^2}{\sigma^2}\right) \tag{6.40}$$

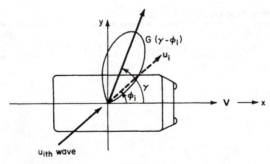

Figure 6.5 Received E field signal coordinates for a directional antenna.

The CPD of $r_d(t)$ from a directional antenna

If an omnidirectional (whip) antenna is replaced by a directional antenna pointing in the direction γ at the mobile receiver as shown in Fig. 6.5, then the received E field signal E_d is

$$E_d = X_d + jY_d \tag{6.41}$$

where the mobile is traveling in the x direction and where

$$X_d = \sum_{i=1}^{N'} (R_i \cos \xi_i + S_i \sin \xi_i) G(\gamma - \phi_i) \tag{6.42}$$

$$Y_d = \sum_{i=1}^{N'} (S_i \cos \xi_i - R_i \sin \xi_i) G(\gamma - \phi_i) \tag{6.43}$$

$$\xi_i = \beta Vt \cos \phi_i \tag{6.44}$$

and $G(\gamma - \phi_i)$ is the directional-antenna pattern [4]. N' is the number of wave arrivals received by the directional antenna.

From applying the descriptions of the model to Eqs. (6.42) and (6.43), it appears that X_d and Y_d are Gaussian functions regardless of the directional antenna pattern $G(\gamma - \phi_i)$. Also, from Sec. 5.2, the average power of the envelope r_d obtained from a directional antenna is almost the same as that of r_e obtained from an omnidirectional antenna:

$$E[r_d^2] = E[r_e^2] = 2\sigma^2 \tag{6.45}$$

Hence the cpd of r_d can be expressed:

$$P(r_d \leq A) = 1 - \exp\left(-\frac{A^2}{2\sigma^2}\right) \tag{6.46}$$

Though the right-hand side of Eq. (6.46) is the same as that of Eq. (6.19), this only means that the first-order statistics among r_e and r_d are the same. It should be noted that the second-order statistics among these two are not the same, as will be shown in the following sections.

Finding the PDF of angular wave arrival

On the basis of the assumption that the distributions of angles θ and ϕ, as shown in Fig. 5.1, are mutually independent, the pdf of angular wave arrival can be expressed:

$$p_r(\theta, \phi) = p_r(\theta)p_r(\phi)$$

If the signal is received from a directional antenna, its beam pattern can be expressed as $G(\gamma - \phi_i)$ as shown in Fig. 6.5, where $\mathbf{u}_i$ is the unit vector representation of the ith wave, ϕ_i is the angle between vectors $\mathbf{V}$ and $\mathbf{u}$ ($\mathbf{u} \cdot \mathbf{V} = \cos \phi_i$), and γ is the normal direction of the antenna array measured along the positive x axis. Under these given conditions, the received E field of signal E_d is

$$E_d = X_d + jY_d \tag{6.47}$$

From Eqs. (6.42) and (6.43), the following expression is easily derived:

$$\langle X_d^2 \rangle = \langle Y_d^2 \rangle = \frac{N}{2\pi} \int_{-\pi}^{\pi} G^2 \, (\gamma - \phi_i) p_r(\phi_i) \, d\phi_i$$

$$= N \langle G^2(\gamma - \phi_i) \rangle \tag{6.48}$$

and the pdf of angular wave arrival $p_r(\phi)$ can be expressed:

$$p(\gamma) = \int_{-\pi}^{\pi} \delta(\gamma - \phi_i) p_r(\phi_i) \, d\phi_i \tag{6.49}$$

where $\delta(\ )$ is the delta function.

In the case where the antenna beamwidth is very narrow, then $\delta(\gamma - \phi_i)$ can be replaced in Eq. (6.49) by $G^2(\gamma - \phi)$. By comparing Eq. (6.49) with Eq. (6.48), an approximation of $p(\phi)$ can be obtained:

$$p(\phi) \approx \frac{2\pi}{N} \, \langle X_d^2 \rangle \approx \frac{\pi}{N} \, \langle r_d^2 \rangle$$

where ϕ replaces γ to preserve consistency within the notation. In Eq. (6.45), the term r_d was representative of the signal envelope.

In the case where the antenna beamwidth is not narrow, then $\delta(\gamma - \phi_i)$ should be replaced by $G^N(\gamma - \phi_i)$; likewise, where N can become a large value. Under these conditions, $p_r(\phi)$ can be expressed

$$p_r(\phi) = \pi \langle G^N(\gamma - \phi) \rangle \tag{6.50}$$

and $\langle G^N(\gamma - \phi) \rangle$ can be calculated for $N = 4$ and $N = 6$:

$$\langle G^4(\gamma - \phi) \rangle = \frac{1}{6N} \, \{ \langle r_d^4 \rangle - 2 \langle r_d^2 \rangle^2 \}$$

$$\langle G^6(\gamma - \phi) \rangle = \frac{1}{30N} \, \{ \langle r_d^6 \rangle - 9 \langle r_d^4 \rangle \langle r_d^2 \rangle + 12 \langle r_d^2 \rangle^3 \}$$

Lee's method [4] can be used to find the general expression for $\langle G^N(\gamma - \phi) \rangle$ in terms of r_d.

6.4   Level-Crossing Rate (LCR) [3]

The lcr of $r_0(t)$ or $r(t)$ is a second-order statistic. As shown in Eq. (2.73), the level-crossing rate (lcr) at a given level A is

$$n(r_0 = A) = \int_0^\infty \dot{r}_0 p(r_0 = A, \dot{r}_0) \, d\dot{r}_0 \qquad (6.51)$$

where the slope of signal $r_0, \dot{r}_0$, is a derivative of r_0 with respect to time, as

$$\dot{r}_0 = \frac{dr_0}{dt}$$

Since Eq. (6.51) involves time, the lcr is a second-order statistic. The lcr for each of three field components received from an omnidirectional antenna and the lcr of a signal received from a directional antenna are derived in the following sections.

The LCR of $r_e(t)$ from an E field signal

The E field signal expressed in Eq. (6.10) contains real and imaginary parts:

$$E_z = X_1 + jY_1$$

where the envelope of E_z is $r_e(t)$, as expressed in Eq. (6.17). The joint probability density function $p(r_e, \dot{r}_e)$ is obtained first; then Eq. (6.51) is applied to obtain the lcr or $r_e(t)$.

To obtain $p(r_e, \dot{r}_e)$, all N values of R and S in Eqs. (6.11) and (6.12) are assumed to be time-independent; then, the derivatives of Eqs. (6.11) and (6.12) with respect to time are:

$$\dot{X}_1 = \beta V \sum_{i=1}^N (-R_i \sin \xi_i + S_i \cos \xi_i) \cos (\phi_i - \alpha) \qquad (6.52)$$

$$\dot{Y}_1 = \beta V \sum_{i=1}^N (-S_i \sin \xi_i - R_i \cos \xi_i) \cos (\phi_i - \alpha) \qquad (6.53)$$

where ξ_i is expressed:

$$\xi_i = \beta V t \cos(\phi_i - \alpha) \qquad (6.54)$$

Since the time averages of $\sin \xi_i$ and $\cos \xi_i$ are

$$\langle \sin \xi_i \rangle = \langle \cos \xi_i \rangle = 0$$

the mean values and variances of $\dot{X}_1$ and $\dot{Y}_1$ are:

$$E[\dot{X}_1] = \beta V \sum_{i=1}^{N} E[(R_i\langle\sin\xi_i\rangle + S_i\langle\cos\xi_i\rangle)\cos(\phi_i - \alpha)] = 0$$

$$E[\dot{Y}_1] = \beta V \sum_{i=1}^{N} E[(-S_i\langle\sin\xi_i\rangle - R_i\langle\cos\xi_i\rangle)\cos(\phi_i - \alpha)] = 0$$

$$v^2 = E[\dot{X}_1^2] = (\beta V)^2 NE[\cos^2(\phi_i - \alpha)] = (\beta V)^2 \frac{N}{2} \qquad (6.55)$$

and thus

$$E[\dot{Y}_1^2] = E[\dot{X}_1^2] = v^2 \qquad (6.56)$$

By following the same steps as were used to obtain Eq. (6.16), the following is derived:

$$\langle X_1 Y_1 \rangle = \langle X_1 \dot{Y}_1 \rangle = \langle Y_1 \dot{X}_1 \rangle = \langle Y_1 \dot{Y}_1 \rangle = \langle \dot{X}_1 \dot{Y}_1 \rangle = \langle X_1 \dot{X}_1 \rangle = 0 \qquad (6.57)$$

From the central limit theorem, it follows that X_1, Y_1, $\dot{X}_1$, and $\dot{Y}_1$ are four independent random variables that are distributed normally as the value N approaches infinity.

The probability density function of four independent real random variables X_1, Y_1, $\dot{X}_1$, and $\dot{Y}_1$ is $p(X_1, Y_1, \dot{X}_1, \dot{Y}_1)$ [5], expressed:

$$p(X_1, Y_1, \dot{X}_1, \dot{Y}_1) = \frac{1}{(2\pi)^2\sigma^2 v^2} \exp\left[-\frac{1}{2}\left(\frac{X_1^2 + Y_1^2}{\sigma^2} + \frac{\dot{X}_1^2 + \dot{Y}_1^2}{v^2}\right)\right] \qquad (6.58)$$

where σ^2 and v^2 are shown in Eq. (6.15) and Eq. (6.56), respectively.

From Eq. (6.10), the following is indicated:

$$E_z = X_1 + jY_1 = r_e e^{j\psi_e}$$

The quantity r_e is the envelope and ψ_e is the phase, both of them slow variable functions of time. Then

$$X_1 = r_e \cos\psi_e$$
$$Y_1 = r_e \sin\psi_e \qquad (6.59)$$

$$\dot{X}_1 = \dot{r}_e \cos\psi_e - r_e\dot{\psi}_e \sin\psi_e$$
$$\dot{Y}_1 = \dot{r}_e \sin\psi_e + r_e\dot{\psi}_e \cos\psi_e \qquad (6.60)$$

The jacobian of the transformation from $(X_1, Y_1, \dot{X}_1, \dot{Y}_1)$ space to $(r_e, \psi_e, \dot{r}_e, \dot{\psi}_e)$ space is $|J| = r_e^2$ [see Eq. (2.49)]. Therefore, the change of variables makes it possible to express the probability density in the form:

$$p(r_e, \psi_e, \dot{r}_e, \dot{\psi}_e) = r_e^2 p(X_1, Y_1, \dot{X}_1, \dot{Y}_1) = r_e^2 q(r_e, \psi_e, \dot{r}_e, \dot{\psi}_e)$$

$$= \frac{r_e^2}{(2\pi)^2 \sigma^2 v^2} \exp\left[-\frac{1}{2}\left(\frac{r_e^2}{\sigma^2} + \frac{r_e^2 \dot{\psi}_e^2 + \dot{r}_e^2}{v^2}\right)\right] \quad (6.61)$$

where $q(r_e, \psi_e, \dot{r}_e, \dot{\psi}_e)$ is the density obtained by substituting the values r_e, ψ_e, etc., for X_1, Y_1, etc., obtained from Eq. (6.59) and its time derivative Eq. (6.60). To obtain $p(r_e, \dot{r}_e)$, the probability density of the envelope and its rate of change, ψ_e and $\dot{\psi}_e$, must be integrated over their respective ranges (0 to 2π and $-\infty$ to ∞). Then, from Eq. (6.61), the following is obtained:

$$p(r_e, \dot{r}_e) = \frac{r_e}{\sqrt{2\pi v^2 \sigma^2}} \exp\left[-\frac{1}{2}\left(\frac{r_e^2}{\sigma^2} + \frac{\dot{r}_e^2}{v^2}\right)\right] \quad (6.62)$$

In obtaining the lcr, it should be noted that the expression on the right of Eq. (6.62) is independent of t. Hence, the expected number of level crossings $n(A)$ at a given signal amplitude $r_e = A$ can be obtained from Eq. (6.51) by using $p(A, \dot{r}_e)$ in Eq. (6.62), as follows:

$$n(r_e = A) = \int_0^\infty \dot{r}_e\, p(A, \dot{r}_e)\, d\dot{r}_e = \frac{v}{\sigma}\frac{A}{\sqrt{2\pi\sigma^2}} \exp\left(-\frac{A^2}{2\sigma^2}\right) \quad (6.63)$$

The variance of r_e is

$$\langle r_e^2 \rangle = \langle X_1^2 \rangle + \langle Y_1^2 \rangle = 2N = 2\sigma^2$$

The normalized level R can be defined:

$$R = \frac{A}{\sqrt{E[r_e^2]}} = \frac{A}{\sqrt{2\sigma^2}} \quad (6.64)$$

Substituting the various values of σ^2 and v^2 from Eqs. (6.15) and (6.55) into Eq. (6.63) and applying the relationships of Eq. (6.64) gives the following:*

$$n\left(\frac{r_e}{\sqrt{2\sigma^2}} = R\right) = \frac{\beta V}{\sqrt{2\pi}} R \exp(-R^2) \quad (6.65)$$

The function of Eq. (6.65) is plotted in Fig. 6.6, where the abscissa is in decibels ($20 \log R$) and the ordinate is ($\sqrt{2\pi}/\beta V)n(R)$).

The normalized $r_0(t)$ can be found from Eq. (6.22). Inserting Eq. (6.22) into Eq. (6.65) gives the following:*

---

* These equations first appeared in the literature in 1967. See Ref. 3.

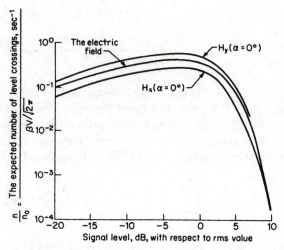

Figure 6.6   Plot of lcr curves for E_z, H_x, and H_y mobile radio signals [3].

$$n(r_0 = R_0) = \frac{\beta V}{2\sqrt{2}} R_0 \exp\left(-\frac{\pi}{4} R_0^2\right) \tag{6.66}$$

The LCR of r_{h_x} and r_{h_y} from magnetic-field components

H_x and H_y, expressed in Eqs. (6.32) and (6.33), respectively, contain real and imaginary parts. The derivatives with respect to time are:

$$\left.\begin{array}{c}\dot{X}_2 \\ \dot{X}_3\end{array}\right\} = \beta V \sum_{i=1}^{N} \left\{\begin{array}{c}\sin\phi_i \\ -\cos\phi_i\end{array}\right\} \cos(\phi_i - \alpha)\,(-R_i \sin\xi_i + S_i \cos\xi_i) \tag{6.67}$$

$$\left.\begin{array}{c}\dot{Y}_2 \\ \dot{Y}_3\end{array}\right\} = \beta V \sum_{i=1}^{N} \left\{\begin{array}{c}\sin\phi_i \\ -\cos\phi_i\end{array}\right\} \cos(\phi_i - \alpha)\,(-S_i \sin\xi_i - R_i \cos\xi_i) \tag{6.68}$$

The means, variances, and covariances of the four Gaussians X, Y, $\dot{X}$, and $\dot{Y}$ of each field component are:

$$E[X_k] = E[Y_k] = E[\dot{X}_k] = E[\dot{Y}_k] = 0 \qquad k = 2, 3 \tag{6.69}$$

$$\sigma_k^2 = E[X_k^2] = E[Y_k^2] = \frac{N}{2} = \frac{\sigma^2}{2} \qquad k = 2, 3 \tag{6.70}$$

$$v_2^2 = E[\dot{X}_2^2] = E[\dot{Y}_2^2] = v^2 \frac{\cos^2\alpha + 3\sin^2\alpha}{4} \tag{6.71}$$

$$v_3^2 = E[\dot{X}_3^2] = E[\dot{Y}_3^2] = v^2 \frac{3\cos^2\alpha + \sin^2\alpha}{4} \tag{6.72}$$

and similarly, the following can be obtained:

$$E[X_k Y_k] = E[X_k \dot{X}_k] = E[X_k \dot{Y}_k] = E[\dot{X}_k Y_k]$$

$$= E[Y_k \dot{Y}_k] = E[\dot{X}_k \dot{Y}_k] = 0 \qquad k = 2, 3$$

The probability density of the envelopes of H_x and H_y and their rates of change $p(r_h, \dot{r}_h)$ are then obtained by using the same procedure that was used in deriving $p(r_e, \dot{r}_e)$ in Eq. (6.62). Moreover, $p(r_h, \dot{r}_h)$ has the same form as $p(r_e, \dot{r}_e)$ in Eq. (6.62), but with different values of σ_k and v_k. Hence, the lcr of r_h, $n(r_h = A)$, where r_h represents r_{h_x} or r_{h_y}, has the same form as $n(r_e = A)$, shown in Eq. (6.63), but with different values of σ_k and v_k. Substituting Eqs. (6.70) through (6.72) into Eq. (6.63) gives the following:

$$n(r_{h_x} = A) = \frac{\beta V}{\sqrt{2\pi}} \sqrt{1 - \tfrac{1}{2} \cos 2\alpha} \, \frac{A}{\sqrt{N}} \exp\left(-\frac{A^2}{N}\right) \qquad (6.73)$$

and

$$n(r_{h_y} = A) = \frac{\beta V}{\sqrt{2\pi}} \sqrt{1 + \tfrac{1}{2} \cos 2\alpha} \, \frac{A}{\sqrt{N}} \exp\left(-\frac{A^2}{N}\right) \qquad (6.74)$$

When r_{h_x} and r_{h_y} are normalized by $2\sigma^2 = 2N$, then:

$$n\left(\frac{r_{h_x}}{\sqrt{2\sigma^2}} = R\right) = \frac{\beta V}{\sqrt{2\pi}} \sqrt{2 - \cos 2\alpha} \, R \exp\left(-2R^2\right) \qquad (6.75)$$

$$n\left(\frac{r_{h_y}}{\sqrt{2\sigma^2}} = R\right) = \frac{\beta V}{\sqrt{2\pi}} \sqrt{2 + \cos 2\alpha} \, R \exp\left(-2R^2\right) \qquad (6.76)$$

The functions of Eqs. (6.75) and (6.76) are plotted in Fig. 6.6 for $\alpha = 0$; the angle α has already been illustrated in Fig. 6.3.

The LCR of $r_d(t)$ from a directional antenna [4]

The received E field signal E_d from a directional antenna (see Fig. 6.5) contains real and imaginary parts, as shown in Eq. (6.41). The time derivatives of these parts are:

$$\dot{X}_d = \beta V \sum_{i=1}^{N'} (-R_i \sin \xi_i + S_i \cos \xi_i) \cos \phi_i \, G(\gamma - \phi_i)$$

$$\dot{Y}_d = \beta V \sum_{i=1}^{N'} (-S_i \sin \xi_i - R_i \cos \xi_i) \cos \phi_i \, G(\gamma - \phi_i) \qquad (6.77)$$

where

$$\xi_i = \beta V t \cos \phi_i \qquad (6.78)$$

Assuming that the antenna pattern is $G(\gamma - \phi_i)$ for an M-element directional array and that M is an even number, then:

$$G(\gamma - \phi_i) = \frac{\sin (M\xi)}{\sin \xi} = 2 \sum_{m=1}^{M/2} \cos [(2m - 1)\xi]$$

$$-\frac{\pi}{2} + \gamma \le \phi_i \le \frac{\pi}{2} + \gamma \qquad (6.79)$$

$$G(\gamma - \phi_i) = 0 \qquad \text{elsewhere}$$

where

$$\xi = \frac{\beta d}{2} \sin (\gamma - \phi_i) \qquad (6.80)$$

and

$$G^2(\gamma - \phi_i) = 4 \left[\frac{M}{4} + \sum_{m=1}^{M-1} \left(\frac{M - m}{2} \right) \cos (2m\xi_i) \right] \qquad (6.81)$$

In this case, the following relationships hold true:

$$\sigma_d^2 = E[X_d^2] = E[Y_d^2] \qquad (6.82)$$

$$v_d^2 = E[\dot{X}_d^2] = E[\dot{Y}_d^2] \qquad (6.83)$$

$$E[X_d] = E[Y_d] = E[X_d Y_d] = E[\dot{X}_d \dot{Y}_d] = 0 \qquad (6.84)$$

However, the following terms may not equal zero:

$$v_{12}^2 = E[X_d \dot{Y}_d] = -E[Y_d \dot{X}_d] \ne 0 \qquad (6.85)$$

Hence, the joint probability density function $p(X_d, Y_d, \dot{X}_d, \dot{Y}_d)$ becomes:

$$p(X_d, Y_d, \dot{X}_d, \dot{Y}_d) = \frac{1}{(2\pi)^2 |\mu|^{1/2}} \exp \left\{ -\frac{1}{2|\mu|^{1/2}} \right.$$

$$\left. \times [v_d^2(X_d^2 + Y_d^2) + \sigma_d^2(\dot{X}_d^2 + \dot{Y}_d^2) - 2 v_{12}^2(X_d \dot{Y}_d - Y_d \dot{X}_d)] \right\} \qquad (6.86)$$

where

$$|\mu|^{1/2} = \sigma_d^2 v_d^2 - v_{12}^4 \qquad (6.87)$$

By applying Eqs. (6.59) and (6.60) to Eq. (6.86) and changing the variables, the following equations can be derived:

$$p(r_d, \dot{r}_d, \psi_d, \dot{\psi}_d) = \frac{1}{(2\pi)^2 |\mu|^{1/2}} \exp \left\{ - \frac{1}{2|\mu|^{1/2}} \right.$$

$$\left. \times [v^2 r_d^2 + \sigma^2(\dot{r}_d^2 + r_d^2 \dot{\psi}_d^2) - 2v_{12}^2 r_d^2 \dot{\psi}_d] \right\} \quad (6.88)$$

Then

$$p(r_d, \dot{r}_d) = \int_{-\infty}^{\infty} \int_0^{2\pi} p(r_d, \dot{r}_d, \psi_d, \dot{\psi}_d) \, d\psi_d \, d\dot{\psi}_d$$

$$= \frac{r_d}{\sqrt{2\pi} |\mu|^{1/2} \sigma_d^2} \exp \left(- \frac{r_d^2}{2\sigma_d^2} \right) \exp \left(- \frac{\sigma_d^2}{2|\mu|^{1/2}} \dot{r}_d^2 \right) \quad (6.89)$$

and the lcr of r_d is

$$n(r_d = A) = \int_0^{\infty} \dot{r}_d p(A, \dot{r}_d) \, d\dot{r}_d$$

$$= \frac{A}{\sqrt{2\pi \sigma_d^2}} \sqrt{\frac{v_d^2}{\sigma_d^2} - \frac{v_{12}^4}{\sigma_d^4}} \exp \left(- \frac{A^2}{2\sigma_d^2} \right) \quad (6.90)$$

For example, if a two-element antenna array is used, then Eqs. (6.79) and (6.81) become:

$$G(\gamma - \phi_i) = 2 \cos \xi_i \quad (6.91)$$

and

$$G^2(\gamma - \phi) = 4[\tfrac{1}{2} + \tfrac{1}{2} \cos 2\xi_i] = 2 + 2 \cos (2\xi_i) \quad (6.92)$$

Since the angle of wave arrival ϕ_i is uniformly distributed, i.e.,

$$p(\phi_i) = \frac{1}{2\pi}$$

it is easy to show that:

$$\sigma_d^2 = E[X_d^2] = E[Y_d^2] = N' \, E[G^2(\gamma - \phi_i)] = N'[1 + J_0(\beta d)] \quad (6.93)$$

$$v_d^2 = E[\dot{X}_d^2] = E[\dot{Y}_d^2] = (\beta V)^2 N' \, E[G^2(\gamma - \phi_i) \cos^2 \phi_i]$$

$$= \frac{(\beta V)^2}{2} N'[1 + J_0(\beta d) + \cos (2\gamma) J_2(\beta d)] \quad (6.94)$$

and

$$v_{12}^2 = E[X_d \dot{Y}_d] = E[\dot{X}_d Y_d]$$

$$= (\beta V) N' \cos \gamma \, J_1(\beta d) \qquad (6.95)$$

Substituting Eqs. (6.93), (6.94), and (6.95) into Eq. (6.90) gives the following:

$$n(r_d = A) = \frac{\beta V}{\sqrt{2\pi}} \sqrt{1 + \frac{\cos 2\gamma \, J_2(\beta d)}{1 + J_0(\beta d)} - \frac{2 \cos^2 \gamma \, J_1^2(\beta d)}{[1 + J_0(\beta d)]^2}}$$

$$\times R_d \exp(-R_d^2) \qquad (6.96)$$

where

$$R_d = \frac{A}{\sqrt{2 \, \sigma_d^2}}$$

In the case where the antenna spacing $d = \pi/2$ and the antenna pointing direction $\gamma = 90°$, then:

$$n(r_d = A) = 0.5636 \, \frac{\beta V}{\sqrt{2\pi}} \, R_d \exp(-R_d^2) \qquad (6.97)$$

The directional antenna therefore has a reduced fading rate when compared with that of Eq. (6.65), but its relative fading rate is linearly proportional to the fading rate for an omnidirectional whip antenna.

Example 6.1 The envelope of an E field signal, at a frequency of 850 MHz, is received by a mobile unit traveling at 15 mi/h. Figure E6.1 illustrates 1200 positive level crossings within a 2-min interval, referenced to an arbitrary reference level A. Assuming that the signal is Rayleigh-distributed, what is the average power $E[r^2]$ in decibels above reference level A?

solution The level-crossing rate can be expressed:

$$n = \frac{1200}{2 \times 60} = 10 \text{ crossings per second}$$

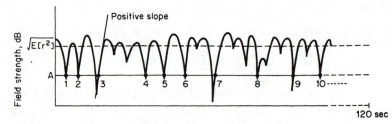

Figure E6.1   Reference lcr data for Example 6.1.

Since it is known that $V = 15$ mi/h = 22 ft/s and that $\lambda = 1.158$ ft, the normalized value becomes:

$$n_0 = \frac{\beta V}{\sqrt{2\pi}} = 47.6 \text{ crossings per second}$$

and the ratio can be expressed:

$$\frac{n}{n_0} = 0.21$$

Figure 6.6 can then be used to find the decibel level of average power above reference level A, which is 15 dB, and which is written:

$$\left.\frac{A}{\sqrt{E[R^2]}}\right|_{dB} \approx -15 \text{ dB}$$

Example 6.2 On the basis of the parameters given in Fig. E6.2, find the level R_{max} at which the maximum value of the probability density function occurs, that is, $p(r_e = R_{max}) = \max$. Also, find the $1\sigma_{r_e}$ of Rayleigh standard deviation levels above and below the average power in the E field envelope, in decibels.

solution The maximum value of the probability density function, $p(R_{max})$, can be obtained by differentiating Eq. (6.18) with respect to r and then setting the equation equal to zero, as follows:

$$\frac{d}{dr}p(r_e) = \frac{d}{dr}\left[\frac{r}{\sigma^2}\exp\left(-\frac{r^2}{2\sigma^2}\right)\right] = 0$$

$$\frac{r_{max}^2}{2\sigma^2} = \frac{1}{2}$$

Let

$$R_{max} = \frac{r_{max}}{\sqrt{2}\sigma}$$

$$R_{max} = -3 \text{ dB}$$

The $1\sigma_{r_e}$ standard deviation levels above and below the average power in the E field can be found from Eq. (6.24), where the value of σ_{r_e} is:

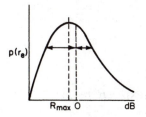

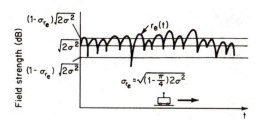

Figure E6.2 Reference data for Example 6.2.

$$\sigma_{r_e} = \frac{\sqrt{4-\pi}}{2} \sqrt{2\sigma^2} = 0.463 \sqrt{2\sigma^2}$$

$$\sqrt{2\sigma^2} + \sigma_{r_e} = 1.463 \sqrt{2\sigma^2} \approx 3.3 \text{ dB with respect to } \sqrt{2\sigma^2}$$

$$\sqrt{2\sigma^2} - \sigma_{r_e} = 0.5367 \sqrt{2\sigma^2} \approx -5.4 \text{ dB with respect to } \sqrt{2\sigma^2}$$

6.5   Calculating the Average Duration of Fades [3]

As previously derived in Eq. (2.75), the average duration of fades is expressed:

$$\bar{t}(r = A) = \frac{P(r \leq A)}{n(r = A)} \tag{6.98}$$

From the values of $P(r \leq A)$ and $n(r = A)$ described in Secs. 6.3 and 6.4, respectively, Eq. (6.98) is obtained, to express the average duration of fades for different field components. The average duration of fades for an E field signal, r_e, can be obtained by inserting Eqs. (6.19) and (6.63) into Eq. (6.98) [3] to obtain*

$$\bar{t}(r_e = A) = \frac{P(r_e \leq A)}{n(r_e = A)} = \frac{\sqrt{2\pi}}{\beta V} \frac{1}{R} [\exp (R^2) - 1] \tag{6.99}$$

where

$$R = \frac{A}{\sqrt{2\sigma^2}}$$

The function of Eq. (6.99) is plotted in Fig. 6.7.

The average duration of fades for magnetic-field components, r_h, can be obtained by inserting Eqs. (6.40) and (6.75) into Eq. (6.98), to obtain

$$\bar{t}(r_{h_x} = A) = \frac{P(r_{h_x} \leq A)}{n(r_{h_x} = A)}$$

$$= \frac{\sqrt{2\pi}}{\beta V} \frac{1}{\sqrt{2 - \cos 2\alpha}} \frac{1}{R} [\exp (2R^2) - 1] \tag{6.100}$$

Similarly,*

$$\bar{t}(r_{h_y} = A) = \frac{P(r_{h_y} \leq A)}{n(r_{h_y} = A)}$$

$$= \frac{\sqrt{2\pi}}{\beta V} \frac{1}{\sqrt{2 + \cos 2\alpha}} \frac{1}{R} [\exp (2R^2) - 1] \tag{6.101}$$

---

* These equations first appeared in the literature in 1967. See Ref. 3.

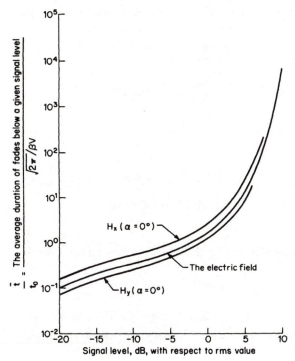

Figure 6.7 Average duration of fades for E_z, H_x, and H_y mobile-radio signals [3].

where

$$R = \frac{A}{\sqrt{2\sigma^2}} \qquad (6.102)$$

Equations (6.100) and (6.101) for $\alpha = 0$ are plotted in Fig. 6.7.

Example 6.3 To estimate the average duration of fades from the raw signal data shown in Fig. E6.3.1, it is first necessary to find the average signal level, using the method described in Example 6.1. For this example, the number of signal level crossings is 55 in a 4-h period, referenced to the −5-dB level. The −5-dB level was chosen because it is relatively easy to count the number of upward or down-

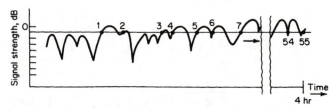

Figure E6.3.1 Reference data for Example 6.3.

ward crossings by observation and still obtain a sufficient number of crossings to meet the requirements of statistical analysis.

Assuming that the signal data are Rayleigh-fading, then Eqs. (6.65) and (6.99) can be applied to obtain the following expressions:

$$n = AR \exp(-R^2) = \frac{55}{4 \times 60 \times 60} = 0.0038 \text{ s}^{-1} \tag{E6.3.1}$$

$$\bar{t} = \frac{1}{AR} [\exp(R^2) - 1] \tag{E6.3.2}$$

where

$$20 \log R = -5 \text{ dB} \quad \text{or} \quad R = 10^{-5/20} \tag{E6.3.3}$$

Substituting Eq. (E6.3.3) into Eq. (E6.3.1) gives the following:

$$A = 0.00922$$

and by substituting the value of A into Eq. (E6.3.2), the following is obtained:

$$\bar{t} = \frac{1}{0.00922R} [\exp(R^2) - 1] \tag{E6.3.4}$$

The function of Eq. (E6.3.4) is plotted in Fig. E6.3.2 and $\bar{t} = 71.7$ s at $R = -5$ dB.

If the operating frequency is given as 30 MHz, then the vehicle speed can be found as:

$$\frac{\beta V}{\sqrt{2\pi}} = A$$

or

$$V = \frac{\sqrt{2\pi}}{\beta} A = 3.68 \times 10^{-2} \text{ m/s}$$

6.6   Envelope Correlation of the Mobile Received Signal Based on Time Separation

The correlation function $R(r_1, r_2)$ obtained at the mobile-receiver location is different from that obtained at the base-station location and therefore must be described separately.

The E field can be expressed:

$$E_z = X(t) + jY(t) \tag{6.103}$$

where

$$r_1(t) = \sqrt{X^2(t) + Y^2(t)} = \sqrt{X_1^2 + Y_1^2} \tag{6.104}$$

and

$$r_2(t) = \sqrt{X^2(t+\tau) + Y^2(t+\tau)} = \sqrt{X_2^2 + Y_2^2} \tag{6.105}$$

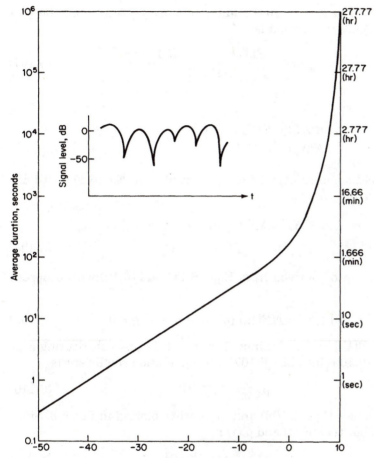

Figure E6.3.2   Estimate of the average duration of fades.

Assuming that the signal envelope received by a mobile unit is a Rayleigh-distributed random variable r and the signal phase ψ is uniformly distributed, then the joint probability distribution $p(r_1, r_2, \psi_1, \psi_2)$ can be found, where the values of r and ψ are at times t and $t + \tau$ [6]. The envelope correlation function $R(\tau)$ can be obtained from the following equation:

$$R(\tau) = \int_0^{\infty} r_1 r_2 p(r_1, r_2, \psi_1, \psi_2) \, dr_1 \, dr_2 \, d\psi_1 \, d\psi_2 \qquad (6.106)$$

Since Eq. (6.106) is very difficult to solve and involves two complete elliptic integrals [7], a very close approximation can be derived from Booker, Ratcliff, and Shinn [8], as follows. If two random signals r_1 and

r_2, as indicated in Eqs. (6.104) and (6.105), are present, then the normalized correlation function is

$$\rho_r(\tau) = \frac{(E[X_1 X_2])^2 + (E[X_1 Y_2])^2}{(E[X_1^2])^2} \tag{6.107}$$

provided

$$\left. \begin{array}{l} E[X_m^2] = E[Y_m^2] \\ E[X_m] = E[Y_m] = 0 \end{array} \right\} \quad m = 1, 2$$

For an E field signal, $E[X_1 X_2]$ can be derived from Eq. (6.11) with different subscripts [9]:

$$E[X_1 X_2] = NE[\cos \{\beta V \tau \cos (\phi_i - \alpha)\}]$$
$$= N J_0(\beta V \tau) = N \rho(\tau) \tag{6.108}$$

and $E[X_1 Y_2]$ can be derived from Eqs. (6.11) and (6.12) with different subscripts:

$$E[X_1 Y_2] = NE[\sin \{\beta V \tau \cos (\phi_i - \alpha)\}] = 0 \tag{6.109}$$

Also, where $E[X_1^2] = N$ is obtained from Eq. (6.14), by substituting Eqs. (6.108) and (6.109) into Eq. (6.107), the correlation coefficient is

$$\rho_r(\tau) = J_0^2(\beta V \tau) \tag{6.110}$$

The functions for Eqs. (6.108) and (6.110) are plotted in Fig. 6.8. Since the relation between $R_r(\tau)$ and $\rho_r(\tau)$ is

$$\rho_r(\tau) = \frac{R_r(\tau) - (E[r])^2}{R_r(0) - (E[r])^2} \tag{6.111}$$

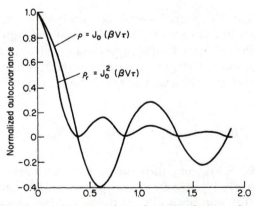

Figure 6.8 Linear and envelope correlations of a mobile received signal.

and from Eqs. (6.21) and (6.20)

$$E[r] = \frac{\sqrt{\pi}}{2} \sqrt{2N}$$

$$R_r(0) = E[r_1^2] = 2N$$

then the correlation function can be expressed:

$$R_r(\tau) = (E[r])^2 + \{R_r(0) - (E[r])^2\}\rho_r(\tau) \tag{6.112}$$

In order to calculate the unit total power, it is first necessary to apply Eq. (6.112) to obtain the following:

$$\frac{R_r(\tau)}{R_r(0)} = \frac{\pi}{4} + \left(1 - \frac{\pi}{4}\right)\rho_r(\tau) = \frac{\pi}{4} + \left(1 - \frac{\pi}{4}\right)J_0^2(\beta V\tau) \tag{6.113}$$

Example 6.4 The autocorrelation coefficient with a time delay τ obtained from the envelope of an E field signal is:

$$\rho_{r_e}(\tau) = J_0^2(\beta V\tau) \tag{E6.4.1}$$

This can also be considered a cross-correlation of two signals r_1 and r_2 which have been received by two antennas separately mounted on top of the mobile vehicle, as shown in Fig. E6.4. Under these conditions, Eq. (E6.4.1) becomes:

$$\rho_{r_1 r_2}(d) = J_0^2(\beta d) \tag{E6.4.2}$$

Find the distance d_1 between the two colocated mobile antennas such that the two received signals are uncorrelated, or such that

$$J_0^2(\beta d_1) = 0 \tag{E6.4.3}$$

solution When the two received signals from the dual mobile antennas are uncorrelated, the distance d_1 between the two antennas is

$$\rho_{r_1 r_2}(d) = J_0^2(\beta d) = 0 \tag{E6.4.4}$$

The value of $\beta d \approx 2.4$ can be obtained from the Bessel function tables in published reference books, and the value of d_1 can be calculated as follows:

$$d_1 = \frac{2.4\lambda}{2\pi} = 0.38\lambda \tag{E6.4.5}$$

Also, the value of $d_1 = 0.38\lambda$ can be obtained by reference to Fig. 6.8.

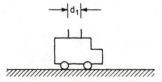

Figure E6.4 Mobile with dual mounted antennas.

6.7 Envelope Correlation of the Mobile
Received Signal Based on Time
and Space Separation [11]

Envelope correlation of mobile received signals based on time and space separation is a useful parameter in analyzing mobile-radio problems. If a signal consisting of multipath vertically polarized waves is received by an M-type space-diversity antenna array and an M-branch combining mobile receiver, then the received signal at time t_1, the sum of the individual signal amplitudes from the M individual antenna branches, can be expressed:

$$\varepsilon_1(t_1; d_1, d_2, d_3, \ldots, d_M) = r_1(t_1; d_1) + r_2(t_1; d_2) + \cdots + r_M(t_1; d_M)$$

$$= \sum_{m=1}^{M} r_m(t_1; d_m) \tag{6.114}$$

After time t_2, the signal becomes:

$$\varepsilon_2(t_2; d_1', d_2', d_3', \ldots, d_M') = \sum_{m=1}^{M} r_m(t_2; d_m') \tag{6.115}$$

The correlation of ε with both time and space separation is:

$$R_\varepsilon(\tau; \{d_m - d_n'\}) = \langle \varepsilon_1 \varepsilon_2 \rangle$$

where $\{d_m - d_n'\}$ represents a set of spacings between different space separations d_m and d_n', where both m and n are values from 1 to M.

The normalized correlation function can then be derived, as follows:

$$\rho_\varepsilon(\tau; \{d_m - d_n'\}) = \frac{R_\varepsilon(\tau; \{d_m - d_n'\}) - m_\varepsilon^2}{\sigma_\varepsilon^2} \tag{6.116}$$

where

$$m_\varepsilon = \langle \varepsilon_1(0; d_1, d_2, \ldots, d_M) \rangle$$

$$\sigma_\varepsilon^2 = \langle \varepsilon_1^2(0; d_1, d_2, \ldots, d_M) \rangle - m_\varepsilon^2$$

In the case where all values of d_m are fixed and $d_m = md$ and where d is the spacing between adjacent elements, then $d_m - d_n' = (m - n)d$, and the normalized correlation is:

$$\rho_\varepsilon(\tau \mid \{(m - n)d\}) = \frac{R_\varepsilon(\tau \mid \{(m - n)d\}) - m_\varepsilon^2}{\sigma_\varepsilon^2} \tag{6.117}$$

For a linear M-element array [10], ρ_e becomes:

$\rho_\varepsilon \left(\tau \,|\, \{(m-n)d\} \right)$

$$= \frac{M\rho_{11} + (M-1)(\rho_{12} + \rho_{21}) + (M-2)(\rho_{13} + \rho_{31}) + \cdots + \rho_{1M} + \rho_{M1}}{M\rho_{11}^0 + (M-1)(\rho_{12}^0 + \rho_{21}^0) + (M-2)(\rho_{13}^0 + \rho_{31}^0) + \cdots + \rho_{1M}^0 + \rho_{M1}^0} \quad (6.118)$$

where

$$\rho_{mn} = \rho_{mn}(\tau; (m-n)d) \qquad \text{and} \qquad \rho_{mn}^0 = \rho_{mn}(0; (m-n)d)$$

and

$$\rho_{mn}(\tau; (m-n)d) = J_0^2[\beta\sqrt{(V\tau)^2 + (m-n)^2 d^2 - 2(m-n)V\tau d \cos \alpha}] \quad (6.119)$$

The value of $E[X_m X_n]$ can be inserted into Eq. (6.107), since it is known that $E[X_m Y_n] = 0$. The parameter V is the speed of the mobile unit, and α is the direction of travel, as shown in Fig. 6.9(a). The normalized correlation for Eq. (6.118) is shown in Fig. 6.9(b), for values of $\alpha = 0$ and $\alpha = 90°$. Note that the correlation function drops faster for $\alpha = 90°$ than it does for $\alpha = 0$ [11].

6.8 Envelope Correlation of the Mobile Received Signal Based on Frequency and Time Separation

Time-delay spread and coherence bandwidth were discussed briefly in Chap. 1, Secs. 1.5 and 1.6. Both of these terms can be predicted from the envelope correlation of a mobile received signal, on the basis of frequency separation. For this discussion, a model which is more general than the model expressed in Eq. (6.8) can be used.

Assuming that the ith wave arrives at angle ϕ_i with a delay time T_i, then the sum of N waves becomes

$$E_z = \sum_{i=1}^{N} a_i \exp j\,[\omega t + \beta V t \cos (\phi_i - \alpha) - \omega T_i]\, G(\phi_i) \quad (6.120)$$

Where $G(\phi_i)$ is the horizontal pattern of the antenna, and the variable ϕ is uniformly distributed, $p(\phi_i) = 1/2\pi$. The delay time T_i can be approximated by an exponential distribution [12–16]:

$$p(T) = \frac{1}{\Delta}\, e^{-T/\Delta} \quad (6.121)$$

where Δ is the time-delay spread.

The correlation of $r = |E_z|$, based on a frequency separation of $\Delta\omega$ and a time separation of $\tau = t_1 - t_2$, can be expressed:

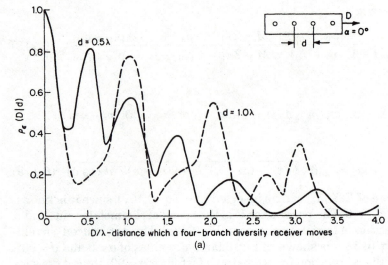

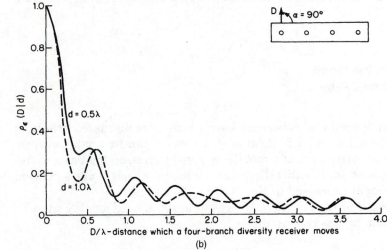

Figure 6.9 Normalized autocorrelation functions for a four-branch diversity receiver: (a) for a mobile receiver moving at $\alpha = 0$; (b) for a mobile receiver moving at $\alpha = 90°$.

$$R_r(\Delta\omega, \tau) = \langle r_1(\omega_1, t_1) r_2(\omega_2, t_2) \rangle$$

$$= \int_0^\infty r_1 r_2 \, p(r_1, r_2) \, dr_1 \, dr_2 \qquad (6.122)$$

where

$$p(r_1, r_2) = \int_0^\infty \int_0^\infty p(r_1, r_2, \theta_1, \theta_2) \, d\theta_1 \, d\theta_2$$

By following the same steps of derivation shown in Eqs. (6.103) through (6.110), the following hold true:

$$E[X_1X_2] = N \int_0^\infty \int_0^{2\pi} \cos(\beta V\tau \cos\phi_i - \Delta\omega T)\, G(\phi_i)p(\phi_i)\,p(T)\,d\phi_i\,dT$$

$$= N\,\frac{J_0(\beta V\tau)}{\sqrt{1 + (\Delta\omega)^2\Delta^2}} \tag{6.123}$$

$$E[X_1Y_2] = 0 \tag{6.124}$$

$$E[X_1^2] = N \int_0^\infty \int_0^{2\pi} G(\phi_i)p(\phi_i)p(T)\,d\phi_i\,dT \tag{6.125}$$

where $\Delta\omega = \omega_1 - \omega_2$ and $p(T)$ is as shown in Eq. (6.121). Substituting Eqs. (6.123) through (6.125) into Eq. (6.107) shows that the envelope correlation on frequency and time separation is:

$$\rho_r(\Delta\omega, \tau) = \frac{J_0^2(\beta V\tau)}{1 + (\Delta\omega)^2\Delta^2} \tag{6.126}$$

The function of Eq. (6.126) is plotted in Fig. 6.10 for different time separations, where $f_m = V/\lambda$ is the maximum fading frequency. For τ set to 0 (no time separation), the coherence bandwidth B_c associated with a particular $\Delta\omega$, $(\Delta\omega)_c = 2\pi B_c$, can be found from two different criteria [17, 18]:

$$\rho_r(B_c, 0) = 0.5 \qquad \text{or} \qquad \rho_r(B_c, 0) = \frac{1}{e} = 0.3678 \tag{6.127}$$

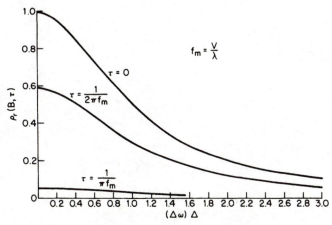

Figure 6.10 Plot of envelope correlation coefficient vs. the product of frequency separation $\Delta\omega = \omega_2 - \omega_1$ and time-delay spread Δ.

The first criterion is chosen for use in this book. By substituting the first criterion into Eq. (6.126), B_c is obtained as follows:

$$B_c = \frac{(\Delta\omega)_c}{2\pi} = \frac{1}{2\pi\Delta} \tag{6.128}$$

For example, a time delay of 0.25 μs requires a coherence bandwidth B_c of 0.636 MHz. Another definition of coherence bandwidth may be derived from the phase correlation based on frequency separation, discussed in Chap. 7.

6.9 Envelope Correlation of the Base-Station Received Signal Based on Space Separation [19]

At the base station, the signal from the mobile unit is usually confined to a small angular sector, as shown in Fig. 6.11. It can be assumed that the ith incoming wave is incident at an angle α with a probability density expressed in the form:

$$p(\phi_i) = \frac{Q}{\pi}\left[\cos^n(\phi_i - \alpha) + U\right] \qquad -\frac{\pi}{2} + \alpha \le \phi_i \le \frac{\pi}{2} + \alpha \tag{6.129}$$

where $\alpha = \langle\phi_i\rangle$, n is an even integer, and U is a value smaller than unity and can therefore be treated as a noiselike background signal associated with the major incoming signal. To simplify the model, let $U = 0$ in the following calculation. Q is a constant that can be found by substituting Eq. (6.129) into the following equation:

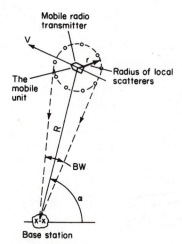

Figure 6.11 Model of base-station received signal.

$$\int_{-\pi/2 + \alpha}^{\pi/2 + \alpha} p(\phi_i)\, d\phi_i = 1 \qquad (6.130)$$

For any value of n, the incoming wave beamwidth BW can be represented by $\cos^n x$. See Ref. 19. For $n = 10^3$, the 3-dB BW = 3°. If Eq. (6.129) is known, then Eqs. (6.108) and (6.109) can be solved:

$$E[X_1 X_2] = NE[\cos \{\beta V\tau \cos (\phi_i - \alpha)\}] \qquad (6.131)$$

and

$$E[X_1 Y_2] = NE[\sin \{\beta V\tau \cos (\phi_i - \alpha)\}] \qquad (6.132)$$

Then by substituting Eqs. (6.131) and (6.132) into Eq. (6.107), and knowing $E[X^2] = N$, the correlation coefficient of two base-station signal envelopes can be obtained. The cross-correlation of the envelope of two signals r_1 and r_2 received from two antennas can be expressed as:

$$\rho_r(\tau) = (E[\cos \{\beta V\tau \cos (\phi_i - \alpha)\}])^2 + (E[\sin \{\beta V\tau \cos (\phi_i - \alpha)\}])^2 \quad (6.133)$$

Where

$$E[f(\phi_i)] = \int_{-\pi/2 + \alpha}^{\pi/2 + \alpha} f(\phi_i) p(\phi_i)\, d\phi_i$$

$$= \frac{Q}{\pi} \int_{-\pi/2 + \alpha}^{\pi/2 + \alpha} f(\phi_i) \cos^n (\phi_i - \alpha)\, d\phi_i \qquad (6.134)$$

Eq. (6.133) can be calculated numerically [19], as shown in Fig. 6.12. The correlation coefficient versus antenna spacing for different angles α with the beamwidth of the incoming signal equal to 0.4° is shown in Fig. 6.12(a). The correlation coefficient versus antenna spacing for different angles α with the beamwidth of the incoming signal equal to 3° is shown in Fig. 6.12(b). The theoretical and experimental correlations versus antenna spacing for the broadside propagation case are shown in Fig. 6.10(c). At this particular experimental setup, the data match to the BW = 0.5° curve (or $n = 3 \times 10^4$).

The radius of effective scatterers surrounding the mobile unit

The radius r to the scatterers surrounding the mobile unit is defined for the idealized model shown in Fig. 6.11 as

$$r = \frac{R \times \text{BW}}{2}$$

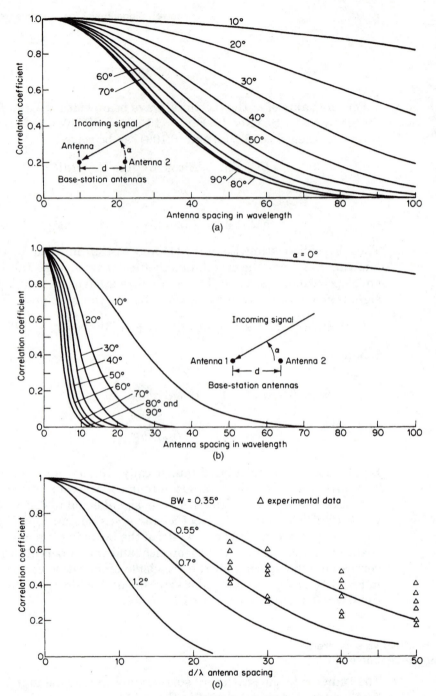

Figure 6.12 Cross-correlation of signal envelopes from two base-station antennas: (*a*) correlation coefficient vs. antenna spacing for different angles α with BW of incoming signal equal to 0.4°; (*b*) correlation coefficient vs. antenna spacing for different angles α with BW of incoming signal equal to 3°; (*c*) theoretical and experimental correlations vs. antenna spacing for the broadside propagation case.

where R is the propagation distance from the transmitter to the receiver, and BW is the beamwidth of the incoming signal. For example, if BW = 0.5° and R = 3 mi, then, r = 69 ft. The value of r gives a rough idea of how large an area surrounding the mobile unit is effective for scattering.

6.10   Power-Spectrum Analysis [21]

The power spectrum $S_{r_e}(f)$ of the signal envelope for an E field signal can be derived by taking the Fourier transform of Eq. (6.113), as shown in Eq. (2.98):

$$S_{r_e}(f) = \int_{-\infty}^{\infty} R(\tau)e^{-j\omega\tau}\, d\tau = \frac{\pi}{4}\,\delta(f) + \left(1 - \frac{\pi}{4}\right)\mathscr{S}_{r_e}(f) \qquad (6.135)$$

where $\mathscr{S}_{r_e}(f)$ is obtained by taking the Fourier transform of the correlation coefficient $\rho_{r_e}(\tau)$, and where $\delta(f)$ is an impulse; then:

$$\mathscr{S}_{r_e}(f) = \int_{-\infty}^{\infty} J_0^2(\beta V\tau)e^{-j\omega\tau}\, d\tau = \frac{K(\sqrt{1 - (f/f_0)^2})}{\pi^2 f_0} \qquad (6.136)$$

where $f_0 = 2V/\lambda$, and $K(\ )$ is the complete elliptic integral. The function for Eq. (6.136) is plotted in Fig. 6.13. Note that the power density drops at $f = f_0$ (that is, at $v = 1$), which is twice the maximum Doppler frequency ($f_0 = 2f_m$).

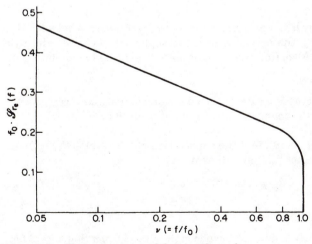

Figure 6.13   Power spectrum of the signal envelope r_e.

Problem Exercises

1. On the basis of the relationship between r and x in Eq. (P6.1.1), express the Rayleigh variable r, as derived from the uniform random variable x $(0 \leq x \leq 1)$ [20].

$$r = (-2 \ln x)^{1/2} \tag{P6.1.1}$$

2. Given two independent random variables x_1 and x_2 with a correlation ρ, find the two new correlated random variables y_1 and y_2.

3. What is the asymptotic form of Eq. (6.19) when level A is much smaller than the rms value of the signal?

4. If the received signal power is –80 dBm and the fade margin is 12 dB, what percentage of the received signal is above threshold level?

5. Find the level-crossing rate for the 4-branch signal received by the 4-element linear array shown in Fig. 6.9, as a function of the direction of the vehicle in motion [11].

6. Prove Eq. (6.119), the ρ_{mn} for a linear multiple-element array.

7. On the basis of the following assumptions, what is the correlation coefficient $\rho(\tau)$ and the separation τ for the mobile received signal? Transmission frequency is 850 MHz. Vehicular speed is 40 km/h. Mean signal power is –80 dBm. Mean signal voltage is –81 dBm. Correlation function $R(\tau)$ for τ separation is –82 dBm.

8. Given a correlation of ρ, explain why the physical separation of two co-located base-station antennas is larger for in-line propagation than for broadside propagation.

9. The cpd for a 2-branch signal is a function of the correlation ρ between the two branches. Assuming that $\rho = 0.7$, find the separation between the two base-station antennas. Also, find the separation between the two mobile-unit antennas.

10. What is the level-crossing rate at the rms value of a signal received by a mobile antenna with a 13° beamwidth traveling at a speed of 30 km/h?

11. On the basis of Eq. (6.50), find the expectation of $G^8(\gamma - \phi)$, $\langle G^8(\gamma - \phi) \rangle$, in terms of the signal envelope r_d, that will express $p_r(\phi)$.

$$p_r(\phi) = \pi \langle G^N(\gamma - \phi_i) \rangle \tag{P6.11.1}$$

References

1. W. C. Y. Lee and Y. S. Yeh, "On the Estimation of the Second-Order Statistics of Log Normal Fading in the Mobile Radio Environment," *IEEE Trans. Comm.*, vol. 22, no. 6, June 1974, pp. 869–873.

2. E. N. Gilbert, "Energy Reception for Mobile Radio," *Bell System Technical Journal,* vol. 44, October 1965, pp. 1779–1803.
3. W. C. Y. Lee, "Statistical Analysis of the Level Crossings and Duration of Fades of the Signal from an Energy Density Mobile Radio Antenna," *Bell System Technical Journal,* vol. 46, February 1967, pp. 417–448.
4. W. C. Y. Lee, "Finding the Approximate Angular Probability Density Function of Wave Arrival by Using a Directional Antenna," *IEEE Trans. Anten. Prop.,* vol. 21, no. 3, May 1973, pp. 328–334.
5. W. B. Davenport, Jr. and W. L. Rost, *Random Signals and Noise,* McGraw-Hill, New York, 1958, p. 151.
6. S. O. Rice, "Mathematical Analysis of Random Noise," *Bell System Technical Journal,* vol. 24, January 1945, pp. 46–156.
7. J. L. Lawson and G. E. Uhlenbeck, *Threshold Signals,* McGraw-Hill, New York, 1950, p. 62.
8. H. G. Booker, J. A. Ratcliff, and D. H. Shinn, "Diffraction from an Irregular Screen with Applications to Ionosphere Problems," *IEEE Trans. Royal Society (London),* vol. 262, sec. A., 1950, p. 579.
9. Edward A. Erdelyi, *Tables of Integral Transforms,* vol. 1, McGraw-Hill, New York, 1954, p. 46.
10. W. C. Y. Lee, "An Extended Correlation Function of Two Random Variables Applied to Mobile Radio Transmission," *Bell System Technical Journal,* vol. 48, December 1969, pp. 3423–3440.
11. W. C. Y. Lee, "A Study of the Antenna Array Configuration of an M-Branch Diversity-Combining Mobile Radio Receiver," *IEEE Trans. Veh. Tech.,* vol. 20, November 1971, pp. 93–104.
12. T. Aulin, "A Modified Model for the Fading Signal of a Mobile Radio Channel," *IEEE Trans. Veh. Tech.,* vol. 28, August 1979, pp. 183–203.
13. H. Hashemi, "Simulation of the Urban Radio Propagation Channel," *IEEE Trans. Veh. Tech.,* vol. 28, August 1979, pp. 213–225.
14. H. F. Schmid, "A Prediction Model for Multipath Propagation of Pulse Signals at VHF and UHF Over Irregular Terrain," *IEEE Trans. Anten. Prop.,* vol. 18, March 1970, pp. 253–258.
15. H. Suzuki, "A Statistical Model for Urban Radio Propagation," *IEEE Trans. Comm.,* vol. 25, July 1977, pp. 673–679.
16. G. L. Turin, et al., "A Statistical Model for Urban Multipath Propagation," *IEEE Trans. Veh. Tech.,* February 1972, pp. 1–8.
17. W. C. Jakes, *Microwave Mobile Communications,* Wiley, New York, 1974, p. 51.
18. P. A. Bello and B. D. Nelin, "The Effect of Frequency Selective Fading on the Binary Error Probabilities of Incoherent and Differentially Coherent Matched Filter Receivers," *IEEE Trans. Comm. Syst.,* vol. 11, June 1963, pp. 170–186.
19. W. C. Y. Lee, "Effects on Correlation between two Mobile Radio Base-Station Antennas," *IEEE Trans. Comm.,* vol. 21, November 1973, pp. 1214–1224.
20. G. E. P. Box and M. E. Muller, "A Note on the Generation of Random Normal Derivates," *American Mathematical Statistics,* vol. 29, 1958, pp. 610–611.
21. W. C. Y. Lee, "Theoretical and Experimental Study of the Properties of the Signal from an Energy Density Mobile Radio Antenna," *IEEE Vehicular Group Conference Record,* December 1–2, 1966, Montreal, Quebec, Canada, pp. 121–127.

Received-Signal Phase Characteristics

7.1 Random Variables Related to Mobile-Radio Signals

A signal $s_0(t)$ received at the mobile unit can be expressed:

$$s_0(t) = m(t)s(t) \qquad (7.1)$$

The long-term-fading factor $m(x)$ is extracted from $s_0(t)$ and the resultant can be expressed:

$$s(t) = r_0 e^{j\psi(t)} \qquad (7.2)$$

where $r_0(t)$ and $\psi(t)$ are the envelope and phase terms, respectively. The characteristics of $r_0(t)$ have been discussed in Chap. 6. The phase and time derivative of $\psi(t)$, $\dot{\psi}(t) = d\psi(t)/dt$, is the random FM that is described in this chapter.

Assume that $a_j(t)$ is the jth wave arrival, then $s(t)$ represents the sum of all wave arrivals, as has been shown in Eq. (6.10):

$$s(t) = \sum_{j=1}^{N} a_j(t) = X_1 + jY_1 \qquad (7.3)$$

where X_1 and Y_1 are as defined in Eqs. (6.11) and (6.12), respectively.

Hence $\psi_1(t)$ can be defined:

$$\psi_1(t) = \tan^{-1} \frac{Y_1}{X_1} \qquad (7.4)$$

The terms X_1 and Y_1 of the signal $s(t)$ are two independent Gaussian variables with zero mean and a variance of σ^2. This means that:

$$E[X_1 Y_1] = 0 \qquad (7.5)$$

and

$$E[X_1^2] = E[Y_1^2] = \sigma^2 \tag{7.6}$$

But when two signals $s_1 = s(t)$ and $s_2 = s(t + \tau)$, expressed

$$s_1(t) = X_1 + jY_1 = r_1 e^{j\psi_1} \tag{7.7}$$

and $\qquad s_2(t) = X_2 + jY_2 = r_2 e^{j\psi_2} \tag{7.8}$

are correlated, then $E[X_1X_2]$ and $E[X_1Y_2]$ are not necessarily zero. Consequently, it is first necessary to find the covariance matrix for these random variables, and then the characteristics of $\psi(t)$ and $\dot{\psi}(t)$ can be introduced.

Finding the covariance of random variables

Since the ergodic process is always applied to random variables in the mobile-radio environment, then Eq. (2.80) can also be used here:

$$E[s_1 s_2^*] = \mathcal{R}(\tau) = \lim_{T \to \infty} \frac{1}{2T} \int_{-T}^{T} s(t)s^*(t - \tau) \, dt \tag{7.9}$$

From Eq. (7.9), the following expressions are obtained:

$$R_c(\tau) = E[X_1X_2] = E[Y_1Y_2]$$

$$= \lim_{T \to \infty} \frac{1}{2T} \int_{-T}^{T} X(t)X(t - \tau) \, dt \tag{7.10}$$

$$R_s(\tau) = E[X_1Y_2] = -E[Y_1X_2]$$

$$= \lim_{T \to \infty} \frac{1}{2T} \int_{-T}^{T} X(t)Y(t - \tau) \, dt \tag{7.11}$$

Equations (7.10) and (7.11) have been used in Chap. 6 for calculating the covariances of random variables. Also $E[s_1 s_2^*]$ can be expressed as in Eq. (2.99):

$$E[s_1 s_2^*] = R(\tau) = \int_{-\infty}^{\infty} S(f)e^{j\omega\tau} \, df \tag{7.12}$$

where $S(f) \, df$ is the average power that lies in the frequency range f, $f + df$. $S(f)$ is often given in order to specify the spectrum of a given noise. Since the signals s_1 and s_2 shown in Eqs. (7.7) and (7.8) contain four random variables—X_1, Y_1, X_2, and Y_2—then the following expressions of covariances can be derived from Eq. (7.12):

$$R_c(\tau) = E[X_1 X_2] = E[Y_1 Y_2]$$

$$= 2 \int_0^\infty S(f) \cos 2\pi f\tau \, df \qquad (7.13)$$

$$R_{cs}(\tau) = E[X_1 Y_2] = -E[Y_1 X_2]$$

$$= 2 \int_0^\infty S(f) \sin 2\pi f\tau \, df \qquad (7.14)$$

Following Rice's notations [1],

$$R_c'(\tau) = E[X_1 \dot{X}_2] = E[Y_1 \dot{Y}_2] = -E[\dot{X}_1 X_2] = -E[\dot{Y}_1 Y_2] \qquad (7.15)$$

$$R_{cs}'(\tau) = E[X_1 \dot{Y}_2] = E[\dot{Y}_1 X_2] = -E[\dot{X}_1 Y_2] = -E[X_1 \dot{Y}_2] \qquad (7.16)$$

$$R_c''(\tau) = -E[\dot{X}_1 \dot{X}_2] = -E[\dot{Y}_1 \dot{Y}_2] \qquad (7.17)$$

$$R_{cs}''(\tau) = E[\dot{Y}_1 \dot{X}_2] = -E[\dot{X}_1 \dot{Y}_2] \qquad (7.18)$$

When $\tau = 0$, the moments can be found from these correlation functions:

$$b_n = (2\pi)^n \int_{-fm}^{fm} S(f) f^n \, df \qquad (7.19)$$

Thus,

$$b_0 = E[X_i^2] = E[Y_i^2] = R_c(0) = \sigma^2$$

$$E[X_i Y_i] = R_{cs}(0) = 0$$

$$E[X_i \dot{X}_i] = E[Y_i \dot{Y}_i] = R_c'(0) = 0 \qquad (7.20)$$

$$b_1 = E[X_i \dot{Y}_i] = -E[\dot{X}_i Y_i] = R_{cs}'(0)$$

$$b_2 = E[\dot{X}_i^2] = E[\dot{Y}_i^2] = -R_c''(0)$$

$$E[\dot{X}_i \dot{Y}_i] = -R_{cs}''(0) = 0$$

$$b_n = \begin{cases} (-1)^{n/2} R_c^{(n)}(0) & n \text{ even} \qquad (7.21) \\ (-1)^{(n+3)/2} R_{cs}^{(n)}(0) & n \text{ odd} \qquad (7.22) \end{cases}$$

Power spectra of a signal $s(t)$

$E[s_1 s_2^*]$ can also be obtained from $S(f)$ in Eq. (7.12). Sometimes it is much easier to calculate $E[s_1 s_2^*]$ by using $S(f)$ than by time-averaging from Eq. (7.9) as described in Chap. 6. Therefore, the expression of $S(f)$ must be determined.

Assume that all waves are traveling in the horizontal plane and that the antenna is pointing in a horizontal direction with an azimuthal antenna beam angle of γ. The power contributed to the the received signal by waves arriving in the horizontal plane within the angle $d\phi$ is the power arriving in that angular interval that would normally be received by an isotropic antenna of the same polarization:

$$S_s(\phi)\, d\phi = Ap(\phi)G^2(\phi - \gamma)\, d\phi \qquad (7.23)$$

where A is a constant that will be defined later on, $p(\phi)$ is the angular distribution of wave arrival, and $G(\phi - \gamma)$ is the antenna pattern. The power spectral density $S_s(f)$ is [2]

$$S_s(f) = S_s(\phi)\left|\frac{d\phi}{df}\right| \qquad (7.24)$$

and from Eq. (1.17), the Doppler frequency (denoted by f) is

$$f = \frac{V}{\lambda}\cos\phi = f_m \cos\phi \qquad (7.25)$$

where ϕ is the angle of wave arrival with respect to the direction of vehicle travel and f_m is the maximum Doppler shift frequency. Then, by taking the derivative of Eq. (7.25), the following is obtained:

$$d\phi = -\left[f_m\sqrt{1 - \left(\frac{f}{f_m}\right)^2}\right]^{-1} df \qquad (7.26)$$

Substituting Eq. (7.26) into Eq. (7.24) and combining the two angles $\pm\phi$ gives the Doppler shift f, and the power spectral density $S_s(f)$ of $s(t)$ is

$$S_s(f) = \frac{S_s(\phi) + S_s(-\phi)}{f_m\sqrt{1 - (f/f_m)^2}} \qquad (7.27)$$

where

$$\phi = \left|\cos^{-1}\frac{f}{f_m}\right| \qquad 0 \le \phi \le \pi \qquad (7.28)$$

The power spectral density $S_X(f)$ of the real part of $s(t)$ can be easily obtained:

$$S_X(f) = \frac{S_X(\phi) + S_X(-\phi)}{f_m\sqrt{1 - (f/f_m)^2}} \qquad (7.29)$$

where

$$S_X(\phi)\, d\phi = p(\phi)G^2(\phi - \gamma)\, d\phi \qquad (7.30)$$

and, to satisfy Eq. (7.23),

$$A = 1 \tag{7.31}$$

The power spectral density expressed in Eq. (7.29) is that of the real part X of signal $s(t)$. The power spectral density $S_Y(f)$ of the imaginary part Y of signal $s(t)$ then becomes

$$S_Y(f) = S_X(f) \tag{7.32}$$

On the basis of the preceding calculations, the average power $S(f)$ can be rewritten:

$$S(f) = S_Y(f) = S_X(f)$$

$$= \frac{p(\phi)G^2(\phi - \gamma) + p(-\phi)G^2(-\phi - \gamma)}{f_m\sqrt{1 - (f/f_m)^2}}$$

$$= \frac{G^2(\phi - \gamma)}{\sqrt{f_m^2 - f^2}} \, [p(\phi) + p(-\phi)] \tag{7.33}$$

In most instances where the antenna patterns $G(\phi - \gamma)$ are symmetrical around γ, the following relationship holds true:

$$G^2(\phi - \gamma) = G^2(-\phi - \gamma) \tag{7.34}$$

The average power relationships expressed in Eq. (7.33) are used repeatedly in subsequent discussions.

The following examples are based on the assumption that all angles of incident waves are uniformly distributed:

$$p(\phi) = p(-\phi) = \frac{1}{2\pi} \tag{7.35}$$

1. Vertical monopole, $G^2(\phi - \gamma) = \frac{3}{2}$:

$$S_s(f) = \frac{3}{2\pi\sqrt{f_m^2 - f^2}}$$

$$S(f) = S_X(f) = S_Y(f) = \frac{1}{2}S_s(f) \tag{7.36}$$

2. Vertical loop in a plane perpendicular to vehicle motion, $\gamma = 90°$ in Fig. 6.5:

$$G^2(\phi - \gamma) = \frac{3}{2}\cos^2(\phi - \gamma) \tag{7.37}$$

Then

$$S_s(f) = \frac{3}{2\pi f_m^2} \sqrt{f_m^2 - f^2} \tag{7.38}$$

3. Vertical loop in plane of vertical motion, $\gamma = 0$, as shown in Fig. 6.5:

$$G^2(\phi - \gamma) = \tfrac{3}{2} \cos^2 (\phi - \gamma) \tag{7.39}$$

Then

$$S_s(f) = \frac{3f^2}{2\pi f_m^2 \sqrt{f_m^2 - f^2}} \tag{7.40}$$

Equations (7.36), (7.38), and (7.40) are plotted in Fig. 7.1 [2].

The covariance $E[s_1 s_2^*]$ can be obtained by Eq. (7.12), and the average power is

$$E[s_1 s_1^*] = R(0) = 2 \int_0^\infty S(f)\, df \tag{7.41}$$

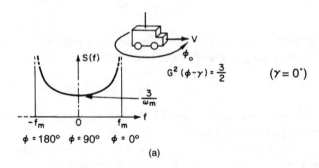

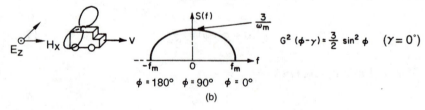

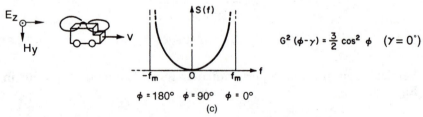

Figure 7.1　Power spectral density functions for different field components.

Example 7.1 Mobile reception of vertically polarized FM transmissions is affected by the power spectral density, the angular distribution of wave arrival, and various random FM variables that affect correlation-coefficient responses. Other factors, such as carrier frequency and vehicular speed, must be taken into consideration, since they too can have an adverse affect on reception. Doppler effects and click noise are typical symptoms of poor FM reception in the mobile-radio environment.

Assuming that the vertically polarized antenna beam is directed along the path of a vehicle in motion ($\gamma = 0$) where

$$G^2(\phi) = \begin{cases} G & \text{for } -\dfrac{\Phi}{2} \le \phi \le \dfrac{\Phi}{2} \quad \text{and} \quad \pi - \dfrac{\Phi}{2} \le \phi \le \pi + \dfrac{\Phi}{2} \\ 0 & \text{otherwise} \end{cases} \tag{E7.1.1}$$

and assuming that the angular distribution of wave arrival is uniformly distributed:

$$p(\phi) = \frac{1}{2\pi} \tag{E7.1.2}$$

then what is the range of the multipath frequency?

Figure E7.1.1 illustrates the direction of the vehicle in motion with respect to the angular arrival of the vertically polarized wave. Figure E7.1.2 illustrates the parameters for determining the range of the multipath frequency, where $f_1 \le f_d \le f_2$.

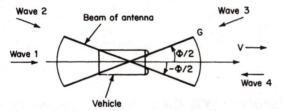

Figure E7.1.1   Model for analysis, Example 7.1.

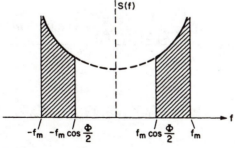

Figure E7.1.2   Parameters for determining multipath frequency range.

solution From Eq. (7.25), the Doppler frequency f is:

$$f = \frac{V}{\lambda} \cos \phi = f_m \cos \phi \qquad 0 \leq \phi \leq \pi \tag{E7.1.3}$$

Therefore, based on the parameters of Fig. E7.1.1, the range of the multipath Doppler frequency can be determined as follows:

1. When an incoming wave arrives from an angle of 0 (see wave 1 in Fig. E7.1.1),

$$f = f_m = \frac{V}{\lambda} \cos 0 = \frac{V}{\lambda} \tag{E7.1.4}$$

2. When an incoming wave arrives from $\Phi/2$,

$$f = f_m \cos \frac{\Phi}{2} \tag{E7.1.5}$$

3. When an incoming wave arrives from $180° - \Phi/2$,

$$f = f_m \cos \left(\pi - \frac{\Phi}{2} \right) = -f_m \cos \frac{\Phi}{2} \tag{E7.1.6}$$

4. When an incoming wave arrives from $180°$,

$$f = f_m \cos \pi = -f_m \tag{E7.1.7}$$

5. Hence, the range of the Doppler frequency is

$$-f_m \leq f_d \leq -f_m \cos \frac{\Phi}{2}$$

$$+f_m \cos \frac{\Phi}{2} \leq f_d \leq f_m \tag{E7.1.8}$$

Covariance matrix

A covariance matrix is always associated with a joint probability density function. If the given random process is real, the joint probability density function of N random variables x_n in general can be expressed:

$$p(x_1, \ldots, x_N) = \frac{\exp \left[-\frac{1}{2|\Lambda|} \sum_{m=1}^{N} \sum_{n=1}^{N} |\Lambda|_{nm}(x_n - m_n)(x_m - m_m) \right]}{(2\pi)^{N/2} |\Lambda|^{1/2}} \tag{7.42}$$

where the set of x_n is defined

$$\{x_n\} = \{X_1, Y_1, X_2, Y_2, \ldots, X_n, Y_n\}$$

and where

$$m_n = E[x_n] \tag{7.43}$$

The elements of the covariance matrix $[\Lambda]$ are:

$$[\Lambda] = \begin{bmatrix} \lambda_{11} & \lambda_{12} & \cdots & \cdots & \cdots & \lambda_{1N} \\ \lambda_{21} & \cdots & \cdots & \cdots & \cdots & \cdots \\ \cdots & \cdots & \cdots & \cdots & \cdots & \cdots \\ \lambda_{N1} & \lambda_{N2} & \cdots & \cdots & \cdots & \lambda_{NN} \end{bmatrix} \quad (7.44)$$

where

$$\begin{aligned} \lambda_{nm} &= E[(x_n - m_n)(x_m - m_m)] \\ &= E[x_n x_m] - m_n E[x_m] - m_m E[x_n] + m_n m_m \\ &= E[x_n x_m] - m_n m_m \end{aligned}$$

$|\Lambda|_{nm}$ is the cofactor of the element λ_{nm} in the determinant $|\Lambda|$ of the covariance matrix. In this case, $N = 4$ and the mean value of x_n is $E[X_i] = E[Y_i] = 0$, and the covariance matrix can be expressed:

$$\begin{aligned} [\Lambda] &= \begin{bmatrix} E[X_1 X_1] & E[X_1 Y_1] & E[X_1 X_2] & E[X_1 Y_2] \\ E[Y_1 X_1] & E[Y_1 Y_1] & E[Y_1 X_2] & E[Y_1 Y_2] \\ E[X_2 X_1] & E[X_2 Y_1] & E[X_2 X_2] & E[X_2 Y_2] \\ E[Y_2 X_1] & E[Y_2 Y_1] & E[Y_2 X_2] & E[Y_2 Y_2] \end{bmatrix} \\ &= \begin{bmatrix} \sigma_1^2 & 0 & R_c(B, \tau) & R_{cs}(B, \tau) \\ 0 & \sigma_1^2 & -R_{cs}(B, \tau) & R_c(B, \tau) \\ R_c(B, \tau) & -R_{cs}(B, \tau) & \sigma_2^2 & 0 \\ R_{cs}(B, \tau) & R_c(B, \tau) & 0 & \sigma_2^2 \end{bmatrix} \quad (7.45) \end{aligned}$$

where $R_c(B, \tau)$ is the correlation of X_1 and X_2 separated by a frequency B and a time τ, and where $R_{cs}(B, \tau)$ is the correlation of X_i and Y_j separated by a frequency B and a time τ, and where $i \neq j$.

For the case where there is no frequency separation, $B = 0$, $R_c(0, \tau) = R_c(\tau)$, and $R_{cs}(0, \tau) = R_{cs}(\tau)$.

When the matrix elements of Eq. (7.45) have been obtained, then the cofactors of the covariance matrix can be found and inserted into Eq. (7.42); thus, the joint probability density function for a given set of random variables is acquired.

7.2 Phase-Correlation Characteristics

Phase-correlation characteristics of $\psi(t)$

There are several characteristics that can be used to describe $\psi(t)$.

The terms X and Y in Eq. (7.4) are the real and imaginary components of the mobile-radio signal $s(t)$. These two components are Gaussian relationships, as shown in Eqs. (7.5) and (7.6). Then, the distribution of the random process $\psi(t)$ can be found from Eq. (2.40):

$$p(\psi) = \begin{cases} \dfrac{1}{2\pi} & 0 \le \psi \le 2\pi \\ 0 & \text{otherwise} \end{cases} \tag{7.46}$$

Hence, ψ is uniformly distributed, as shown in Fig. 7.2.

When there are two phase angles, ψ_1 and ψ_2, of the signals s_1 and s_2, then the distribution of ψ_1 and ψ_2, assuming the two signals are correlated, is:

$$p(\psi_1, \psi_2) = \int_0^\infty \int_0^\infty p(r_1, r_2, \psi_1, \psi_2)\, dr_1\, dr_2 \tag{7.47}$$

where $p(r_1, r_2, \psi_1, \psi_2)$ is similar to, but not the same as, the form shown in Eq. (6.88).

Assuming that

$$E[X_1^2] = E[Y_1^2] = E[X_2^2] = E[Y_2^2] = \sigma^2 \tag{7.48}$$

then $p(r_1, r_2, \psi_1, \psi_2)$ can be derived from $p(X_1, Y_1, X_2, Y_2)$, which was developed in Eq. (7.42) and can be written as follows:

$$p(X_1, Y_1, X_2, Y_2) = \frac{1}{4\pi^2 |\Lambda|^{1/2}}$$

$$\times \exp\left\{ -\frac{1}{2|\Lambda|^{1/2}} \left[\begin{array}{l} \sigma^2(X_1^2 + Y_1^2 + X_2^2 + Y_2^2) \\ -2R_c(\tau)(X_1X_2 + Y_1Y_2) \\ -2R_{cs}(\tau)(X_1Y_2 - Y_1X_2) \end{array} \right] \right\} \tag{7.49}$$

The jacobian transformation based on Eq. (2.49) is

$$|J| = \frac{\partial(X_1, Y_1, X_2, Y_2)}{\partial(r_1, \psi_1, r_2, \psi_2)} = r_1 r_2 \tag{7.50}$$

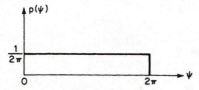

Figure 7.2 Uniform distribution of the random process $p(\psi)$.

It therefore follows that

$$p(r_1, \psi_1, r_2, \psi_2) = |J| p(X_1, Y_1, X_2, Y_2)$$

$$= \begin{cases} \dfrac{r_1 r_2}{4\pi^2 |\Lambda|^{1/2}} \exp\left\{-\dfrac{1}{2|\Lambda|^{1/2}}\begin{bmatrix} \sigma^2(r_1^2 + r_2^2) \\ -2R_c(\tau)r_1 r_2 \cos(\psi_2 - \psi_1) \\ -2R_{cs}(\tau)r_1 r_2 \sin(\psi_2 - \psi_1) \end{bmatrix}\right\} \\ \qquad\qquad\qquad\qquad r_1, r_2 > 0 \text{ and } 0 \le \psi_1, \psi_2 \le 2\pi \\ 0 \qquad\qquad\qquad\qquad \text{otherwise} \end{cases} \qquad (7.51)$$

Substituting Eq. (7.51) into Eq. (7.47) gives the following [3]:

$$p(\psi_1, \psi_2) = \begin{cases} \dfrac{|\Lambda|^{1/2}}{4\pi^2\sigma^4}\left[\dfrac{(1-Q)^{1/2} + (\pi - \cos^{-1} Q)}{(1-Q^2)^{3/2}}\right] & 0 \le \psi_1, \psi_2 \le 2\pi \\ 0 & \text{otherwise} \end{cases} \qquad (7.52)$$

where

$$Q = \frac{R_c(\tau)}{\sigma^2} \cos(\psi_2 - \psi_1) + \frac{R_{cs}(\tau)}{\sigma^2} \sin(\psi_2 - \psi_1) \qquad (7.53)$$

If there is no correlation between s_1 and s_2, then $R_c(\tau)$ and $R_{cs}(\tau)$ are zero. Thus, $Q = 0$, from Eq. (7.53), and $|\Lambda|^{1/2} = \sigma^4$, from Eq. (7.45). When these two values are inserted into Eq. (7.52), $p(\psi_1, \psi_2)$ becomes:

$$p(\psi_1, \psi_2) = \frac{1}{4\pi^2} = p(\psi_1)p(\psi_2) \qquad (7.54)$$

which is what is normally expected.

Phase correlation between frequency and time separation

In expressing the phase correlation between frequency separations $\Delta\omega$ and time separations τ, the phase correlation can be expressed:

$$R_\psi(\Delta\omega, \tau) = \int_0^{2\pi} \int_0^{2\pi} \psi_1 \psi_2 p(\psi_1, \psi_2)\, d\psi_1\, d\psi_2 \qquad (7.55)$$

where $p(\psi_1, \psi_2)$ is as shown in Eq. (7.52) and the covariance matrix elements are obtained from Eq. (7.44), but with Q expressed differently from Eq. (7.53), as follows:

$$Q = \frac{R_c(\Delta\omega, \tau)}{\sigma^2} \cos(\psi_2 - \psi_1 - \phi) + \frac{R_{cs}(\Delta\omega, \tau)}{\sigma^2} \sin(\psi_2 - \psi_1 - \phi) \qquad (7.56)$$

where

$$\phi = \tan^{-1}(-\Delta\omega \, \Delta) \tag{7.57}$$

$$R_c(\Delta\omega, \tau) = \sigma^2 \left[\frac{J_0^2(\omega_m\tau)}{1 + (\Delta\omega)^2\Delta^2} \right]^{1/2} \tag{7.58}$$

and $$R_{cs}(\Delta\omega, \tau) = 0$$

can be obtained from Eqs. (6.123) through (6.125). The integral of Eq. (7.55) cannot be carried out exactly, but an approximation can be shown [4]:

$$R_\psi(\Delta\omega, \tau) = \pi^2 \left[1 + \Gamma(\rho_c, \phi) + 2\Gamma^2(\rho_c, \phi) + \frac{1}{24}\,\Omega(\rho_c) \right] \tag{7.59}$$

where $\rho_c = \rho_c(\Delta\omega, \tau)$ is the normalized correlation $\rho_c = R_c(\Delta\omega, \tau)/\sigma^2$, which equals $\sqrt{\rho_r}$ in Eq. (6.126), expressed as:

$$\Gamma(\rho_c, \phi) = \frac{1}{2\pi}\sin^{-1}(\rho_c \cos \phi) \tag{7.60}$$

and

$$\Omega(\rho_c) = \frac{6}{\pi^2}\sum_{n=1}^{\infty}(\rho_c)^{2n}\frac{1}{n^2} \qquad \Omega(1) = 1 \tag{7.61}$$

The function $\Omega(\rho_c)$ is plotted in Fig. 7.3. Equation (7.59) can be solved by using the information contained in Fig. 7.3.

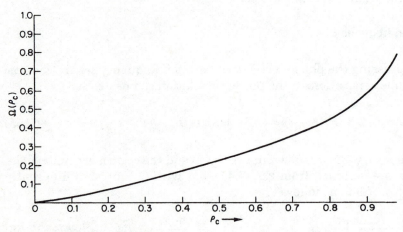

Figure 7.3 Plot of function $\Omega(\rho_c)$.

If it is assumed that $p(\psi) = 1/2\pi$, then the mean average values of ψ_1 and ψ_2 are:

$$\langle \psi_1 \rangle = \langle \psi_2 \rangle = \pi \quad \text{and} \quad \langle \psi_1^2 \rangle = \langle \psi_2^2 \rangle = \frac{4\pi^2}{3} \tag{7.62}$$

and therefore the correlation coefficient can be expressed

$$\rho_\psi(\Delta\omega, \tau) = \frac{R_\psi - E[\psi_1]E[\psi_2]}{E[\psi_1^2] - E^2[\psi_1]} = \frac{3}{\pi^2}\,[R_\psi(\Delta\omega, \tau) - \pi^2] \tag{7.63}$$

The function of Eq. (7.63) is plotted in Fig. 7.4.

The coherence bandwidth $B_c = (\Delta\omega)_c/2\pi$ at values of $\tau = 0$ and $\rho_\psi = 0.5$ can be found from the data given in Fig. 7.4, where:

$$\rho_\psi(B_c, 0) = 0.5 \tag{7.64}$$

and

$$B_c = \frac{1}{4\pi\Delta} \tag{7.65}$$

The coherence bandwidth, according to the phase correlation, is one-half of the envelope correlation value previously shown in Eq. (6.128).

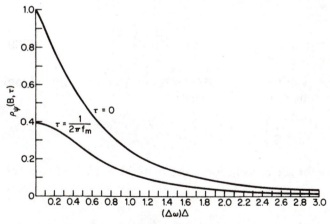

Figure 7.4 Plot of phase correlation coefficient vs. the product of frequency separation and time-delay spread.

7.3 Characteristics of Random FM

Probability distribution of random FM

The term $p(r, \dot{r}, \psi, \dot{\psi})$ has been defined in Eq. (6.61), and therefore the probability density function (pdf) of a random FM signal $\dot{\psi}$, $p(\dot{\psi})$, can be derived from $p(r, \dot{r}, \psi, \dot{\psi})$ in combination with the following integrals:

$$p(\dot{\psi}) = \int_0^\infty dr \int_{-\infty}^\infty d\dot{r} \int_0^{2\pi} p(r, \dot{r}, \psi, \dot{\psi}) \, d\psi$$

$$= \frac{\sigma}{2v} \left(1 + \frac{\sigma^2}{v^2} \dot{\psi}^2 \right)^{-3/2} \tag{7.66}$$

From Eq. (6.15) and Eq. (6.55), σ^2 and v^2 for an E field signal are expressed:

$$\sigma^2 = E[X^2] = N \tag{7.67}$$

and

$$v^2 = E[\dot{X}^2] = E[\dot{Y}^2] = \frac{(\beta V)^2}{2} N \tag{7.67}$$

Then,

$$p(\dot{\psi}) = \frac{1}{\sqrt{2}\beta V} \left[1 + \frac{2}{(\beta V)^2} \dot{\psi}^2 \right]^{-3/2} \tag{7.68}$$

The correlation coefficient $p(\dot{\psi})$ of Eq. (7.68) is plotted in Fig. 7.5(a).

The distribution of random FM can be expressed:

$$P(\dot{\psi} \le \dot{\Psi}) = \int_{-\infty}^{\dot{\Psi}} p(\dot{\psi}) \, d\dot{\psi}$$

$$= \frac{1}{2} \left[1 + \frac{\sigma}{v} \dot{\Psi} \left(1 + \frac{\sigma^2}{v^2} \dot{\Psi} \right)^{-1/2} \right] \tag{7.69}$$

By substituting the results of Eq. (7.67) for σ and v of an E field signal into Eq. (7.69), the following is obtained:

$$P(\dot{\psi} \le \dot{\Psi}) = \frac{1}{2} \left[1 + \frac{\sqrt{2}}{\beta V} \dot{\Psi} \left(1 + \frac{2}{(\beta V)^2} \dot{\Psi}^2 \right)^{-1/2} \right] \tag{7.70}$$

The response curve for the correlation coefficient for $P(\dot{\psi} \le \dot{\Psi})$ is plotted in Fig. 7.5(b).

Power spectrum of random FM

The power spectrum of random FM can be obtained from the autocorrelation function described by Rice [1] and expressed:

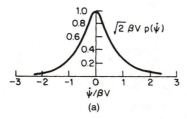

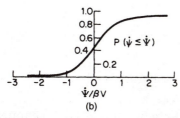

Figure 7.5 (a) Characteristics of $p(\dot\psi)$, Eq. (7.68); (b) characteristics of $P(\dot\psi \le \dot\Psi)$, Eq. (7.70).

$$R_\psi(\tau) = E[\psi(t)\psi(t-\tau)]$$

$$= -\frac{1}{2}\left\{\left[\frac{R_c'(\tau)}{R_c(\tau)}\right]^2 - \left[\frac{R_c''(\tau)}{R_c(\tau)}\right]\right\}\ln\left\{1 - \left[\frac{R_c(\tau)}{R_c(0)}\right]^2\right\} \qquad (7.71)$$

Assume that the angular wave arrival is uniformly distributed, that is, that $p(\phi_i) = 1/2\pi$. Then the expression of $R_c(\tau)$ in Eq. (7.71) can be found from Eq. (6.108) for the case of the E_z field. The first and second derivatives, $R_c'(\tau)$ and $R_c''(\tau)$, can be obtained as follows:

$$\frac{R_c(\tau)}{R_c(0)} = \frac{E[X_1 X_2]}{E[X_1^2]} = J_0(\beta V\tau) \qquad (7.72)$$

$$\frac{R_c'(\tau)}{R_c(\tau)} = \frac{1}{R_c(\tau)}\frac{dR_c(\tau)}{d\tau} = -\beta V\frac{J_1(\beta V\tau)}{J_0(\beta V\tau)} \qquad (7.73)$$

$$\frac{R_c''(\tau)}{R_c(\tau)} = \frac{1}{R_c(\tau)}\frac{d^2 R_c(\tau)}{d\tau^2} = (\beta V)^2\left[\frac{J_1(\beta V\tau)}{\beta V\tau J_0(\beta V\tau)} - 1\right] \qquad (7.74)$$

The power spectrum can be expressed as the Fourier transform of $R_\psi(\tau)$ as follows:

$$S_\psi(f) = \int_{-\infty}^{\infty} R_\psi(\tau)e^{-j\omega\tau}\,d\tau = 2\int_0^{\infty} R_\psi\cos\omega\tau\,d\tau \qquad (7.75)$$

Equation (7.75) can be integrated either approximately or numerically. Figure 7.6 shows the curve obtained by integrating the autocorrelation function over three regions in the time domain [2].

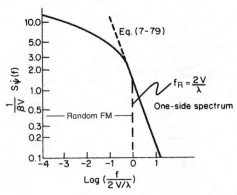

Figure 7.6 Power spectrum of random FM.

For mobile speeds of up to 60 mi/h at UHF channel frequencies, and with an audio band of from 300 Hz to 3000 Hz, an asymptotic form can be used as follows [2]:

$$S_{\dot\psi}(f) = \frac{v^2/\sigma^2 - v_{12}^4/\sigma^4}{f} \tag{7.76}$$

On the basis of the model shown in Fig. 6.3 and described in Sec. 6.2, the following E field signal relationships are true:

$$\sigma^2 = E[X_1^2] = E[Y_1^2] = E[X_2^2] = E[Y_2^2] = N \tag{7.77}$$

$$v^2 = E[\dot X_1^{\,2}] = E[\dot Y_1^2] = E[\dot X_2^2] = E[\dot Y_2^2] = \frac{(\beta V)^2}{2} N \tag{7.78}$$

$$v_{12}^2 = E[X_1 \dot Y_2] = -E[X_2 \dot Y_1] = 0$$

and therefore

$$S_{\dot\psi}(f) = \frac{(\beta V)^2}{2f} \tag{7.79}$$

The corresponding baseband output noise due to random FM in an audio band (W_1, W_2) is:

$$E[\dot\psi^2] = \int_{W_1}^{W_2} S_{\dot\psi}(f)\, df = \frac{(\beta V)^2}{2} \ln \frac{W_2}{W_1} \tag{7.80}$$

Hence, the average power of the random FM is a function of vehicular speed V and the audio bandwidth (W_1, W_2).

In Fig. 7.6, the power spectrum of the random FM drops very drastically as soon as reaching the frequency

$$f_m = \frac{2V}{\lambda}$$

Therefore, we can use this criterion to calculate the impact of the random FM versus the mobile speed. The criterion of considering the impact of random FM is

$$\text{Impact of random FM} > f_R = \frac{2V}{\lambda} \tag{7.81}$$

The signal will be distorted by the random FM if Eq. (7.81) is met. When $V = 0$, the random FM disappears. In the data transmission, we have to transmit the data rate higher than f_R. Thus, f_R is the lower limit. For the voice transmission, as long as the random FM does not exceed 300 Hz, the low-pass filter will filter it out.

Example 7.2 Assuming that the carrier frequency is 850 MHz and the audio band is 300 to 3000 Hz, what is the average power ratio between two random FM signals, one received while the mobile unit is traveling at 30 mi/h, and the other at 15 mi/h?

solution The baseband receiver output noise due to random FM is found from Eq. (7.80):

$$N_{\text{rfm}} = \frac{(\beta V)^2}{2} \ln \frac{W_2}{W_1} \tag{E7.2.1}$$

Let $V_1 = 30$ mi/h and $V_2 = 15$ mi/h, where N_{rfm_1} is due to V_1 and N_{rfm_2} is due to V_2. Then the following expression is valid:

$$\frac{N_{\text{rfm}_1}}{N_{\text{rfm}_2}} = \frac{V_1^2}{V_2^2} = \left(\frac{30}{15}\right)^2 = 4 \approx 6 \text{ dB} \tag{E7.2.2}$$

Therefore, an average power ratio of 6 dB per octave between the two random FM signals is found when the velocity of the mobile receiver is changing.

Level-crossing rate (LCR) of random FM

In order to obtain the lcr of random FM, it is first necessary to find

$$p(r, \dot{r}, \ddot{r}, \psi, \dot{\psi}, \ddot{\psi}) = |J| p(X, \dot{X}, \ddot{X}, Y, \dot{Y}, \ddot{Y}) \tag{7.82}$$

where $\dot{r}$ and $\dot{\psi}$ are first derivatives and $\ddot{r}$ and $\ddot{\psi}$ are second derivatives with respect to time. From Eq. (7.81) the following expression is derived:

$$p(r, \psi, \dot{\psi}, \ddot{\psi}) = \int_{-\infty}^{\infty} d\dot{r} \int_{-\infty}^{\infty} p(r, \dot{r}, \ddot{r}, \psi, \dot{\psi}, \ddot{\psi}) \, d\ddot{r}$$

The lcr of the random FM can then be found [1]:

$$N(\psi = \dot{\Psi}) = \int_{-\pi}^{\pi} d\psi \int_{0}^{\infty} dr \int_{0}^{\infty} \dot{\psi} p(r, \psi, \dot{\Psi}, \dot{\psi}) \, d\dot{\psi}$$

$$= \frac{\sqrt{b_4/b_2 - b_2/b_0 + 4\dot{\Psi}^2}}{2\pi(1 + (b_0/b_2)\,\dot{\Psi}^2)} \tag{7.83}$$

where b_0, b_2, and b_4 are as shown in Eqs. (7.20) through (7.22).

For the case where the noise is constant, with a value of η across the frequency band from $f - \beta/2$ to $f + \beta/2$, then the terms b_0, b_2, and b_4 can be found by applying Eq. (7.19), as follows:

$$b_0 = \beta\eta$$

$$b_2 = \frac{\pi^2\beta^2 b_0}{3}$$

$$b_4 = \frac{\pi^4\beta^4 b_0}{5}$$

By substituting the above values into Eq. (7.83), the following expression is obtained:

$$N(\psi = \dot{\Psi}) = \frac{\beta}{1 + z^2} \frac{(1 + 5z^2)^{1/2}}{15} \tag{7.84}$$

where

$$z = \sqrt{\frac{b_0}{b_2}} \, \dot{\Psi} = \frac{\sqrt{3}\,\dot{\Psi}}{\beta\pi} \tag{7.85}$$

The function of Eq. (7.84) is plotted in Fig. 7.7, where a peak value is shown at $z = 0.8$ or at $\dot{\Psi} = 1.45\beta$.

Example 7.3 Assuming a noise bandwidth of 20 kHz, what is the maximum level-crossing rate for the random FM signal, and what is the reference level at which the number of crossings are counted?

solution Based on the data plotted in Fig. 7.7, the maximum lcr is:

$$\frac{N(\dot{\Psi})}{\beta} = 0.322 \qquad \text{at} \qquad \dot{\Psi} = 1.45\beta \tag{E7.3.1}$$

or

$$N(\dot{\Psi}) = 6.44 \times 10^3 \text{ s}^{-1} \qquad \text{at} \qquad \dot{\Psi} = 2.9 \times 10^4 \tag{E7.3.2}$$

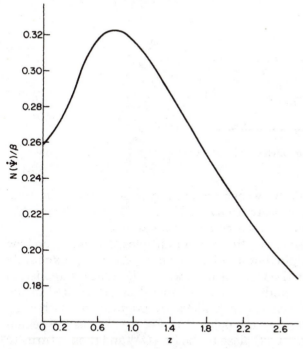

Figure 7.7   Level-crossing rate of random FM.

7.4   Click-Noise Characteristics

As the mobile unit moves through a mobile-radio field, the received signal fluctuates in both amplitude r and phase ψ. In a conventional FM discriminator, the fluctuations in phase ψ due to multipath fading generate noise in the signal. Such noise is called "random FM noise." The random FM noise is essentially concentrated at frequencies less than 2 to 3 times the maximum Doppler frequency. Outside this range the power spectrum falls off at a rate of $1/f$. Davis [5] has found that random FM noise due to fading can be considered a secondary effect in the reception of mobile-radio signals.

Another kind of noise, called "click noise," may be found at the output of the FM discriminator.

Click noise

To understand the click-noise phenomenon described by Rice [6], refer to Fig. 7.8, which shows the signal phasor construction of $\tilde{R} = Re^{j\psi_R}$. Phasor $\tilde{a} = re^{j\psi_r}$ represents the fading signal, and phasor $\tilde{n} = ne^{j\psi_n}$ corresponds to the additive noise. Since the fading is very slow compared with the noise, the resultant vector $\tilde{R}$ appears to rotate rapidly about

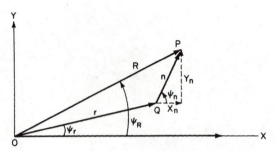

Figure 7.8 Signal phasor plot of click-noise characteristics.

point Q, and it occasionally sweeps around the origin when $r < n$, thus
causing ψ_R to increase or decrease by 2π. Figure 7.9 shows the mechanism by which such excursions produce impulses in ψ_R.

   The resultant impulses have different amplitudes, depending on how
close P comes to the origin, but all impulses have areas approximately
equal to $\pm 2\pi$ rad. When the waveform shown in Fig. 7.9(b) is applied to
a low-pass filter, corresponding but wider impulses are excited in the
output and are heard as clicks. Such clicks are produced only when ψ_R
changes by $\pm 2\pi$. When $r > n$, the point P, in Fig. 7.8, leaves the region
Q, cuts across the segment $\overline{OQ}$ close to the line $\overline{OX}$, and then returns to
the region Q. As ψ_R rapidly changes by approximately $\pm\pi$ during the
sweep past line $\overline{OX}$, the resulting pulse ψ_R has little low-frequency content and therefore produces only a small excursion, which is different
from the condition of $r < n$. In essence, click noise is closely associated
with the threshold behavior of the FM discriminator when it is
responding to multipath fading.

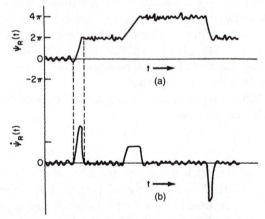

Figure 7.9 Waveforms for phasor function, showing
click noise.

Above the threshold, it can be assumed that the fading signal is much larger than the additive noise, so that most of the time $n \ll r$. The vector $\tilde{R}$ is rotating rapidly around the point Q most of the time. Occasionally, the vector $\tilde{R}$ (or point P) may encircle the origin and generate clicks; however, the potential for generating clicks becomes less as the amplitude of r increases. The vector $\tilde{R}$, rotating rapidly about the point Q but never encircling the origin, provides a term ψ_s such that the total ψ_R can be written:

$$\psi_R = \psi_s + 2\pi(Z_+ - Z_-) \tag{7.86}$$

where the term $2\pi(Z_+ - Z_-)$ represents an additional factor necessary to account for the positive and negative clicks in the time interval T that are present at, above, and below threshold. The terms Z_+ and Z_- are the sums of positive and negative impulses, respectively, that occur at random times and are assumed to be independent of one another. They are regarded as Poisson processes associated with average click rates. The average positive click rate can be expressed:

$$N_+ = \frac{Z_+}{T} \tag{7.87}$$

and the average negative click rate can be expressed as:

$$N_- = \frac{Z_-}{T} \tag{7.88}$$

Furthermore, the click rates are also assumed to be independent of ψ_s. When the conditions for the generation of a click occur, i.e., when

$$n > r$$

and $\qquad\qquad \pi < \psi_n < \pi + \Delta\psi$ $\qquad\qquad$ (7.89)

the average click rate can be expressed [7]:

$$N_+ = N_-$$
$$= \frac{1}{T} \int_0^T dt \int_0^\infty p(r)\, dr \int_r^\infty dn \int_0^\infty \dot\psi_n\, p(n,\, \psi_n = \pi,\, \dot\psi_n)\, d\dot\psi_n \tag{7.90}$$

where

$$p(n,\, \psi_n,\, \dot\psi_n) = \int_{-\infty}^\infty p(n,\, \dot n,\, \psi_n,\, \dot\psi_n)\, d\dot n \tag{7.91}$$

Note that Eq. (7.90) is similar to Eq. (2.72), the expression for the level-crossing rate. To solve Eq. (7.91), it is first necessary to obtain a value for $p(n,\, \dot n,\, \psi_n,\, \dot\psi_n)$.

Procedure for obtaining $p(n, \dot{n}, \psi_n, \dot{\psi}_n)$

The noise phasor as shown in Fig. 7.8 can be expressed:

$$\tilde{n} = n e^{j\psi_n} = X_n + jY_n \qquad (7.92)$$

Since X_n and Y_n are Gaussian noise terms, the joint probability density function of X_n and Y_n and their derivatives $p(X_n, Y_n, \dot{X}_n, \dot{Y}_n)$ can be found by transforming the variables $p(n, \dot{n}, \psi_n, \dot{\psi}_n)$ as follows:

$$p(n, \dot{n}, \psi_n, \dot{\psi}_n) = |J| p(X_n, Y_n, \dot{X}_n, \dot{Y}_n) \qquad (7.93)$$

Average click rate

The term $p(n, \dot{n}, \psi_n, \dot{\psi}_n)$ of Eq. (7.93) can be found from Eq. (6.61) with σ_n^2 and v_n^2 specified for the noise terms instead of the fading signal and can be written:

$$\sigma_n^2 = E[X_n^2] = E[Y_n^2] \qquad (7.94)$$

and

$$v^2 = E[\dot{X}_n^2] = E[\dot{Y}_n^2] \qquad (7.95)$$

Therefore, $p(n, \psi_n, \dot{\psi}_n)$ can be obtained by inserting Eq. (7.93) into Eq. (7.91):

$$p(n, \psi_n, \dot{\psi}_n) = \frac{n \exp(-n^2/2\sigma_n^2)}{2\pi\sigma_n^2} \frac{\exp(-\dot{\psi}_n^2/2v_n^2)}{\sqrt{2\pi v_n^2}} \qquad (7.96)$$

In actual practice, the Rayleigh-fading distribution is truncated at some point, because when the output receiver noise exceeds some predetermined value, the circuit will be either cut off or switched to a better path. Assuming that a minimum level of r_0 is set, based on the receiver noise, when the signal envelope $r(t)$ is held above the minimum level r_0, no switching will occur. The probability density function of r therefore follows a truncated Rayleigh curve density as shown in Fig. 7.10:

$$p(r) = (1 - e^{-r_2^0/2\sigma^2})\delta(r - r_0) + \frac{r}{\sigma^2} e^{-r^2/2\sigma^2} u(r - r_0) \qquad (7.97)$$

where $\delta(r - r_0)$ is an impulse function and $u(r - r_0)$ is a unit-step function. The term σ^2 is the variance of r.

Substituting Eq. (7.96) and Eq. (7.97) into Eq. (7.90) gives the following:

$$N_+ = N_- = \frac{v_n^2}{\sigma_n^2} \frac{1}{8\pi(\sigma^2/\sigma_n^2)} \left\{ \mathrm{erfc}\left[\frac{1}{\sqrt{2}}\left(\frac{r_0}{\sigma}\right)\left(\frac{\sigma}{\sigma_n}\right)\right] \right.$$
$$\left. + \sqrt{\frac{2}{\pi}}\left(\frac{r_0}{\sigma}\right)\left(\frac{\sigma}{\sigma_n}\right)e^{(-1/2)(r_0/\sigma)^2(\sigma/\sigma_n)^2} \right\} \qquad (7.98)$$

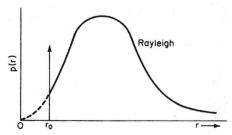

Figure 7.10 Plot of a truncated Rayleigh curve at r_0.

where erfc ( ) is the complementary error function. Equation (7.98) involves σ^2/σ_n^2, which represents the carrier-to-noise ratio (cnr), defined:

$$cnr = \frac{\sigma^2}{\sigma_n^2} \qquad (7.99)$$

The relationship of Eq. (7.98) is plotted in Fig. 7.11 for specific values of r_0/σ.

7.5   Simulation Models

Many radio-propagation models have been proposed that can be used to predict the amplitude and phase of radio signals propagated within a mobile-radio environment [8.20]. These models can be classified into two general categories. The first category deals only with multipath-fading phenomena, whereas the second category deals with both multipath- and selective-fading phenomena. The following paragraphs describe some of the features associated with these two general categories of simulation models.

Rayleigh multipath fading simulator

(A) Hardware simulator.   Based on analyses of the statistical nature of a mobile fading signal and its effects on envelope and phase, a fading simulator can be configured either from hardware or a combination of hardware and software. Figure 7.12 shows a simple hardware configuration for a Rayleigh multipath fading simulator that consists of two independent Gaussian noise generators (GNG), two variable low-pass filters (VLPF), and two balanced mixers (BM). The cutoff frequency of the low-pass filter is selected on the basis of the frequency f_b and the assumed average speed of the mobile vehicle V. The output of the simulator represents the envelope and phase of a Rayleigh-fading signal.

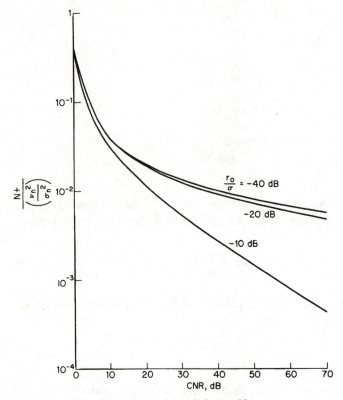

Figure 7.11   Plot of average positive click rate N_+.

The mathematical expression is as follows. The two noise sources after passing their corresponding VLPF (variable low-pass filter) are expressed in Eqs. (1.46) and (1.47) of Chap. 1 as:

$$n_c(t) = \sum_{k=-M/2}^{M/2} A_k \cos (2\pi k \, \Delta f t + \theta_k) \tag{7.100}$$

$$n_s(t) = \sum_{k=-M/2}^{M/2} A_k \sin (2\pi k \, \Delta f t + \theta_k) \tag{7.101}$$

when the bandwidth B is determined by the fading frequency f_b as

$$B = f_b = M \cdot \Delta f \tag{7.102}$$

and the fading frequency f_b in the simulator is determined by

$$f_b = \frac{V}{\lambda} \tag{7.103}$$

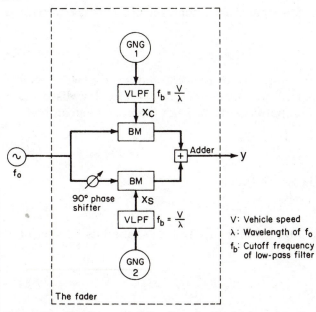

Figure 7.12 Hardware configuration of a Rayleigh multipath-fading simulator.

The output signal $y(t)$ after the adder is

$$y(t) = n_c \cos \omega_0 t + n_s \sin \omega_0 t \qquad (7.104)$$

The amplitude $A_t(t)$ and phase $\psi_t(t)$ are expressed in Eq. (1.48) as

$$A_t(t) = [n_c^2(t) + n_s^2(t)]^{1/2} \qquad (7.105)$$

$$\psi_t(t) = \tan^{-1} \frac{n_s(t)}{n_c(t)} \qquad (7.106)$$

$A_t(t)$ is the Rayleigh amplitude, the characteristics of which are expressed in Secs. 6.3–6.5, and $\psi(t)$ is the random phase expressed in Sec. 7.2. $\dot\psi(t)$ is the random FM expressed in Sec. 7.3, which can be obtained by taking the derivative of Eq. (7.106).

(B) Software simulator. In a software-configured simulator [10], the model described in Sec. 6.2 is based upon, and Eqs. (6.10) through (6.13) are used to simulate, a Rayleigh-fading signal. In both Eg. (6.11) and Eq. (6.12), there are N Gaussian random variables for R_i and S_i. Each of them has zero mean and variance one. Since a uniform angular distribution is assumed for N incoming waves, we let $\phi_i = 2\pi \, i/N$. To simplify the simulator model, the direction α of the moving vehicle can be set to $\alpha = 0°$ without loss generosity, then the parameter ξ_i of the ith incoming wave is

$$\xi_i = \beta V t \cos \phi_i \tag{7.107}$$

We have found that summing six or more incoming waves will perform a Rayleigh fading signal. We use nine evenly angular distributed waves, each of them is separated by 40° from the next wave arrival.

$$\phi_i = \frac{2\pi i}{9} \qquad i = 1, 9 \tag{7.108}$$

Let the parameter t in Eq. (7.107) be

$$t = k \cdot \Delta t \tag{7.109}$$

and Δt is the sample interval that can be set as small as 10 sampling points in every wavelength.

$$V \cdot \Delta t = \frac{\lambda}{10} \tag{7.110}$$

Now Eq. (6.11) and Eq. (6.12) become

$$X_k = \sum_{i=1}^{N=9} \left[R_i \cos \left(k \frac{2\pi}{10} \cos \phi_i \right) + S_i \sin \left(k \frac{2\pi}{10} \cos \phi_i \right) \right] \tag{7.111}$$

$$Y_k = \sum_{i=1}^{N=9} \left[S_i \cos \left(k \frac{2\pi}{10} \cos \phi_i \right) - R_i \sin \left(k \frac{2\pi}{10} \cos \phi_i \right) \right] \tag{7.112}$$

The kth sample point of a Rayleigh envelope is:

$$r_k = \{X_k^2 + Y_k^2\}^{1/2} \tag{7.113}$$

Equation (7.113) is the simulated Rayleigh fading signal plotted in Fig. 7.13 as the increment number k started from one. For our demonstration, a frequency of 850 MHz, the wavelength of which is 0.353 m, is given, and the sampling interval is $\lambda/10 = 0.0353$ cm. Each second will have m samples depending on the velocity of the mobile unit. The following table lists the mobile speeds that correspond to the number of samples per second.

V (km/h)	V(λ/s)	# of samples/sec (m)
25	19.6	196
50	39.3	393
75	59	590
100	78.7	787

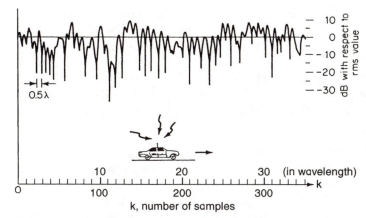

Figure 7.13 A simulated fading signal generated as if the mobile unit is moving with a speed V.

For any different mobile speed, the Rayleigh fading curve shown in Fig. 7.13 remains unchanged. Only the number of samples on the x-axis is rescaled according to the different mobile speed conditions indicated above.

The application of the software fading simulator (fader) [19] is described as follows. Since the random FM generally does not have an impact on the received signal, only the Rayleigh fading is considered. The modulated signal at the baseband is first digitized at its data rate R_b and then sent through the software fading simulator. For those data bits N_1 above the threshold level of the simulator (fader), no errors occur. Those data bits N_2 below the threshold will cause the errors depending on the states M of the modulation ($M = 2^L$). The error probability for each bit is

$$\frac{M-1}{M}$$

Assume that the total number of bits is N; if so, the BER (bit error rate) can be obtained by

$$\text{BER} = \frac{N_1 + N_2 \left(\dfrac{M-1}{M} \right)}{N} \tag{7.114}$$

The speed of the vehicle should not have an impact on the BER (see Sec. 16.14). We may verify this statement by using this simulator. The simulator also can evaluate the WER (word error rate) or FER (frame error rate) by grouping a sequence of bits called a frame or word. If

FEC (forward error correction) can correct 2 errors in a frame, then the frame has no error unless 3 or more errors occur. The simple signal process can generate the FER through this software fader very easily.

Other similar software configuration simulations are shown by Smith [20] and Arredondo [21]. The main advantage in a software-configured simulator is the ability to quickly change the operational parameters by merely making changes in the software program.

Multipath- and selective-fading simulator

A simulator of multipath and selective fading can also be configured either from hardware or a combination of hardware and software. The hardware version uses an arrangement of components similar to those shown in Fig. 7.12; the components are duplicated several times, and a delay line is added to each assembly [22]. The number of assemblies and delay lines is determined by the kind of environment that is being studied. The several delayed output signals from the Rayleigh-fading simulators are summed together to form a multipath- and selective-fading signal.

The software version [13, 17] is based on a model developed by Turin [17], in which the mobile-radio channel is represented as a linear filter with a complex-valued impulse response expressed:

$$h(t) = \sum_{k=0}^{\infty} a_k \delta(t - t_k) e^{j\psi_k} \tag{7.101}$$

where $\delta(t - t_k)$ is the delta function at time t_k, and ψ_k and a_k are the phase and amplitude, respectively, of the kth wave arrival. The terms a_k, ψ_k, and t_k are all random variables. During transmission of a radio signal $s(t)$, the channel response convolves $s(t)$ with $h(t)$. Figure 7.14 shows the mathematical simulation generated by the software for this

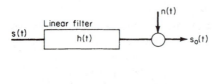

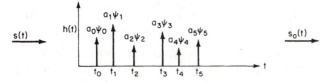

Figure 7.14 Mathematical model for multipath- and selective-fading simulation.

model [14]. The simulator enables evaluation of mobile-radio system performance without the need for actual road testing, since it simulates the propagation medium for mobile-radio transmissions.

Problem Exercises

1.   A mobile unit is traveling at a speed of 40 km/h while receiving a mobile-radio transmission at a frequency of 850 MHz. Assuming that the time-delay spread for the medium is 0.5 µs, what is the correlation coefficient for a frequency change from 850 MHz to 850.1 MHz in a time interval of 0.1 s?

2.   If the time-delay spread is measured and found to be 3 µs and the required phase correlation coefficient is 0.5, what is the coherent bandwidth for a time delay of $\tau = 0$?

3.   Which parameter has the greater effect on a random FM signal, velocity, or audio band? If the upper limit of the audio band is increased from 10 to 20 times over the lower limit, how much of a reduction in vehicle speed is required to maintain the same random FM level?

4.   In receiving a mobile-radio transmission at 850 MHz, a Doppler frequency of 20 to 60 Hz is observed. What is the beamwidth of the mobile antenna, and how fast is the mobile unit traveling?

5.   Prove that the power spectrum of click noise when the frequency is small compared with the reciprocal of pulse duration is $8\pi^2(N_+ + N_-)$.

6.   Give the proof for Eq. (7.52).

7.   Let $R_c(\tau) = NJ_0(\beta V\tau)$. On the basis of Eqs. (7.20) through (7.22), derive the values for terms b_0, b_2, and b_4, and insert these values into Eq. (7.83). What is the resultant level-crossing rate (lcr)?

References

1. S. O. Rice, "Statistical Properties of a Sine Wave plus Random Noise," *Bell System Technical Journal,* vol. 27, January 1948, pp. 109–157.
2. M. J. Gans, "A Power-Spectral Theory of Propagation in the Mobile-Radio Environment," *IEEE Trans. Veh. Tech,* vol. 21, no. 1, February 1972, pp. 27–38.
3. W. F. Davenport, Jr., and W. L. Root, *Random Signals and Noise,* McGraw-Hill, New York, 1958, pp. 152 and 164.
4. W. C. Jakes, *Microwave Mobile Communications,* Wiley, New York, 1974, p. 53.
5. B. R. Davis, "FM Noise with Fading Channels and Diversity," *IEEE Trans. Comm.,* vol. 19, December 1971, pp. 1189–1200.
6. S. O. Rice, "Noise in FM Receivers," in M. Rowenblatt (ed.), *Proceedings of the Symposium on Time Series Analysis,* Wiley, New York, 1963, chap. 25.
7. D. L. Schilling, E. A. Nelson, and K. K. Clarke, "Discriminator Response to an FM Signal in a Fading Channel," *IEEE Trans. Comm.,* vol. 15, April 1967, pp. 252–263.
8. J. F. Ossana, Jr., "A Model for Mobile Radio Fading Due to Building Reflections: Theoretical and Experimental Fading Waveform Power Spectra," *Bell System Technical Journal,* vol. 43, November 1964, pp. 2935–2971.

9. E. N. Gilbert, "Energy Reception for Mobile Radio," *Bell System Technical Journal,* vol. 44, October 1965, pp. 1779–1803.
10. W. C. Y. Lee, "Statistical Analysis of the Level Crossings and Duration of Fades of the Signal from an Energy Density Mobile Radio Antenna," *Bell System Technical Journal,* vol. 46, February 1967, pp. 417–448.
11. R. Clarke, "A Statistical Theory of Mobile Radio Reception," *Bell System Technical Journal,* vol. 47, July–August 1968, pp. 957–1000.
12. T. Aulin, "A Modified Model for the Fading Signal of a Mobile Radio Channel," *IEEE Trans. Veh. Tech.,* vol. 28, August 1979, pp. 182–203.
13. H. Hashemi, "Simulation of the Urban Radio Propagation Channel," *IEEE Trans. Veh. Tech.,* vol. 28, August 1979, pp. 213–225.
14. G. L. Turin, "Communication through Noisy, Random Multipath Channels," *IRE Conv. Rec. pt. 4,* 1956, pp. 154–166.
15. G. L. Turin et al., "A Statistical Model for Urban Multipath Propagation," *IEEE Trans. Veh. Tech.,* vol. 21, February 1972, pp. 1–9.
16. H. Suzuki, "A Statistical Model for Urban Radio Propagation," *IEEE Trans. Comm.,* vol. 25, no. 7, July 1977, pp. 673–680.
17. G. L. Turin, "Simulation of Urban Radio Propagation and of Urban Radio Communication System," *Proc. Int. Symp. Anten. Prop.,* Sendai, Japan, 1978, pp. 543–546.
18. H. F. Schmid, "A Prediction Model for Multipath Propagation of Pulse Signals at VHF and UHF over Irregular Terrain," *IEEE Trans. Anten. Prop.,* March 1970, pp. 253–258.
19. W. C. Y. Lee, "A Computer Simulation Model for the Evaluation of Mobile Radio Systems in the Military Tactical Environment," *IEEE Transaction on Vehicular Technology,* vol. VT-32, September 1983, pp. 177–190.
20. J. D. Smith, "A Multipath Fading Simulator for Mobile Radio," *IEEE Trans. Veh. Tech.,* vol. 24, November 1975, pp. 39–40.
21. G. Arredondo, W. Chiss, and E. Walker, "A Multipath Fading Simulator for Mobile Radio," *IEEE Trans. Veh. Tech.,* vol. 22, November 1973, pp. 241–244.
22. E. L. Caples, K. E. Massad, and T. R. Minor, "A UHF Channel Simulator for Digital Mobile Radio," *IEEE Trans. Veh. Tech.,* vol. 29, May 1980, pp. 281–289.

Modulation Technology

8.1 System Application

Chapters 6 and 7 presented studies of the characteristics of fading based on envelope and phase. In this chapter, modulation schemes that can be used to counter the effects of multipath fading and time-delay spread in mobile-radio communication are discussed. In amplitude-modulated (AM) signaling, the information is contained in the envelope. During severe multipath fading, most of the amplitude information may be wiped out. One possible solution is to use a type of angle modulation. In FM signaling the noise or interference amplitude increases linearly with frequency; therefore the power spectrum of the noise will vary proportionately with the square of the frequency. This parabolic response on the power spectrum reflects the fact that input noise components that are close to the carrier frequency are suppressed.

Digital modulation techniques are also discussed in this chapter, and the difference in error rates between nonfading and fading cases is shown. Spread-spectrum modulation techniques are also discussed, since these modulation techniques have military as well as civilian applications. Finally, a new concept of using a modified SSB system is briefly introduced.

8.2 FM for Mobile Radio

Frequency modulation (FM) can improve the baseband signal-to-noise ratio without necessitating an increase in transmitting power. Also, FM can capture the signal from the interference even when the signal-to-interference ratio is small. Assume that a signal transmitted from a base station is

$$s_t(t) = Ae^{j(\omega_c t + \mu(t))} \tag{8.1}$$

where A is a constant and $\mu(t)$ is defined:

$$\mu(t) = \int_0^t v(t')\, dt' \tag{8.2}$$

and $v(t')$ is the actual message. Then the received signal at the mobile-unit antenna (see Fig. 8.1), as shown in Eq. (6.8), can be expressed:

$$s'(t) = \left(\sum_{i=1}^N \tilde{a}_i e^{-j\beta Vt\,\cos\,(\phi_i - \alpha)} \right) s_t(t) \tag{8.3}$$

The term $s'(t)$ can also be expressed:

$$s'(t) = [r(t)e^{j\psi_r(t)}]Ae^{j\mu(t)}e^{j\omega_c t} \tag{8.4}$$

The relationships for the parameters shown in Eqs. (8.3) and (8.4) were previously expressed in Eq. (1.20). On the assumption that Fig. 8.2 is a block diagram of an FM receiver typically used in a mobile-radio unit, then the signal at the input of the IF filter is

$$s'(t) = [Q(t)Ae^{j\mu(t)} + n_c + jn_s]e^{j\omega_c t} \tag{8.5}$$

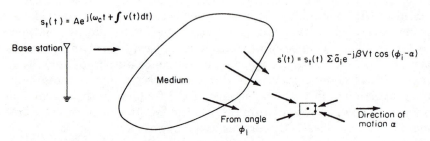

Figure 8.1   Received signal at the mobile antenna.

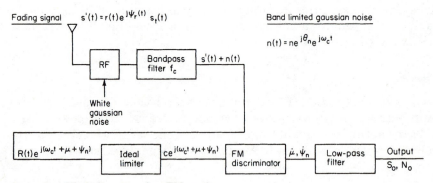

Figure 8.2   Block diagram of an FM receiver.

where

$$Q(t) = r(t)e^{j\psi_r(t)} \qquad (8.6)$$

Nonfading case

For a nonfading case, let $Q(t) = 1$ in Eq. (8.5). Upon reception by an FM system, the baseband signal-to-noise ratio of this signal can be improved without increasing the transmitted power, but rather by increasing the frequency deviation of the modulation and consequently increasing the IF bandwidth. The nonfading case can be analyzed for two situations: (1) signal present and (2) no signal present.

Signal present. The special case of Eq. (8.5), $Q(t) = 1$, is illustrated in Fig. 8.3 and can be expressed:

$$(Ae^{j\mu(t)} + n_c + jn_s)e^{j\omega_c t} = Re^{j[\mu(t) + \psi_n]}e^{j\omega_c t} \qquad (8.7)$$

where the derivative of μ, $\dot{\mu}$, is the message and the derivative of ψ_n, $\dot{\psi}_n$, is the noise at the output of the discriminator. Then, logarithmic differentiation of Eq. (8.7) with respect to time gives:

$$\frac{\dot{R}}{R} + j(\dot{\mu} + \dot{\psi}_n) = \frac{j\dot{\mu}Ae^{j\mu} + \dot{n}_c + j\dot{n}_s}{Ae^{j\mu} + n_c + jn_s} \qquad (8.8)$$

The four Gaussian variables n_c, n_s, $\dot{n}_c$, and $\dot{n}_s$ are independent when $S(f)$ is symmetrical about f_c, as previously discussed in Chap. 7, Sec. 7.1. Then, as described by Rice [1],

$$E[\dot{R}R^{-1} + j(\dot{\mu} + \dot{\psi}_n)]$$

$$= \overline{\iiiint_{-\infty}^{\infty}} \frac{j\dot{\mu}A \exp j\mu + \dot{n}_c + j\dot{n}_s}{A \exp j\mu + n_c + jn_s} \, p(n_c, n_s, \dot{n}_c, \dot{n}_s) \, dn_c \, dn_s \, d\dot{n}_c \, d\dot{n}_s$$

$$= j\dot{\mu}\left[1 - \exp\left(-\frac{A^2}{2b_0}\right)\right] \qquad (8.9)$$

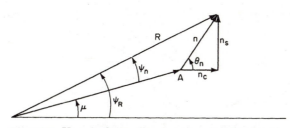

Figure 8.3 Vector relationships between signal and noise.

where

$$b_0 = E[n_c^2] = E[n_s^2] \tag{8.10}$$

As a result of balancing the two sides of Eq. (8.9), the following is obtained:

$$E[\dot{R}R^{-1}] = 0 \tag{8.11}$$

and

$$E[\dot{\mu} + \dot{\psi}_n] = \dot{\mu}(1 - e^{-\gamma}) \tag{8.12}$$

where γ is the cnr, as shown in Eq. (8.9) and expressed:

$$\gamma = \frac{A^2}{2b_0} \tag{8.13}$$

and $\dot{\mu}$, shown in Eq. (8.12), is the input signal after the discriminator. If the output of the discriminator is denoted by $v_o(t)$, then Eq. (8.12) becomes:

$$v_o(t) = \dot{\mu}(t)(1 - e^{-\gamma}) \tag{8.14}$$

Equations (8.14) and (8.12) both indicate that the presence of noise reduces the output signal by multiplying it by a factor of $1 - e^{-\gamma}$. The output baseband signal power then becomes:

$$S_o = E[v_o^2(t)] = (1 - e^{-\gamma})^2 E[\dot{\mu}^2(t)]$$
$$= (1 - e^{-\gamma})^2 S_i \tag{8.15}$$

where S_i is the input signal power to the discriminator, which must be specified before S_o of Eq. (8.15) can be evaluated.

From Carson's rules, the transmission bandwidth is:

$$B = 2(W + f_d) = 2W(1 + \beta_i) \tag{8.16}$$

where W is the maximum modulating frequency, analogous to the modulation index β_i of worst-case tone modulation, and f_d is the maximum deviation, expressed:

$$f_d = \frac{B - 2W}{2}$$

The input power S_i can be expressed:

$$S_i = \alpha(2\pi f_d)^2 \tag{8.17}$$

where α is less than 1. If S_i in Eq. (8.15) is 10 dB less than $(2\pi f_d)^2$, then $\alpha = 0.1$. The input power S_i can then be expressed:

$$S_i = \frac{4\pi^2}{10} f_d^2 = \frac{\pi}{10} (B - 2W)^2 \qquad (8.18)$$

By substituting Eq. (8.18) into Eq. (8.15), S_o/W^2 versus γ for different values of $B/2W$ can be plotted, as shown in Fig. 8.4.

Obtaining the baseband noise spectrum. The derivative of $\psi_n(t)$, $\dot{\psi}_n(t)$, represents the noise current, where $\dot{\psi}_n(t) = \dot{\psi}_R(t)$ is shown in Fig. 8.3 for $\mu(t) = 0$. The one-sided power spectrum is

$$S_{\dot{\psi}_n}(f) = 4 \int_0^\infty R_{\dot{\psi}_n}(\tau) \cos 2\pi f\tau \, d\tau \qquad (8.19)$$

where $R_{\dot{\psi}_n}(\tau)$ is the autocorrelation, expressed:

$$R_{\dot{\psi}_n}(\tau) = E[\dot{\psi}_n(t) \, \dot{\psi}_n(t + \tau)] \qquad (8.20)$$

and where $\dot{\psi}_n(t)$ can be obtained from $\psi_n(t)$, since

$$\psi_n = \tan^{-1} \frac{n_s}{A + n_c} \qquad (8.21)$$

or

$$\sec^2 \psi_n = 1 + \left(\frac{n_s}{A + n_c}\right)^2 \qquad (8.22)$$

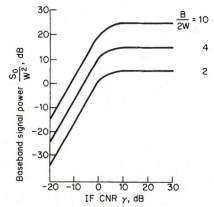

Figure 8.4 Plot of baseband signal power S_o/W^2 vs. IF carrier-to-noise ratio γ for different $B/2W$.

As a result of differentiating Eq. (8.21) and substituting the terms of Eq. (8.22) into Eq. (8.21), $\dot{\psi}_n$ becomes:

$$\dot{\psi}_n = \frac{(A + n_c)\dot{n}_s - n_s\dot{n}_c}{\sec^2\psi_n(A + n_c)^2} = \frac{(A + n_c)\dot{n}_s - n_s\dot{n}_c}{(A + n_c)^2 + n_s^2} \tag{8.23}$$

and the correlation of $\dot{\psi}_n(t)$, $E[\dot{\psi}_n(t)\dot{\psi}_n(t + \tau)]$, turns out to be an eightfold integral [2], written:

$$R_{\dot{\psi}_n}(\tau) = \int_{-\infty}^{\infty} dn_{c1} \int_{-\infty}^{\infty} dn_{s1}$$

$$\cdots \int_{-\infty}^{\infty} d\dot{n}_{s2}\, p(n_{c1}, n_{s1}, \dot{n}_{c1}, \dot{n}_{s1}, \ldots, \dot{n}_{s2})\, \dot{\psi}_{n1}\dot{\psi}_{n2} \tag{8.24}$$

where $\dot{\psi}_{n1}$ and $\dot{\psi}_{n2}$ are as defined in Eq. (8.23). Unfortunately Eq. (8.24) increases the complexity, but an approximation can be ascertained.

The noise output of the FM discriminator can be approximated [2] by the overall baseband noise spectrum, assuming a Gaussian-shaped IF filter with a bandwidth B of

$$H_{IF}(f) = e^{-\pi(f - f_c)^2/B^2} \tag{8.25}$$

With this type of filter, Rice [1] divided the one-sided baseband noise spectrum $S_B(f)$ into three components:

$$S_B(f) = S_1(f) + S_2(f) + S_3(f) \tag{8.26}$$

Obtaining $S_1(f)$.   It is assumed that $S_1(f)$ has the shape of the output noise spectrum when the carrier is very large. The same factor, used in Eq. (8.15) to derive the output baseband signal power may be applied to obtain the output noise spectrum:

$$S_1(f) = (1 - e^{-\gamma})^2 S_{\dot{\psi}_n}(f) \tag{8.27}$$

The input noise spectrum $S_{\dot{\psi}_n}(f)$ in Eq. (8.27) can be obtained when $\gamma \gg 1$, as follows (see Fig. 8.3). First, ψ_n is expressed as:

$$\psi_n = \tan^{-1} \frac{n_s(t)}{A + n_c(t)} = \frac{n_s(t)}{A} \tag{8.28}$$

when $\mu = 0$ as shown in Fig. 8.3. Differentiating Eq. (8.28) yields:

$$\dot{\psi}_n = \frac{\dot{n}_s(t)}{A}$$

The one-sided spectrum of $n_s(t)$ is

$$S_{n_s}(f) = \eta[H_{\text{IF}}(f_c - f) + H_{\text{IF}}(f_c + f)] \tag{8.29}$$

where $H_{\text{IF}}(f_c - f)$ is the shape of the IF filter and η is as shown in Eq. (1.40). The power spectrum of the derivative of a stationary stochastic process, which is $(2\pi f)^2$ times the spectrum of the process itself [2], is derived from Eq. (8.29) and is expressed:

$$S_{\psi_n}(f) = \left(\frac{2\pi f}{A}\right)^2 \eta[H_{\text{IF}}(f_c - f) + H_{\text{IF}}(f_c + f)] \tag{8.30}$$

Hence, when $\gamma \gg 1$, the baseband noise spectrum is approximately parabolic in shape and proportional to f^2, as shown in Eq. (8.30).

Substituting Eq. (8.25) into Eq. (8.30) and then Eq. (8.30) into Eq. (8.27) yields:

$$S_1(f) = 2 \left[\frac{2\pi f(1 - e^{-\gamma})}{A}\right]^2 \eta e^{-(\pi f^2/B^2)}$$

$$= \frac{[2\pi f(1 - e^{-\gamma})]^2}{B\gamma} e^{-(\pi f^2/B^2)} \tag{8.31}$$

where the cnr is expressed as:

$$\gamma = \frac{A^2}{2b_0} = \frac{A^2}{2\eta B}$$

Obtaining $S_2(f)$ and $S_3(f)$.   It is assumed that $S_2(f)$ has the same spectrum shape as the output noise spectrum when the carrier is absent; $S_3(f)$ is a correction term that predominates in the threshold region of γ. Rice [2] gives the following equation for $S_2(f)$:

$$S_2(f) = \sqrt{2}\pi B\, e^{-\gamma} \sum_{n=1}^{\infty} n^{-3/2} e^{-\pi f^2/2nB^2} \tag{8.32}$$

and an even more complex expression for $S_3(f)$. Therefore, $S_2(f)$ and $S_3(f)$ require special numerical evaluation. As an alternate, the following approximation can be used: In the frequency range from 0 to f_a, where $f_a < B$, the sum of the spectral components $S_2(f) + S_3(f)$ may be accurately approximated by an empirical relationship developed by Davis [3]:

$$S_2(f) + S_3(f) \approx 8\pi B e^{-\gamma}[2(\gamma + 2.35)]^{-1/2} \tag{8.33}$$

Equation (8.33) approximates Rice's exact analysis. Substituting Eqs. (8.31) and (8.33) into Eq. (8.26) gives the following:

$$S_B(f) = \frac{[2\pi f(1 - e^{-\gamma})]^2}{B\gamma} e^{-(\pi f^2/B^2)} + \frac{8\pi B e^{-\gamma}}{\sqrt{2(\gamma + 2.35)}} \qquad (8.34)$$

Equation (8.34) is plotted in Fig. 8.5. The approximate solution agrees quite well with Rice's exact analysis for $f < \frac{1}{3}\sqrt{\pi/2}B$. The total noise output of the rectangular baseband filter is then:

$$N_t(\gamma) = \int_0^W S_B(f)\, df$$

$$= \frac{a(1 - e^{-\gamma})^2}{\gamma} + \frac{8\pi B W e^{-\gamma}}{\sqrt{2(\gamma + 2.35)}} \qquad (8.35)$$

where

$$a = \frac{(2\pi)^2}{B} \int_0^W f^2 e^{-\pi f^2/B^2}\, df = \frac{(2\pi)^2}{B} \int_0^W f^2 \sum_{n=0}^{\infty} \frac{(-\pi f^2/B^2)^n}{n!} df$$

$$= \frac{4\pi^2 W^3}{3B} \left[1 - \frac{6\pi}{10} \left(\frac{W}{B}\right)^2 + \cdots \right] \qquad (8.36)$$

Equation (8.35) is plotted in Fig. 8.6.

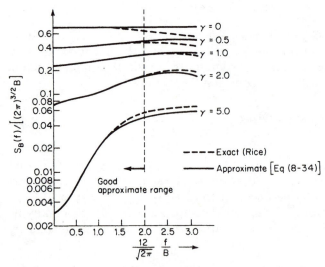

Figure 8.5 One-sided baseband noise spectrum $S_B(f)$. (*From Ref. 3.*)

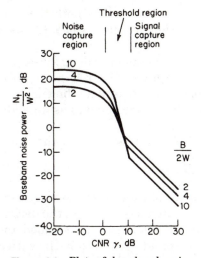

Figure 8.6 Plot of baseband noise power vs. carrier-to-noise ratio.

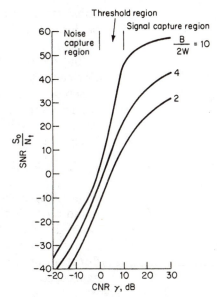

Figure 8.7 Plot of signal-to-noise ratio as a function of carrier-to-noise ratio.

The signal-to-noise ratio (snr) at the baseband output is defined:

$$\text{snr} = \frac{S_o}{N_t} = \frac{S_i(1 - e^{-\gamma})^2}{N_t}$$

and is plotted in Fig. 8.7. In Fig. 8.7, note that for an FM system the threshold effect is very prominent in a nonfading case.

For $\gamma \gg 1$, S_o/N_t can be obtained as follows:

$$\frac{S_o}{N_t} = \frac{S_i(1 - e^{-\gamma})^2}{a(1 - e^{-\gamma})^2/\gamma} = \frac{S_i}{a}\,\gamma = 3\alpha\beta_i^2\,\frac{B}{W}\,\gamma \tag{8.37}$$

where S_i and a are obtained from Eqs. (8.17) and (8.36), respectively, and inserted into Eq. (8.37) to obtain the final result. β_i is the modulation index.

Rayleigh-fading case

In a Rayleigh-fading case, if the multipath rate is much slower when compared with the baseband bandwidth [3, 4], or if the multipath duration is much larger than the reciprocal signal bandwidth, then the fading envelope of the FM signal is considered to be flat, and therefore the power density is the same over the entire signal spectrum. The signal-to-noise ratio of the FM discriminator output can then be found by the quasi-static approximation, from which both the signal and noise can be determined by the carrier-to-noise ratio γ. It was previously shown in the nonfading case.

Rapid phase changes accompany the deep fades, producing a random FM component at the output of the FM receiver. Furthermore, the baseband output signal is suppressed as the cnr decreases. This rapid random suppression of the signal appears as an additional noise component in the baseband output.

The received fading FM signal is given by the equation

$$s(t) = R(t) \exp\left[j(\omega_c t + \mu(t) + \psi_r + \psi_n)\right]$$

where $R(t)$ is the fading amplitude, ψ_r is the random phase change due to fading, ψ_n is the Gaussian noise, and $\dot{\mu}(t)$, the derivative of $\mu(t)$, is the desired modulating signal. $\dot{\mu}(t)$ can be assumed to be a zero-mean Gaussian process, ideally band-limited to W Hz, with an rms frequency duration of $\sqrt{S_i}/2\pi$, where S_i is the power of the modulating signal, expressed:

$$E[\dot{\mu}^2(t)] = E[v^2(t)] = S_i \tag{8.38}$$

The envelope $R(t)$ is Rayleigh-distributed, with average power expressed:

$$E[R^2(t)] = P_{av} \tag{8.39}$$

The carrier-to-noise ratio (cnr) is denoted by $\gamma(t)$, as follows:

$$\gamma(t) = \frac{R^2(t)}{2\sigma_t^2} \tag{8.40}$$

where σ_t^2 is the mean square value of the combined Gaussian noise ψ_r and ψ_n after IF filtering. $R(t)$ varies slowly with time and has an exponential density function with parameter γ_o, which is the average cnr based on the following relationships:

$$p_\gamma(\alpha) = \frac{1}{\gamma_o} \exp\left(-\frac{\alpha}{\gamma_o}\right) \tag{8.41}$$

$$\gamma_o = \frac{P_{av}}{2\sigma_t^2} \tag{8.42}$$

The total phase relationship expressed by the term $\psi_R(t)$ at the input of the IF filter can be shown as:

$$\psi_R(t) = \mu(t) + \psi_r(t) + \psi_n(t) \tag{8.43}$$

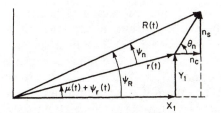

Figure 8.8 Vector relationships among phases at the IF filter input.

as illustrated in Fig. 8.8. The time derivative of $\psi_R(t)$, $\dot\psi_R(t)$, becomes:

$$\dot\psi_R(t) = \dot\mu(t) + \dot\psi_r(t) + \dot\psi_n(t) + 2\pi(Z_+ - Z_-) \tag{8.44}$$

The term $2\pi(Z_+ - Z_-)$ in Eq. (8.44) is derived from Eq. (7.86).

In an ideal FM detector, the bandwidth B is wide enough that it does not distort the signal. Then B has to be:

$$B = 2\left(W + \frac{\sqrt{10S_i}}{2\pi}\right) \tag{8.45}$$

which is derived from Eq. (8.18). The output of an ideal FM detector is $\dot\psi_o(t)$, which can be obtained from the input $\dot\psi_R(t)$. The terms $\dot\mu(t)$ and $\dot\psi_n(t)$ in Eq. (8.44), after the signal passes through the IF detector, will be suppressed by the factor $1 - e^{-\gamma}$, as shown in Eqs. (8.14) and (8.15). The terms ψ_r and $2\pi(Z_+ - Z_-)$ remain unchanged.

Let $g(t) = 1 - e^{-\gamma(t)}$, which can also be written as $g(t) = E[g] + g(t) - E[g]$. Then, after IF detection, $\dot\psi_o(t)$ can be expressed [5]:

$$\dot\psi_o(t) = g(t)[\dot\mu(t) + \dot\psi_n(t)] + \dot\psi_r(t) + 2\pi(Z_+ - Z_-)$$

$$= E[g]\dot\mu(t) + \{g(t) - E[g(t)]\}\dot\mu(t) + \dot\psi_r(t)$$

$$+ g(t)\dot\psi_n(t) + 2\pi(Z_+ - Z_-) \tag{8.46}$$

Let

$$g = g(t) = 1 - e^{-\gamma} \tag{8.47}$$

The expectation for $E[g]$ is

$$E[g] = \int_0^\infty gp(\gamma)\,d\gamma = \int_0^\infty \frac{(1 - e^{-\gamma})e^{-\gamma/\gamma_o}}{\gamma_o}\,d\gamma = \frac{\gamma_o}{1 + \gamma_o} \tag{8.48}$$

The terms ψ_r and ψ_n are the random phase characteristics caused by multipath fading and Gaussian noise, respectively, as shown in Fig.

8.8. The terms Z_+ and Z_- are the sums of positive and negative impulses, respectively, with average occurrence rates N_+ and N_-, as expressed in Eqs. (7.87) and (7.88). The first term in Eq. (8.46) is the desired signal and the other four are noise terms: (1) the noise due to suppression of the signal (N_{sn}), (2) the noise due to random FM caused by fading (N_{rfm}), (3) the usual Gaussian noise present due to the additive noise at the input (N_{gn}), and (4) the click noise (N_{cn}).

Taking the Fourier transform of the spectrum function of $\psi_o(t)$ and integrating from 0 to W gives the output power:

$$E[\psi_o^2] = \int_0^W S_{\psi_o}(f)\, df \tag{8.49}$$

where $S_{\psi_o}(f)$ is the one-sided power spectrum as

$$S_{\psi_o}(f) = 4 \int_0^\infty R_{\psi_o}(\tau) \cos 2\pi f\tau\, d\tau \tag{8.50}$$

Where $E[\psi_o^2(t)]$ consists of many components, we obtain the five relationships in Eqs. (8.51) through (8.55):

$$\overline{S}_o = (E[g])^2 S_i \tag{8.51}$$

where $\overline{S}_o$ is the signal output power.

$$\overline{N}_{sn} = E[(g - E[g])^2] S_i \tag{8.52}$$

where $\overline{N}_{sn}$ is the output power of the noise caused by signal suppression.

$$N_{rfm} = \int_0^W S_{\psi_r}(f)\, df \tag{8.53}$$

where N_{rfm} is the random FM noise power and $S_{\psi_r}(f)$ is the power spectrum found in Eq. (7.75), which is independent of the cnr γ.

$$\overline{N}_{gn} = \int_0^\infty \int_0^W g S_B(f) p(\gamma)\, df\, d\gamma$$

$$= \int_0^\infty \int_0^W g[S_1(f) + S_2(f) + S_3(f)] p(\gamma)\, df\, d\gamma \tag{8.54}$$

where $\overline{N}_{gn}$ is the cnr noise output power and the general expression for S_B is shown in Eq. (8.26). A general approximation for S_B is shown in Eq. (8.34).

For the case of $\gamma \gg 1$, $S_1(f)$ of Eq. (8.27) plays a major role in Eq. (8.54); then

$$\overline{N}_{\mathrm{gn}} = \int_0^\infty \int_0^W gS_1(f)p(\gamma)\,df\,d\gamma = \int_0^\infty \int_0^W \frac{g^2}{2\gamma}\,(2\pi f)^2 [H_{\mathrm{IF}}(-f) + H_{\mathrm{IF}}(f)]\,df$$

$$\overline{N}_{\mathrm{cn}} = 4\pi^2 (N_+ + N_-)(2W) \tag{8.55}$$

where $\overline{N}_{\mathrm{cn}}$ is the click-noise power and N_+ and N_- are as shown in Eq. (7.98).

The detected snr at the output of the discriminator can be computed:

$$\mathrm{snr} = \frac{\overline{S}_o}{\overline{N}_{\mathrm{sn}} + N_{\mathrm{rfm}} + \overline{N}_{\mathrm{gn}} + \overline{N}_{\mathrm{cn}}} \tag{8.56}$$

Equation (8.56) is a general expression.

The click noise N_{cn} can be ignored, since this effect is significant only in the vicinity of the FM threshold as opposed to well below or well above the threshold. Also, it produces relatively small changes in the noise output.

In mobile communications, random FM is caused by the interference between waves of different Doppler frequency arriving at a mobile antenna from various directions. Assuming a uniform angle of arrival, the random FM output noise, as shown in Eq. (7.80), is:

$$E[\dot{\psi}_r^2] = \frac{(\beta V)^2}{2} \ln \frac{W_2}{W_1} \tag{8.57}$$

The random FM output noise shown in Eq. (8.57) is independent of the cnr, γ. The signal power S_o at the output of the discriminator is suppressed by the factor $1 - e^{-\gamma}$. When $\gamma \gg 1$, as shown in Eq. (8.15), then the expression for S_o can be obtained from Eq. (8.18):

$$S_o = S_i = \frac{\pi^2}{10}(B - 2W)^2 \tag{8.58}$$

Then the limiting output snr can be expressed:

$$\mathrm{snr}_{\mathrm{rfm\ only}} = \frac{S_o}{E[\dot{\psi}_r^2]} = \frac{(B - 2W)^2}{20(V/\lambda)^2 \ln\,(W_2/W_1)} \tag{8.59}$$

where $W \approx W_2 - W_1$. In the audio band, the values $W_1 = 300$ Hz and $W_2 = 3$ kHz are assumed.

Since the snr_{rfm} varies on account of the vehicle speed, it may be treated separately from the snr that is not associated with random FM and expressed:

$$\text{snr}_{\text{no rfm}} = \frac{\overline{S}_o}{\overline{N}_{\text{sn}} + \overline{N}_{\text{gn}}} \tag{8.60}$$

and

$$\text{snr}_{\text{rfm only}} = \frac{(B - 2W)^2}{20(V/\lambda)^2 \ln 10} = \text{snr}_{\text{rfm}} \tag{8.61}$$

The term $\overline{N}_{\text{gn}}$ in Eq. (8.60) can be obtained by averaging Eq. (8.35) over γ:

$$\overline{N}_{\text{gn}} = \int_0^\infty p(\gamma) N_t(\gamma) \, d\gamma = \frac{a}{\gamma_o} \ln \frac{(1 + \gamma_o)^2}{(1 + 2\gamma_o)}$$

$$+ 8BW \sqrt{\frac{\pi}{2\gamma_o(\gamma_o + 1)}} \, e^{2.35(\gamma_o + 1)/\gamma_o} \, \text{erfc} \, (\sqrt{2.35(\gamma_o + 1)/\gamma_o}) \tag{8.62}$$

where the value a is defined in Eq. (8.36), erfc $(x) = 1 - \text{erf} \, (x)$, and erf (x) is as follows:

$$\text{erf} \, (x) = \frac{2}{\pi} \int_0^x e^{-y^2} \, dy \tag{8.63}$$

The term $\overline{N}_{\text{sn}}$ in Eq. (8.60) can be obtained by carrying out Eq. (8.52), where g in Eq. (8.52) is defined in Eq. (8.47):

$$\overline{N}_{\text{sn}} = \int_0^\infty p(\gamma)(g - E[g]) S_i \, d\gamma = S_i \left(\frac{1}{2\gamma_o + 1} - \frac{1}{(\gamma_o + 1)^2} \right) \tag{8.64}$$

The two signal-to-noise terms, $\text{snr}_{\text{no rmf}}$ and $\text{snr}_{\text{rfm only}}$, are plotted in Fig. 8.9. Note that when the output snr is below the value of the snr_{rfm}, then output snr increases linearly with the cnr. When the output snr reaches the value of the snr_{rfm}, increasing cnr has no effect on the output snr.

The sharp threshold shown in Fig. 8.7, which was due to the capture, is no longer apparent in Fig. 8.9. With increasing transmitter power as γ increases, the output signal-to-noise ratio in decibels increases approximately linearly with the described value of γ_o. When $\gamma \gg 1$, the threshold crossings of the noise elements are less frequent, there is less signal suppression, and thus the quadrature noise is suppressed.

N_{rfm} in the noise term does not change as γ increases. Hence, when $\gamma \gg 1$, the $\text{snr}_{\text{rfm only}}$, due to the effects of random FM, is the limiting snr. For example, with a carrier frequency of 900 MHz and a vehicle veloc-

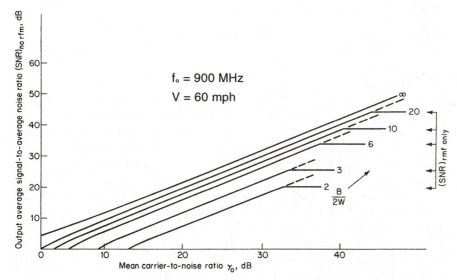

Figure 8.9 Plot of output signal-to-noise ratio vs. average carrier-to-noise ratio. (*From Ref. 3.*)

ity of 60 mi/h, the maximum Doppler frequency shift is $f_m = 80$ Hz. The limiting $\text{snr}_{\text{rfm only}}$ for $f_m = 80$ Hz and $W = 3$ kHz is illustrated in Fig. 8.9 for various values of bandwidth ratio $B/2W$.

Example 8.1   In mobile communication, random FM is caused by the interference between waves of different Doppler frequency arriving at a mobile antenna from various directions. The limiting output snr, shown in Eq. (8.61), is due to random FM only. Under these conditions, the snr cannot be improved by increasing the cnr. Assuming that the maximum modulation frequency is 3 kHz, the vehicular speed is 45 mi/h, and the carrier frequency is 850 MHz, what is the ratio $B/2W$ for $\text{snr}_{\text{rfm}} \geq 30$ dB?

solution   The ratio $B/2W$ for $\text{snr}_{\text{rfm}} \geq 30$ dB can be found by the equation

$$\text{snr}_{\text{dB}} = 10 \log \frac{(B - 2W)^2}{20(V/\lambda)^2 \ln 10} \tag{E8.1.1}$$

where

$$20 \left(\frac{V}{\lambda} \right)^2 \ln 10 = (386.8)^2 \tag{E8.1.2}$$

Then B can be derived from Eq. (E8.1.1) as follows:

$$\frac{(B - 2W)^2}{20 \left(\dfrac{V}{\lambda} \right)^2 \ln 10} = 10^3 \tag{E8.1.3}$$

or

$$B = 6W = 18 \text{ kHz} \qquad\qquad (E8.1.4)$$

8.3 Digital Modulation

When digital modulation techniques are used, speech signals must be coded into a format suitable for digital transmission. This can be accomplished by using PCM, DPCM, and delta modulation [6, 7]. A digitally controlled companding scheme [8, 9] can be used to obtain a good telephone link with a bit rate of 32 kb/s, which appears to meet CCITT specifications. For bit rates as low as 20 kb/s, other types of modulation schemes have been reported that are effective in providing intelligible speech signals. Where companded delta modulation is used, a rate of 32 kb/s yields a baseband snr of at least 30 dB over a dynamic range of 45 dB. Figure 8.10 illustrates the snr for various bit rates and input signal levels [8]. Recently, using linear predictive coding (LPC) requires only 2.4 kb/s. The description of LPC is shown in Sec. 14.5.

Sampled speech data at a bit rate f_s can be transmitted over a radio channel by either AM, PM, or FM. From the viewpoints of power economy and system simplicity, only two-level systems are considered. Also, the filter shaping is performed at the transmitter, whereas the receiving filter is used only for channel selection purposes.

Nonfading case

There are five selection schemes that are worthy of discussion: (1) coherent binary AM (PSK), (2) FM with discrimination, (3) coherent FSK,

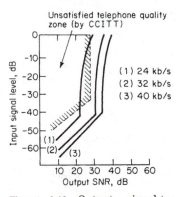

Figure 8.10 Output signal-to-noise ratio for varying bit rates and input signal levels for delta modulation.

(4) noncoherent FSK, and (5) differential PSK. These are discussed in the following paragraphs.

Coherent binary AM (PSK). In coherent binary AM, also called "phase-shift keying" (PSK) detection, the transmitted waveform is expressed:

$$x(t) = \sum_{k=-\infty}^{\infty} a_k h(t - kT) \cos (2\pi f_c t) \qquad (8.65)$$

where f_c is the carrier frequency, f_s is the sampling frequency, $T = 1/f_s$ is the sampling interval, the a_k's are information digits, and $h(t)$ is a baseband waveform band-limited to bandwidth B. The model for coherent binary AM signaling is shown in Fig. 8.11. The received waveform is expressed:

$$y(t) = \mathrm{Re} \left[\left(\sum_{k=-\infty}^{\infty} a_k h(t - kT) + \tilde{n} \right) e^{j\omega_c t} \right] \qquad (8.66)$$

where $\tilde{n} = n_c + jn_s$. The power of $\tilde{n}$ is controlled by the filter bandwidth, where:

$$E[n_c^2] = E[n_s^2] = \sigma_n^2 \qquad (8.67)$$

When $a_k = \pm 1$, the signaling is either binary AM or two-phase PM. The received waveforms can be detected coherently by multiplying Eq. (8.66) by $\cos \omega_c t$ and passing the waveform through a low-pass filter to obtain the baseband output:

$$z(t) = \sum_{k=-\infty}^{\infty} a_k h(t - kT) + n_c(t) \qquad (8.68)$$

At the sampling instant, $t = kT$, and

$$z(kT) = a_k + n_c(kT) \qquad (8.69)$$

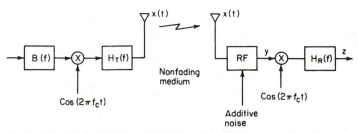

Figure 8.11 Model for coherent binary AM signaling.

Since the likelihood exists that n_c can be either positive or negative, the decision threshold is set at zero. At $z(kT) > 0$, let $a_k = A$. Then, error probability is:

$$P_e = P\{A + n_c < 0\} = P\{n_c < -A\}$$

$$= \int_{-\infty}^{-A} p_{n_c}(x)\,dx = \frac{1}{2}\left[1 + \text{erf}\left(-\frac{A}{\sqrt{2}\sigma_n}\right)\right] = \frac{1}{2}\,\text{erfc}\,(\sqrt{\gamma}) \quad (8.70)$$

where the snr at the sampling instant is

$$\gamma = \frac{A^2}{2\sigma_n^2} \quad (8.71)$$

Coherent binary AM (or coherent PSK) is rated for best performance over the other systems [10, 11]. The probability of error for a coherent binary AM system, Eq. (8.70), is plotted in Fig. 8.12. However, coherent detection requires carrier phase recovery, and in the mobile-radio environment, where rapid phase fluctuations are common, this is not a simple task.

FM with discrimination.   It is interesting to compare the error rate for a digital signal using FM with discriminator detection with that of a digital signal using coherent FSK detection.

It should be noted that FM with discriminator detection, where the peak frequency deviation $f_d = 0.35 f_s$ and the IF bandwidth equals f_s, has

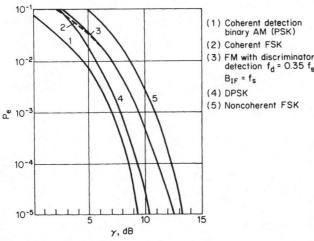

Figure 8.12   Comparisons of error probability between coherent binary AM, coherent FSK, FM with discriminator detection, DPSK, and noncoherent FSK.

almost the same bit-error rate as is found in coherent FSK detection. This can be seen by comparing the curves [10, 12] shown in Fig. 8.12.

Coherent FSK. The error rate for coherent FSK detection has an analytic solution, which can be applied, as follows. One filter, at frequency ω_1, has an output containing signal and noise, expressed:

$$y_1(t) = \text{Re} \left\{ \left[\sum_{k=-\infty}^{\infty} a_k h(t - kT) + \tilde{n}_1 \right] e^{j\omega_1 t} \right\} \tag{8.72}$$

The other filter, at frequency ω_2, has an output containing only noise, which can be expressed:

$$y_2(t) = \text{Re} \left[\tilde{n}_2 e^{j\omega_2 t} \right] \tag{8.73}$$

The coherent detection process, using reference signals $\cos \omega_1 t$ and $\sin \omega_2 t$, respectively, gives the corresponding detector outputs shown in the following equations, where the sampling instant $t = kT$:

$$v_1(kT) = a_k + n_{c_1}(kT) \tag{8.74}$$

$$v_2(kT) = n_{s_2}(kT) \tag{8.75}$$

Assuming that a_k is positive, $a_k = A$. Then the probability of an error occurs when $v_1 < v_2$. This relationship can be expressed:

$$p_e = \text{prob}\ (v_1 < v_2) = \text{prob}\ (a_k + n_{c_1} < n_{s_2}) \tag{8.76}$$

The noise components n_{s_2} and n_{c_1} at their respective filter outputs are independent zero-mean Gaussian variables with variance equal to σ^2. Therefore their difference, $n = n_{s_2} - n_{c_1}$, is also a zero-mean Gaussian variable, with a variance of $2\sigma^2$. Then, from Eq. (8.70), the following expression is obtained:

$$P_e = \int_{-\infty}^{-A} p_n(x)\, dx = \frac{1}{2}\,\text{erfc} \left(\frac{1}{\sqrt{2}} \frac{A}{\sqrt{2\sigma^2}} \right) = \frac{1}{2}\,\text{erfc} \left(\sqrt{\frac{\gamma}{2}} \right) \tag{8.77}$$

where the snr is expressed:

$$\gamma = \frac{A^2}{2\sigma^2}$$

The coherent FSK noise characteristics of Eq. (8.77) are plotted in Fig. 8.12.

Noncoherent FSK. The noncoherent FSK selection scheme uses envelope detection rather than phase detection. A pair of bandpass filters tuned to different frequencies corresponding to a "mark" and a "space," respectively, are used to distinguish between the two signal envelopes r_1 and r_2. The probability distribution function of the envelope for r_1, containing signal A, is a Rician function [2], expressed:

$$p(r_1) = \frac{r_1}{\sigma^2} \exp\left(-\frac{r_1^2 + A^2}{2\sigma^2}\right) I_0\left(\frac{r_1 A}{\sigma^2}\right) \qquad 0 \le r_1 < \infty \qquad (8.78)$$

where $I_0()$ is a zero-order modified Bessel function. The distribution of r_1 is usually called Rician after S. O. Rice [2] and will be shown in Fig. 15.12. Then, at the same instant of time, the other filter passes only noise, and the pdf of envelope r_2, for the other filter output, is

$$p(r_2) = \frac{r_2}{\sigma^2} \exp\left(-\frac{r_1^2}{2\sigma^2}\right) \qquad 0 \le r_2 < \infty \qquad (8.79)$$

Thus, the probability of error can be expressed:

$$P_e = \text{prob } (r_1 < r_2) = \int_{r_1=0}^{\infty} p(r_1) \int_{r_2=r_1}^{\infty} p(r_1)\, dr_2\, dr_1$$

$$= \frac{1}{2} \exp\left(-\frac{A^2}{\sigma^2}\right) = \frac{1}{2} \exp\left(-\frac{\gamma}{2}\right) \qquad (8.80)$$

The function of Eq. (8.80) is plotted in Fig. 8.12.

Coherent detection in a DPSK system. A DPSK system can be used effectively when there is enough stability in the timing of the oscillator—and within the mobile-radio medium—to ensure that there will be negligible changes in phase from one information bit to the next. Figure 8.13 illustrates a coherent phase detector that can be used within a DPSK system, where the input pulse and the appropriately delayed previous input pulse are both fed to the phase detector.

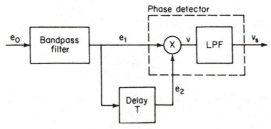

Figure 8.13 DPSK model employing phase-coherent detection with time delay.

TABLE 8.1   Message-Code Processing in a DPSK System

Message operation	Message coding								
Input message	1	0	1	1	0	1	0	0	
Encoded message	1 → 1	→ 0	→ 0	→ 0	→ 1	→ 1	→ 0	→ 1	
Transmitted phase e_0	0	0	π	π	π	0	0	π	0
Phase-comparison output	+	−	+	+	−	+	−	−	
Output message	1	0	1	1	0	1	0	0	

Table 8.1 illustrates the operations that typically occur in the type of DPSK system shown in Fig. 8.13 for an input message assumed to be 10110100. The encoded message becomes 110001101. Assuming that the first two marks (mark = 1V) are transmitted, both signals, e_1 and e_2, at the receiver have the same phase. However, when the third digit, a space (space = 0V), is transmitted, the phases of the two signals e_1 and e_2 are opposing. The two inputs to the phase detector shown in Fig. 8.13 can be expressed in the following form:

$$e_1(t) = [u(t) + n_{c_1}(t)] \cos \omega_0 t + n_{s_1}(t) \sin \omega_0 t$$

$$= \text{Re} \ [z_1 \exp \ (j\omega_0 t)]$$

and
$$e_2(t) = [u(t) + n_{c_2}(t)] \cos \omega_0 t + n_{s_2}(t) \sin \omega_0 t$$

$$= \text{Re} \ [z_2 \exp \ (j\omega_0 t)]$$

where n_{c_1}, n_{c_2}, n_{s_1}, and n_{s_2} are all zero-mean, mutually uncorrelated noise with a variance $\sigma^2/2$ and $u(t)$ is the signal. The following relationships also apply:

$$z_1 = u(t) + n_{c_1}(t) - jn_{s_1}(t) \tag{8.81a}$$

$$z_2 = u(t) + n_{c_2}(t) - jn_{s_2}(t) \tag{8.81b}$$

and therefore

$$w_1 = \frac{z_1 + z_2}{2} \quad \text{and} \quad w_2 = \frac{z_1 - z_2}{2}$$

Then the pdf of $|w_1|$ can be shown as a Rician distribution, as expressed in Eq. (8.78), and the pdf of $|w_2|$ is a Rayleigh distribution, as expressed in Eq. (8.79). The probability of error is given when the two signals z_1 and z_2 are detected as:

$$P_e = \text{prob} \left(\text{Re} \, [z_1 z_2^*] < 0 \right)$$

$$= \text{prob} \left(|w_1| < |w_2| \right)$$

The result of the above equation is similar to that obtained for Eq. (8.80):

$$P_e = \frac{1}{2} \exp \left(-\frac{u^2}{2\sigma^2} \right) = \frac{1}{2} \exp \left(-\gamma \right) \qquad (8.82)$$

The function of Eq. (8.82) is plotted in Fig. 8.12.

Of the five types of selection schemes described in the preceding paragraphs, the coherent binary AM (PSK) system provides the best overall performance. The noncoherent FSK system is the least desirable from the performance viewpoint; however, it has the advantages of simplicity in that it does not require phase correlation as a basis for coherent detection.

Fading case (mobile-radio environment)

If the time-delay spread of the mobile environment is small in comparison with the inverse of the signaling bandwidth, the received signal, then, will be corrupted only by fading.

Recent developments in speech digitization [6] have prompted an examination of digital coding as a possibility for mobile radiotelephone, which conventionally employs analog techniques for speech transmission. If a mobile-radio "control signal" is digital, and if the speech is handled digitally as well, it is simple to interleave the voice and control bits. Digital methods also offer the possibilities of inexpensive coder-decoder implementation, straightforward speech encryption (by bit scrambling), and efficient signal regeneration. The greatest incentive for the use of digital speech is to find a properly designed digital code that will provide a good resistance to multipath fading in the mobile-radio environment.

The following discussion applies only to digital voice signals; digital control signals are not discussed here, but will be described in Chap. 11.

The error rate of each detection system is increased by multipath fading. The snr γ in Eqs. (8.70) and (8.77) is a varying quantity on account of the effect of the fading. Also γ is proportional to the square of the Rayleigh-fading envelope, r^2, which can be obtained by letting $\gamma = r^2/(2\sigma_n^2)$ and by using Eq. (2.53).

$$p_\gamma(\gamma) = \frac{1}{\gamma_0} e^{-\gamma/\gamma_0}$$

where

$$\gamma_o = E[\gamma]$$

The average error rate may be obtained as follows:

$$\langle P_e \rangle = \int_0^\infty p_\gamma(\gamma) P_e \, d\gamma \tag{8.83}$$

For coherent binary AM detection, the average error rate is:

$$\langle P_e \rangle = \int_0^\infty \frac{1}{\gamma_o} e^{-\gamma/\gamma_o} \frac{1}{2} \, \mathrm{erfc} \, (\sqrt{\gamma}) \, d\gamma = \frac{1}{2} \left(1 - \frac{1}{\sqrt{1 + 1/\gamma_0}} \right)$$

For coherent FSK or FM with discrimination, the average error rate is:

$$\langle P_e \rangle = \int_0^\infty \frac{1}{\gamma} e^{-\gamma/\gamma_o} \frac{1}{2} \, \mathrm{erfc} \left(\sqrt{\frac{\gamma}{2}} \right) \, d\gamma$$

$$= \frac{1}{2} \left(1 - \frac{1}{\sqrt{1 + 2/\gamma_0}} \right)$$

For noncoherent FSK, the average error rate is:

$$\langle P_e \rangle = \int_0^\infty \frac{1}{\gamma} e^{-\gamma/\gamma_o} \frac{1}{2} \exp \left(-\frac{\gamma}{2} \right) \, d\gamma = \frac{1}{2 + \gamma_o}$$

For DPSK, the average error rate is:

$$\langle P_e \rangle = \int_0^\infty \frac{1}{\gamma} e^{-\gamma/\gamma_o} \frac{1}{2} \exp \, (-\gamma) \, d\gamma = \frac{1}{2 + 2\gamma_o}$$

The comparisons among the five digital modulation systems over both fading and nonfading channels are plotted in Fig. 8.14. In comparing Fig. 8.14 with Fig. 8.12, it can be seen that the error rate for each type of system increases substantially in a mobile-radio environment.

Example 8.2   For a given average error rate, the coherent binary AM detection process is approximately 3 dB better than the coherent FSK detection process. In coherent binary AM detection, the decision threshold is set at zero. The signal $z = A + n_c$ is in error when:

$$A + n_c < 0 \qquad \text{or} \qquad A < -n_c \tag{E8.2.1}$$

and

$$E[n_c^2] = \sigma_n^2 \tag{E8.2.2}$$

In coherent FSK detection, the decision is made only when the output signal component is below the output noise component, that is when

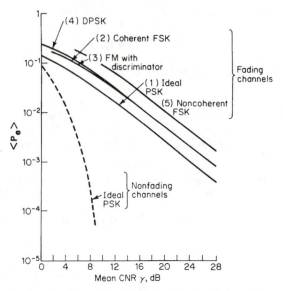

Figure 8.14 Comparisons among five digital modulation systems over fading channels.

$$A + n_{c_1} < n_{s_1} \quad \text{or} \quad A < n_{s_1} - n_{c_1} \quad \text{(E8.2.3)}$$

and

$$E[(n_{s_1} - n_{c_1})^2] = 2\sigma_n^2 \quad \text{(E8.2.4)}$$

By comparing the results of Eq. (E8.2.2) with those of Eq. (E8.2.4), it is apparent that the coherent binary AM detection process is always at least a 3-dB improvement over the coherent FSK detection process.

8.4   Constant Envelope Modulation

In the mobile radio, since the multipath fading distorts the amplitude of the carrier, the signal is sent by modulating the phase or frequency of the carrier, which has no impact on the amplitude. We call those modulations *constant envelope modulations;* that is, no signal is modulated on the amplitude. The distortion of carrier amplitude by other factors such as fading or nonlinear amplification will not affect the signal. Therefore, it is possible to use a nonlinear amplifier. Although, the nonlinear amplifier can distort the amplitude but not the phase, as we have mentioned in the inverse-sine modulation method in Sec. 12.5. But the intermodulation introduced by the AM-PM conversion in a nonlinear amplifier can be taken care of by a few particular modulations.

Constant envelope modulation can be a linear or a nonlinear modulation in digital mobile systems. In general, constant envelope modulation is used for the nonlinear modulation (FM) in analog mobile systems.

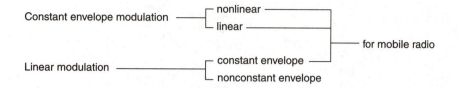

QPSK (quadrature phase shift keying)

QPSK is used for increasing the modulation efficiency from the BPSK. In BPSK (binary PSK) as mentioned previously, one symbol phase 0 or 180° at the modulation stage represents one bit. The modulation is one bit per second per Hz. In QPSK, one symbol (one of four phases) at the modulation stage represents two bits. The modulation efficiency is double. The possible combination of two bits is mapped according to the gray code the adjacent symbols (phase states) offer by one bit 00, 01, 11, 10. Therefore, the advantage is that a single symbol error corresponds to single bit error.

For QPSK, four phases are used, 0°, ±90°, 180°. In QPSK, half of the bit stream goes to the I multiplier, which has phases 0 or 180°, and the other half goes to the Q multiplier, which has +90° or −90°. This QPSK can be treated as two BPSKs. The modulator circuit and the signal space diagram are shown in Fig. 8.15(a).

The even bit stream d_I ($d_0, d_2, d_4 \ldots$) and odd bit stream d_Q ($d_1, d_3, d_5 \ldots$) the QPRK signal is:

$$s(t) = \frac{1}{\sqrt{2}} \, d_I(t) \cos (\omega_0 t + \phi_0) + \frac{1}{\sqrt{2}} \, d_Q(t) \sin (\omega_0 t + \phi_0) \quad (8.84)$$

where $\phi_0 = 45°$. It is necessary to modulate both the I and Q channels during each bit interval to retain the constant envelope of $s(t)$. Equation (8.84) also can be expressed as

$$s(t) = \cos [\omega_0 t + \phi(t)] \quad (8.85)$$

where $\phi(t) = 0°, \pm90°$, and 180°. The carrier phase changes every symbol interval $T_s = 2T_b$, where T_b is the bit interval.

In eliminating the leak of signal energy into the adjacent channels, the pulse shaping is done at the base band to provide the proper RF filtering at the transmitter output. When a QPSK signal is under filtering to reduce the spectral side lobes, the constant envelope characteristics are lost. Occasionally, the 180° phase shift can cause the envelope to go to zero instantly. Since the linear amplifier should be used to preserve the signal fidelity, the less efficient amplification from the linear amplifier is observed.

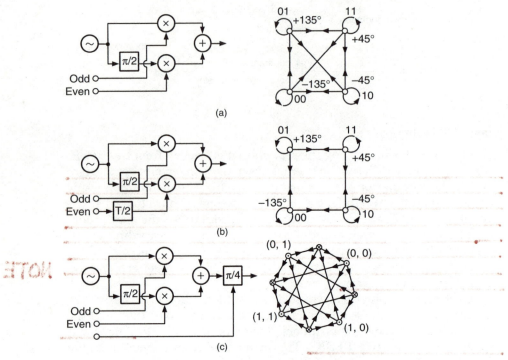

Figure 8.15   Modulator circuits and signal-space diagrams of (a) conventional QPSK, (b) offset QPSK, and (c) π/4 shift. (*After Hirade et al., Ref. 2, p. 14.*)

OQPSK (offset QPSK)

Like QPSK, the unfiltered OQPSK signal has a constant envelope. The timing of the pulse streams $d_I(t)$ and $d_Q(t)$ is shifted and offset by $T_s/2$. Therefore, Eq. (8.84) can be used for the expression of OQPSK as

$$s(t) = \frac{1}{\sqrt{2}} d_I(t) \cos(\omega_0 t + \phi_0) + \frac{1}{\sqrt{2}} d_Q\left(t + \frac{T_s}{2}\right) \sin(\omega_0 t + \phi_0) \quad (8.86)$$

where the two streams offset by $T_s/2$ can be expressed as

$$
\begin{array}{ccccccc}
& \overbrace{}^{T_s} & \overbrace{}^{T_s} & \overbrace{}^{T_s} & \cdots \\
d_I(t) = d_0 & d_0 & d_2 & d_2 & d_4 & d_4 \\
d(Q) = & d_1 & d_1 & d_3 & d_3 & d_5 & d_5 \cdots \\
& \underbrace{}_{T_s} & \underbrace{}_{T_s} & \underbrace{}_{T_s}
\end{array}
\quad (8.87)
$$

Therefore, two streams do not change status at the same time and eliminate the 180° phase change. The envelope will not go to zero as it does with QPSK. In this case, the nonlinear amplification can be used to provide more efficient amplification. The power spectrum of OQPSK is shown in Fig. 8.15(b).

The power spectral of the OQPSK modulation can be obtained by taking Fourier transform of $s(t)$ from Eq. (8.86) as

$$S(f) = \int_0^{T_b} s(t)e^{-j2\pi ft}\,dt$$

$$= 2T\left[\frac{\sin \pi\,(f-f_0)T_s}{\pi\,(f-f_0)T_s}\right]^2 \tag{8.88}$$

where $T_s = 2T_b$. Equation (8.88) is plotted in Fig. 8.16.

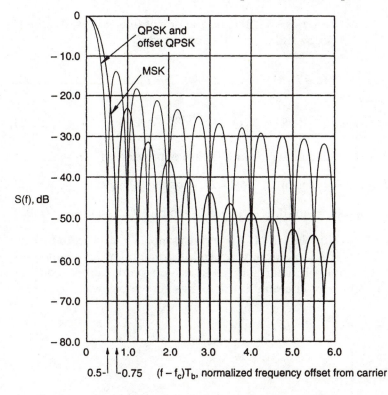

$$S_{\text{GMSK}}(f) = \frac{8PT_b[1-\cos 4\pi\,(f-f_0)T_b]}{\pi^2[1-16T_b^2(f-f_c)^2]^2}$$

$$S_{\text{OQPSK}}(f) = 2PT_b\left[\frac{\sin 2\pi\,(f-f_0)T_b}{2\pi(f-f_0)T_b}\right]^2$$

Figure 8.16 Normalized power spectral densities of unfiltered QPSK, OQPSK, and MSK systems. Because the modulated spectrum is symmetrical around the carrier frequency, only the upper sideband is shown. See Eqs. (8.88) and (8.105).

π/4-DQPSK

There is another modulation scheme called π/4-DQPSK which belongs to the class of QPSK. The modulation architecture of π/4-QPSK and π/4-DQPSK systems is essentially the same as shown in Fig. 8.15(c).

In a differential encoding in DQPSK, symbols are represented as the relative changes in phase rather than an absolute phase.

The two carriers in quadrature with each other generate the π/4-DQPSK waveform.

$$s(t) = s_I + s_Q$$

$$= d_I(t - T_s/2) \cos (\omega_0 t + \phi_0) + d_Q(t - T_s/2) \sin (\omega_0 t + \phi_0) \quad (8.89)$$

where $d_I(t - T_s/2)$ and $d_Q(t - T_s/2)$ can take one of five values: 0, +1, −1, $1/\sqrt{2}$, or $-1/\sqrt{2}$. Also, d_I and d_Q can be expressed as

$$d_I(t - T_s/2) = \sum_{k=0}^{N-1} I_k \, p(t - kT_s - T_s/2) \quad (8.90)$$

$$d_Q(t - T_s/2) = \sum Q_k \, p(t - kT_s - T_s/2) \quad (8.91)$$

where $p(x)$ is the pulse shape. I_k and Q_k are expressed as

$$I_k = \cos \theta_k = \cos (\theta_{k-1} + \phi_k) = I_{k-1} \cos \phi_k - Q_{k-1} \sin \phi_k \quad (8.92)$$

$$Q_k = \sin \theta_k = \sin (\theta_{k-1} + \phi_k) = I_{k-1} \sin \phi_k + Q_{k-1} \cos \phi_k \quad (8.93)$$

ϕ_k is the phase shift based on the input symbols s_I and s_Q to form a pair. Gray code is used in mapping a two-bit symbol. The adjacent symbols differ in a single bit. Therefore, most two-bit symbol errors contain only a single bit error. The combination is as shown below:

s_I	s_Q	ϕ_k
1	1	π/4
0	1	3π/4
0	0	−3π/4
1	0	−π/4

The reason for choosing π/4-DQPSK is based on our choice of having a power-efficient modulation or a spectral-efficient modulation. For a power-efficient modulation, the dynamic power range of the nonlinearly amplifier (NLA) is applied to increase power efficiency. For example, an RF amplifier for operating in a linear range has to take 6 dB output back off (OBO). This means that a 4-watt amplifier can only operate at 1 watt maximum power. Therefore, a gain of 6 dB by operating at a nonlinearly amplified range for an NLA shows the power efficiency. In a power-efficient modulation, spectral regrowth is a problem that reduces the spectrum efficiency. For a modulation with spectral efficiency, we have to use a linear amplifier to reduce the spectral regrowth. However, different modulation schemes have different spectral regrowth natures.

QPSK, due to the instantaneous 180° phase shift, leads to a significant spectral regrowth and thus has a low spectral efficiency. In a $\pi/4$-DQPSK system, its instantaneous phase transitions are limited to $\pm 135°$, the spectral regrowth is reduced. Therefore, it is a better modulation scheme than QPSK. In OQPSK its instantaneous phase transitions are limited to $\pm 90°$. The spectral regrowth is the lowest of all three.

Comparing $\pi/4$-DQPSK with OQPSK, the signal of $\pi/4$-DQPSK can be easily differentially demodulated, and the scheme has an ease of hardware implementation. The demodulation of OQPSK is much harder. Therefore, $\pi/4$-DQPSK with a linear amplifier is used to achieve the spectral efficiency for the cellular and PCS systems. One scheme called FQPSK can be used to increase the power efficiency yet stop the spectral growth [13].

Subcarrier QPSK

The subcarrier QPSK is proposed [14] to carry a weak signal on top of the regular QPSK signal. This weak signal need only be transmitted at a very short distance. The regular QPSK signal is generally transmitted at a long distance. The QPSK signal will be modified as follows:

$$s(t) = \frac{1}{\sqrt{2}} d_I(t) \cos (\omega_0 t + \phi_0) + \frac{1}{\sqrt{2}} d_Q(t) \sin (\omega_0 t + \phi_0)$$

$$+ \frac{a}{\sqrt{2}} d_I'(t) \cos (\omega_0 t + \phi_0 + \Delta\phi_0) + \frac{a}{\sqrt{2}} d_Q'(t) \sin (\omega_0 t + \phi_0 + \Delta\phi_0) \quad (8.94)$$

where $d_I'(t)$ and $d_Q'(t)$ are the different bits of I and Q information. The amplitude a is very small; $a \ll 1$. The phase difference $\Delta\phi_0$ from ϕ_0 is also very small. This subcarrier QPSK signal can be detected at a short distance without being corrupted by the mobile radio multipath medium. It does naturally generate a certain noise level for the QPSK signal. The noise level can be calculated and tolerated in the QPSK system.

The application of the subcarrier system is very valuable. Many times, the mobile transmitter can send two signals: one to the base station (a regular QPSK signal) and one to a nearby repeater or receiver such as a location finder.

GMSK

GMSK stands for *Gaussian filtered minimum shift keying*. It is a type of constant envelope FSK where the frequency modulation is a carefully handled phase modulation. The constant amplitude of the GMSK signal makes it suitable for use with high-efficiency amplifiers [15]. The wave forms started from a data stream to the final MSK wave form are shown in Fig. 8.17. In Fig. 8.17, there are two frequency versions—high frequency and low frequency—to provide two frequencies for sending the data information. The relationship between the data bits and frequencies is shown below [16]:

| Digital Input | | MSK Output | |
| Bit Value | | Frequency | Sense |
Odd Bit	Even Bit	High or Low	+ or −
1	1	High	+
−1	1	Low	−
1	−1	Low	+
−1	−1	High	−

When sense is positive, the carrier frequency remains unchanged. When sense is negative, the carrier frequency is upside down.

The resulting MSK waveform shown in Fig. 8.17 has a relatively smooth phase transition from one frequency to the other. Such a smooth transition can reduce the spectral regrowth. In QPSK family modulations, the phase transitions are discontinuous. Therefore, the power spectrum density of OQPSK has shown that the spectral regrowth is wider than that of GMSK (see Fig. 8.16). The minimum shift keying

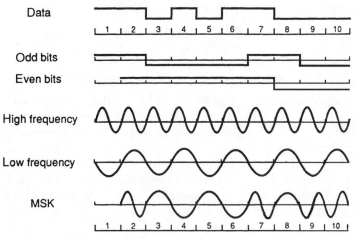

Figure 8.17   Generating minimum-shift keying.

means the minimum tone-space shift keying. We have to use the criterion that two waveforms be orthogonal:

$$\int_0^T \cos(2\pi f_1 t + \phi) \cdot \cos(2\pi f_2 t)\, dt = 0 \qquad (8.95)$$

where $1/T$ is the separation in hertz between two frequencies f_1 and f_2. Equation (8.95) can be solved by applying the condition $f_1 + f_2 \gg 1$, and

$$\frac{\sin 2\pi(f_1 + f_2)T}{2\pi(f_1 + f_2)} \approx \frac{\cos 2\pi(f_1 + f_2)T}{2\pi(f_1 + f_2)} \approx 0 \qquad (8.96)$$

The result of Eq. (8.95) becomes

$$\cos\phi \cdot \sin 2(f_1 - f_2)T + \sin\phi \cdot [\cos 2\pi(f_1 - f_2)T - 1] \approx 0 \qquad (8.97)$$

For a noncoherent FSK, $\phi \neq 0$, Eq. (8.97) becomes

$$f_1 - f_2 = \frac{1}{T} \qquad (8.98)$$

For a coherent FSK or MSK, $\phi = 0$. Equation (8.97) becomes

$$f_1 - f_2 = \frac{1}{2T} \qquad (8.99)$$

Equation (8.99) is the minimum spacing requirement shown in Fig. 8.18. Since all the continuous-phase FSK schemes have a low modulation index, they intrinsically have constant envelope properties, unless severe band pass filtering is introduced to the modulator output. In MSK, the frequency shift precisely increases or decreases the phase by 90° every T seconds. Thus, the signal waveform is

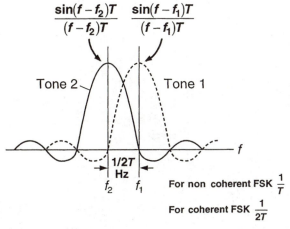

Figure 8.18 MFSK—minimum tone spacing for FSK.

$$s(t) = \sin\left(\omega_0 t + 2\pi \int_0^t s_i \, dt + \frac{n\pi}{2}\right) \qquad 0 < t < T \qquad (8.100)$$

where

$$s_i = \begin{cases} s_1 = \dfrac{1}{4T} & \text{for data bit 1} \\[2ex] s_2 = \dfrac{1}{4T} & \text{for data bit 0} \end{cases} \qquad (8.101)$$

Equation (8.100) also can be expressed in a different form as

$$s(t) = \sin\left(\omega_0 t + \frac{n\pi}{2} \pm \frac{\pi t}{2T}\right) \qquad (8.102)$$

or

$$s(t) = \cos\left(\pm\frac{\pi t}{2T}\right)\sin\left(\omega_0 t + \frac{n\pi}{2}\right) + \sin\left(\pm\frac{\pi t}{2T}\right)\cos\left(\omega_0 t + \frac{n\pi}{2}\right) \quad (8.103)$$

When we compare Eq. (8.103) with Eq. (8.86), we find that the two equations are very similar. In fact, the phase-modulation waveforms of the *I*- and *Q*-channel modulations of OQPSK are modulated by sine and cosine waveforms; thus, the output will be identical to that of MSK. The GMSK is the Gaussian low-pass filter added to MSK, as shown in Fig. 8.19. The power spectrum density function of GMSK can be obtained by taking the signal *s(t)* through a filter, the frequency response of which is

$$H(f) = \exp\left\{-\left(\frac{f}{B}\right)^2 \frac{\ln 2}{2}\right\} \qquad (8.104)$$

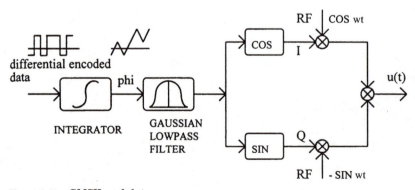

Figure 8.19   GMSK modulator.

The power spectrum density $S(f)$ of $s(t)$ can be obtained by taking the Fourier transform of Eq. (8.102), which is the same process as shown in Eq. (8.88). We can get the power spectrum density function $S_G(f)$ of GMSK by applying Eq. (1.41). The result is

$$S_G(f) = |H(f)|^2 \cdot S(f)$$

$$= \frac{8T_b[1 - \cos 4\pi(f - f_0)T_b]}{\pi^2[1 - 16T_b^2(f - f_0)^2]^2} \tag{8.105}$$

The power spectrum density functions of GMSK with different filter bandwidth is shown in Fig. 8.20. The B_bT is the normalized bandwidth of a Gaussian filter. The narrower the Gaussian filter, the less the spectral regrowth, but the bit error performance is degraded, as shown in Fig. 8.20. As compared with the power spectrum densities between OQPSK and GMSK, GMSK has less spectral regrowth (see Fig. 8.20), but OQPSK has better BER performance (see Fig. 8.21).

8.5   Nonconstant Envelope Modulation

In the nonconstant envelope modulation, the information can be modulated on the amplitude of carrier frequency such as AM (amplitude mod-

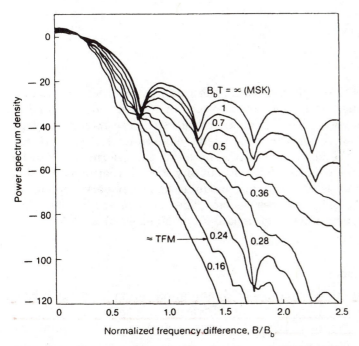

Figure 8.20 Power spectrum density functions of GMSK. *(After Hirade et al., Ref. 28, p. 15.)*

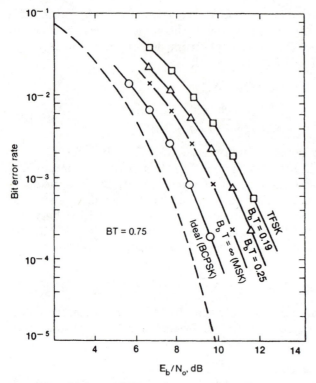

Figure 8.21 Measured BER performance. (*After Hirade et al., Ref. 28, p. 17.*)

ulation) and SSB (single sideband), the common modulation schemes belonging to this family. The power spectrum efficiency of this kind of modulation is always higher than other kinds such as constant envelope modulation. However, in the mobile radio environment, the multipath fading distorts the signal envelope. Therefore, the information can be very hard to receive undistorted without any special treatment. For this reason, the nonconstant envelope modulation is not very popular in mobile radio or mobile wireless systems. The following section describes one of the nonconstant envelope modulations called QAM [17, 24, 25, 26].

QAM (quadrature amplitude modulation)

QAM signaling can be viewed as a combination of amplitude shift keying (ASK) and phase shift keying (PSK). It can also be viewed as an amplitude shift keying in two dimensions. In an *M*-ary QAM signal, the signal in each state *I* is

$$s_i(t) = R_e \left[v_i(t) e^{j2\pi f_c t} \right]$$

$$= Aa_i \cos (2\pi f_c t) + Ab_i \sin (2\pi f_c t) \qquad 0 < t < T \qquad (8.106)$$

where A is a constant, $\sqrt{2}A$ representing the amplitude of the lowest state, and (a_i, b_i) is a pair of identifying states in the constellation of the ith state. The 16-QAM constellation diagram is shown in Fig. 8.22.

The probability of error for QAM can be determined from the probability of error for PAM. For the probability of error of a $\sqrt{M}$-ary PAM is expressed as

$$P_{\sqrt{M}} = 2 \left(1 - \frac{1}{\sqrt{M}} \right) Q \left(\sqrt{\frac{3}{M-1} \frac{S_{av}}{N_0}} \right) \qquad (8.107)$$

where S_{av}/N_0 is the average SNR per symbol of an M-QAM, and

$$S_{av} = 2 \left(\frac{A^2 \sum_{1}^{\sqrt{M}} a_i^2}{\sqrt{M}} \right) \qquad (8.108)$$

The probability of a symbol error for an M-ary QAM is

$$P_M = 1 - (1 - P_{\sqrt{M}})^2 \qquad (8.109)$$

When $P_{\sqrt{M}}$ is a small value:

$$P_M = 2P_{\sqrt{M}} - P_{\sqrt{M}}^2 \approx 2P_{\sqrt{M}} \qquad (8.110)$$

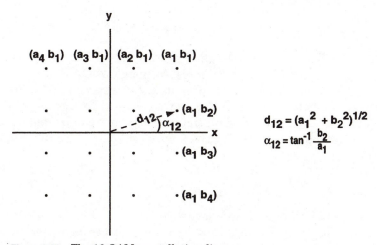

Figure 8.22 The 16-QAM constellation diagram.

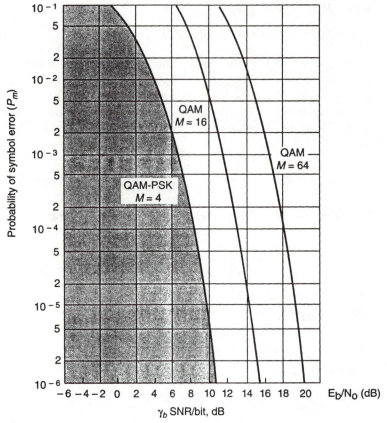

Figure 8.23 Probability of a symbol error for QAM and PSK in the 10^{-1} to 10^{-6} range. (From Proakis, 1989.) For $M = 4$, the performance of QPSK is the same as that of 4-QAM.

Figure 8.23 shows the bit error rate for an M-QAM on nonfading and Rayleigh fading channels. Although M-QAM is a power-efficient modulation, the bit error probability increases as M increases. Besides, the envelope of M-QAM is not a constant. It means that the partial information is carried on the envelope and can be distorted while the envelope is distorted by the Gaussian noise or Rayleigh fading.

8.6   OFDM Modem

Orthogonal frequency division multiplexing (OFDM) is a modulation technique where source symbols are transmitted in parallel by applying to a large number of orthogonal subcarriers. The OFDM signal $s(t)$ can be expressed

$$s(t) = R_e \{v(t)e^{j2\pi f_c t}\} \tag{8.111}$$

The complex envelope $v(t)$ of an OFDM signal $s(t)$ can be expressed [20]

$$v(t) = \sum_{k=1}^{K} v_k(t) = A \sum_{k=0}^{K} \sum_{n=0}^{N-1} x_{k,n}\phi_n(t - kT) \tag{8.112}$$

where A is the carrier amplitude and k is the kth block of N serial source symbols.

The N orthogonal waveforms $\{\phi_n(t)\}$ in Eq. (8.112) are chosen as

$$\phi_n(t) = h_a(t) \exp\left\{\frac{j2\pi\left(n - \dfrac{N-1}{2}\right)t}{T}\right\} \tag{8.113}$$

and $h_a(t)$ can be a rectangular amplitude shaping pulse, $h_a(t) = U_T(t)$. The source symbol block $\underline{x}_k$ at time $t = kT_s$ can be expressed as

$$\underline{x}_k = \{x_{k,0}, x_{k,1}, x_{k,2}, x_{k,3}, \cdots\cdots x_{k,N-1}\}$$

A block of N serial source symbols has a symbol duration T_s, where T_s can be less than the rms time delay spread Δ of the medium ($T_s < \Delta$). The block of N serial source can be converted to a block of N parallel modulated symbols, each in a duration of $T = NT_s$. Let the block length N be chosen so that $NT_s \gg \Delta$. Since the symbol rate ($1/T$) on each subcarrier is much less than the serial source rate

$$\left(\frac{1}{T} < \frac{1}{T_s}\right)$$

the effects of delay spread are greatly reduced, and we can even try to eliminate the equalizer. The source symbol block $\underline{x}_k$ in time domain is applied to N subcarrier and converted to $\underline{X}_k$ in frequency domain as shown in Fig. 8.24. The N transmitted signals are of equal energy and

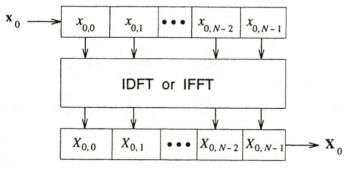

Figure 8.24   Block diagram of OFDM transmitter using IDFT or IFFT.

equal duration, and the signal subcarriers are separated by $1/T$ hertz, making the signals orthogonal among themselves. The medium dispersion will still cause consecutive modulation symbol blocks to overlap, and creates a so-called residual intersymbol interference (ISI). It can be reduced by introducing the guard intervals between the blocks. The source symbol duration with guard interval is

$$T'_s = \frac{T_s}{1 + \dfrac{G}{N}} \tag{8.114}$$

where G should be greater than the time delay spread Δ.

The attractive advantage of using OFDM is that the modulation can be expressed in a discrete frequency domain after going through the inverse fast Fourier transform (IFFT). Let the source symbol block at the first block, $k = 0$, the complex envelope

$$v_o(t) = A \sum_{n=0}^{N-1} x_{o,n} \exp \left\{ \frac{j2\pi nt}{NT_s} \right\} \qquad 0 < t < T \tag{8.115}$$

The complex envelope $v_o(t)$ is sampled at $t = kT_s$ and forms a sequence through IFFT operation.

$$X_{o,k} = v(kT_s) = A \sum_{n=0}^{N-1} x_{o,n} \exp \left\{ \frac{j2\pi nk}{N} \right\} \tag{8.116}$$

Then, $X_{o,k}$ passes a *D/A* converter and is subcarrier-modulated. OFDM is a means of providing power-efficient modulation for an OFDM signal expressed in Eq. (8.111), the power spectrum density (PSD) of its complex envelope is

$$S(f) = \frac{A^2}{T} \sigma_x^2 \sum_{n=0}^{N-1} \left| H\left(f - \frac{1}{T}\left(n - \frac{N-1}{2} \right) \right) \right|^2 \tag{8.117}$$

where

$$\sigma_x^2 = \frac{1}{2} E[|x_{k,n}|^2]$$

and

$$H(f) = F^{-1}[h(t)]$$

The PSD of OFDM with $N = 4$ and $N = 32$ is shown in Fig. 8.25. For large N, the PSD becomes more flat at

$$f = \frac{1.6}{T_s} \left(\text{or } \frac{1.6N}{T} \right)$$

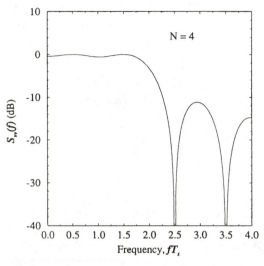
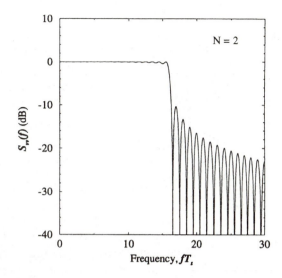

Figure 8.25   Psd of OFDM with $N = 4$ and $N = 32$.

Therefore, as the block length N is large, the PSD of OFDM approaches that of single-carrier modulation. This is power-efficient modulation.

8.7   Spread-Spectrum Systems

A spread-spectrum system is used to spread a signal over a frequency band that is much wider than the minimum bandwidth required to transmit the information being sent. Typically, a multimegahertz bandwidth is used to transmit a voiceband signal of a few kilohertz. A spread-spectrum system can be used to improve spectrum efficiency in mobile-radio communications.

In the newly allocated 850-MHz mobile-telephone frequency spectrum, a band of 40 MHz (825 to 845 and 870 to 890 MHz) is assigned to mobile communications, as was mentioned in the introduction to this book. One-half of the frequency band is used for transmitting and the other half for receiving, with a one-way transmission bandwidth consisting of a 20-MHz continuous band. Communications channels can be evenly assigned a particular frequency for conventional use in a cellular planned system, or a coded waveform covering the total 20-MHz band can be assigned to a particular user. Codes that can be used include the following different kinds of modulation:

1. **Direct-sequence-modulated system**—Modulation of a carrier by a digital-code sequence whose bit rate is much higher than the information signal bandwidth.

2. **Frequency-hopping system**—Carrier-frequency shifting in discrete increments, in a pattern governed by a code sequence which determines the order of frequency usage.

3. **Time-hopping system**—The timing of transmission—e.g., low-duty-cycle or short-duration—is governed by a code sequence.

4. **Time/frequency-hopping system** [18]—The code sequence determines both the transmitted frequency and the transmission bit rate.

5. **"Chirp" or pulse-FM modulation system**—A carrier is swept over a wide band during a given pulse interval.

Among these five types of modulation, the time/frequency-hopping system is a primary choice for mobile-radio use. In this type of system, each mobile unit is assigned a unique set of time- and frequency-coded waveforms that are used for both transmission and reception. These coded waveforms have large time-bandwidth products, so that interfering signals are easily suppressed. This type of system is relatively complicated; however with present-day advanced technology, such systems are already under development. A comprehensive discussion of the codes for this type of system available for mobile-communication use is contained in Cooper and Nettleton [18]; it is known either as a continuous-phase discrete-frequency-modulated (CPDFM) signaling system, or as a frequency-hopped differential phase-shift keying (FH-DPSK) system.

8.8   Frequency-Hopped Differential Phase-Shift Keying (FH-DPSK) Systems

The frequency-hopped differential phase-shift keying (FH-DPSK) system is a spread-spectrum method of communication that can provide mobile-radiotelephone services to a large number of urban customers. Each FH-DPSK signal is a sinusoidal, constant-envelope, continuous-phase signal divided into discrete time intervals—each of duration t_1—called "chips." Over a waveform interval of duration $T = Lt_1$, a specific sequence of L different tones or chips is assigned and no frequency assignment is repeated (L is the number of different frequencies). The signal is periodic with period T. The frequency assignment takes the following form:

$$f_i^k = f_c + a_i^k f_1$$

where f_i^k is the frequency shift from the carrier f_c assigned to the ith time chip of the kth waveform of the set, a^k is the integer from the kth code of a code set, $i = 1, \ldots, L$, and f_1 is the fundamental frequency channel before hopping. Hence, there are L distinct signal waveforms, each with L time chips and each with a bandwidth of approximately L/t_1.

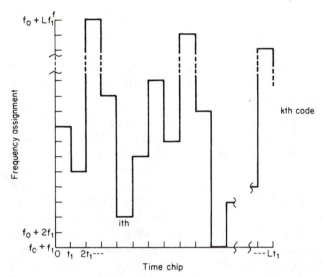

Figure 8.26 Representative signaling in the time-frequency plane.

A representative signal in the time-frequency plane is shown in Fig. 8.26. The "continuous-phase" property ensures that the waveform chip will form a continuous joint at the edges with other chips and therefore will contain an integer of cycles.

The figure of merit representing system and/or spectral efficiency can be expressed [19]:

$$\eta = M \frac{R_b}{B} \qquad (8.118)$$

where M is the number of users simultaneously served by the system, B is the one-way transmission bandwidth occupied by the system, and R_b is the information-transmission rate, defined:

$$R_b = \frac{\log_2 L}{T} = \frac{\log_2 L}{Lt_1} \qquad (8.119)$$

Each mobile unit receives its own L-tone sequence, as well as $M - 1$ interfering sequences. Assuming all transmitters are uncorrelated and $M \gg 1$, then the received interference is equivalent to white Gaussian noise over the system bandwidth B. The frequency-managing properties of the code are as follows:

1. If the codes are synchronized and if the frequency slots are nonoverlapping, the signals are all mutually orthogonal. A set of L orthogonal words is defined:

$$\sum_{j=1}^{L} w_{k,j} w_{l,j} = L\delta_{k,l}$$

where $w_{k,j}$ is the jth bit of the kth word and $\delta_{k,l}$ is the Kronecker delta function. However, synchronization is difficult to acquire in a mobile-communication medium.

2. If synchronization is not employed, the code used must be such that no two codes have more than one frequency coincidence for any time shift. This type of code can be verified by shifting any given row in the code set cyclically to either the right or the left. This property is useful in a mobile-communication application, since no other type of synchronization is needed and therefore any number of users can transmit independently with minimal mutual interference. The major limitation of this type of code is that all signals must be received with equal power. In actual practice, this is difficult to achieve and usually requires additional design effort.

Signal and noise representation

The signal waveform of each set of L chips may use frequency-hopped "carriers" for a biphase code message; thus, the kth code in the ith chip period is

$$s_i^k(t) = A\sin(\omega_i t + \theta_i)$$
$$= A\sin[2\pi(f_c + a_i^k f_1)t + \theta_i] \qquad (i-1)t_1 < t < it_1$$

where a_i^k is the ith integer from the kth code, $\omega_i = 2\pi(f_c + a_i^k f_1)$, A is the signal amplitude, and $\theta_i = 0$ or π is constant in the ith chip period and represents the transmitted message.

Figure 8.27 illustrates a receiver model with differential phase-reference functional capabilities.

Within any given area, there are typically M mobile units roaming about and $M-1$ interference sequences to contend with. $M-1$ interference sequences can be treated as narrowband conventional noise; the noise in the ith chip period can be expressed [20]:

$$n_i(t) = I_{c,i}(t)\cos(\omega_i t) + I_{s,i}(t)\sin(\omega_i t) \qquad (8.120)$$

where $I_{c,i}$ and $I_{s,i}$ are the in-phase and quadrature components of the noise.

For a time interval t_i, the ith chip (at frequency ω_i) develops its maximum power at the output of the appropriate matched filter. The output, the sum of the signal plus interference, may be written (see Fig. 8.27) as:

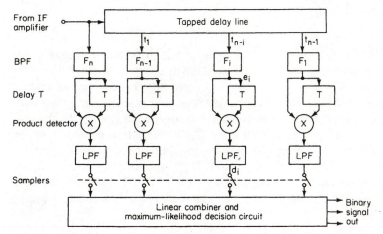

Figure 8.27 Receiver model with differential phase reference.

$$v_i(t) = A \sin [\omega_i t + \theta_i(t)] + I_{s,i}(t) \sin (\omega_i t) + I_{c,i}(t) \cos (\omega_i t) \quad (8.121)$$

where $\theta_i(t)$ is the phase of the ith chip (0 or π). The interference envelopes, $I_{s,i}(t)$ and $I_{c,i}(t)$, are independent Gaussian random variables with zero mean and a variance of σ^2.

$$\sigma_i^2 = \langle I_{s,i}^2 \rangle = \langle I_{c,i}^2 \rangle \quad (8.122)$$

The output of the low-pass filter for the ith detector (see Fig. 8.27) is

$$d_i(t) = 2\langle v_i(t)v_i(t-T) \rangle = A^2 w_{l,i} + N_i(t) \quad (8.123)$$

where the interference components $N_i(t)$ are expressed:

$$N_i(t) = \pm AI_{s,i}(t) \pm AI_{s,i}(t-T)$$
$$+ I_{s,i}(t)I_{s,i}(t-T) + I_{c,i}(t)I_{c,i}(t-T) \quad (8.124)$$

and where the plus and minus signs in Eq. (8.124) are chosen as follows:

$\theta(t-T)$	$\theta(t)$	Sign
0	0	+ +
π	0	− +
0	π	+ −
π	π	− −

and
$$\langle N_j(t)N_k(t) \rangle = 0 \qquad \text{for } j \neq k$$

8.9   Error Rate and System Efficiency of an FH-DPSK System

A typical DPSK system uses a receiver with differential-phase error circuitry to overcome the effects of multipath signal losses while maintaining phase coherence when hopping from one frequency to another. Figure 8.27 illustrates this type of receiver, where each time chip contains differential phase-modulated information with respect to the corresponding time chip. For a frequency of 800 MHz and a vehicle speed as high as 100 mi/h (160 km/h), a waveform duration of 200 µs (a hopping rate of 5 kHz) may be assumed. During this time interval, the vehicle traverses only 0.35 in (0.89 cm), or 0.03 λ. It is therefore reasonable to assume that a given waveform can provide a reliable phase reference.

Single-active-base-station intracell interference

To further analyze the effects of tone-sequence interference on reception, consider a mobile-radio system consisting of only one, centrally located, active station serving M mobile users within a given service area. Each mobile unit receives its own tone sequence and $M - 1$ interfering tone sequences. The received interference can be expressed:

$$\sigma_I^2 = \langle I_{s,i}^2 \rangle = \langle I_{c,i}^2 \rangle = \frac{A^2}{2} \frac{M-1}{Bt_1} \qquad (8.125)$$

where $1/t_1$, the reciprocal of the chip pulse width, is the noise bandwidth for each bandpass filter, the system bandwidth is B, and A is the amplitude of the tone sequence of the desired mobile unit's address. The kth output of the combiner (shown in Fig. 8.27) in response to the lth word being transmitted is represented by $C_{k,l}$ and defined:

$$C_{k,l} = LA^2 \delta_{k,l} + \sum_{j=1}^{L} w_{k,j} N_j \qquad (8.126)$$

where $w_{k,j}$ is the jth bit of the kth word, $\delta_{k,l}$ is the Kronecker delta function, and N_j is as expressed in Eq. (8.124) by changing i to j. A word-detection error occurs only when the "wrong" combiner output $C_{k,l}$ exceeds the "right" combiner output $C_{l,l}$, which means that

$$C_{l,l} - C_{k,l} = LA^2 + 2 \sum_{\{j\}} w_{l,j} N_j < 0 \qquad (8.127)$$

where $\sum_{\{j\}}$ indicates the summation over the $L/2$ values of j, and where $w_{l,j}$ and $w_{k,j}$ disagree. Equation (8.127) is written in the form of a signal plus noise, with a total signal-to-interference ratio of

$$\gamma = \frac{L^2 A^4}{4 \sum_{\{j\}} \langle N_j^2 \rangle} \tag{8.128}$$

Substituting Eq. (8.124) into Eq. (8.128) yields the following expression:

$$\gamma = \frac{L \gamma_f^2}{2\gamma_f + 1} \tag{8.129}$$

γ_f expresses the signal-to-interference ratio at the output of each bandpass filter and can be written:

$$\gamma_f = \frac{A^2}{2\sigma_I^2}$$

$$= \frac{Bt_1}{M-1} = \frac{B}{M-1} \frac{\log_2 L}{LR_b} \tag{8.130}$$

where σ_I is as shown in Eq. (8.125) and t_1 is as shown in Eq. (8.119).

Example 8.3   Given the condition that there are 40 mobile users within an area served by an FH-DPSK system and the coding consists of 32 orthogonal words, each with its own unique tone, then with a system bandwidth of $B = 20$ MHz and a signaling rate of $R_b = 32$ kb/s, find the total signal-to-interference ratio.

solution   The signal-to-interference ratio (s.i.r.) at the output of each bandpass filter is:

$$\gamma_f = \frac{B}{M-1} \frac{\log_2 L}{LR_b} = \frac{20 \times 10^6}{39} \frac{\log_2 32}{32 \times 32 \times 10^3} = 2.5 \approx 4 \text{ dB} \tag{E8.3.1}$$

Then the total s.i.r. is:

$$\gamma = \frac{L \gamma_f^2}{2\gamma_f + 1} = 33.3 \approx 15 \text{ dB} \tag{E8.3.2}$$

Example 8.4   What is the s.i.r. range of γ_f within which a word-detection error can occur when a wrong combiner output exceeds the right combiner output?

solution   A word-detection error can occur when the conditions of Eq. (8.127) are present, expressed as follows:

$$(LA^2)^2 < \left(2 \sum_{\{j\}}^{L} w_{l,j} N_j \right)^2 = 4 \sum_{\{j\}}^{L} N_j^2 \tag{E8.4.1}$$

When Eq. (E8.4.1) is inserted into Eq. (8.128), the value of γ should be less than 1.

The following expression is obtained from Eq. (8.129) with $\gamma < 1$:

$$\frac{L\gamma_f^2}{2\gamma_f + 1} < 0 \qquad (E8.4.2)$$

In solving Eq. (E8.4.2), the following is obtained:

$$\gamma_f < \frac{1}{L} + \sqrt{1/L^2 + 1/L} \qquad (E8.4.3)$$

Where $L = 32$, $\gamma_f < 0.21$ (or -6.77 dB). Therefore, when γ_f is less than -6.77 dB for $L = 32$, a word-detection error can occur.

Since $C_{l,l}$ and $C_{k,l}$ are Gaussian terms, then the difference $C_{l,l} - C_{k,l}$ is also Gaussian. The probability of a word-detection error ($C_{l,l} - C_{k,l} < 0$) is

$$P = Q(\sqrt{\gamma}) \triangleq \frac{1}{\sqrt{2\pi}} \int_\gamma^\infty e^{-x^2/2} \, dx$$

There are $L - 1$ independent ways that an error can occur in detecting a given word. Consequently, the system bit-error probability is:

$$P_b = \frac{1}{2} \left[1 - (1 - P)^{L-1} \right] \approx \frac{L-1}{2} Q(\sqrt{\gamma}) \qquad (8.131)$$

For $L = 32$, as suggested by Cooper and Nettleton [18], the nonfading bit-error probability is as shown in Fig. 8.28.

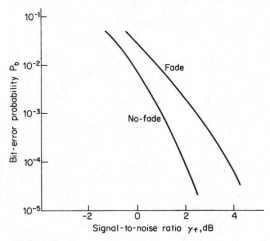

Figure 8.28 Bit-error probability as a function of signal-to-noise ratio at the matched filters for $L = 32$ (Rayleigh model).

For good-quality speech, where $P_b = 10^{-3}$, the signal-to-noise ratio at the output of each bandpass filter is $\gamma_f = 1$ dB. This shows that the signal-to-noise ratio can be lowered by increasing the bandwidth.

Multiple-base-station intercell interference

In a high-capacity mobile-radio system employing multiple base stations, only adjacent cells typically cause interference (intercell interference). Nonadjacent cells in the system, because of high attenuation by the radio medium (power falls off faster than the range of $1/r^3$ to $1/r^4$), do not generally cause interference. The worst interference occurs when a mobile unit approaches the corner boundaries of a cell, where the interfering frequencies from adjacent base stations are almost equidistant from the mobile unit. Each base station serves M mobile units, and the signal strength conforms to the propagation rules.

Figure 8.29 shows a typical cellular layout for a mobile-radio system [19]. The mobile unit at the location of the cell corner in Fig. 8.29 is in the area of worst-case interference. The total contribution of signal energy from a cell Y_i transmitted at its base station X to a mobile unit is

$$P_{Y_i} = \frac{P_t}{\pi R_s^2} \int_0^{2\pi} \int_0^{R_s} \frac{r^{\beta + 1}}{(D_i^2 + r^2 - 2D_i r \cos \theta)^{\beta/2}} \, dr \, d\theta \qquad (8.132)$$

where $\beta = 3$ or 4, depending on the radio medium.

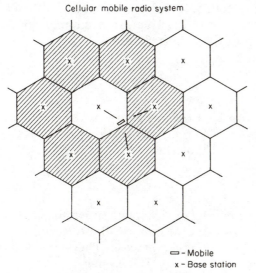

Cellular mobile radio system

⊂⊃ - Mobile
x - Base station

Figure 8.29 Cellular layout for a mobile-radio system, worst-case interference.

According to the rules of propagation, P_t is the total power, R_s is the cell radius, and D_i is the distance to the ith base station. The term dr $d\theta$ represents the integration factor for cell Y_i.

Dividing by P_t and summing the total contributions from all $N - 1$ interfering cells yields the signal-to-interference ratio at cell X base station. If each cell contains $M \gg 1$ users, the ratio of signal power for one mobile to total interference at each matched filter output can be expressed:

$$\hat{\gamma}_f = \frac{P_t}{M \sum\limits_{i=1}^{N-1} P_{Y_i}}$$ (8.133)

where N is the total number of cells in the service area. Adding the interference contributions from the dozen base stations nearest a given cell corner allows the signal-to-interference ratio for a mobile unit located in a cell corner to be derived from Eq. (8.133):

$$\hat{\gamma}_f = 0.26\gamma_f \qquad \text{worst case}$$ (8.134)

The s.i.r. for a randomly located mobile unit within the cell can be calculated as follows [18]:

$$\hat{\gamma}_f = 0.5\gamma_f \qquad \text{average case}$$ (8.135)

where γ_f, given by Eq. (8.130), is the s.i.r. obtained with M mobile units in an isolated cell. $\hat{\gamma}_f$ is the s.i.r. from $N = 12$ cells, where each cell contains M mobile units. Then, in a manner similar to that used in Eq. (8.131), the following is obtained:

$$P_b' = \frac{L-1}{2} Q(\sqrt{\hat{\gamma}})$$

where there are $L - 1$ independent ways to make an error while detecting a given word. The relation between $\hat{\gamma}$ and $\hat{\gamma}_f$ is the same as the relation between γ and γ_f as shown in Eq. (8.129).

Multipath-fading environments

In a multipath-fading environment, it can be assumed that individual signals fade; however, the sum of all interfering signals does not fade, because of the averaging process. Therefore, the output of the ith matched filter [see Eq. (8.121)] becomes:

$$v_j(t) = r_j(t)A \sin [\omega_i t + \phi_i(t)] + I_{s,i}(t) \sin \omega_i t + I_{c,i}(t) \cos \omega_i t$$ (8.136)

where $r_j(t)$ is a Rayleigh-distributed random variable with $p(r_j) = 2r_j e^{-r_j^2}$.

The effective signal-to-noise ratio [analogous to the quantity measured by Eq. (8.129)] can be expressed as follows [19]:

$$\gamma' = Z \frac{\gamma_f}{1 + (L/4)(1/Z\gamma_f)} \tag{8.137}$$

where

$$Z = \sum_{\{j\}} r_j^2(t) \quad j = 1, \ldots, L \qquad \text{(fading)} \tag{8.138}$$

$$= 1/2 \qquad \qquad \text{(nonfading)}$$

The expression for $\{j\}$ is the summation over the L values of j from 1 to L.

Z is a random variable that has a chi-square distribution, with L degrees of freedom and a mean value of $L/2$; it is defined:

$$Z = \frac{1}{2} \chi^2 \tag{8.139}$$

where the probability density function for χ^2 is:

$$p(\chi^2) = \frac{\chi^{N-2}}{2^{L/2}\Gamma\left(\dfrac{L}{2}\right)} \exp\left(-\frac{\chi^2}{2}\right) \tag{8.140}$$

For a given L and γ_f, γ' can be obtained from Eq. (8.137). Let γ' express the function of the slowly varying random variable χ^2. The average P_b of Eq. (8.131) can then be expressed:

$$\overline{P}_b = \int_0^\infty p(\chi^2)P_b \, d\chi^2 = \frac{L-1}{2} \int_0^\infty p(\chi^2)Q(\sqrt{\gamma'(\chi^2)}) \, d\chi^2 \tag{8.141}$$

Equation (8.141) is applicable to an isolated cell. When intercell interference is present, γ_f from Eq. (8.137) should be replaced by $\hat{\gamma}_f$, which is shown in Eq. (8.134).

It should be noted that Eq. (8.141) can only be integrated numerically, as shown in Fig. 8.28 for the fading case.

8.10 Spectrum Efficiency and Number of Channels Per Cell

In this section, the spectrum-efficiency merits of two different systems will be compared; the first is the FM frequency-reuse system, and the second is the frequency-hopping DPSK System. The figure of merit η is shown in Eq. (8.118) and is the basis for these spectrum-efficiency comparisons.

The FM frequency-reuse system used by the Advanced Mobile Phone Service (AMPS) reuses assigned frequencies within a 12.5-MHz, one-way transmission spectrum bandwidth W at approximately 850 MHz [21, 22]. The available bandwidth can be defined:

$$W = 2^k \, \Delta f = Q \Delta f \qquad (8.142)$$

where Q is the total number of available channels, each having a channel width of Δf. In actual practice, the total number of available channels, Q, is subdivided into C frequency subsets that are each assigned to specified cells; e.g., subset C_i could be assigned to a particular cell and also be reused by another cell that is far enough away to preclude mutual interference at that subset of operating channel frequencies.

The spectral efficiency of the AMPS system is:

$$\eta = \frac{R_b}{C \Delta f} \qquad (8.143)$$

In an isolated-cell situation, the optimal value of C is unity, because there is no interference from neighboring cells.

Example 8.5   In an isolated-cell situation, where $R_b = 32$ kb/s and $\Delta f = 30$ kHz, then $\eta = 1.06$. In an intercell situation where omnidirectional antennas are used [21], $C = 12$ cells, and the spectral efficiency is $\eta = 0.088$. Therefore, the average number of usable channels per cell, M_{fm}, would be:

$$M_{fm} = \frac{\mathcal{L}B}{\Delta f C} = \begin{cases} 444 & \text{isolated-cell case} \\ 37 & \text{intercell case} \end{cases}$$

where $\mathcal{L}$ is the multitrunk loading factor (usually 0.666), $B = 20$ MHz, and $\Delta f = 30$ kHz.

In the frequency-hopping system of Cooper and Nettleton [22] the number of channels per cell, M_{fd}, depends on the spectral efficiency η, which depends on the bit rate R_b per user, as was shown in Eq. (8.118). R_b is expressed in Eq. (8.130) in terms of the s.i.r., γ_f. For the condition where $M \gg 1$, the spectral efficiency η is expressed [19]:

$$\eta = \frac{M R_b}{B} = \frac{M}{B} \frac{B \log_2 L}{(M-1) L \gamma_f} \approx \frac{\log_2 L}{L \gamma_f} \qquad (8.144)$$

Example 8.6   For an isolated cell in a fading environment, the value of s.i.r. $\gamma_f = 2.4$ dB can be found from Fig. 8.28, where $L = 32$ and $P_b = 10^{-3}$. The value of s.i.r. $\hat{\gamma}_f$ for the intercell case in a fading environment can be found from Eq. (8.135) as $\hat{\gamma}_f = 0.5 \gamma_f$. Under these conditions, the spectral efficiency η is:

$$\eta = \begin{cases} \dfrac{\log_2 32}{32} \times \dfrac{1}{1.73} = 0.09 & \text{isolated-cell case} \\ \dfrac{0.09}{2} = 0.045 & \text{intercell case} \end{cases}$$

TABLE 8.2   Spectrum Efficiency of FH-DPSK Mobile-Radio System Using Omnidirectional Base-Station Antennas

Systems	Isolated-cell multipath fading		Cellular-network multipath fading	
	η	Number of channels per cell	η	Number of channels per cell
AMPS	1.06	444	0.088	37
FH-DPSK	0.09	56	0.045	28

$L = 32, P_b = 10^{-3}$

The number of channels per cell can be expressed:

$$M_{\text{fd}} = \eta \frac{B}{R_b} = \begin{cases} 56 & \text{isolated-cell case} \\ 28 & \text{intercell case} \end{cases}$$

Table 8.2 compares the spectral efficiency and number of channels per cell for the AMPS and FH-DPSK systems. From Table 8.2, the number of channels per cell is significantly larger for AMPS than it is for the FH-DPSK system. Changes in system parameters and basic assumptions could cause a corresponding change in the results; however, the important thing is to understand the process for making this type of analysis.

8.11 Spread Spectrum Modulation—Direct Sequence (DS) [27]

The spread spectrum can provide the redundancy of the message bits while in transmission. The two direct sequence techniques can be illustrated in Fig. 8.30. The first one is shown in Fig. 8.30(a). The data $x(t)$ transmitted at a rate R_b are modulated, first by carrier f_0 and then by a spreading code $G(t)$ to form a DS signal $s_t(t)$ with a chip rate R_c, which takes a DS bandwidth B_{ss}. Then the DS signal $s_t(t - T)$, after a propagation time delay T, is received and goes through a correlator that consists of two functions: multiplier and averager. The correlator uses the same prestored spreading code $G(t)$ to despread the DS signal. Then the despread signal $x(t - T)$ is covered. The second DS technique is to spread first, then modulate, as shown in Fig. 8.30(b). These two DS

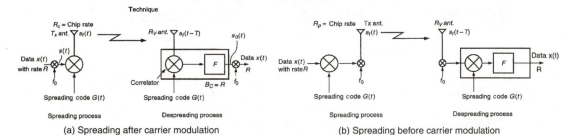

(a) Spreading after carrier modulation   (b) Spreading before carrier modulation

Figure 8.30   Basic spread—spectrum technique. Tx Ant.—transmitting antenna; Rv Ant.—receiving antenna.

techniques provide the same DS signal $s_t(t)$. The following analysis uses the technique from Fig. 8.30(a).

Let $x(t)$ be a data stream modulated by a binary phase shift keying (BPSK) so that

$$s_t(t) = x(t) \cos (2\pi f_0 t) \qquad (8.145)$$

where $x(t) = \pm 1$ and its data rate $= R_b$. At the transmitting end, the spread sequence $G(t)$ modulation also uses BPSK:

$$G(t) = \pm 1 \qquad (8.146)$$

with a chip rate R_c. The spread signal is

$$s_t(t) = x(t) \, G(t) \cos (2\pi f_0 t) \qquad (8.147)$$

At the receiving end, $s_t(t - T)$ is received after T seconds of propagation delay. The despreading process takes place. The signal $s_o(t - T)$ coming out from the correlator is

$$s_o(t - T) = E\{x(t - T) \cdot G(t - T), G(t - \hat{T}) \cos [2\pi f_0 (t - T)]\}$$
$$= x(t - T) \cdot E\{G(t - T) \cdot G(t - \hat{T})\} \cdot \cos [2\pi f_0 (t - T)] \qquad (8.148)$$

where $\hat{T}$ is the estimated propagation delay generated in the receiver. Since $G(t) = \pm 1$ and when $T = \hat{T}$

$$E\{G(t - T) \cdot G(t - \hat{T})\} = 1 \qquad \text{at } T = \hat{T} \qquad (8.149)$$

Then $s_o(t - T)$ shown in Eq. (8.148) becomes

$$s_o(t - T) = x(t - T) \cos [2\pi f_0 (t - \tau)] \qquad (8.150)$$

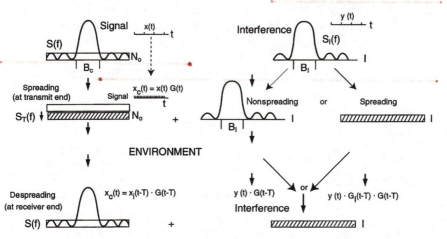

Figure 8.31 Spread spectrum. The interference source could have a different $G_I(t)$ to do the spreading and end the same result.

The data stream $x(t - T)$ is recovered after demodulated by the carrier frequency f_0.

Reduction of interference by a DS signal

The reduction of interference by a DS signal is shown in Fig. 8.31. After spreading $s(t)$ with a desired $G(t)$, the output $S_t(t - T)$ is transmitted out while the interference in the air is either a narrowband signal $S_I(t - T)$ or a DS signal $S_I(t - T)$ $G_I(t - T)$ with a difference $G_I(t)$. The two interference signals received can be expressed as

$$s_I(t - T) = E\{y(t - T) \cdot G(t - T) \cdot G_I(t - T) \cos [2\pi f_0(t - T)]\} \quad (8.151)$$

or

$$s_I(t - T) = E\{y(t - T) \cdot G(t - \hat{T}) \cos [2\pi f_0(t - T)]\} \quad (8.152)$$

In Eq. (8.151), the $s_I(t - T)$ is very weak because $E[G(t - \hat{T}) \, G_I(t - T)] \approx 0$.

In Eq. (8.152), the $s_I(t - T)$ is weak because $y(t - T) \cdot G(t - \hat{T})$ is an energy-spread signal, just as is $x(t) \cdot G(t - T)$ in the air before despreading. Therefore, a DS signal can recover its desired signal $x(t)$ and weaken the interferent signals.

Correlators and rake receivers

The correlator receiver provides despreading processes in spread spectrum modulation. Assuming the message $x(t)$ with a data rate R_b is spread by a pseudonoise code $G(t)$ to become a chip rate R_c before transmitting and despreading by the same $G(t)$ after receiving. The message $x(t)$ is recovered. In this case, the chip rate for this individual correlator should be considered. In a strong Rician environment, the correlator receiver with a large number of chips per data symbol should be used.

The rake receiver combines the outputs of N-correlators. The rake receiver, while equivalent to an equal-gain diversity combination, cannot be guaranteed to perform better than a single correlator under any circumstance. Usually in Rayleigh fading, or if the channel contains a strong specular component, a low-chip-rate rake receiver provides better results than a correlator receiver [30].

8.12 Modified Single-Sideband (SSB) System

One of the prime objectives of modern technology is to find more effective ways to use the existing mobile-radio frequency bands with enhanced efficiency. An efficient technique which is a radical departure from the spread-spectrum concept was proposed by Lusignan [23], who proposed a single-sideband, amplitude-modulated transmission that uses frequency companding to reduce the bandwidth and amplitude companding to enhance the signal-to-noise performance. This system is referred to as a single-sideband-frequency/amplitude (SSB-F/A) system.

The SSB-F/A system employs a radio-frequency bandwidth of 1.7 kHz with a suggested channel separation of 2.0 to 2.5 kHz. This channel separation is only one-tenth of the 25-kHz FM separation currently in use. The amplitude compandor compresses the dynamic range of the speech signal prior to transmission and expands the signal back to its original range upon reception. This technique (SSB-A) reduces power yet the snr remains the same. The frequency compandor compresses the frequencies of a voice waveform prior to transmission and expands them upon reception. This technique reduces the bandwidth while retaining the same signal-to-noise ratio. The amplitude compandor function is described in Chap. 13. The differences in channel separation for various modulation schemes are as follows:

Modulation scheme	Channel separation
Conventional FM	15 kHz
Amplitude-compandored SSB(SSB-A)	3.0–3.5 kHz
Frequency/amplitude-compandored SSB(SSB-F/A)	2.0–2.5 kHz

Figure 8.32 shows the bandwidth-versus-power relationship for the different modulation schemes. Three sets of data are shown, corresponding to values of 40, 30, and 20 dB of signal-to-noise ratio, respectively. It should be noted that when the required signal-to-noise ratio increases, the required transmitting power also increases, regardless of the modulation scheme used.

The functions plotted in Fig. 8.32 show that the SSB-F/A scheme requires less transmission power than either the SSB-A or the conventional FM modulation scheme to achieve the same signal-to-noise benefits. Lusignan's road-test studies [23] proved that the SSB-F/A scheme provides better cochannel interference isolation and better adjacent-

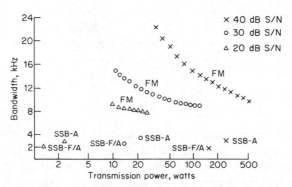

Figure 8.32 Bandwidth-vs.-power comparisons for systems using different modulation schemes. (*From Ref. 18.*)

channel-interference rejection than was possible in using conventional FM. The major problem with the SSB-F/A modulation scheme is the requirement for frequency stability. Hence, specially designed circuits, are required to compensate for short-term jitter and long-term frequency drifting. In the mobile-radio environment, Doppler frequencies as high as 100 Hz are typical for mobile speeds of 110 km/h, for operation in the 800- to 900-MHz band. Therefore specially designed circuits are also required to compensate for the Doppler frequency shift due to the motion of the vehicle. The SSB-F/A modulation scheme may be suitable for air-to-ground communications but not mobile communications. Also, the use of SSB-F/A has shown that there is no improvement in spectrum efficiency over the use of FM modulation in the cellular (frequency reuse) systems [29].

Problem Exercises

1. Derive Eq. (8.9).

2. In Fig. 8.9, if the vehicular speed is changed to 100 km/h and $W = 6$ kHz at an operating frequency of 450 MHz, what are the limiting lines due to the random FM signal?

3. The baseband-signal power bandwidth, $B = 8W$, has been satisfied when the IF carrier-to-noise ratio is cnr $\gamma = +10$ dB. Assuming that $W = 3$ kHz and B increases to $B = 30W$, how much lower can the value of γ be increased while maintaining the same level of baseband signal power?

4. Give the proof for Eq. (8.95).

5. Prove that the following fourth-order matrix can be used to form a set of four orthogonal words [14].

$$\begin{pmatrix} 1 & 1 & 1 & 1 \\ -1 & -1 & 1 & 1 \\ -1 & 1 & 1 & -1 \\ 1 & -1 & 1 & -1 \end{pmatrix}$$

Note A set of N orthogonal words can be developed by the first N Walsh functions.

6. Give the proof for Eq. (8.43).

7. Give the proof for Eq. (8.93).

8. Assuming the case of an isolated cell where the total s.i.r. γ is 15 dB, for an FH-DPSK system using a set of 16 orthogonal words, what is the s.i.r. γ_f at the output of the matched filters, and what is the average bit-error probability during the fading condition?

References

1. S. O. Rice, "Noise in FM Receivers," in M. Rowenblatt (ed.), *Proceedings of the Symposium on Time Series Analysis,* Wiley, New York, 1963, chap. 25.

2. S. O. Rice, "Statistical Properties of a Sine Wave plus Random Noise," *Bell System Technical Journal,* vol. 27, January 1948, pp. 109–157.

3. B. R. Davis, "FM Noise with Fading Channels and Diversity," *IEEE Trans. Comm.,* vol. 19, no. 6, December 1971, pp. 1189–1200.

4. D. L. Schilling, E. A. Nelson, and K. K. Clarke, "Discriminator Response to an FM Signal in a Fading Channel," *IEEE Trans. Comm.,* vol. 15, no. 2, April 1967, pp. 252–263.

5. H. H. Park, Jr. and C. Chayavadhanangkur, "Effect of Fading on FM Reception with Co-Channel Interference," *IEEE Trans. Aerospace and Electronic Syst.,* vol. 13, March 1977, pp. 127–132.

6. N. S. Jayant, "Digital Coding of Speech Waveform—PCM, DPCM and DM Quantizers," *Proc. IEEE,* May 1974, pp. 611–632.

7. N. S. Jayant, R. W. Schafer, and M. R. Karim, "Step-Size-Transmitting Differential Coders for Mobile Telephone" in N. S. Jayant (ed.), *Waveform Quantization and Coding,* IEEE Press, 1976, pp. 527–531.

8. J. A. Greefkes and K. Riemens, "Code Modulation with Digitally Controlled Companding for Speech Transmission," *Philips Tech. Rev.,* vol. 31, no. 11/12, pp. 335–353.

9. J. A. Greefkes, "A Digitally Companded Delta Modulation Modem for Speech Transmission," *Proc. IEEE Int. Conf. Comm.,* June 1970, pp. 7-33 to 7-48.

10. M. Schwartz, W. R. Bennett, and S. Stein, *Communication Systems and Techniques,* McGraw-Hill, New York, 1966, p. 299.

11. W. C. Jakes, *Microwave Mobile Communications,* Wiley, New York, 1974, p. 229.

12. T. T. Tjhung and P. H. Wittke, "Carrier Transmission of Binary Data in a Restricted Band," *IEEE Trans. Comm.,* vol. 18, no. 4, August 1970, pp. 295–304.

13. K. Feher, *Wireless Digital Communication,* Prentice Hall PTR, New Jersey, 1995, pp. 456–485.

14. Mike Fitts, private communication.

15. R. Steele, *Mobile Radio Communications,* Pentech Press, London, 1992.

16. S. M. Redl, M. K. Weber, and M. W. Oliphant, *An Introduction to GSM,* Artech House Publishers, Boston, 1995, pp. 135–141.

17. B. Sklar, *Digital Communications,* Prentice Hall, 1988.

18. G. R. Cooper and R. W. Nettleton, "A Spread-Spectrum Technique for High-Capacity Mobile Communications," *IEEE Trans. Veh. Tech.,* vol. 27, November 1978, pp. 264–275.

19. P. S. Henry, "Spectrum Efficiency of a Frequency-Hopping DPSK Spread-Spectrum Mobile Radio System," *IEEE Trans. Veh. Tech.,* vol. 28, November 1979.

20. M. Schwartz, W. R. Bennett, and S. Stein, *Communication Systems and Techniques,* McGraw-Hill, New York, 1966, p. 121.

21. F. H. Blecher, "Advanced Mobile Phone Service," *IEEE Trans. Veh. Tech.,* vol. 29, May 1980, pp. 238–242.

22. W. C. Y. Lee, *Mobile Cellular Telecommunications: Analog and Digital Systems,* Chap. 2, 2d ed., McGraw-Hill Book Co., 1995.

23. R. W. Wilmotte and B. B. Lusignan, "Spectrum-Efficient Technology for Voice Communications," UHF Task Force Report FCC/OPP UTF 78-01 (PB 278340), FCC, February 1978.

24. G. L. Stuber, *Principles of Mobile Communication,* Kluwer Academic Publishers, Boston, 1996.

25. T. S. Rappaport, *Wireless Communications, Principles and Practice,* Prentice Hall PTR, New Jersey, 1996.

26. J. G. Proakis, *Digital Communications,* McGraw-Hill, 1989.

27. M. K. Simon, Jim K. Omura, R. A. Schaltz, and B. K. Levitt, *Spread Spectrum Communications Handbook,* part 4, chapter 6, McGraw-Hill, 1994.

28. K. Hirade and K. Murota, "A Study of Modulation for Digital Mobile Telephony," *29th IEEE Vehicular Technology Conference Record,* Arlington Heights, Illinois, March 27–30, 1979, pp. 13–19.

29. W. C. Y. Lee, *Mobile Cellular Telecommunications: Analog and Digital Systems,* 2d ed., McGraw-Hill Book Co., 1995, pp. 407–411.

30. D. Noneaker and M. B. Pursley, "Rake Reception for a CDMA Mobile Communication System with Multipath Fading," *Code Division Multiple Access Communications,* Kluwer Academic, 1995, pp. 183–201.

Diversity Schemes

9.1 Functional Design of Mobile-Radio Systems—Diversity Schemes

A diversity scheme is a method that is used to develop information from several signals transmitted over independently fading paths. The objective is to combine the multiple signals and reduce the effect of excessively deep fades. The combining of signals will be described in the next chapter. Diversity schemes can minimize the effects of fading, since deep fades seldom occur simultaneously during the same time intervals on two or more paths. Two uncorrelated fading signals received via independently fading paths are shown in Fig. 9.1.

Since the chance of having two deep fades from two uncorrelated signals at any instant is rare, the effect of the fades can be reduced by combining them. There are two general types of diversity schemes. One is called the "macroscopic diversity scheme" and the other is the "microscopic diversity scheme." The macroscopic diversity scheme is used for combining two or more long-term lognormal signals, which are obtained via independently fading paths received from two or more different antennas at different base-station sites. The microscopic diversity scheme is used for combining two or more short-term Rayleigh signals, which are obtained via independently fading paths received from two or more different antennas but only at one receiving cosite.

9.2 Macroscopic Diversity Scheme— Applied on Different-Sited Antennas

As was discussed previously in Chaps. 3 through 5, the long-term lognormal fading which the mobile-radio signal undergoes in a shadow region causes the average power to drop over a long period of time. The

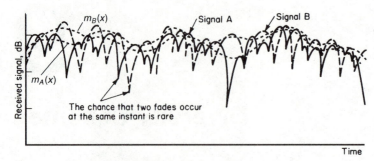

Figure 9.1 Uncorrelated fading signals.

undulation of the terrain contour determines the duration of envelope variations during long-term fading. The roughness of the surface along the propagation path determines the range of variation of the long-term fading. The macroscopic diversity scheme defeats shadowing and other terrain effects by using transmitted and received base-station signals at two different geographical sites, as shown in Fig. 9.2.

The same transmitted signal received simultaneously at two different base-station sites can be used to determine which site is better for communication with the mobile unit. In Fig. 9.2, the improved mobile-radio communication with site A occurs while the mobile unit is traveling along section A of the road; and with site B it occurs while the mobile unit is traveling along section B of the road. Assume that the average power terms of the two received signals at the two respective sites are

$$P_{r_A} = m_A(x) \qquad\qquad (9.1)$$

$$P_{r_B} = m_B(x) \qquad\qquad (9.2)$$

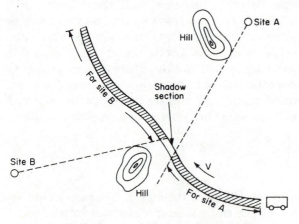

Figure 9.2 Using base stations at two different sites.

here $m_A(x)$ and $m_B(x)$ are lognormal-distributed local means, as described in Chap. 6. In the macroscopic diversity scheme, the site which serves the strongest path is selected. If

$$m_A(x) > m_B(x) \qquad \text{use site A} \tag{9.3}$$

This is the only combining technique that is used in dealing with two signals in a macroscopic diversity scheme.

The macroscopic diversity scheme can be used for any arbitrary number of sites, as needed. Macroscopic diversity is also called "multiple-base-station diversity."

9.3 Microscopic Diversity Schemes—Applied on Cosited Antennas

The microscopic diversity scheme is used when two or more uncorrelated short-term Rayleigh signals are received, with the same long-term fading experienced on those signals. Identical long-term fading means that all of the signal paths have to follow the same terrain contour. Therefore, different short-term fading signals are received from different antennas at the same antenna site. The term "cosite" can be applied to both the mobile unit and the base station. There are basically six methods of achieving diversity reception. The differences between the diversity schemes applied at the base station and those applied at the mobile unit are described in Sec. 9.4.

9.4 Space Diversity

In Chap. 1, it was shown that the signal received at the mobile unit is obtained by summing up the incoming waves from all directions. From the expression shown in Eq. (1.16), signals are received at different instants as the mobile unit moves with a speed V. Since the distance $x = Vt$, the instantaneous location x_1 corresponds to time t_1, and x_2 corresponds to time t_2. It can also be shown that the instantaneous signal strengths received at x_1 and x_2 are different.

If the receiver has multiple antennas, spaced sufficiently far apart that their received signals fade independently, then they can be used for diversity reception. The necessary space separation required to obtain two uncorrelated signals must be determined.

At the mobile unit

In Sec. 6.6 of Chap. 6, the correlation of received signals at the mobile-unit location was described. The correlation coefficient $\rho_r(d)$ for a distance separation d can be obtained from $\rho_r(\tau)$ for a time separation τ and a constant speed V, which was shown in Eq. (6.110):

$$\rho_r(d) = J_0^2(\beta V\tau) = J_0^2(\beta d) \tag{9.4}$$

For a uniform angular distribution of wave arrival, the first null of $J_0^2(\beta d)$ is at $d = 0.4\lambda$, as shown in Fig. 9.3. After the first null, the correlation coefficient starts to increase again; however, measurements show that the distance associated with the first null of $\rho_r(d)$ is about 0.8λ in suburban areas [1, 2]. This may be due to a lack of uniform angular distribution of wave arrival. The separation of d becomes less for the first null of $\rho_r(d)$ measured in an urban area.

Usually a separation of 0.5λ can be used to obtain two almost uncorrelated signals at the mobile unit. The fact is that as long as the correlation coefficient is less than 0.2, the two signals are considered to be uncorrelated. This is explained in more detail in Chap. 10.

At the base station

The correlation between two signals received at a base station has been described in Sec. 6.9 of Chap. 6. Figure 6.12 shows the correlation versus antenna spacing for different directions and different beamwidths of incoming signals. The antenna spacing between the mobile and base-station antennas becomes larger if the beamwidth of incoming signals is small. Hence, the antenna spacing shown in Fig. 6.12(a) is larger than the antenna spacing in Fig. 6.12(b). The experimental data also support this contention [3]. Besides, when the base-station antenna height is increased, the correlation is decreased, provided that the base-station antenna spacing remains unchanged. Since there are many parameters involved, a new design parameter is defined as follows [4]:

$$\eta = \frac{\text{antenna height}}{\text{antenna spacing}} \tag{9.5}$$

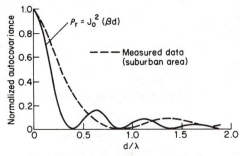

Figure 9.3 Autocorrelation coefficient vs. spacing for uniform angular distribution.

which simplifies the design criterion. This parameter is useful in selecting an antenna spacing to achieve a desired signal decorrelation between the two antennas. For a broadside case, the correlation ρ is plotted versus η in Fig. 9.4. The correlation versus η is shown in Fig. 9.5 for various angles of signal arrival with respect to the normal incident waves from the two antennas. Note that the data in both Fig. 9.4 and Fig. 9.5 were obtained at a frequency of 850 MHz [3].

The empirical curve shown in Fig. 9.4 can be expressed from a formula as follows:

$$\rho = 0.2 + 0.7 \log_{10} \frac{\eta}{2} \tag{9.6}$$

$$\eta = 2 \cdot 10^{[(\rho - 0.2)/0.7]} \tag{9.6a}$$

As an example, given $h = 100$ ft and $\rho = 0.7$ for the broadside case, the antenna separations can be found by obtaining $\eta = 11$ from Fig. 9.4. Then,

$$\eta = \frac{h}{d} = \frac{100 \text{ ft}}{d} = 11 \tag{9.5a}$$

$$d = 9 \text{ ft} \qquad \text{(or } 8\lambda \text{ at 850 MHz)}$$

The antenna separation at the base-station antenna is therefore 9 ft for this case.

Though the curves shown in Figs. 9.4 and 9.5 are obtained at a distance of 3 mi, these same curves can be used for distances greater than 3 mi, since the radius of effective local scatterers at the location of the

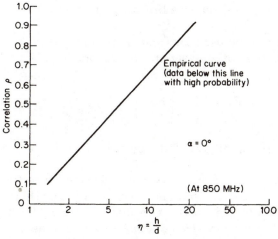

Figure 9.4 Correlation vs. antenna height and spacing (broadside case) in a suburban area.

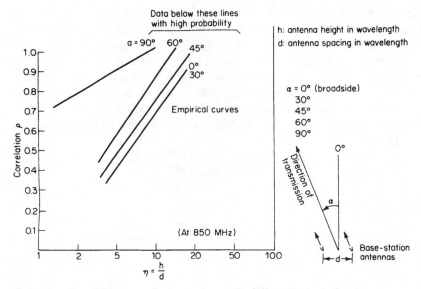

Figure 9.5   Correlation vs. η for two antennas in different orientation.

mobile unit is roughly fixed. When those effective local scatterers surrounding the mobile unit as it travels are farther than 3 mi away, the required antenna spacing at the base station tends to increase. Another factor that must be considered is the presence of scatterers along the propagation path. These scatterers will cause the signal received by the two base-station antennas to become less correlated as the propagation-path length increases, and the required antenna spacing tends to reduce. These two factors counteract each other; hence, the curves of Figs. 9.4 and 9.5 can be used for a link path of approximately 3 mi or greater.

Figures 9.4 and 9.5 can also be used for frequencies other than 850 MHz, simply by converting the antenna separation d from 850 MHz to the new corresponding frequency d':

$$d' = d \times \left(\frac{850}{f}\right) = \frac{h}{\eta} \times \left(\frac{850}{f}\right)$$

where f is the frequency in megahertz. This formula is valid for $f \geq$ 100 MHz.

Space diversity is not recommended at lower frequencies, since the physical separation between the two antennas becomes large and therefore is impractical. For example, given the parameters of $h = 100$ ft, $\eta = 10$ (which corresponds to $\rho = 0.7$ for $\alpha = 0$ in Fig. 9.4), and a frequency of $f = 100$ MHz, the separation d' is 85 ft. A separation of 96 ft between the two base-station antennas is obviously impractical.

Example 9.1 Figure E9.1 shows the parameters for a base station employing space-diversity reception and two mobile units transmitting from directions of $\alpha = 0$ and $\alpha = 90°$, respectively. For transmission frequency $f_c = 850$ MHz, a correlation coefficient of 0.8 between the two received signals, and an antenna separation d of 10 ft, what should the height of the antenna be?

solution To find the optimal base-station antenna height for transmission direction $\alpha = 0$, the parameter $\eta = 15$ is obtained from Fig. 9.5, where

$$\eta = \frac{\text{antenna height}}{\text{antenna spacing}} = \frac{h}{d} = \frac{h}{10}$$

and therefore

$$h = \eta \times d = 15 \times 10 \text{ ft} = 150 \text{ ft}$$

Similarly, to find the optimal base-station antenna height for transmission direction $\alpha = 90°$, the parameter $\eta = 2$ is obtained from Fig. 9.5; then:

$$h = \eta \times d = 2 \times 10 \text{ ft} = 20 \text{ ft}$$

This example has shown an important fact. Lowering two antenna heights decreases the correlation but weakens the strengths of two received signals. The former improves the performance of diversity and the latter reduces the average snr. Therefore, a trade-off study has to be made by designers.

Example 9.2 On the basis of the parameters $\alpha = 0$ to $\alpha = 90°$, what procedure can be used to ensure that all three base-station antennas in a multiple-antenna array will have equal and minimum heights with a signal correlation of approximately 0.8 between any two of the three antennas?

solution It is only necessary to consider the parameter $\alpha = 90°$ to find the equal and minimum height of a three-antenna multiple array. A triangular configuration should be used, as shown in Fig. E9.2. A mobile unit can be located at any given angle α ($0 \leq \alpha \leq 360°$) and at any given distance r, and usually r is much greater than d ($r \gg d$), as shown in Fig. E9.2. The actual location of the mobile antenna with respect to the base station can be chosen from one of the following three sectors:

Sector A——$-60° \leq \alpha \leq +60°$, where antennas 1 and 2 provide the required correlation.

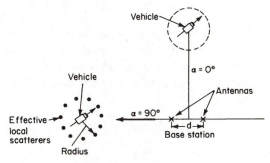

Figure E9.1 Model for Example 9.1.

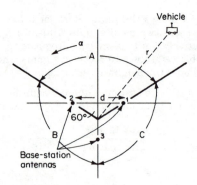

Figure E9.2 Parameters for Example 9.2.

Sector B—$+60° \le \alpha \le 180°$, where antennas 2 and 3 provide the required correlation.

Sector C—$180° \le \alpha \le 300°$, where antennas 1 and 3 provide the required correlation.

Hence, $\alpha = 60°$ and $\rho_r = 0.8$ should be used to find the parameter $\eta = 9$ from Fig. 9.5, and the minimum antenna height is then:

$$h = \eta \times d = 9 \times d$$

Where d is given as 10 ft as in Example 9.1, then h can be calculated as follows:

$$h = 9 \times 10 \text{ ft} = 90 \text{ ft}$$

9.5 Field-Component Diversity

In Sec. 5.5 of Chap. 5, the characteristics of field components were described. The components E_z, H_x, and H_y are shown in Eqs. (5.20), (5.21), and (5.22), respectively. They are uncorrelated at any given instant while being received at either the mobile unit or the base station. However, the power relationship among these three components at the time of reception is [5, 6, 7]:

$$|E_z^2| = |H_x^2| + |H_y^2| \qquad V^2/m^2 \tag{9.7}$$

Because of the pattern differences between the loop and dipole antennas, the three components are uncorrelated and can be used to form three kinds of diversity. One method is to combine them incoherently. The second method is to combine them coherently. The incoherently combined signal is:

$$V_I = E_z + H_x + H_y$$

The coherently combined signal is:

$$V_{II} = |E_z| + |H_x| + |H_y|$$

The third method is the energy-diversity expression that was initially suggested by J. R. Pierce, and subsequently verified by analysis and measurements performed by Lee [5, 6, 7]:

$$V_{\text{III}}^2 = |E_z|^2 + |H_x|^2 + |H_y|^2 \rightarrow \text{constant}$$

All three kinds of field-component diversity can be realized by using an energy-density antenna described in Sec. 5.5.

9.6 Polarization Diversity

Signals transmitted in either horizontal or vertical electric fields are uncorrelated at both the mobile and base-station receivers, as shown in Fig. 9.6.

In Sec. 5.6 of Chap. 5, the mechanism for receiving two polarized signals was described. Suppose that the vertically polarized signal is

$$\Gamma_{11} = \sum_{i=1}^{N} a_i e^{j\psi_i} e^{-j\beta Vt \cos \phi_i} \tag{9.8}$$

and the horizontally polarized signal is

$$\Gamma_{22} = \sum_{i=1}^{N} a'_i e^{j\psi'_i} e^{-j\beta Vt \cos \phi_i} \tag{9.9}$$

where a_i and ψ_i are the amplitude and phase, respectively, of each wave path indicated in Eq. (9.8) and a'_i and ψ'_i are their counterparts in Eq. (9.9). V is the velocity of the vehicle, and ϕ_i is the angle of the ith wave arrival. Although these two polarized signals arrive at the receiver from the same number of incoming waves, it is not difficult to see from Eqs. (9.8) and (9.9) that Γ_{11} and Γ_{22} are uncorrelated, because of their different amplitudes and phases. These uncorrelated properties are found at both the mobile and base-station receivers.

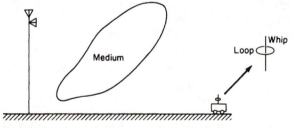

Figure 9.6 Horizontal and vertical polarization diversity signals.

Polarization diversity at the base

We described in Sec. 5.6 a base station polarization antenna. This antenna is suggested to be used at the base station. The signal transmitted at the mobile unit is supposedly from a tilted whip antenna. The two components, either E_v and E_h or $E_{-45°}$ and $E_{45°}$, can only make one component of two be a strong one. If the difference between the two received signals is greater than 2 dB, the diversity gain is diminished. If the mobile unit signal is totally vertically polarized while receiving, then the polarization diversity at the base can get the best diversity gain by combining two components $E_{-45°}$ and $E_{45°}$. If the mobile unit signal is polarized at 45° from the vertical while receiving, then E_v and E_h will give the best diversity gain. Another thought: using the right and left circular antennas at the base station can always achieve the best diversity gain from the mobile signal at any linear polarization.

9.7   Angle Diversity

At the base station

The angular diversity scheme applied at the base station may have several concerns. They are related to the definition of angular diversity. There are in-time angular diversity and out-of-time angular diversity.

In-time angular diversity.   We try to combine the two received signals at the same time in order to achieve the diversity gain. It is a microscopic diversity. The signal coming from the mobile unit is from a spread angle we called BW as described in Sec. 16.18. The effect of the angle spreading at the base station is very small, less than 2°. The measurement data has shown this fact by taking the difference in signal strength between two signals from two corresponding beams at the base. The beam angle at the base station antenna is 30°. The result is shown in Fig. 9.7. There are two sets of data: one collected from the horizontal spaced antennas (8λ) and one collected from the angular separated antennas (two 30° beams). The differences between the two received signal strengths—spaced or angular—from many mobile runs are shown in Fig. 9.7. We realize that the differences in the two signal strengths from two angular separated beams are so large that the diversity gain is diminished according to the required conditions stated in Sec. 16.8. From the horizontal space antenna, the differences between the two received signal strengths from all the runs are extraordinarily small, which means that the diversity gain is obtained [8, 9].

Out-of-time angular diversity.   In this case, the signal strength of the mobile unit is constantly monitored at the base station by each beam of a multibeam antenna system. The strongest beam is used for the traf-

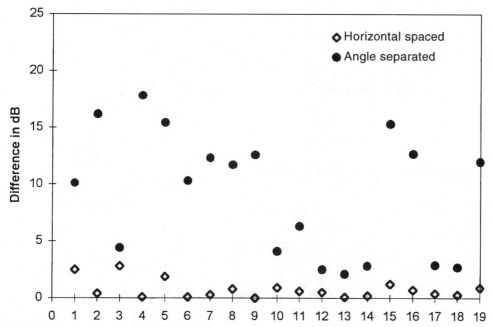

Figure 9.7 Comparison of the difference in signal strengths between two signals from horizontal spaced and angle separated arrangement.

fic link at the time. When the mobile unit is moved to another location, the strongest signal strength at the base will be detected by another beam. Then the system will switch the traffic link to the other beam. This is called the fixed-beam switched smart antenna [8] and has been proven to be a good system [9].

At the mobile unit

If the received signal arrives at the antenna via several paths, each with a different angle of arrival, the signal components can be isolated by means of directive antennas. Each directive antenna will isolate a different angular component. As previously discussed in Sec. 6.4 of Chap. 6, directive mobile antennas pointing in widely different directions can receive scattered waves at the mobile site from all directions and can provide a less severe fading signal. If directional antennas are mounted on the mobile unit, as shown in Fig. 9.8, the signals received from different directive antennas pointing at different angles are uncorrelated.

The signal obtained from the directive antenna J can be expressed:

$$s_J = X_J + jY_J \qquad (9.10)$$

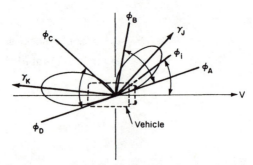

Figure 9.8 Angle-diversity vectors.

where

$$X_J = \sum_{i=1}^{N} (R_i \cos \xi_i + S_i \sin \xi_i) G(\gamma_J - \phi_i) \qquad (9.11)$$

$$Y_J = \sum_{i=1}^{N} (S_i \cos \xi_i - R_i \sin \xi_i) G(\gamma_J - \phi_i) \qquad (9.12)$$

$$\xi = \beta V t \cos \phi_i \qquad (9.13)$$

$G(\gamma_J - \phi_i)$ is the pattern for directive antenna J pointing at an angle γ_J, ϕ_i is the angle of the ith wave arrival, and N is the total number of scattered waves. Let:

$$G(\gamma_J - \phi_j) = 1 \qquad \phi_A \leq \phi_j \leq \phi_B$$

Then Eq. (9.11) and Eq. (9.12) become:

$$X_J = \sum_{j=1}^{N_j} (R_j \cos \xi_j + S_j \sin \xi_j) \qquad (9.14)$$

$$Y_J = \sum_{j=1}^{N_j} (S_j \cos \xi_j - R_j \sin \xi_j) \qquad (9.15)$$

where

$$\xi_j = \beta V t \cos \phi_j \qquad \phi_A \leq \phi_j \leq \phi_B \qquad (9.16)$$

$$N_J = N \frac{\phi_B - \phi_A}{360°} \qquad (9.17)$$

The statistics of R_j and S_j are described from the theoretical model in Sec. 6.2 of Chap. 6. It is easy to show that the signal $s_k = X_k + jY_k$ received from another directive antenna k can be expressed:

$$X_K = \sum_{k=1}^{N_K} (R_k \cos \xi_k + S_k \sin \xi_k) \tag{9.18}$$

$$Y_K = \sum_{k=1}^{N_K} (S_k \cos \xi_k - R_k \sin \xi_k) \tag{9.19}$$

$$\xi_k = \beta V t \cos \phi_k \tag{9.20}$$

where

$$\phi_C \leq \phi_k \leq \phi_D \tag{9.21}$$

$$N_k = N \frac{\phi_D - \phi_C}{360°} \tag{9.22}$$

On the basis of the theoretical model, it is easy to show that

$$E[X_J Y_J] = E[X_J X_K] = E[X_J Y_K] = E[Y_J Y_K] = 0 \tag{9.23}$$

Hence, the signal received at the mobile site from two different directive antennas is uncorrelated; that is, $E[s_J s_K^*] = 0$.

9.8   Frequency Diversity

Two mobile-radio signals separated by two carrier frequencies far apart are possibly independent. Since frequency diversity is statistically dependent on the separation of two carrier frequencies, the criterion for achieving frequency diversity is of primary importance.

Statistical dependence on frequency

The variation of the transmission coefficient of an ith wave has been expressed in Eq. (6.120):

$$E_z^i = \tilde{a}_i \exp [j(\omega_c - \beta V \cos \phi_i)(t - \Delta t) - j\omega_c T_i]$$

where $\beta V \cos \phi_i$ is the Doppler shift, Δt is the time increase corresponding to vehicular travel, and T_i is the time delay of the ith wave. A resultant signal is obtained by summing up all of the waves:

$$s(t) = \sum_{i=1}^{N} \tilde{a}_i \exp\left[j(\omega_c - \beta V \cos \phi_i)(t - \Delta t) - j\omega_c T_i\right]$$

$$= X(t) + jY(t) \tag{9.24}$$

As time t increases, the phase of the signal changes proportionately to $\beta V t \cos \phi_i$, so that the correlation of the time variations, $\Delta t = t_1 - t_2$, is determined by the Doppler frequency spectrum $S(f)$ as:

$$R_s(\Delta t) = \int_{-\infty}^{\infty} S(f) e^{j2\pi f_c \Delta t} \, df \tag{9.25}$$

As the transmitted frequency f_c varies, the phase of each wave changes relative to T_i, so that the correlation of the frequency variations $\Delta f = f_1 - f_2$ is determined by the various delay distributions $p(T)$ which relates to the frequency correlation as:

$$R_s(\Delta f) = \int_{-\infty}^{\infty} p(T) \exp\left(j2\pi \, \Delta f \, T\right) dT \tag{9.26}$$

There are three models (see Fig. 9.9) that can be used to express the time-delay distribution:

1. Maxwell-distribution model:

$$p_1(T) = \begin{cases} \dfrac{C \sqrt{2/\pi} \, (3 - 8/\pi)^{3/2}}{\Delta^3} \, (\Delta t)^2 \exp\left[-\dfrac{(3 - 8/\pi) \, T^2}{2\Delta^2}\right] & T \geq 0 \\ 0 & T < 0 \end{cases} \tag{9.27}$$

where C is the total power received by the antenna, T is the delay relative to a directive path from the base station to the mobile unit, and Δ is the standard deviation of the time delays:

$$\Delta = \sqrt{\langle T^2 \rangle - \langle T \rangle^2}$$

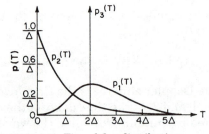

Figure 9.9 Time-delay distributions.

2. Exponential-distribution model:

$$p_2(T) = \begin{cases} \dfrac{C}{\Delta} \exp\left(-\dfrac{T}{\Delta}\right) & T \geq 0 \\ 0 & T < 0 \end{cases} \qquad (9.28)$$

3. Approximation model:

$$p_3(T) = \frac{C}{2}\,[\delta(T) + \delta(T - 2\Delta)] \qquad (9.29)$$

Of the three models, the exponential-distribution model, which was shown in Eq. (6.121), is the simplest and most frequently used model.

Correlation versus frequency

To find the correlation of the transmission coefficient at two different frequencies, the Fourier transformation of $p(T)$ can be taken, as follows:

$$R(\Delta f) = E[s^*(f_c, t)s(f_c - \Delta f, t)]$$

$$= \int_{-\infty}^{\infty} p(T)\,\exp\,[j2\pi\,\Delta f\,\Delta t]\,d(\Delta t) \qquad (9.30)$$

Let

$$g(\Delta f) = \tfrac{1}{2}\,\mathrm{Re}\,\{E[s^*(f_c, t)s(f_c - \Delta f, t)]\} \qquad (9.31)$$

$$h(\Delta f) = \tfrac{1}{2}\,\mathrm{Im}\,\{E[s^*(f_c, t)s(f_c - \Delta f, t)]\} \qquad (9.32)$$

then the normalized correlation coefficient is

$$\rho(\Delta f) = \frac{|E[s^*(f_c, t)s(f_c - \Delta f, t)]|}{E[s^*(f_c, t)s(f_c, t)]}$$

$$= \frac{|g^2(\Delta f) + h^2(\Delta f)|^{1/2}}{g^2(0)} \qquad (9.33)$$

The correlation coefficients ρ_1, ρ_2, and ρ_3 for the time-delay distributions of Eqs. (9.27), (9.28), and (9.29), respectively, are:

$$\rho_1(\Delta f) = e^{-b^{3/2}}\sqrt{(1 - b^2)^2 + \frac{8}{\pi}\,b_1^2 F_1^2\left(-\frac{1}{2};\frac{3}{2};\frac{b^2}{2}\right)} \qquad (9.34)$$

$$\rho_2(\Delta f) = \frac{1}{\sqrt{1 + (2\pi\,\Delta f\,\Delta)^2}} \qquad (9.35)$$

$$\rho_3(\Delta f) = |\cos\,(2\pi\,\Delta f\,\Delta)| \qquad (9.36)$$

where

$$b = \frac{2\pi \, \Delta f \, \Delta}{\sqrt{3 - 8/\pi}} \tag{9.37}$$

and where $_1F_1$ is the confluent hypergeometric function, which can be expressed in terms of the modified Bessel functions of the first kind:

$$_1F_1(\nu + \tfrac{1}{2};\, 2\nu + 1;\, z) = 2^{2\nu}\Gamma(\nu + 1)z^{-\nu}e^{z/2}I_\nu\!\left(\frac{z}{2}\right)$$

The three correlation-coefficient functions, Eqs. (9.34) to (9.36), are plotted in Fig. 9.10. The exponential-distribution model is the most widely accepted of the three models since it fits the measured data well [10].

Coherence bandwidth

The coherence bandwidth of a channel is the maximum width of the band in which the statistical properties of the transmission coefficients of two CW signals are strongly correlated. Coherence bandwidth can be defined as that frequency separation for which the magnitude of the normalized complex correlation coefficient first drops below a certain value A, where A is always smaller than 1. Then

$$\rho(\Delta f_A) \leq A \tag{9.38}$$

where $B_c = \Delta f_A$ is the coherence bandwidth. A typical assumed value for A would be 0.5 for either envelope correlations—as shown in Eq. (6.127)—or phase correlations—as shown in Eq. (7.64).

Two frequencies
with noncoherent bandwidth

If a channel is frequency-selective, two widely spaced frequencies will fade independently. This means that the two frequencies have non-

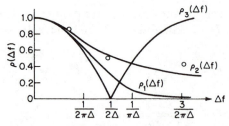

Figure 9.10   Correlation vs. frequency separation (circles are measured points). *(From Ref. 8.)*

coherent bandwidth. Therefore, if the same message—voice or digital signal—is simultaneously transmitted on multiple frequencies, the fading signals are statistically independent at the time of reception. Since frequency selectivity is largely determined by the time-delay spread between the various multipath components, the greater the multipath time-delay spread, the smaller the noncoherent bandwidth between the two CW frequencies. The expression for coherent bandwidth was previously shown in Eqs. (1.53), (6.128), and (7.65).

Two frequencies with coherent bandwidth

When a pilot frequency is transmitted together with a normal signal, the pilot frequency does not carry any of the message information; however, when the two frequencies are received, their fading characteristics are essentially the same, providing the frequency separation is close enough to remain within the coherent bandwidth. The separation between the pilot and signal frequencies must be within the coherent bandwidth in order to correct for amplitude and phase changes that are caused by the transmission coefficient.

On the basis of the exponential distribution shown in Fig. 9.10, the coherence bandwidth can be approximated by defining $\rho_r = \rho_2^2(\Delta f) = 0.5$, as was previously discussed in Sec. 6.8. The coherence bandwidth $B_c = \Delta f = 1/(2\pi\Delta)$ obtained from Fig. 9.10 is identical to Eq. (6.128).

> **Example 9.3**   In order to find the coherence bandwidth for a pilot frequency f_p associated with a signal frequency f_c in a mobile-radio environment, it is first necessary to calculate the time-delay spread between the two frequencies under consideration.
>
> The time-delay spread measured in different mobile environments can be expressed:
>
> $$\Delta \approx 5 \text{ μs} \quad B_c = 31.8 \text{ kHz} \quad \text{large cities} \tag{E9.3.1}$$
>
> $$\Delta \approx 0.5 \text{ μs} \quad B_c = 318 \text{ kHz} \quad \text{suburban areas} \tag{E9.3.2}$$
>
> On the basis of these measured data, the pilot frequency for a large-city mobile-radio environment would be:
>
> $$f_c - 31 \text{ kHz} \leq f_p \leq f_c + 31 \text{ kHz} \tag{E9.3.3}$$
>
> For conventional FM, the transmission bandwidth of a signal is about 25 kHz, and specially designed circuits are needed to separate f_p from f_c.

9.9   Time Diversity

Time-diversity reception techniques are primarily applicable to the transmission of digital data over a fading channel. In time diversity, the same data are sent over the channel at time intervals of the order of the reciprocal of the baseband fade rate $f_b = 2f_m$. In mobile radio, the reciprocal fade rate can be expressed:

$$\tau_s \geq \frac{1}{2f_m} = \frac{1}{2(V/\lambda)} \tag{9.39}$$

For vehicle speeds of 60 mi/h and transmission frequencies of 1 GHz, a time-diversity separation of 5 ms is required for the two signals; the time separation increases as the fade rate decreases. Multiple diversity channels can be provided by successively transmitting the signal sample in each time slot. The sampling rate for voice transmission of a single channel is 2×4 kHz $= 8$ kHz. For M-branch diversity, the sampling rate must be $M \times 8$ kHz, since the transmission delay spread is usually less than 20 μs, which is much less than the inverse of the sampling rate:

$$f_s < \frac{1}{\Delta} \tag{9.40}$$

Hence, the sampling rate f_s is not limited by the time-delay spread. However, the minimum time separation between samples shown in Eq. (9.39) for diversity application may cause a serious problem, since f_m is a Doppler frequency, expressed:

$$f_m = \frac{V}{\lambda} \tag{9.41}$$

When the vehicle is stationary, $V = 0$, and thus $f_m = 0$. This means that the time separation τ_s is infinite. Therefore, the advantages of time diversity are lost when the vehicle is not moving. This is in sharp contrast to other diversity schemes, in which the branch separation is not a function of vehicle speed and thus the two diversity signals are independent over any value of V. In the mobile-radio case, the velocity V is in the range:

$$0 \leq V \leq 100 \text{ mi/h} \tag{9.42}$$

and therefore, any diversity scheme that is used should be effective over the entire range of V.

Problem Exercises

1.   Two correlated signals are received at the mobile unit via two colocated antennas separated by 0.5λ. When the mobile unit is parked, what is the probability that the two received signals will fade together?

2. Given a required correlation coefficient of 0.7 or less, between two co-located base-station receiving antennas at a height of 100 ft, what is the required antenna separation at a frequency of 400 MHz?

3. Prove that field components E_z, H_x, and H_y are statistically independent in a mobile-radio environment. (Refer to the methods described in Chap. 6.)

4. Prove that two mobile-radio received signals arriving from two different angular directions are uncorrelated.

5. Given a frequency correlation of 0.4, plot the frequency separation Δ_f versus the delay spread Δ.

6. Make a quantitative comparison between frequency diversity and time diversity.

7. If the coherent bandwidth is based on a phase correlation of 0.85, what is the coherent bandwidth in terms of delay spread?

8. How is a pilot frequency used to correct signal fades within a diversity-designed mobile-radio receiver?

References

1. W. C. Y. Lee, "Antenna Spacing Requirement for a Mobile Radio Base-Station Diversity," *Bell System Technical Journal,* vol. 50, July–August 1971, pp. 1859–1874.
2. S. B. Rhee and G. I. Zysman, "Results of Suburban Base-Station Spatial Diversity Measurements on the UHF Band," *IEEE Trans. Comm.,* vol. 22, October 1974, pp. 1630–1634.
3. W. C. Y. Lee, "Effects on Correlation between Two Mobile Radio Base-Station Antennas," *IEEE Trans. Comm.,* vol. 21, November 1973, pp. 1214–1224.
4. W. C. Y. Lee, "Mobile Radio Signal Correlation versus Antenna Height and Spacing," *IEEE Trans. Veh. Tech.,* vol 25, August 1977, pp. 290–292.
5. E. N. Gilbert, "Energy Reception on Mobile Radio," *Bell System Technical Journal,* vol. 44, October 1965, pp. 1779–1803.
6. W. C. Y. Lee, "Statistical Analysis of the Level Crossings and Duration of Fades of the Signal from an Energy Density Mobile Radio Antenna," *Bell System Technical Journal,* vol. 46, February 1967, pp. 416–440.
7. W. C. Y. Lee, "An Energy Density Antenna Model for Independent Measurement of the Electric and Magnetic Fields," *Bell System Technical Journal,* vol. 46, September 1967, pp. 1587–1599.
8. "Cellwave Antenna, a Multi-Beam Antenna," Cellwave, Mariboro, New Jersey.
9. W. C. Y. Lee, "An Optimum Solution for the Switching-Beam Antenna System," 3d Workshop on Smart Antenna in Wireless Mobile Communications, Conference Record, Stanford University, July 25–26, 1996.
10. M. J. Gans, "Power-Spectral Theory of Propagation in the Mobile-Radio Environment," *IEEE Trans. Veh. Tech.,* vol. 21, February 1972, pp. 27–38.

Combining Technology

10.1   Combining Techniques
for Macroscopic Diversity

Signal performance that is degraded by severe fading can be improved by increasing transmitter power, antenna size and height, etc., but these solutions are costly in mobile-radio communications and sometimes impractical. The alternative is to use simultaneous or selective diversity-combining transmission over several channels, to reduce the probability of excessively deep fades at the receiving end. In Chap. 9, several schemes for providing signal diversity at the receiving end were described. In this chapter, the combining techniques for macroscopic diversity and microscopic diversity are analyzed.

In macroscopic diversity, as previously discussed in Sec. 9.2 of Chap. 9, only the local means of the received signal are considered. The local means may vary as a result of long-term fading when the mobile unit is traveling in an extensive area and when the terrain contour in that area is not flat. Diversity reception provides the advantage of being able to receive two signals whose local mean fades rarely occur below a certain level during the same time intervals.

Selective diversity combining

As previously discussed in Sec. 9.1 of Chap. 9, selective diversity combining is chosen primarily to reduce long-term fading. Reducing the effects of long-term fading by combining two signals received from two different-sited transmitting antennas is possible because the local means of the two signals at any given time interval are rarely the same. It will be shown later that to effectively reduce fading requires the combining of two fading signals that have equal mean strengths [1]. Also, if two noncolocated base-station transmitters are not very stable, the phase jittering generated in each of the transmitters will degrade the combined signal. Hence, the technique of selective diversity combining can be used effectively, because it is really only a selection between two signals, rather than a combination of two signals.

Improving the average SNR by selective combining

Assume that there are M different signals obtained, by using a macroscopic diversity scheme, and the local mean of the kth signal, $m_k(t)$, is expressed in decibels and denoted by w_k, as shown in Fig. 10.1. The k signals are distinguishable during reception either by their frequency, their angle of arrival, or their time-division multiplexing differences, which are detectable. Where $w_k(t) = 10 \log m_k(t)$ has a lognormal distribution, the probability that the local mean in decibels, $w_k(t)$, is less than a level A, in decibels, is:

$$P(w_k(t) \leq A) = \int_{-\infty}^{A} \frac{1}{\sqrt{2\pi}\sigma_{w_k}} \exp\left[-\frac{(w_k - \mu_{w_k})^2}{2\sigma_{w_k}^2}\right] dw_k$$

$$= \frac{1}{2} + \frac{1}{2} \operatorname{erf}\left(\frac{A - \mu_{w_k}}{2\sigma_{w_k}}\right) \tag{10.1}$$

where μ_{w_k} and σ_{w_k} are the mean and the standard deviation in decibels of the long-term signal $w_k(t)$. If all M long-term signals are uncorrelated, then the probability that a selectively combined long-term signal $w(t)$ will be less than a level A is as follows [2]:

$$P(w(t) \leq A) = \prod_{k=1}^{M} P(w_k(t) \leq A) \tag{10.2}$$

where $\prod$ is a sign for multiplication and $w_k(t)$ is the long-term signal of the kth branch, measured in decibels. It can therefore be assumed that the means of the long-term signals, μ_w and μ_{w_k}, are the same: $\mu_w = \mu_{w_k}$.

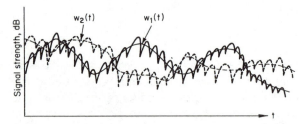

Figure 10.1   Combining long-term-fading signals $w_1(t)$ and $w_2(t)$ for macroscopic diversity.

The function for Eq. (10.2) is plotted in Fig. 10.2, for M equal to 2, 3, and 4. Assuming $\sigma_w = 8$ dB, and comparing the percentage of the signal below the mean level μ_w, Fig. 10.2 shows a probability of 45 percent that the signal will be below 0 dB for a single base station, a probability of 22 percent that it will be below 0 dB for a two-base-station diversity signal, and a probability of 10 percent that it will be below 0 dB for a three-base-station diversity signal. The percentage of probability shows significant improvement as the number of base stations increases.

10.2 Combining Techniques for Microscopic Diversity

In microscopic diversity, which deals with short-term fading, the principle is to obtain a number of signals with equal mean power through the use of diversity schemes. If the individual mean powers of the various signals are unequal, a degree of degradation that is proportional to the differences in mean power will result [1]. Three methods of linear diversity combining are described: selective combining, maximum-ratio combining, and equal-gain combining. These methods of linear diversity combining involve relatively simple weighted linear sums of

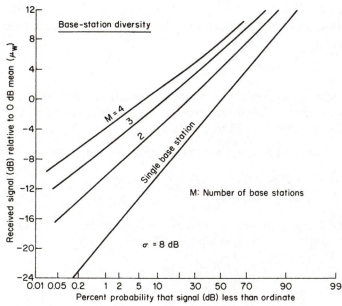

Figure 10.2 Performance of selective combining in macroscopic diversity.

multiple received signals. Let the combined output of the complex envelope $v(t)$ be expressed:

$$v(t) = \sum_{k=1}^{M} \alpha_k v_k(t) = \sum_{k=1}^{M} \alpha_k[s_k(t) + n_k(t)] \tag{10.3}$$

where $v_k(t)$ is the complex envelope of the kth branch and α_k is the weight constant for that branch. For analog transmission, linear diversity combining provides an effective method that is applicable not only for reducing the fading but also for distortionless reception. For digital data transmission, the distortion of the signal may not be a problem, since the signaling-error rate is based on optimal-level decisions.

10.3 Predetection and Postdetection Combining

Diversity receivers can be classified into two basic types—predetection and postdetection—each using one of three combining techniques, such as selective switched, maximum-ratio, or equal-gain combining. The advantage of using predetection combining because of nonlinear detectors can be demonstrated as follows, for two received signals $s_1(t)$ and $s_2(t)$ for the two respective branches:

$$s_1(t) = r_1(t)e^{j(\omega_c t + \psi_1)} \tag{10.4}$$

$$s_2(t) = r_2(t)e^{j(\omega_c t + \psi_2)} \tag{10.5}$$

with noise terms $n_1(t)$ and $n_2(t)$ mutually independent. A square-law detector is assumed.

Using predetection

Each signal is cophased at the IF frequency and combined before detection. Let the noise power of two branches be the same; the snr of the combined signal can be expressed in two cases.

Case 1—$r_1(t) = r_2(t)$ and $n_1 = n_2$ (two identical branches):

$$\frac{S}{N} = \frac{\langle (r_1(t) + r_2(t))^2 \rangle}{\langle (n_1 + n_2)^2 \rangle} = \frac{4\langle r_1^2(t) \rangle}{2n_1^2} = 2\frac{\langle r_1^2(t) \rangle}{n_1^2} \tag{10.6}$$

Case 2—$\langle r_1(t)r_2(t) \rangle = \langle r_1(t) \rangle \langle r_2(t) \rangle$ (two uncorrelated equal-strength branches) with $\langle r_1(t) \rangle = \langle r_2(t) \rangle$ and $\langle r_1^2(t) \rangle = \langle r_2^2(t) \rangle$:

$$\frac{S}{N} = \frac{\langle (r_1(t) + r_2(t))^2 \rangle}{\langle (n_1 + n_2)^2 \rangle}$$

$$= \frac{4\langle r_1^2(t) \rangle}{2n_1^2} = 2\frac{\langle r_1^2(t) \rangle}{n_1^2} \tag{10.7}$$

Using postdetection

Each signal is combined with another signal after detection at the baseband level. Let the noise power of two branches be the same; the snr of the combined signal can be expressed in two cases:

Case 1—$s_1(t) = s_1(t)$ (two identical branches):

$$\frac{S}{N} = \frac{\langle s_1^2(t)\rangle + \langle s_2^2(t)\rangle}{\langle n_1^2(t)\rangle + \langle n_2^2(t)\rangle} = \frac{\langle s_1^2(t)\rangle}{\langle n_1^2\rangle} \tag{10.8}$$

Case 2—$\langle s_1^2(t)\rangle = \langle s_1^2(t)\rangle$ (two equal-strength branches):

$$\frac{S}{N} = \frac{\langle s_1^2(t)\rangle}{\langle n_1^2\rangle} \tag{10.9}$$

Apparently Eq. 10.8 and Eq. 10.9 are identical.

In comparing Eq. (10.6) with Eq. (10.8) for case 1, notice that the predetection method shows an snr gain of 3 dB over the postdetection method. Comparing Eq. (10.7) with Eq. (10.9) for case 2, the snr obtained by using the predetection method is still 3 dB higher than that obtained by using the postdetection method. In predetection, because of the complexity of in-phase addition, the selective combining technique which has no such complexity, is probably the simplest of all diversity-combining methods. In postdetection, the selective combining technique, which selects detected envelopes, is probably the simplest of all diversity-combining methods. Note that there is no essential difference between the two kinds of detection if a linear detector is used.

10.4   Selective Diversity Combining

Selective diversity combining is the least complicated of the three types of linear combining. The other two types of linear combining are equal-gain and maximum-ratio combining. The algorithm for the selective diversity-combining technique is based on the principle of selecting the best signal among all of the signals received from different branches, at the receiving end. The differences between the different signals obtained from the various diversity schemes were described in Chap. 9. The resultant signal from selectively combining two individual signals is shown in Fig. 10.3. For selective combining, let α_m of Eq. (10.3) denote the index of a channel for which the carrier-to noise ratio $\gamma_m \geq \gamma_k$; then

$$\alpha_k = \begin{cases} 1 & k = m \\ 0 & k \neq m \end{cases} \tag{10.10}$$

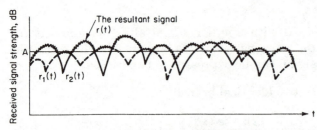

Figure 10.3 Plot of two selectively combined short-term fading signals.

The probability density function of an individual signal envelope r_k, Rayleigh-distributed with a mean power of $\sigma_{r_k}^2$, is:

$$p(r_k) = \frac{2r_k}{\sigma_{r_k}^2} \exp\left(-\frac{r_k^2}{\sigma_{r_k}^2}\right)$$

(10.11)

The cnr (carrier-to-noise ratio) can be defined:

$$\gamma_k = \frac{\text{instantaneous signal per branch}}{\text{mean noise power per branch}}$$

$$= \frac{r_k^2}{2N_k}$$

(10.12)

where $2N_k$ is defined as the average power of the complex envelope $n_k(t)$ of the additive noise in the kth receiver, expressed as:

$$2N_k = \langle |n_k(t)|^2 \rangle$$

(10.13)

as shown in Eq. (1.42). Then Γ_k is the ratio of the mean signal power per branch to the mean noise power per branch and is expressed:

$$\Gamma_k = \langle \gamma_k \rangle = \frac{\langle r_k^2 \rangle}{2N_k} = \frac{\sigma_{r_k}^2}{2N_k}$$

(10.14)

and therefore, Eq. (10.11) becomes

$$p(\gamma_k) = \frac{1}{\Gamma_k} e^{-\gamma_k/\Gamma_k}$$

(10.15)

The probability that cnr γ_k in one branch is less than or equal to a given level x is:

$$P(\gamma_k \leq x) = \int_0^x p(\gamma_k)\, d\gamma_k = 1 - e^{-x/\Gamma_k}$$

(10.16)

Then the probability that the resultant cnr γ will be below level x is equal to γ_k in all M branches that are simultaneously below or equal to a level x.

$$P[\gamma \leq x] = P[\gamma_1, \gamma_2, \ldots, \gamma_M \leq x]$$

$$= \prod_{k=1}^{M} \text{prob } (\gamma_k \leq x)$$

$$= \prod_{k-1}^{M} \left[1 - \exp\left(-\frac{x}{\Gamma_k}\right) \right] \tag{10.17}$$

The following paragraphs describe four special cases that are of particular interest in developing a rationale for applying the selective combining technique.

Case 1.   All $\Gamma_k = \Gamma$

Assume that all Γ_k are equal to a value Γ—i.e., that the mean cnr's over the short-term fading interval are equal among all the diversity branches; then:

$$P(\gamma \leq x) = \left[1 - \exp\left(-\frac{x}{\Gamma}\right) \right]^{M} \tag{10.18}$$

The function for Eq. (10.18) is plotted in Fig. 10.4, where the $M = 1$ curve represents the Rayleigh distribution of the cnr in each diversity branch. The percentage of the total time interval during which a signal is below any given level is called the "outage rate" at that level. Notice that the resultant signal for combining two branch signals is greatly improved at the low outage rate, and although signal improvement

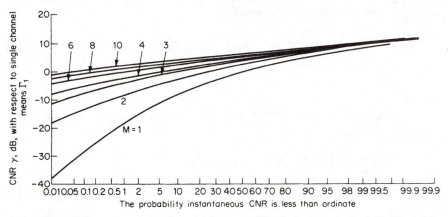

Figure 10.4   Performance curves for selectively combined microscopic-diversity Rayleigh-fading signals.

increases as the number of branches becomes greater, the rate of improvement decreases. The greatest improvement, as shown in Fig. 10.4, is obtained in going from single-branch to two-branch combining.

Case 2. $\gamma \ll \Gamma_k, k = 1, \ldots, M$

Since the approximation is

$$\exp(-\alpha) \approx 1 - \alpha \qquad \alpha \ll 1 \tag{10.19}$$

Eq. (10.17) becomes:

$$P(\gamma \leq x) = \frac{x^M}{\displaystyle\prod_{k=1}^{M} \Gamma_k} \tag{10.20}$$

Case 3. $\Gamma_M < \gamma \ll (\Gamma_1, \ldots, \Gamma_{M-1})$

If $\Gamma_M < \gamma \ll (\Gamma_1, \ldots, \Gamma_{M-1})$, then, the cnr γ is well below the mean cnr values for $M - 1$ branches, and Eq. (10.17) becomes:

$$P(\gamma \leq x) = \left[1 - \exp\left(-\frac{x}{\Gamma_M}\right) \right] \left(\frac{x^{M-1}}{\displaystyle\prod_{k=1}^{M} \Gamma_k} \right) \tag{10.21}$$

and for $\Gamma_M \ll x$, Eq. (10.21) becomes

$$P(\gamma \leq x) = \frac{x^{M-1}}{\displaystyle\prod_{k=1}^{M} \Gamma_k} \tag{10.22}$$

Comparing Eqs. (10.22) and (10.20) confirms that there is essentially only an $(M - N)$-fold diversity action when the mean cnr of N branches is well below the level of interest; i.e.,

$$P(\gamma \leq x) = \frac{x^{M-N}}{\displaystyle\prod_{k=1}^{M-N} \Gamma_k} \qquad \Gamma_{M-N+1}, \Gamma_{M-N+2}, \ldots, \Gamma_M \ll x \tag{10.23}$$

Case 4.   Correlated signals

When calculating Eq. (10.17) with correlated signals, the following expressions are obtained:

$$P(\gamma \leq x) = P(\gamma_1, \gamma_2, \ldots, \gamma_M \leq x)$$

$$= \int_0^x d\gamma_1 \int_0^x d\gamma_2 \cdots \int_0^x p(\gamma_1, \gamma_2, \ldots, \gamma_M) \, d\gamma_M \tag{10.24}$$

where the joint probability density function $p(\gamma_1\gamma_2 \cdots \gamma_M)$ is obtained from

$$p(\gamma_1 \cdots \gamma_M) = |J| p(r_1, r_2, \ldots, r_M) \tag{10.25}$$

$|J|$ is the jacobian transform expressed in Eq. (2.49), and r_k is the signal envelope of the kth branch. The joint probability density function of $r_1 \cdots r_k$ can be obtained as follows:

$$p(r_1, r_2, \ldots, r_M) = \int_{-\pi}^{\pi} d\psi_1 \int_{-\pi}^{\pi} d\psi_2 \cdots \int_{-\pi}^{\pi} d\psi_M \, p(r_1, \psi_1, \ldots, r_M; \psi_M)$$

$$(10.26)$$

where r_i and ψ_i are the envelope and phase of a complex signal $s_i(t)$, expressed:

$$s_i(t) = r_i e^{j\psi_i}$$

$$= X_i + jY_i$$

and $p(r_1, \psi_1, \ldots, r_M, \psi_M)$ is obtained from the joint probability density functions of the X_i's and Y_i's, which are Gaussian variables. For 2-branch selective combining, $p(r_1, \psi_1, r_2, \psi_2)$ is shown in Eq. (7.51). By applying Eq. (10.26), the following is obtained [2]:

$$p(r_1, r_2) = \frac{r_1 r_2}{\sigma_{x_1}^2 \sigma_{x_2}^2 (1 - |\rho|^2)} I_0 \left[\frac{|\rho| r_1 r_2}{(1 - |\rho|^2)\sigma_{x_1}\sigma_{x_2}} \right.$$

$$\left. \times \exp \left[-\frac{\sigma_{x_2}^2 r_1^2 + \sigma_{x_1}^2 r_2^2}{2\sigma_{x_1}^2 \sigma_{x_2}^2 (1 - |\rho|^2)} \right] \right. \qquad (10.27)$$

where

$$|\rho|^2 = \frac{R_c^2(\tau) + R_{cs}^2(\tau)}{\sigma_{x_1}^2 \sigma_{x_2}^2} = 4 \frac{R_c^2(\tau) + R_{cs}^2(\tau)}{\sigma_{r_1}^2 \sigma_{r_2}^2} \qquad (10.28)$$

and $R_c(\tau)$ and $R_{cs}(\tau)$ are defined in Eq. (7.45). ρ is the correlation coefficient. The signal envelope correlation coefficient $\rho_r(\tau)$, with a time separation of τ as the mobile unit travels at a speed V, is expressed:

$$\rho_r(\tau) = |\rho(\tau)|^2 = J_0^2(\beta V t) \qquad (10.29)$$

as was previously shown in Eq. (6.110).

Since $|J|$ of Eq. (10.25) is defined

$$|J| = \left(\frac{\partial r_1}{\partial \gamma_1}\right)\left(\frac{\partial r_2}{\partial \gamma_2}\right) = \frac{\sigma_{x_1}^2 \sigma_{x_2}^2}{r_1 r_2 \Gamma_1 \Gamma_2} \qquad (10.30)$$

and the value of γ_k is:

$$\gamma_k = \frac{r_k^2 \Gamma_k}{\sigma_{r_k}^2} = \frac{r_k^2 \Gamma_k}{2\sigma_{x_k}^2} \qquad \text{or} \qquad \frac{\gamma_k}{\Gamma_k} = \frac{r_k^2}{2\sigma_{x_k}^2}$$

then Eq. (10.27) becomes:

$$p(\gamma_1, \gamma_2) = \frac{1}{\Gamma_1\Gamma_2(1 - |\rho|^2)} I_0\left(\frac{2\sqrt{\gamma_1\gamma_2/\Gamma_1\Gamma_2}|\rho|}{1 + |\rho|^2}\right) \exp\left(-\frac{\gamma_1/\Gamma_1 + \gamma_2/\Gamma_2}{1 - |\rho|^2}\right)$$

(10.31)

By letting $\Gamma_1 = \Gamma_2$, and by inserting Eq. (10.31) into Eq. (10.24), the following is obtained:

$$P(\gamma \leq x) = 1 - e^{-x/\Gamma}\left[1 - Q(a, b) + Q(b, a)\right]$$ (10.32)

where

$$Q(a, b) = \int_b^\infty e^{-(1/2)(a^2 + x^2)}I_0(ax)x \, dx$$

$$a = \sqrt{\frac{2x}{\Gamma(1 + |\rho|^2)}}$$

$$b = \sqrt{\frac{2x}{\Gamma(1 - |\rho|^2)}}$$ (10.33)

For the case of $x \ll \Gamma$,

$$P(\gamma \leq x) = \frac{x^2}{\Gamma^2(1 - |\rho|^2)}$$ (10.34)

The function for Eq. (10.32) is shown in Fig. 10.5.

Selective combining is very difficult to implement, because a floating threshold level is needed. Therefore, switched combining is a practical alternative, based on a fixed threshold level and a practical combining technique. It will be described in Sec. 10.5.

Example 10.1 Antenna-pattern ripples, caused by a multiple-array antenna configuration at the base station, produce unequal average cnr values in M branches of a diversity-received signal during any given time interval. This phenomenon was described previously in Chap. 5. The resulting degradation in performance is significantly greater when two unequal power branches are combined than it is when two equal power branches are combined.

When the selective combining technique is used to combine two unequal power branches, the cumulative probability distribution (cpd) for a combined cnr γ can be calculated by substituting Eq. (10.31) into Eq. (10.24) to obtain the following result [2]:

$$\text{prob}(\gamma \leq x) = 1 - \exp\left(-\frac{x}{\Gamma_1}\right) Q(a, \sqrt{\rho}b)$$

$$- \exp\left(-\frac{x}{\Gamma_2}\right)[1 - Q(\sqrt{\rho}a, b)]$$ (E10.1.1)

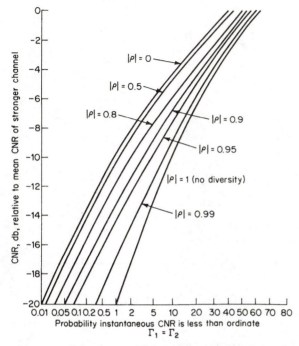

Figure 10.5   Selective combining of two correlated signals.

where ρ_r is as expressed in Eq. (10.29).

$$a = \sqrt{\frac{2x}{\Gamma_2(1 + \rho_r)}} \qquad \text{(E10.1.2)}$$

$$b = \sqrt{\frac{2x}{\Gamma_1(1 - \rho_r)}} \qquad \text{(E10.1.3)}$$

The function of Eq. (E10.1.1) is plotted in Fig. E10.1, for values of $\Gamma_1 \neq \Gamma_2$, and the curves show that performance degradation increases as the ratio of the average power in the signal branches increases.

10.5   Switched Combining

Since, as was shown in the preceding section, selective combining is an impractical technique for mobile-radio communication, a more practical technique known as "switched combining" is described in the following paragraphs. Assuming that two independent Rayleigh signals $r_1(t)$ and $r_2(t)$ are received from two respective diversity branches, the resultant carrier envelope $r(t)$, then, can be obtained by using a switch-and-stay strategy. The strategy is to stay with the signal envelope $r_1(t)$ or $r_2(t)$ until the envelope drops below a predetermined switching

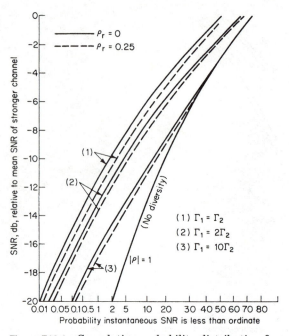

Figure E10.1 Cumulative probability distribution for unequal power branches using selective diversity-combining techniques.

threshold A, and then to switch to the stronger of the two signals. The parameters for switched combining can be defined:

$$q_A = P(r \leq A) = \int_0^A p(r)\, dr = 1 - e^{-A^2/2\sigma_x^2} \tag{10.35}$$

$$p_A = P(r > A) = \int_A^\infty p(r)\, dr = e^{-A^2/2\sigma_x^2} = 1 - q_A \tag{10.36}$$

where

$$E[r^2] = 2\sigma_x^2$$

In most cases, the switching will occur at a fixed threshold level A, as shown in Fig. 10.6. Then, to find the cumulative probability distribution below any arbitrary level B, the following expression is used [3]:

$$P(r \leq B) = \begin{cases} P(r \leq B \mid r = r_1) \\ \qquad\quad \text{or} \\ P(r \leq B \mid r = r_2) \end{cases} \tag{10.37}$$

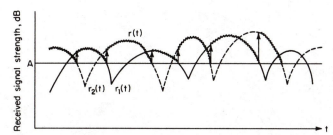

Figure 10.6 Switching threshold level and discontinuous nature of a switched-combined signal.

Since r_1 and r_2 are statistically indistinguishable, then:

$$P(r \le B) = \begin{cases} P(A < r_1 < B, \text{ or both } r_1 < A \text{ and } r_1 < B) & B > A \\ P(r_1 < A \text{ and } r_1 < B) & B \le A \end{cases} \quad (10.38)$$

where

$$P(A < r_1 < B, \text{ or both } r_1 < A \text{ and } r_1 < B) = \int_A^B p(r_1) \, dr_1$$

$$+ \left[\int_0^A p(r_1) \, dr_1 \right] \left[\int_0^B p(r_1) \, dr_1 \right] = q_B - q_A + q_A q_B \quad (10.39)$$

and where q_A and q_B are as defined in Eq. (10.35). The condition $P(r_1 < A$ and $r_1 < B)$ can be expressed:

$$P(r_1 < A \text{ and } r_1 < B) = \int_0^A p(r_1) \, dr_1 \int_0^B p(r_1) \, dr_1$$

$$= q_A q_B \quad (10.40)$$

Hence, Eq. (10.38) becomes:

$$P(r \le B) = \begin{cases} q_B - q_A + q_A q_B & B > A \\ q_A q_B & B \le A \end{cases} \quad (10.41)$$

The distribution of r for several values of switching threshold A is shown in Fig. 10.7. As illustrated in Fig. 10.7, the improvement is obtained above the threshold level. Below the threshold level, the combined signal follows the Rayleigh characteristics. The switched-combined signal always performs worse than the selectively combined signal, except at the threshold level, where performance is equal.

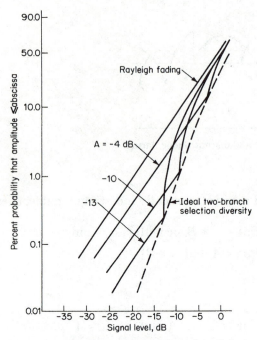

Figure 10.7   Performance of a 2-branch switched-combined signal with various threshold levels.

10.6   Maximal-Ratio Combining

In maximal-ratio combining, M signals are weighted for optimum performance and are cophased before being combined. In predetection maximal-ratio combining, each signal is cophased at the IF level, as previously described in Sec. 10.3. The maximal-ratio combining technique can also be applied during postdetection of the received signal; a gain control is required following each detection. The complex envelope of each diversity branch, at the input of the IF receiving circuit, can be expressed:

$$v_k(t) = s_k(t) + n_k(t) = a_0(t)u_k(t) + n_k(t)$$

where $a_0(t)$ is a time-varying signal and $u_k(t)$ is due to multipath fading with no specular component. The predetection linear combining results in a complex output envelope, expressed as:

$$v(t) = \sum_{k=1}^{M} \alpha_k v_k(t) = \sum_{k=1}^{M} \alpha_k[s_k(t) + n_k(t)] \tag{10.42}$$

where

$$v(t) = s(t) + n(t)$$

and where the resultant signal and noise $s(t)$ and $n(t)$, respectively, are expressed:

$$s(t) = \sum_{k=1}^{M} \alpha_k s_k(t) = a_0(t) \sum_{k=1}^{M} \alpha_k u_k(t) \tag{10.43}$$

$$n(t) = \sum_{k=1}^{M} \alpha_k n_k(t) \tag{10.44}$$

Without loss generality, let $\langle a_0^2(t) \rangle = 1$ over a period of $2T$ for a unit mean square envelope. The cnr is:

$$\gamma = \frac{1}{2} \frac{|s(t)|^2}{\eta} = \frac{1}{2} \frac{\left| \sum_{k=1}^{M} \alpha_k u_k \right|^2 \langle a_0^2(t) \rangle}{\sum_{k=1}^{M} |\alpha_k|^2 \langle n_k^2 \rangle} = \frac{1}{2} \frac{\left| \sum_{k=1}^{M} \alpha_k u_k \right|^2}{\sum_{k=1}^{M} |\alpha_k|^2 \eta_k} \tag{10.45}$$

As a result of applying the Schwarz inequality to complex-valued numbers, the following relationship holds:

$$\left| \sum_{k=1}^{M} \alpha_k u_k \right|^2 \leq \left(\sum_{k=1}^{M} \frac{|u_k|^2}{\eta_k} \right) \left(\sum_{k=1}^{M} |\alpha_k|^2 \eta_k \right) \tag{10.46}$$

Substituting Eq. (10.46) into Eq. (10.45) yields:

$$\gamma = \frac{1}{2} \frac{\left| \sum_{k=1}^{M} \alpha_k u_k \right|^2}{\sum_{k=1}^{M} |\alpha_k|^2 \eta_k} \leq \frac{1}{2} \sum_{k=1}^{M} \frac{|u_k|^2}{\eta_k} \tag{10.47}$$

To obtain the maximum in Eq. (10.47), it is necessary to apply an equals sign on Eq. (10.46). The equals sign is used if and only if:

$$\alpha_k = K \frac{u_k^*}{\eta_k} \tag{10.48}$$

for each value of k, where K is any arbitrary complex number. Equation (10.48) indicates that the optimum weight for each branch has a magnitude proportional to the conjugate of the fading signal and inversely proportional to the branch noise-power level. Equation (10.47) can be rewritten to express the maximal-ratio output:

$$\gamma = \sum_{k=1}^{M} \frac{\frac{1}{2} |u_k|^2}{\eta_k} = \sum_{k=1}^{M} \gamma_k \tag{10.49}$$

where

$$\gamma_k = \frac{1}{2} \frac{u_k u_k^*}{\eta_k} \tag{10.50}$$

Equation (10.49) is a χ distribution. The term "maximal-ratio" was used by Brennan [4] to define the sum of the instantaneous cnr's for individual branches in a multibranched diversity system, as expressed by Eq. (10.49).

Example 10.2   Two signals, u_1 and u_2, represent two multipath-fading signals in a mobile-radio environment. What method can be used to find the correlation between the two complex Gaussian signals u_1 and u_2, $\langle u_1 u_2^* \rangle$, from two branches separated by a distance d?

solution   Let $u_1 = X_1 + Y_1$ and $u_2 = X_2 + jY_2$. Then:

$$\langle u_1 u_2^* \rangle = (\langle X_1 X_2 \rangle + \langle Y_1 Y_2 \rangle) + j(\langle Y_1 X_2 \rangle - \langle X_1 Y_2 \rangle) \tag{E10.2.1}$$

Applying the model described in Eqs. (6.11) and (6.12) of Chap. 6 yields the following expressions:

$$X_k = \sum_{i=1}^{N_k} (R_i \cos \xi_i + S_i \sin \xi_i) \tag{E10.2.2}$$

$$Y_k = \sum_{i=1}^{N_k} (S_i \cos \xi_i - R_i \sin \xi_i) \tag{E10.2.3}$$

where

$$\xi_i = \beta x_k \cos (\phi_i - \alpha) \tag{E10.2.4}$$

The value k denotes two different signals ($k = 1, 2$), and N_k is the number of waves within a given interval. Other parameters appearing in Eqs. (E10.2.1) through (E10.2.4) have been described in Sec. 6.2 of Chap. 6.

Since

$$E[R_i R_j] = E[S_i S_j] = \delta_{ij} \quad \text{and} \quad E[R_i S_j] = 0 \tag{E10.2.5}$$

then where ϕ_i is the angle of the ith wave arrival and a uniform distribution is assumed, the following relationships are apparent:

$$\langle X_1 X_2 \rangle = \langle Y_1 Y_2 \rangle = N_1 N_2 \int_0^{2\pi} \cos [\beta d_{12} \cos (\phi_i - \alpha)] \, d\phi_i$$

$$= N_1 N_2 J_0(\beta d_{12}) \tag{E10.2.6}$$

and

$$\langle Y_1 X_2 \rangle = -\langle X_1 Y_2 \rangle$$

$$= N_1 N_2 \int_0^{2\pi} \sin [\beta d_{12} \cos (\phi_i - \alpha)] \, d\phi_i = 0 \tag{E10.2.7}$$

where $d_{12} = x_2 - x_1$ is the separation between the two signal branches. Therefore,

$$\langle u_1 u_2^* \rangle = 2N_1 N_2 J_0(\beta d_{12}) \tag{E10.2.8}$$

where $J_0(\beta d_{12}) = J_0(\beta d_{21})$, in accordance with the Bessel function property $J_0(-x) = J_0(x)$. The general expression is therefore

$$J_0(\beta d_{jk}) = J_0(\beta d_{kj}) \tag{E10.2.9}$$

**Calculating the probability
density distribution**

Let each of the M jointly distributed complex Gaussians $\{z_k\}$ be defined:

$$z_k = \frac{u_k}{\sqrt{\eta_k}} = x_k + jy_k$$

Then the distribution of M pairs of real Gaussian variables $\{x_k, y_k\}$ can be written in the form:

$$p(\{x_k, y_k\}) = \frac{1}{(2\pi)^M \, |\Lambda|^{1/2}} \exp \{-\tfrac{1}{2}([Z]^t - \langle [Z]^t \rangle)$$

$$\times [\Lambda]^{-1}([Z^*] - \langle [Z^*] \rangle)\} \tag{10.51}$$

where $[Z]$ is a column in a matrix, with its elements $\{z_k\}$, $[\Lambda]$ representing the $M \times M$ covariance matrix, as shown in Eq. (7.44) and expressed:

$$[\Lambda] = \tfrac{1}{2}\{[Z^*] - \langle [Z^*] \rangle\} \{[Z]^t - \langle [Z]^t \rangle\} \tag{10.52}$$

$[Z^*]$ and $[Z]^t$ are the complex conjugate and transpose matrices, respectively, and the total snr γ from Eq. (10.49) becomes:

$$\gamma = \tfrac{1}{2}[Z^*]^t[Z] \tag{10.53}$$

The characteristic function of γ can be found from:

$$\Phi_\gamma(s) = E[\exp(js\gamma)] \tag{10.54}$$

Then the probability density function, as an inverse Fourier transform, can be obtained as follows:

$$p(\gamma) = \int_{-\infty}^{\infty} \exp(-js\gamma) \, \Phi_\gamma(s) \, ds \tag{10.55}$$

From Turin [5], Eq. (10.54) can be derived as follows:

$$\Phi_\gamma(s) = \frac{1}{\det([I] + s[\Lambda])} \tag{10.56}$$

The covariance matrix $[\Lambda]$, shown in Eq. (10.52), can be written in another form:

$$[\Lambda] = \frac{\langle u_1 u_1^* \rangle}{2\eta_1} [\mathcal{R}] = \Gamma_1 [\mathcal{R}] \qquad (10.57)$$

where Γ_1 is the cnr of the first single branch and $[\mathcal{R}]$ is a normalized covariance matrix of $\{z_k\}$. Each element of $[\mathcal{R}]$, for a mobile-reception case, can be found from Eq. (E10.2.8) and expressed:

$$\mathcal{R}_{jk} = \frac{\langle z_j z_k^* \rangle}{\langle z_1 z_1^* \rangle} = \frac{\sqrt{\Gamma_j \Gamma_k}}{\Gamma_1} J_0(\beta d_{jk}) \qquad (10.58)$$

where d_{jk} is the antenna spacing between the jth branch and the kth branch and β is the wave number. η_j and $\Gamma_j = \frac{1}{2} \langle z_j z_j^* \rangle$ are the noise power and cnr of the jth branch, respectively. Equation (10.56) can be further simplified [5]:

$$\Phi_\gamma(s) = \frac{1}{\displaystyle\prod_{j=1}^{M} (1 + s\lambda_j' \Gamma_1)} \qquad (10.59)$$

where the λ_j''s are the eigenvalues of the matrices $[\mathcal{R}]$, $\lambda_j = \lambda_j' \Gamma_1$. Substituting Eq. (10.59) into Eq. (10.55) yields the probability density function, expressed as:

$$p(\gamma) = \frac{1}{\displaystyle\prod_{j=1}^{M} \lambda_j} \sum_{j=1}^{M} \frac{e^{-\gamma/\lambda_j}}{\displaystyle\prod_{j \neq k}^{M} (1/\lambda_k - 1/\lambda_j)} \qquad (10.60)$$

where the eigenvalues λ_j may be either positive real values or complex conjugate pairs. The cumulative distribution of γ, of the combined signal, can then be expressed as:

$$P(\gamma \leq x) = 1 - \sum_{j=1}^{M} \frac{(\lambda_j)^{M-1} \exp(-x/\lambda_j)}{\displaystyle\prod_{k \neq j}^{M} (\lambda_j - \lambda_k)}$$

Example 10.3 Given a three-branch linear array with equal half-wavelength spacing between adjacent branches, find the eigenvalues λ_j' of a normalized covariance matrix $[\mathcal{R}]$, assuming that the average cnr's of the individual branches are the same ($\Gamma_k = \Gamma$).

solution The elements of the normalized covariance matrix $[\mathcal{R}]$ can be found from Eq. (10.58) and expressed as:

$$[\mathcal{R}] = \begin{bmatrix} 1 & J_0(\beta d_{12}) & J_0(2\beta d_{13}) \\ J_0(\beta d_{21}) & 1 & J_0(\beta d_{23}) \\ J_0(\beta d_{31}) & J_0(\beta d_{32}) & 1 \end{bmatrix} \tag{E10.3.1}$$

Since

$$d_{13} = x_3 - x_1 = 2d_{12} = 2d_{23} = \lambda \quad \text{and} \quad \beta = \frac{2\pi}{\lambda} \tag{E10.3.2}$$

then

$$\beta d_{13} = 2\beta d_{12} = 2\pi \tag{E10.3.3}$$

and where

$$J_0(\beta d_{jk}) = J_0(\beta d_{kj})$$

from Eq. (E10.2.9), then the eigenvalues for the normalized covariance matrix elements of Eq. (E10.3.1) can be found from the equation

$$\left| [\mathcal{R}] - \lambda'_m[I] \right| = 0 \tag{E10.3.4}$$

where $[I]$ is an identity matrix and the λ'_m's are the eigenvalues.

Note λ'_m and λ_m should not be confused with λ, which is the symbol for wavelength.

By substituting the values of Eq. (E10.3.1) into Eq. (E10.3.4), the following covariance matrix is obtained:

$$\begin{vmatrix} 1 - \lambda'_m & J_0(\pi) & J_0(2\pi) \\ J_0(\pi) & 1 - \lambda'_m & J_0(\pi) \\ J_0(2\pi) & J_0(2\pi) & 1 - \lambda'_m \end{vmatrix} = 0 \tag{E10.3.5}$$

Solving Eq. (E10.3.5) and letting $a = J_0(\pi) = -0.3033$ and $b = J_0(2\pi) = 0.2194$ yields the following:

$$(1 - \lambda'_m)^3 + A(1 - \lambda'_m) + B = 0 \tag{E10.3.6}$$

where $A = -(b^2 + 2a^2)$ and $B = 2a^2b$.

Equation (E10.3.6) can be tested as follows:

$$\frac{B^2}{4} + \frac{A^3}{27} < 0 \tag{E10.3.7}$$

which shows that there are three real and unequal roots, which are expressed as:

$$\lambda'_1 = 1 - 2\sqrt{\frac{-A}{3}} \cos\frac{\phi}{3} \tag{E10.3.8}$$

$$\lambda'_2 = 1 - 2\sqrt{\frac{-A}{3}} \cos\left(\frac{\phi}{3} + 120°\right) \tag{E10.3.9}$$

$$\lambda'_3 = 1 - 2\sqrt{\frac{-A}{3}} \cos\left(\frac{\phi}{3} + 240°\right) \tag{E10.3.10}$$

where

$$\cos \phi = -\frac{B/2}{\sqrt{-A^3/27}}$$ (E10.3.11)

Special cases

1.   M branches of uncorrelated signals   When all values of λ_j are equal $(\lambda_j = \Gamma_j = \Gamma)$, then Eq. (10.59) is simplified, and Eq. (10.55) becomes:

$$p(\gamma) = \frac{1}{(M-1)!} \frac{\gamma^{M-1}}{\Gamma^M} \exp\left(-\frac{\gamma}{\Gamma}\right)$$ (10.61)

and the cumulative distribution is:

$$P(\gamma < x) = \int_0^x p(\gamma) \, d\gamma = \frac{1}{(M-1)!\Gamma^M} \int_0^x \gamma^{M-1} \exp\left(-\frac{\gamma}{\Gamma}\right) d\gamma$$ (10.62)

For $M = 2$, the following is obtained:

$$P(\gamma < x) = 1 - e^{-x/\Gamma}\left(\frac{x}{\Gamma} + 1\right)$$ (10.63)

For $M = 3$, the following is obtained:

$$P(\gamma < x) = 1 - e^{-x/\Gamma}\left(\frac{x^2}{2\Gamma^2} + \frac{x}{\Gamma} + 1\right)$$ (10.64)

The values of M in Eqs. (10.63) and (10.64), and other values of M from Eq. (10.62), are plotted in Fig. 10.8, where the greatest degree of improvement in performance occurs in going from a single-branch system to a 2-branch diversity system.

2.   Two branches of correlated signals   When two branches are correlated, then Eq. (7.44) becomes:

$$[\Lambda] = \begin{bmatrix} \Gamma_1 & \rho\sqrt{\Gamma_1\Gamma_2} \\ \rho^*\sqrt{\Gamma_1\Gamma_2} & \Gamma_2 \end{bmatrix}$$ (10.65)

where ρ is the complex correlation coefficient between two Gaussian variables.

   Note   Do not confuse the term ρ with ρ_r, which is the correlation coefficient between two Rayleigh signal envelopes. The relationship between these two terms is $\rho_r = |\rho|^2$, which can be derived from Eq. (6.107).

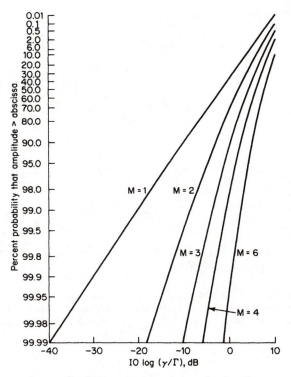

Figure 10.8 Performance curves for maximal-ratio combining within independent channels.

The eigenvalues are the solution to:

$$\begin{vmatrix} \Gamma_1 - \lambda & \rho\sqrt{\Gamma_1\Gamma_2} \\ \rho^* \sqrt{\Gamma_1\Gamma_2} & \Gamma_2 - \lambda \end{vmatrix} = 0 \qquad (10.66)$$

i.e.,

$$\lambda_1 = \tfrac{1}{2}[\Gamma_1 + \Gamma_2 - \sqrt{(\Gamma_1 + \Gamma_2)^2 - 4\Gamma_1\Gamma_2(1 - |\rho|^2)}] \qquad (10.67)$$

$$\lambda_2 = \tfrac{1}{2}[\Gamma_1 + \Gamma_2 + \sqrt{(\Gamma_1 + \Gamma_2)^2 - 4\Gamma_1\Gamma_2(1 - |\rho|^2)}] \qquad (10.68)$$

For $\Gamma_1 = \Gamma_2 = \Gamma$, Eqs. (10.67) and (10.68) become:

$$\lambda_1 = \Gamma(1 - |\rho|) \qquad (10.69)$$

$$\lambda_2 = \Gamma(1 + |\rho|) \qquad (10.70)$$

Then Eq. (10.60) becomes:

$$p(\gamma) = \frac{e^{-\gamma/\lambda_1}}{\lambda_1 - \lambda_2} + \frac{e^{-\gamma/\lambda_2}}{\lambda_2 - \lambda_1}$$

$$= \frac{\exp\left(-\gamma/(1 - |\rho|)\Gamma\right)}{-2|\rho|\Gamma} + \frac{\exp\left(-\gamma/(1 + |\rho|)\Gamma\right)}{2|\rho|\Gamma} \qquad (10.71)$$

and

$$P(\gamma \leq x) = \int_0^x p(\gamma)\, d\gamma = 1 - \frac{1}{2\,|\rho|}\left\{(1 + |\rho|)\exp\left[-\frac{x}{\Gamma(1 + |\rho|)}\right]\right.$$

$$\left. - (1 - |\rho|)\exp\left[-\frac{x}{\Gamma(1 - |\rho|)}\right]\right\} \qquad (10.72)$$

The function for Eq. (10.72) is shown in Fig. 10.9, where ρ^2 is used as the variable instead of $|\rho|$.

Example 10.4   Consider a situation where system design imposes a transmitter power limitation that is 20 dB less than the desired output power level. In such a case, the use of a maximal-ratio diversity-combining scheme would be advan-

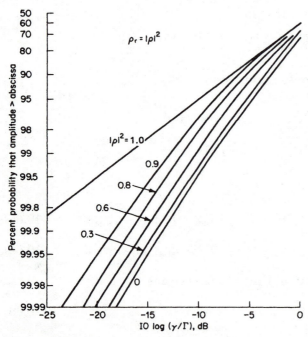

Figure 10.9   Performance curves for maximal-ratio combining with two correlated branches.

tageous in enhancing signal reception. In order to obtain a 20-dB gain at a level where the received signal is below the threshold level only 0.2 percent of the time, how many uncorrelated diversity branches are required?

solution   From Fig. 10.8, the 99.8 percent line, corresponding to 0.2 percent of the time when the signal is below the threshold of reception, is –26 dB with respect to the mean power for a single branch ($M = 1$). In order to obtain the required 20-dB gain, the signal level must remain at –6 dB or higher 99.8 percent of the time. Therefore, a 3-branch maximal-ratio diversity technique is required, as shown in Fig. 10.8.

10.7   Equal-Gain Combining

The maximal-ratio predetection combining technique is an ideal linear diversity-combining technique; however, it requires costly design in receiver circuitry to achieve the correct weighting factors. The selective combining technique selects the strongest signal branch at any given instant of time, but it is also difficult to implement. The switched-combined diversity scheme always provides worse performance than the selective combining diversity scheme. In comparison, equal-gain combining uses a simple phase-locked summing circuit to sum all of the individual signal branches. The equal-gain combining technique still provides incoherent summing of the various noise elements, but it also provides the required coherent summing of all the individual signal branches, as was previously discussed in Sec. 10.2 as another kind of linear diversity combining.

The resultant signal for an equal-gain combining technique is:

$$\gamma = \tfrac{1}{2} \frac{r^2}{\displaystyle\sum_{k=1}^{M} \eta_k} \tag{10.73}$$

where the resultant envelope r is:

$$r = \sum_{k=1}^{M} r_k \tag{10.74}$$

and where r_k is the signal envelope at the kth branch.

Calculating the probability density function

From Eq. (10.27), the probability density function of the resultant signal $r = r_1 + r_2$ can be obtained as follows:

$$p(r) = \int_0^r p(r_1, r_2 = r - r_1) \, dr_1 \tag{10.75}$$

The solution for Eq. (10.75) can be obtained by the numerical integration method. Another method can be used when low values of single branch Γ_{eq}, the mean cnr of an equal-gain combiner, are related to the mean cnr of a maximal-ratio combiner. Under these conditions, an approximation can be obtained [2]:

$$\Gamma_{eq} = g_M \Gamma_{max} = g_M \Gamma \tag{10.76}$$

where

$$g_M = \sqrt[M]{\frac{(M/2)^M \sqrt{\pi}}{(M - 1/2)!}} = \frac{M}{2} \frac{1}{\sqrt[M]{(M - 1/2)!/\sqrt{\pi}}} \tag{10.77}$$

so that for

$$M = 2 \qquad g_2 = 1.16$$
$$M = 3 \qquad g_3 = 1.20$$
$$M = 4 \qquad g_4 = 1.26$$
$$M \gg 1 \qquad g_m = 1.36$$

Then, by substituting the values of Eq. (10.77) into Eq. (10.60), the following expressions are obtained:

$$p_2(\gamma) = \sum_{m=1}^{2} \frac{g_2 \exp(-g_2 \gamma/\lambda_m \Gamma)}{(\lambda_m - \lambda_i)\Gamma} \tag{10.78}$$

$$p_3(\gamma) = \sum_{m=1}^{3} \frac{g_3 \lambda_m \exp(-g_3 \gamma/\lambda_m \Gamma)}{(\lambda_m - \lambda_i)(\lambda_m - \lambda_j)\Gamma} \tag{10.79}$$

$$p_4(\gamma) = \sum_{m=1}^{4} \frac{g_4 \lambda_m^2 \exp(-g_4 \gamma/\lambda_m \Gamma)}{(\lambda_m - \lambda_i)(\lambda_m - \lambda_j)(\lambda_m - \lambda_k)\Gamma} \tag{10.80}$$

For any value of γ, Eqs. (10.78) through (10.80) may generate small errors in relation to the exact solutions.

Cumulative probability distribution

By integrating Eqs. (10.78) through (10.80) for the range of γ from 0 to x, the cumulative probability distribution for the total number of branches can be obtained, as shown in Fig. 10.10. Comparing the performance of the equal-gain combiner with that of the maximal-ratio combiner reveals that equal-gain combiner performance is slightly worse than the performance of the maximal-ratio combiner.

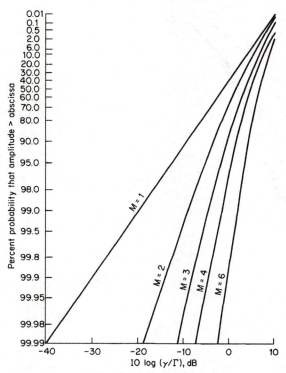

Figure 10.10 Cumulative probability distribution for equal-gain-combined branches.

Example 10.5   To find the cumulative probability distribution (cpd) of two correlated signals from an equal-gain combiner, let ρ_r represent the correlation coefficient between two Rayleigh-fading envelopes. Then an approximation of the cpd can be obtained from Eq. (10.72), as follows:

$$P(\gamma \leq x) = 1 + \frac{(1 - \sqrt{\rho_r})e^{-2ax/\Gamma} - (1 + \sqrt{\rho_r})e^{-2bx/\Gamma}}{2\sqrt{\rho_r}} \qquad \text{(E10.5.1)}$$

where

$$a = \frac{g_2}{2(1 - \sqrt{\rho_r})}$$

$$b = \frac{g_2}{2(1 + \sqrt{\rho_r})}$$

From Eq. (10.77)

$$g_2 = \begin{cases} 1.16 & \rho_r < 1 \\ 1 & \rho_r = 1 \end{cases}$$

The function for Eq. (E10.5.1) is plotted in Fig. E10.5.

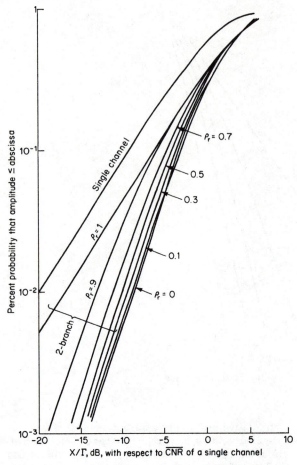

Figure E10.5 Cumulative probability distribution of a 2-branch correlated equal-gain-combining signal.

Calculating the level-crossing rate (LCR)

The complex form of a single branch can be expressed:

$$u_i = X_i + jY_i = r_i e^{j\psi_i} \tag{10.81}$$

and its envelope r_i is expressed:

$$r_i = \sqrt{X_i^2 + Y_i^2} \tag{10.82}$$

and the derivative of the envelope is:

$$\dot{r}_i = \dot{X}_i \cos \psi_i + \dot{Y}_i \sin \psi_i \tag{10.83}$$

The total signal for an M-branched equal-gain combiner can then be expressed:

$$r = \sum_{i=1}^{M} r_i \qquad (10.84)$$

and

$$\dot{r} = \sum_{i=1}^{M} \dot{r}_i \qquad (10.85)$$

The level-crossing rate of r can be calculated from Eq. (2.73):

$$n(r = A) = \int_0^{\infty} \dot{r} p(r = A, \dot{r}) \, d\dot{r} \qquad (10.86)$$

On the basis of the relationships expressed in Eqs. (10.81) through (10.86), it is possible to find the level-crossing rate for a 2-branch equal-gain-combined signal $r(t)$ at either the base-station or mobile-unit receiving location.

Finding the LCR at a base-station site [6, 7]

As the basis for finding the lcr for two signals received by two base-station antennas, it can be assumed that the signal from the mobile transmitter is propagated over an area having uniform scattering properties. Under these conditions, the two signals received by the two base-station antennas are as follows (see Fig. 6.11):

$$e_1 = \sum_{i=1}^{N} A_i \exp{(j\psi_i)} = X_1 + jY_1 \qquad (10.87)$$

$$e_2 = \sum_{i=1}^{N} A_i \exp{j(\psi_i - \beta d \cos \alpha_i)} = X_2 + jY_2 \qquad (10.88)$$

where

$$\psi_i = \omega_c t - \beta V t \cos{(\alpha_i - \gamma)} - \frac{\Omega}{c} \omega_c \cos{(\theta_i - \alpha_i)} \qquad (10.89)$$

and where

d = antenna spacing
V = velocity of mobile unit, which contributes to Doppler shift
γ = angle indicating direction of mobile unit travel
Ω = radius of area of surrounding local scatterers
c = the speed of light
θ_i = angle indicating direction to the ith scatterer at the mobile location

A_i, α_i = amplitude and angle, respectively, at which incoming wave arrives at base-station receiver

N = number of waves arriving at base-station antenna

In choosing a model for calculating the level-crossing rate, line-of-sight propagation is assumed; however, only those waves received by the base-station receiver from scatterers surrounding the mobile unit are considered. Also, the assumption is made that the statistical properties of the scattered signals remain unchanged as the mobile transmitter travels along a given course. Under these conditions, the parameters X_i and Y_i during any given time interval t_1 are Gaussian-distributed, and the angles γ and θ_i are uniformly distributed. The distance between the mobile transmitter and the base-station receiver is assumed to be much greater than the distance from the mobile transmitter to the local scatterers. Therefore, waves reflected from the ith scatterer to antenna A or B are essentially parallel. The absence of local scatterers at the base-station site is also assumed.

On the basis of the preceding assumptions for the lcr model, it is relatively easy to show that the average values for X_1X_2 and X_1Y_2 [each of these four parameters is shown in Eqs. (10.87) and (10.88)] are:

$$R_c = \langle X_1 X_2 \rangle = N \langle \cos (\beta d \cos \alpha_i) \rangle \tag{10.90}$$

$$R_{cs} = \langle X_1 Y_2 \rangle = N \langle \sin (\beta d \cos \alpha_i) \rangle \tag{10.91}$$

$$\sigma_x^2 = \langle X_j^2 \rangle = \langle Y_j^2 \rangle \qquad \text{for } j = 1, 2 \tag{10.92}$$

since the angle at which the incoming wave arrives at the base-station receiver depends on where the mobile unit is located. The angle α_i is typically confined to a small angular sector $\alpha_1 < \alpha_i < \alpha_2$, which was previously described in Sec. 6.9 of Chap. 6, in connection with a probability density model. Equations (10.90) and (10.91) are both nonzero expressions.

By a similar procedure, the derivatives $\dot{X}_1$, $\dot{Y}_1$, $\dot{X}_2$, and $\dot{Y}_2$ are found as shown in the following equations:

$$R_{c'} = \langle \dot{X}_1 \dot{X}_2 \rangle = \langle \dot{Y}_1 \dot{Y}_2 \rangle = \frac{(\beta V)^2}{2} R_c \tag{10.93}$$

$$R_{cs'} = \langle \dot{X}_1 \dot{Y}_2 \rangle = -\langle \dot{X}_2 \dot{Y}_1 \rangle = \frac{(\beta V)^2}{2} R_{cs} \tag{10.94}$$

$$\sigma_{\dot{x}}^2 = \langle \dot{X}_i^2 \rangle = \langle \dot{Y}_i^2 \rangle = \frac{(\beta V)^2}{2} \sigma_x^2 \tag{10.95}$$

$$|\dot{\Lambda}|^{1/2} = |\sigma_{\dot{x}}^4 - R_{c'}^2 - R_{cs'}^2| = \frac{(\beta V^2)}{2} |\Lambda| \tag{10.96}$$

where

$$|\Lambda|^{1/2} = \sigma_x^4 (1 - |\rho|^2) \qquad (10.97)$$

and ρ is the complex correlation coefficient, expressed as:

$$|\rho|^2 = \frac{R_c^2 + R_{cs}^2}{\sigma_x^4} \qquad (10.98)$$

From Eqs. (10.87) and (10.88), the following relationship can also be obtained:

$$\langle X_i \dot{Y}_j \rangle = \langle \dot{X}_i Y_j \rangle = 0 \qquad i, j = 1, 2 \qquad (10.99)$$

Therefore, the joint probability density function for the eight Gaussian parameters $X_1, Y_1, \dot{X}_1, \dot{Y}_1, X_2, Y_2, \dot{X}_2,$ and $\dot{Y}_2$ becomes:

$$p(X_1, Y_1, \dot{X}_1, \dot{Y}_1, X_2, Y_2, \dot{X}_2, \dot{Y}_2)$$

$$= p(X_1, Y_1, X_2, Y_2,)p(\dot{X}_1, \dot{Y}_1, \dot{X}_2, \dot{Y}_2) \qquad (10.100)$$

Equation (10.100) can be simplified by using the following notation:

$$\{X_1\} = (X_1, Y_1, X_2, Y_2) \text{ and } \{\dot{X}_1\} = (\dot{X}_1, \dot{Y}_1, \dot{X}_2, \dot{Y}_2) \qquad (10.101)$$

Then the following expression is obtained:

$$p(\{X_1\}\{\dot{X}_1\}) = p(\{X_1\})p(\{\dot{X}_1\}) \qquad (10.102)$$

Note that both $p(\{X_1\})$ and $p(\{\dot{X}_1\})$ are obtained from Eq. (7.49).

Equation (10.102) can be used to find $p(r_1, r_2, \dot{r})$, by first finding the joint characteristic function for $\Phi(w_1, w_2, w_3)$ and the joint probability function for random variables r_1, r_2 and $\dot{r}$:

$$\Phi(w_1, w_2, w_3) = E[\exp [j(w_1 r_1 + w_2 r_2 + w_3 \dot{r})]]$$

$$= \int_0^\infty \int_0^\infty \exp j(w_1 r_1 + w_2 r_2) \int_0^{2\pi} \int_0^2 \underset{-\infty}{\overset{\infty}{\iiiint}}$$

$$\times \exp (jw_3 \dot{r}) \, p(\{X_1\})p(\{\dot{X}_1\})$$

$$\times d\dot{X}_1 \, d\dot{Y}_1 \, d\dot{X}_2 \, d\dot{Y}_2 \, d\psi_1 \, d\psi_2 \, dr_1 \, dr_2 \qquad (10.103)$$

where $\dot{r}$ can be derived from Eq. (10.85), as a function of ψ_1 and ψ_2:

$$\dot{r} = \dot{X}_1 \cos \psi_1 + \dot{Y}_1 \sin \psi_1 + \dot{X}_2 \cos \psi_2 + \dot{Y}_2 \sin \psi_2 \qquad (10.104)$$

After dX_1, dY_1, $d\dot{X}_2$, $d\dot{Y}_2$, $d\psi_1$, and $d\psi_2$ are integrated, Eq. (10.103) becomes:

$$\Phi(w_1, w_2, w_3) = \int_0^\infty \int_0^\infty [\exp j(w_1 r_1 + w_2 r_2)] \, r_1 r_2 \frac{1}{|\Lambda|^{1/2}}$$

$$\times \exp\left[-\frac{\sigma_x^2}{2|\Lambda|^{1/2}}(r_1^2 + r_2^2) - \frac{w_3^2(\beta V)^2}{2}\sigma_x^2\right]$$

$$\times I_0\left[\left(\frac{r_1 r_2}{|\Lambda|^{1/2}} - w_3^2 \frac{(\beta V)^2}{2}\right)(R_c^2 + R_{cs}^2)^{1/2}\right] dr_1 \, dr_2 \quad (10.105)$$

The probability density can then be obtained from Eq. (10.105), as follows:

$$p(r_1, r_2, \dot{r}) = \frac{1}{(2\pi)^3} \int \int \int \Phi(w_1, w_2, w_3)$$

$$\times e^{-j(w_1 r_1 + w_2 r_2 + w_3 \dot{r})} \, dw_1 \, dw_2 \, dw_3 \quad (10.106)$$

where I_0 is the zero order of the modified Bessel function of the first kind. The probability density function $p(r, \dot{r})$, expressed in Eq. (10.86), can then be obtained as follows:

$$p(r_1 \dot{r}) = \int_0^r p(r_1, r_2 = r - r_1, \dot{r}) \, dr_1 \quad (10.107)$$

By inserting Eq. (10.106) into Eq. (10.107) and then substituting the result of Eq. (10.107) into Eq. (10.86), the level-crossing rate is obtained as follows:

$$n(r = A) = \int_0^\infty \dot{r} p(r = A, \dot{r}) \, d\dot{r} \quad (10.108)$$

The relationship between the instantaneous combined voltage r of a 2-branch received signal and the combined cnr γ is defined in Eq. (10.73) with $\eta_i = \eta$:

$$\gamma = \frac{r^2}{4\eta} \quad (10.109)$$

where η is the mean-squared noise envelope of a single branch, and where the average cnr for a single-branch signal, Γ, can be expressed in terms of the rms value:

$$\Gamma = \frac{\langle r_1^2 \rangle}{2\eta} = \frac{2\sigma_x^2}{2\eta} = \frac{\sigma_x^2}{\eta} \quad (10.110)$$

Combining Eqs. (10.109) and (10.110) yields the following relationship:

$$\frac{\gamma\,(2\ \text{branches})}{\Gamma\,(\text{single branch})} = \tfrac14\left(\frac{r}{\sigma_x}\right)^2 \tag{10.111}$$

Substituting Eq. (10.111), together with the terms $y = w_3(\beta V/\sqrt{2})\,\sigma_x$, $r_1' = r/\sqrt{2}\sigma_x$, and $\rho_r = \rho^2$, into Eq. (10.108) gives the following [6]:

$$\bar{n}(\gamma = x) = \frac{\beta V}{\sqrt{2\pi}}\,\frac{1}{(2\pi)^2}\,\sqrt{\frac{2}{\pi}}\,\exp\left(-\frac{2x/\Gamma}{1-\rho_r}\right)$$

$$\times \int_0^{\sqrt{2x/\Gamma}}\int_0^{\infty}\frac{1}{y^2}\left(\frac{2r_1'(r_1' - \sqrt{2x/\Gamma})}{1-\rho_r}\right)$$

$$\times \exp\left[-\left(\frac{2r_1'(r_1' - \sqrt{2x/\Gamma})}{1-\rho_r} + y^2\right)\right]$$

$$\times I_0\left[\left(\frac{2r_1'(r_1' - \sqrt{2x/\Gamma})}{1-\rho_r} + y^2\right)\sqrt{\rho_r}\right]dy\,dr_1' \tag{10.112}$$

Eq. (10.112) is easily solved numerically, since the integral involving parameter y diminishes very quickly as y itself increases. The function for Eq. (10.112) is plotted in Fig. 10.11.

In Eq. (10.112), ρ_r is the normalized correlation coefficient between the two signals r_1 and r_2, and V is the velocity. For a given value of ρ_r, the base-station antenna separation depends on whether the incoming signal is normal (broadside) or in line with the axis of the two base-station antennas. This relationship was briefly described in Sec. 6.9 of Chap. 6 and in Sec. 9.4 of Chap. 9.

Finding the LCR at a mobile site [8, 9]

At a typical mobile site there are usually more than two signal branches. The methods previously described for finding the lcr of a two-branched diversity combiner are too cumbersome and too complex to be applied where more than two branches of diversity are involved. For this reason, an approximation method is recommended and is described in the following paragraphs.

First, it is necessary to find the derivative of r_j:

$$\dot{r}_j = \frac{X_j\dot{X}_j + Y_j\dot{Y}_j}{(X_j^2 + Y_j^2)^{1/2}} \tag{10.113}$$

The probability density function $p(\dot{Y}_j)$ is Gaussian-distributed with a value of zero mean, as derived from Eq. (6.62) by factoring out $p(r_j)$.

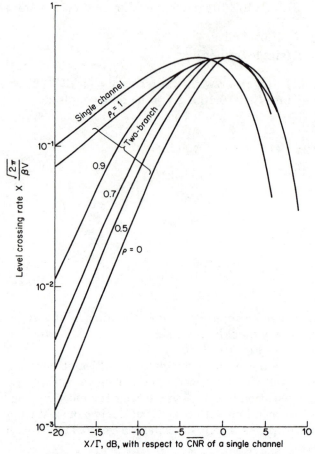

Figure 10.11 Level-crossing rate of a 2-branch equal-gain-combining signal.

Since all of the elements of the set $\{X_j, Y_j, \dot{r}_j\}$ are Gaussian, then $p(X_j, \dot{r}_j)$ is an even-valued function. Term $\dot{r}_j$, as an even-valued function, is expressed in Eq. (10.113). Therefore, the correlation of $X_i \dot{r}_j$ can be expressed as follows:

$$\langle X_i \dot{r}_j \rangle = \iint\limits_{-\infty}^{\infty} X_i \dot{r}_j p(X_i, \dot{r}_j) \, dX_i \, d\dot{r}_j = 0 \qquad (10.114)$$

for any value of i and j.

Where k and l are odd values, then the following relationships are valid:

$$\langle X_i^k \dot{r}_j^{2l} \rangle = \langle X_i^{2k} \dot{r}_j^l \rangle = 0 \qquad (10.115)$$

and

$$\langle X_i^{2k} \dot{r}_j^{2l} \rangle \approx \langle X_i^{2k} \rangle \langle \dot{r}_j^{2l} \rangle \tag{10.116}$$

Equation (10.116) expresses the condition where X^{2k} and $\dot{r}_j^{2l}$ are weak values.

By replacing X_i and Y_i in Eqs. (10.114) through (10.116), similar results are obtained. Hence, the joint probability density function can be expressed as follows:

$$p(X_1, Y_1, X_2, Y_2, \ldots, \dot{r}_1, \dot{r}_2, \ldots)$$
$$= p(X_1, Y_1, \ldots)p(\dot{r}_1, \dot{r}_2, \ldots \mid X_1, Y_1, \ldots)$$
$$\approx p(X_1, Y_1, \ldots)p(\dot{r}_1, \dot{r}_2, \ldots)$$
$$= p(r)p(\dot{r}) \tag{10.117}$$

As a result of inserting Eq. (10.117) into Eq. (10.86) and assuming $p(\dot{r})$ is a Gaussian distribution, the lcr can be expressed as follows:

$$n(r = A) = p(r = A) \int_0^\infty \dot{r}p(\dot{r}) \, d\dot{r}$$

$$= p(r = A)(2\pi)^{-1}\sigma_{\dot{r}} \tag{10.118}$$

where $\sigma_{\dot{r}}$, the standard deviation of the time derivative $\dot{r}$, is a constant value that represents an M-branch linear antenna array, and where the various values for $\sigma_{\dot{r}}$ are:

$$\sigma_{\dot{r}} = [2\langle \dot{r}_1^2 \rangle - 2\langle \dot{r}_1\dot{r}_2 \rangle]^{1/2} \qquad \text{for a 2-branch array}$$

$$= [3\langle \dot{r}_1^2 \rangle + 4\langle \dot{r}_1\dot{r}_2 \rangle + 2\langle \dot{r}_1\dot{r}_3 \rangle]^{1/2} \qquad \text{for a 3-branch array}$$

$$= [4\langle \dot{r}_1^2 \rangle + 6\langle \dot{r}_1\dot{r}_2 \rangle + 4\langle \dot{r}_1\dot{r}_3 \rangle + 2\langle \dot{r}_1\dot{r}_4 \rangle]^{1/2} \qquad \text{for a 4-branch array}$$

$$\tag{10.119}$$

and where

$$\langle \dot{r}_m\dot{r}_n \rangle = -\frac{d^2}{dt^2} \langle r_mr_n \rangle \Big|_{t=0} \tag{10.120}$$

The exact solution for $\langle r_mr_n \rangle$ is:

$$\langle r_mr_n \rangle = 2\sigma_x^2[2E(\sqrt{\rho_{mn}}) - (1 - \rho_{mn})K(\sqrt{\rho_{mn}})] \tag{10.121}$$

where $K(x)$ and $E(x)$ are the complete elliptic functions of the first and second kinds, respectively, and ρ_{mn} is the correlation coefficient between signal envelopes, as shown in Eq. (6.119).

It has been shown that the parameter $\sigma_{\dot{r}}$ is also a function of the direction of motion γ. By inserting Eq. (10.119) into Eq. (10.118), the lcr for a 4-branch equal-gain signal can be calculated for values between $\gamma = 0$ and $\gamma = 90°$. The curves for this condition are plotted in Fig. 10.12, for an antenna spacing of 0.15λ. Note that the lcr is different for different values of γ. The effects of coupling between branches are not included in this section, but are described in Sec. 10.11. Figure 10.12 shows that the antenna elements inclined with the motion ($\gamma = 0$) creates less level crossings in the signal. Thus, the signal performance is better.

10.8   Feed-Combining Techniques

Feed combining consists of feeding of one, or sometimes two, signals into a single receiving channel, splitting the signal in two by a power divider,

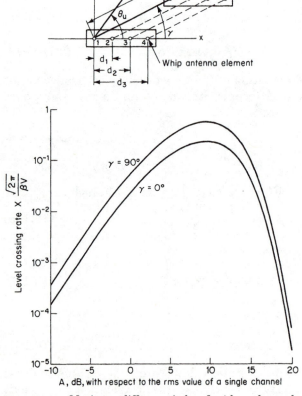

Figure 10.12   Maximum difference in lcr of a 4-branch equal-gain signal between $\alpha = 0$ and $\alpha = 90°$, with antenna spacing of 0.15λ.

and then using a specially designed circuit to combine the two split signals. The purpose of feed combining is to reduce the random FM, or the envelope fading, or both. There are two major feed-combining techniques: feed-forward combining and feedback combining.

10.9 Feed-Forward Combining

A typical circuit that can be used in feed-forward combining is shown in Fig. 10.13 [10].

Use of a nonpilot signal

In feed-forward combining using a nonpilot signal, only one signal is received. In Fig. 10.13, C_1 is a narrowband bandpass filter, whereas C_2 is a normal bandpass filter. The signal at each of eight locations in the feed-forward-combining circuit is labeled with a number for purposes of identification during the following description. At point 1, the input signal can be expressed:

$$s_1 = r(t)e^{j[\omega_c t + \psi_s(t) + \psi_r(t) + \psi_n(t)]} \qquad (10.122)$$

where $r(t)$ is the Rayleigh-fading component, ψ_r is the random phase caused by multipath fading, $\dot{\psi}_r$ is the random FM, ψ_n is the random phase due to system noise, and ψ_s is the message information. For voice transmission, $\dot{\psi}_s$ is 300 to 3000 Hz.

At points 2 and 3, the signal is:

$$s_2 = s_3 = \frac{r(t)}{\sqrt{2}} e^{j(\omega_c t + \psi_s + \psi_r + \psi_n)} \qquad (10.123)$$

and at point 4,

$$s_4 = \frac{r(t)}{\sqrt{2}} \{\exp [j(\omega_0 t + \alpha) + j(\omega_c t + \psi_s + \psi_r + \psi_n)]$$

$$+ \exp [j(\omega_0 t + \alpha) - j(\omega_c t + \psi_s + \psi_r + \psi_n)]\} \qquad (12.124)$$

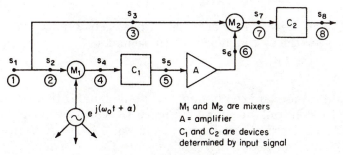

Figure 10.13 Typical feed-forward-combining circuit.

C_1 can be either a narrowband or a normal bandpass filter with a bandwidth of $2\Delta f$, centered at $f_0 - f_c$. If ψ_s contains voice information, then $\dot{\psi}_s$ is in the range of 300 to 3000 Hz and $\Delta f = 100$ Hz, based on the maximum Doppler frequency at the mobile speed of 70 mi/h and the frequency of 850 MHz. Under these conditions, the maximum Doppler frequency is:

$$f_d = \frac{V}{\lambda} \approx 100 \text{ Hz}$$

The random FM power occurring beyond a frequency of $2f_d$ is small, as was shown in Fig. 7.6. Therefore, the filtered signal at point 5 becomes:

$$s_5 = K \frac{r(t)}{\sqrt{2}} e^{j(\omega_0 t - \omega_c t + \alpha - \psi_r - \psi_n)} \tag{10.125}$$

where ψ_n is the average noise power over the range of 0 to 100 Hz, which is negligible.

At point 6, an amplifier A is used to increase the gain of the signal, which now becomes:

$$s_6 = (KA) \frac{r(t)}{\sqrt{2}} \exp\left[j(\omega_0 t - \omega_c t + \alpha - \psi_r)\right] \tag{10.126}$$

At point 7, mixer M_2 provides an output of:

$$s_7 = KA \frac{r^2(t)}{2} \{\exp\left[j(\omega_0 t - \omega_c t + \alpha - \psi_r) + j(\omega_c t + \psi_s + \psi_r + \psi_n)\right]$$

$$+ \exp\left[j(\omega_0 t - \omega_c t + \alpha - \psi_r) - j(\omega_c t + \psi_s + \psi_r + \psi_n)\right]\} \tag{10.127}$$

At point 8, after passing through a bandpass filter centered at f_0, the final signal is:

$$s_8 = KA \frac{r^2(t)}{2} \exp\left[j(\omega_0 t + \alpha + \psi_s + \psi_n)\right] \tag{10.128}$$

where α is a constant phase. The phase term ψ_r has been dropped from Eq. (10.128); however, the system noise ψ_n still remains.

Delayed-signal combining

In delayed-signal feed-forward combining, the input signal is the same as it was in Eq. (10.122), and the circuit of Fig. 10.13 is also used. Up to point 4, the signal behaves in the same way as in Eq. (10.124). At point 5, however, the signal characteristics change, since C_1 in the delayed-signal-combining application is a bandpass filter followed by a time-delay device. The output of the filter therefore is:

$$s_5 = K \frac{r(t)}{\sqrt{2}} \exp j[\omega_0 t - \omega_c t - \psi_s(t - \tau)$$

$$- \psi_r(t - \tau) + \alpha - \psi_n(t - \tau)] \quad (10.129)$$

At point 8, by following the same steps as were shown for pilot-signal combining, the output signal, after passing the amplifier, the mixer, and the bandpass filter, centered at f_0 is

$$s_8 = KA \frac{r^2(t)}{2} \exp j[\omega_0 t + \alpha + \psi_s(t) - \psi_s(t - \tau)$$

$$+ \psi_r(t) - \psi_r(t - \tau) + \psi_n(t) - \psi_n(t - \tau)] \quad (10.130)$$

If time delay t is small enough relative to the rate of change of $\psi_r(t)$, then $\psi_r(t) - \psi_r(t - \tau)$ is negligible. Although the term $\psi_s(t) - \psi_s(t - \tau)$ is one output phase that indicates a reduction in the FM index, it turns out that the detected signal is not necessarily degraded [11]. The maximum allowable delay τ for this system is roughly $\tau \leq 1/B_c$, where B_c is the coherence bandwidth described in Sec. 9.7 of Chap. 9. When $\tau > 1/B_c$, the delay signal will not contribute to the cancellation of random FM.

Use of a pilot signal
with a communication signal [12]

In another type of feed-forward combining, the input at point 1 in Fig. 10.13 is different from that in either of the last two types, in that two signals are present, a communications signal ω_1 and a pilot signal ω_2. The pilot signal does not carry any message information. The difference between ω_1 and ω_2 should be within the coherence bandwidth, so that the phase term ψ_r due to the random FM within the two signals is almost the same, as:

$$\psi_{r_1} \approx \psi_{r_2} \approx \psi_r \quad (10.131)$$

The resultant signal at point 1 consists of the normal communications signal and the pilot signal, which are represented as follows:

$$s_1 = \underbrace{r_1 e^{j(\omega_1 t + \psi_r + \psi_n)}}_{\text{pilot}} + \underbrace{r_2 e^{j(\omega_2 t + \psi_s + \psi_r + \psi_n)}}_{\text{signal}} \quad (10.132)$$

At points 2 and 3, the signal expression is:

$$s_2 = \frac{s_1}{\sqrt{2}} \quad (10.133)$$

By following the same steps as were shown for the other feed-forward cases, the final output signal at point 8, assuming the filter C_2 has a bandpass of $f_0 - f_2 - f_1$, is

$$s_8 = KA \, \frac{r_1 r_2}{2} \, \exp j[(\omega_0 - \omega_2 - \omega_1)t + \alpha - \psi_s] \qquad (10.134)$$

Equation (10.134) shows that the random FM and system noise terms are all canceled.

Alternate-pilot-signal method

Both a pilot signal and a communications signal can be used as inputs to the circuit shown in Fig. 10.14. At point 1, the input signal is:

$$s_1 = r_1 e^{j(\omega_1 t + \psi_r + \psi_n)} + r_2 e^{j(\omega_2 t + \psi_s + \psi_r + \psi_n)} \qquad (10.135)$$

At point 2, the signal at the output of the multiplier is:

$$s_2 = (r_1 e^{\pm j(\omega_1 t + \psi_r + \psi_n)} + r_2 e^{\pm j(\omega_2 t + \psi_s + \psi_r + \psi_n)})^2 \qquad (10.136)$$

At point 3, filter C is a bandpass filter centered at $f_2 - f_1$, and the resultant output signal is:

$$s_3 = r_1 r_2 e^{j[(\omega_2 - \omega_1)t + \psi_s]} \qquad (10.137)$$

The solution for Eq. (10.137) is only a rough approximation, since it does not include the terms for random FM due to multipath fading and noise. The results of Eq. (10.137) are based on the assumption that the fading information carried by the pilot signal is identical to that carried by the communications signal. In reality, the two are not the same.

10.10   Feedback Combining [13] (Granlund Combiner)

The feedback-combining technique, developed by Granlund, is a predetection combining technique that employs feedback as a reference signal in place of a local oscillator, as shown in Fig. 10.15 when the circuit components are as follows. F_1 is a narrowband bandpass filter centered at $f_2 - f_0$ and covering $2\Delta f$. F_2 is a bandpass filter (centered at f_2). L_1 is a limiting amplifier. M_1 and M_2 are mixers.

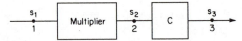

Figure 10.14   Model circuit for the alternate-pilot-signal method.

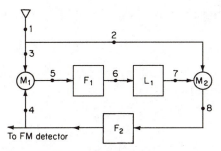

Figure 10.15 Granlund combiner.

In Fig. 10.15, the signals have been labeled at each point and are defined as follows. At point 1, the received input signal is:

$$s_1 = r(t)e^{j[\omega_c t + \psi_s(t) + \psi_r(t) + \psi_n(t)]}$$ (10.138)

At points 2 and 3, the signal is divided into two branches and can be expressed:

$$s_2 = \frac{s_1}{\sqrt{2}}$$ (10.139)

At point 4, the output frequency f_2 from mixer M_2, with phase α and amplitude A, can be expressed:

$$s_4 = Ae^{j(\omega_2 t + \alpha)}$$ (10.140)

After passing the mixer M, the filter F, and the limiting amplifier, the signal at point 7—under the conditions that the amplitude variations have been removed by limiting amplifier L_1 and only the random phase information remains in the signal—is expressed as:

$$s_7 = Ke^{j[(\omega_2 - \omega_c)t + \alpha' - \psi_r(t) - \psi_n(t)]}$$ (10.141)

where K is a constant representing the gain of limiting amplifier L_1.

At point 8, mixer M_2 provides an output signal from input signals s_2 and s_7:

$$s_8 = \frac{Kr}{\sqrt{2}} e^{j[(\omega_2 - \omega_c)t + \alpha' - \psi_r(t) - \psi_n(t)] \pm j[\omega_c t + \psi_s + \psi_r + \psi_n]}$$ (10.142)

At point 4, the final output signal passed through filter F_2 emerges:

$$s_4 = \frac{Kr}{\sqrt{2}} e^{j[\omega_2 t + \alpha' + \psi_s]}$$ (10.143)

Comparing the phase terms of Eqs. (10.140) and (10.143) shows that

$$\alpha = \alpha' + \psi_s \tag{10.144}$$

Since the phase angle of α' after filtering by F_1 is small compared with ψ_s, then it can be assumed that

$$\alpha \approx \psi_s \tag{10.145}$$

Hence, at point 4, the signal becomes:

$$s_4 = \frac{Kr}{\sqrt{2}} \, e^{j[\omega_2 t + \psi_s(t)]} \tag{10.146}$$

Using Eq. (10.146) to check points 5, 6, and 7, we find that Eq. (10.146) is the right solution. At point 4, s_4 contains an amplitude $KV/\sqrt{2}$ and the signal $\psi_s(t)$. It is the output of the Granlund combiner.

10.11   Combining Techniques for Multibranched Antenna Arrays

Because of modern technological advancements, it is possible to fabricate printed-circuit antenna arrays for UHF at a reasonable cost. For a given size of printed-circuit board, the number of array elements contained on each board is primarily determined by the spacing between adjacent elements. When the spacing is sufficiently large, the correlation coefficient between branches approaches 0, and the antenna is configured as an independent-branch array. Under this condition, the advantage of using diversity schemes reaches its maximum potential. However, the size of a printed-circuit board is finite, and therefore there is an optimal number of array elements, say N, that can be placed on the board. In designing an optimal printed-circuit board array, it is necessary to calculate the number of and closeness of spacing between elements so as not to degrade the performance of the combined signal from N branches.

The material in this section is based on the work done by Lee [14, 15]. A significant amount of work has been done in this area by Lee [8, 9, 15].

Of all of the various types of diversity combiners discussed, the maximal-ratio diversity combiner that was defined in Sec. 10.6 yields the maximum carrier-to-noise ratio (cnr) of the combined output signal. When the maximal-ratio diversity combiner is used to combine a large number of branches, the mutual coupling effect due to the mutual impedances among the various branch antennas must be considered. This is particularly valid when the spacing between adjacent antennas is small, i.e., less than 0.5λ. However, even when these mutual coupling effects are disregarded, it has been found that approximately 70 percent correlation between two received signals in their corresponding diversity branches can still provide most of the advantages of the maximal-ratio diversity combining scheme, as was shown in Fig. 10.9. This corresponds to mobile-unit receiving-antenna spacing of only a fraction of one wavelength. Since mobile antennas are spaced close together, their mutual coupling effects should be taken into consideration. When a loading network is attached to the mobile antenna array, the power at the output of the loading network, under optimal conditions, is simply the maximum value of signal power that the antennas can deliver to the receiver. However, an optimal loading network is physically difficult to achieve, and therefore a resistive loading network is usually chosen. A comparison of the effects of the resistive loading network with those of the ideal, nonmutual coupling network, in terms of degraded signal performance, is given in the following paragraphs.

The signals for diversity branches received by their respective antennas can be represented in the form:

$$s_j(t) = u_j(t) \exp j2\pi f_0 t \tag{10.147}$$

where f_0 is a common carrier frequency received by each branch and $u_j(t)$ is the multipath fading for a zero-mean complex Gaussian time process. The incoming signals received by the branch antennas are assumed to be connected to their respective load impedances Z_{L_j}, as shown in Fig. 10.16. It is further assumed that a certain nominal amount of mutual coupling exists between the branch antennas. Z_{jk} denotes the mutual impedance between the jth and kth antennas, and Z_{jj} is the self-impedance of the jth antenna. They are the elements of $[Z]$. The actual and equivalent circuits are shown in Fig. 10.16.

The optimum value for each element within the resistive-load matrix must be determined on the assumption that the load resistances of all branches are the same:

$$[Z_L] = R_L[I] \tag{10.148}$$

where $[I]$ is an identity matrix. Note that each load resistance R_L in Fig. 10.16 is followed by a transformer and a mixer and all of the branch mixers are identical and matched. The output of each mixer is proportional to the input times the insertion loss factor. It is assumed that the output noise is dominated by the receiver front-end noise and therefore the noise incidental to the antennas is negligible.

The spacing between antennas directly determines the amount of mutual coupling. As the antenna spacing increases, the mutual coupling between antennas decreases. With this relationship established, it

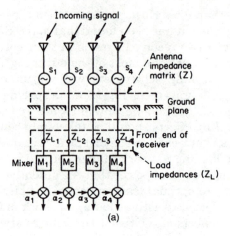

(a)

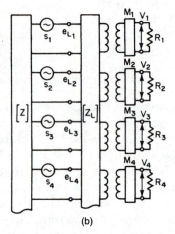

(b)

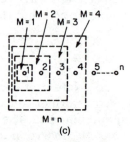

(c)

Figure 10.16 Actual and equivalent circuits of an M-branch maximum-ratio combiner with in-line antenna configurations. (a) Actual circuit of a maximum-ratio combiner. (b) Equivalent circuit of (a). (c) Inline antenna array.

is possible to analyze the mutual coupling effects for a space-diversity combiner with a large number of branches from two points of view: (1) the average carrier-to-noise ratio and (2) the cumulative distribution of the received signal.

CNR analysis method

The total cnr γ can be expressed:

$$\gamma = \frac{1/2\{[u*]^t[C][u]\}}{(1/2R_L)\eta_1} \tag{10.149}$$

where $[u]$ is the column of matrix elements u_i and η_1 is the noise power, which is the same for all branches. The value of the term R_L (load resistance) is as shown in Eq. (10.148), and the general expression for $[C]$ is:

$$[C] = ([Z] + [Z_L])^{-1^t}([Z_L] + [Z_L]^t)([Z] + [Z_L])^{-1} \tag{10.150}$$

The expression for $[C]$ can be simplified when there is no mutual coupling, or when the loading network for the antenna is optimum, as follows. For the case where there is no mutual coupling between antennas,

$$z_{ij} = 0 \qquad i \neq j \tag{10.151}$$

Then

$$[C] = \frac{[I]}{2R_L} \tag{10.152}$$

For the optimum antenna loading-network case,

$$[Z_L] = [Z]^+ \tag{10.153}$$

Then

$$[C] = ([Z] + [Z]^+)^{-1} = \frac{1}{2R_{11}^2}\begin{bmatrix} R_{11} & R_{12} & R_{13} \\ R_{12} & \cdots & \cdots \\ R_{13} & \cdots & \cdots \end{bmatrix} \tag{10.154}$$

where the R_{ij}'s are the real parts of the impedance matrix. The mutual impedance R_{ij} between two $\lambda/4$, normal-sized whip antennas having a length-to-diameter ratio of 36.5 can be found from Tai's results [16, 17] as a function of antenna separation.

Averaging Eq. (10.149) provides the average cnr, where the terms $\langle u_i u_j^* \rangle$ for a mobile-radio environment can be expressed from Eq. (E10.2.6):

$$\langle u_i u_j^* \rangle = 2N_i N_j J_0(\beta d_{ij}) \tag{10.155}$$

where N_i and N_j correspond to the signal levels of the ith and jth branches, respectively. The average cnr for an optimum resistive loading network with mutually independent branches, $\langle\gamma\rangle/\langle\gamma\rangle_{ind}$, is plotted in Fig. 10.17 for a linear array with M branches. For antenna spacing greater than 0.3λ, the power obtained from a resistive-loaded network is only degraded by about 1 dB with respect to the ideal, noncoupling case, regardless of the actual number of branches in the array.

Cumulative-distribution method

The cumulative distribution of the cnr γ for a combined signal was derived in Eq. (10.60). The eigenvalues λ_j are for the product of two matrices $\{2R_{L_1}[\mathcal{R}]*[C]\}$, and the value of $[\mathcal{R}]$ has been shown in Eq. (10.57), where the elements are expressed in Eq. (10.58). Figure 10.18 shows the cumulative distribution for a 4-branch maximal-ratio diversity combiner under two conditions: (1) with no coupling effects and (2) with mutual coupling effects between elements of a linear array. Note that when the antenna spacing d is greater than 0.3λ, then the difference between a signal with coupling and one without coupling effects becomes very small. For antenna spacing less than 0.3λ, the amount of degradation in signal performance can be obtained from the curves plotted in Figs. 10.17 and 10.18.

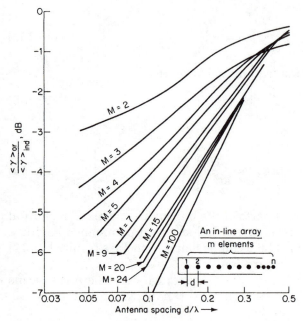

Figure 10.17 Optimum cnr for a resistive loading network consisting of a linear array with M branches.

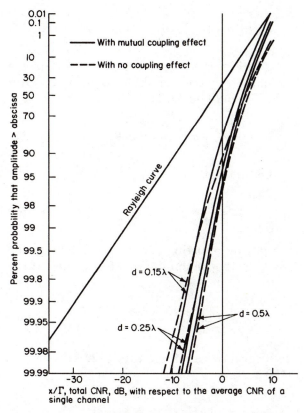

Figure 10.18 Cumulative distribution for a 4-branch maximal-ratio combiner with and without coupling effects between elements of a linear array.

10.12 Reducing Time Delay Spread by Diversity Combining

The diversity combining can reduce the multipath fading as well as the time delay spread. Let's find out the degree of reducing the time delay spread by applying the diversity combining. First we recall the delay spread model in Eq. (1.50), where T_i is the time delay of path i. The probability that T_i in one branch is within the time delay spread Δ is

$$P(T_1 \le \Delta) = 1 - \int_{\Delta}^{\infty} \frac{1}{\Delta} \exp\left(-\frac{T_1}{\Delta}\right) dT_1 = 1 - e^{-1} = 0.632 \quad (10.156)$$

For an M branch, combining all the time delay paths T_i from M branches and reordering the arrival of waves with time scale T_K, the probability that T_K will be within Δ is

$$P(T_K \leq \Delta) = 1 - \left[\int_\Delta^\infty \frac{1}{\Delta} \exp\left(-\frac{T_K}{\Delta}\right) dT_K \right]^M \qquad (10.157)$$

Equation (10.157) is plotted in Fig. 10.19 as a function of M. Since a single branch $P(T_1 \leq \Delta) = 0.632$, we can find the time delay Δ_d after diversity by letting Eq. (10.157) equal 0.632. Then

$$1 - \left[\int_{\Delta_d}^\infty \frac{1}{\Delta} \exp\left(-\frac{T_K}{\Delta}\right) dT_K \right]^M = 0.632 \qquad (10.158)$$

From Eq. (10.158), we may obtain that

$$\Delta_d = \frac{\Delta}{M} \qquad (10.159)$$

Equation (10.159) shows that, by applying diversity combining, the delay spread Δ_d is reduced after the combining by M branches. If the transmission is within

$$R < \frac{1}{2\pi\Delta_d} \qquad (10.160)$$

then no equalizer is needed. For a transmission rate of 24 kilosymbols per second (ksps) and $\Delta = 10$ μs, then

$$\frac{1}{2\pi\Delta} = 16 \text{ kHz}$$

which can have a transmission rate of 16 ksps without ISI. For a two-branch diversity, we let $M = 2$, $\Delta_d = \Delta/2 = 5$ μs.

$$\frac{1}{2\pi\Delta_d} = 32 \text{ kHz}$$

This shows that the transmission rate up to 32 kbps with a two-branch diversity does not need an equalizer.

This exercise tells us that, if we transmit at 24 ksps without diversity, we need an equalizer, since 24 ksps exceeds the limit transmission of 16 ksps. By applying a two-branch diversity, the equalizer can be eliminated.

There are already many advantages of using diversity schemes. Here is another one. Sometimes we may still need an equalizer after implementing the diversity scheme, but in this case the equalizer is more effective when performed in the dispersive medium with the diversity signal and less effective without it.

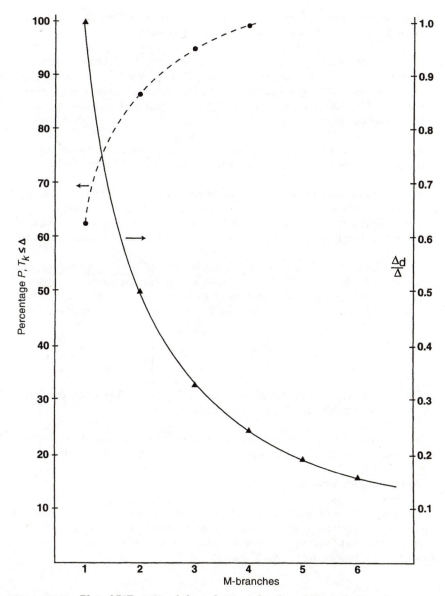

Figure 10.19 Plot of $P(T_1 \leq \Delta)$ and the reduction of Δ_d from M-branch diversity.

Problem Exercises

1. Four types of diversity-combining techniques are described in this chapter: selective combining (Fig. 10.4), switched combining (Fig. 10.7), maximal-ratio combining (Fig. 10.8), and equal-gain combining (Fig. 10.10). On the basis of the corresponding signal levels at which 1 percent of the signal is below the threshold of reception, compare the performance of each diversity-combining technique for 2-branch and 3-branch diversity combining.

2. What correlation coefficient is required between two signals from two diversity branches in order to obtain a gain of 11 dB over the cnr of a single channel for at least 99.7 percent of the time? Compare the resultant correlation coefficient for selective combining (Fig. 10.5) with that obtained by using maximal-ratio combining (Fig. 10.9).

3. Referencing Fig. 10.7, plot an equivalent curve for a switched-combining signal at a threshold level of approximately $A = -6$ dB. Explain the procedure used.

4. Macroscopic diversity techniques can be used successfully when two transmitting antennas are sited at two different locations. What is the rationale for using microscopic diversity techniques for two transmitting antennas located at the same site and separated by a proper spacing between antenna elements?

5. Derive the cumulative probability distribution of the cnr γ for a 3-branch equally spaced linear array with half-wavelength spacing between adjacent branches, using maximal-ratio combining. Assume that a physical triangular antenna configuration is installed and the following matrix of elements is valid:

$$[\mathcal{R}] = \begin{bmatrix} 1 & J_0(\pi) & J_0(\pi) \\ J_0(\pi) & 1 & J_0(\pi) \\ J_0(\pi) & J_0(\pi) & 1 \end{bmatrix}$$

6. Find the average duration of fades for a 2-branch equal-gain combiner with correlated signals, at a base-station site.

7. What are the relative advantages and disadvantages of using feed-forward and feedback combining techniques? List the pros and cons for each type.

8. Verify the process of Eq. (10.112).

9. What is the level of degradation in signal performance as a result of mutual coupling in a 4-branch maximal-ratio combiner, based on an average cnr and the cumulative distribution? Assume that antenna spacing is $d = 0.15\lambda$ and an optimum resistive loading network is used.

References

1. B. B. Barrow, "Diversity Combination of Fading Signals with Unequal Mean Strengths," *IEEE Trans. Comm. Sys.,* vol. 11, no. 1, March 1963.
2. M. Schwartz, W. R. Bennett, and S. Seymour, *Communication Systems and Techniques,* McGraw-Hill, New York, 1966, pp. 435, 470, 476, and 480.
3. A. J. Rustak, Y. S. Yeh, and R. R. Murray, "Performance of Feedback and Switch Space Diversity 900 MHz FM Mobile Radio System with Rayleigh Fading," *IEEE Trans. Comm.* vol. 21, November 1973, pp. 1257–1268.

4. D. G. Brennan, "Linear Diversity Combining Techniques," *Proc. IRE,* vol. 47, June 1959, pp. 1075–1102.
5. G. L. Turin, "The Characteristics and Function of a Hermition Quadratic Form in a Complex Normal Variable," *Biometrika,* vol. 47, June 1960, pp. 199–201.
6. W. C. Y. Lee, "Mobile Radio Performance for a Two-Branch, Equal Gain Combining Receiver with Correlated Signals at the Land Site," *IEEE Trans. Veh. Tech.,* vol. 27, November 1978, pp. 239–243.
7. W. C. Y. Lee, "Effect on Correlation between Two Mobile Radio Base Station Antennas," *IEEE Trans. Comm.,* vol. 21, November 1973, pp. 1214–1224.
8. W. C. Y. Lee, "A Study of the Antenna Array Configuration of an M-Branch Diversity Combining Mobile Radio Receiver," *IEEE Trans. Veh. Tech.,* vol. 20, no. 4, November 1971, pp. 93–104.
9. W. C. Y. Lee, "Level Crossing Rates of an Equal-Gain Predetection Diversity Combiner," *IEEE Trans. Comm.,* vol. 18, no. 4, August 1970, pp. 417–426.
10. C. C. Cutler, R. Kompfner, and L. C. Tillotson, "A Self-Steering Array Repeater," *Bell System Technical Journal,* vol. 42, September 1963, pp. 2013–2032.
11. F. R. Shirley, "FM Suppression in a Mix-On-Self-Loop," *IEEE Trans. Comm.,* vol. 13, no. 4, December 1965, pp. 471–475.
12. W. C. Jakes, *Microwave Mobile Communications,* Wiley, New York, 1974, pp. 464–469.
13. J. Granlund, "Topics in the Design of Antennas for Scatter," MIT Lincoln Laboratory Tech. Rep. 135, November 1956.
14. W. C. Y. Lee, "Effect of Mutual Coupling on a Mobile-Radio Maximum-Ratio Diversity Combiner with a Large Number of Branches," *IEEE Trans. Comm.,* vol. 20, December 1972, pp. 1188–1193.
15. W. C. Y. Lee, "Mutual Coupling Effect on Maximum-Ratio Diversity Combiners and Applications to Mobile Radio," *IEEE Trans. Comm.,* vol. 18, no. 6, December 1970, pp. 779–791.
16. C. T. Tai, "Coupling Antenna," *Proc. IRE,* vol. 36, April 1948, pp. 487–500.
17. J. D. Kraus, *Antenna,* McGraw-Hill, New York, 1950, p. 265.

11

Signal Processes

11.1 Mobile-Radio-System Functional Design—Signaling Problems

Many of the problems commonly associated with the mobile-radio environment have already been discussed in the preceding chapters. To briefly summarize, the natural phenomena that result in excess path loss over that normally occurring as free-space path loss were discussed in Chaps. 3 and 4. Multipath fading and the effects of random FM were discussed in Chaps. 6 and 7. Improving signal reception by increasing transmitted power and/or transmission bandwidth was discussed in Chap. 8. The option of using diversity techniques instead of increasing the transmitted power to improve performance was discussed in Chap. 9. And finally, the techniques for improving performance through the use of diversity combining were described in Chap. 10. In this chapter, the problems encountered in sending and receiving both voice, which uses analog transmission, and control signals, which use digital data transmission, through the mobile-radio environment are discussed. By designing a waveform for the control signal, we can filter the voice and the control signal at the baseband.

Problems relating to signal transmission in the mobile-radio environment are usually associated with the variables of distance and vehicular velocity, the waveforms of the transmitted pulses, and the time-delay spread attributable to the mobile-transmission medium. These problem areas are briefly described in the following paragraphs.

Distance-dependency factors

In Sec. 8.3 of Chap. 8, the error rate was found to be a function of the carrier-to-noise ratio (cnr). The cnr is affected by the distance between

. the transmitting location and the receiving location. If the transmitted power remains fixed, the bit-error rate increases as the distance increases. The bit-error rate is also a function of the signaling rate. The Shannon channel-capacity formula can be used to verify that the bit-error rate will increase as the distance increases:

$$C = B \log_2 \left(1 + \frac{S_c}{N}\right) \tag{11.1}$$

where C is the maximum capacity for a given signaling rate, B is the transmission bandwidth, S_c is the carrier power, and N is the received noise.* To show that the bit-error rate increases as the signal rate decreases with distance, it is necessary to relate S_c in Eq. (11.1) to a function of distance. On the basis of Eq. (4.42), the received power as a function of distance can be expressed:

$$P_d = P_R + \gamma \log \frac{R}{d} \tag{11.2}$$

where R is a distance of 1 mi, P_R is the received power at the 1-mi intercept point, $\gamma = 38.4$ dB per decade is the slope of the path loss in a suburban area, and P_d is the received power of the signal carrier, or the S_c of Eq. (11.1). Hence, the channel capacity for the signaling rate is:

$$C = B \log_2 (1 + 10^{(P_d - 10 \log N)/10}) \tag{11.3}$$

Equation (11.3) indicates the dependency of the signaling rate on path distance and therefore can be used to determine the signaling rate.

Example 11.1.   Given a power $P_R = -61.7$ dBm at the 1-mi intercept point, a slope of path loss of $\gamma = 38.4$ dB per decade, a bandwidth of 25 kHz, and a noise level of -120 dBm, what is the maximum signaling rate that can be used for a communications link of 10 mi?

solution   The received power is derived from Eq. (11.2), as follows:

$$P_d = -61.7 - 38.4 = -100.1 \text{ dBm} \tag{E11.1.1}$$

To obtain the maximum signaling rate for a communications link of 10 mi, Eq. (11.3) is applied as follows:

$$C = 25 \times 10^3 \log_2 (1 + 10^{(-100.1 + 120)/10})$$

$$\doteq 165.6 \text{ kb/s} \tag{E11.1.2}$$

---

* The Shannon channel-capacity formula is applied to a Gaussian noise channel. In the mobile-radio environment, the capacity should be less than Eq. (11.1) except in the case of a Rician-fading signal that approaches a Gaussian as its snr becomes large. Hence Eq. (11.1) serves as an upper limit. The channel capacity in Rayleigh fading environments appears in Sec. 17.1.

Velocity-dependency factors

As previously discussed in Sec. 8.2 of Chap. 8, the vehicular velocity does not directly affect the bit-error rate; however, the word-error rate is influenced by the velocity of the mobile unit, as shown in Fig. 11.1.

The average duration of fades for a single-channel receiver can be obtained from Eq. (6.99) and was shown in Fig. 6.7. The average duration of fades $\bar{t}$ at the rms level is:

$$\bar{t} = 0.036 \text{ s} \qquad V = 15 \text{ mi/h}$$

$$\bar{t} = 0.018 \text{ s} \qquad V = 30 \text{ mi/h}$$

The bursts of errors due to fades in the mobile-radio environment are different from errors caused by Gaussian noise. The duration of fades determines the necessary rate of repetition for transmitting message data, or for coding information.

11.2 Bit-Stream Waveform Analysis

Pulses transmitted in the time domain can be converted into the frequency domain. The spectral energy of a particular waveform of transmitted pulses falls within a certain frequency band at which the pulses can be acquired if there is a proper bandpass filter at the receiving end. Because of this requirement, the waveform characteristics of transmitted pulses are essential parameters for compatible receiver design.

Time-delay-spread dependency factor

The parameters and ranges for time-delay spread in urban and suburban mobile-radio environments were explained in Sec. 1.5 of Chap. 1. The maximum signaling-bit rate can be calculated on the basis of time-delay spread within the mobile-radio medium. An approximate estimate of the maximum bit rate $R_{\max}$ can also be found roughly from the

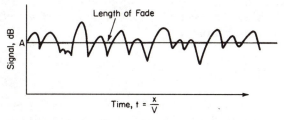

Figure 11.1 Influence of vehicular velocity on mobile-radio signals.

coherent bandwidth of the signal, as was shown in Eqs. (6.128) and (7.65). The relationships based on the maximum bit rate R_{max} can be expressed as follows:

$$R_{max} \leq \begin{cases} \dfrac{1}{2\pi\Delta} & \text{for linear types of modulation, such as ASK} \\[3mm] \dfrac{1}{4\pi\Delta} & \text{for exponential types of modulation, such as FSK and PSK} \end{cases}$$

A more accurate calculation of R_{max} will be found in Sec. 13.10 of Chap. 13. The maximum bit rate R_{max} cannot be increased by increasing power. The use of coding or diversity schemes, or both, compensates for a Rayleigh-fading medium and enables R_{max} to be increased to a value approaching $1/\Delta$. At this signaling rate, $R^\circ_{max} = 1/\Delta$, further increases are not attainable by any known method. Therefore, the following expression is valid:

$$R \leq R_{max} \leq R^\circ_{max}$$
$$\uparrow$$
by coding and diversity

The dependency factors that have been described in the preceding paragraphs are used to determine the signaling-bit rate, the length of the word bits, the pulse waveform, and the message coding. The methods used to achieve bit and word synchronization are described in Sec. 11.3.

Analysis of bit-stream waveforms

The various types of binary PCM waveforms are shown in Fig. 11.2. Note that each waveform has a different power spectral density. The nonreturn-to-zero (NRZ) and biphase Manchester-code waveforms are commonly used in pulsed signaling applications. The power spectral densities of these two waveforms provide a basis for selecting an adequate mobile-radio signaling scheme, and for designing a bandpass filter that can pass the total signal energy. The techniques for determining the power spectral density of a pulsed waveform are described in the following subsection.

Power spectral density of a random data sequence

Assume that a random binary source transmits an elementary signal from the set $\{s_i(t), i = 1, 2\}$, with a probability of P_i ($P_1 = p$ and $P_2 = 1 - p$), at each time interval T_s. Further assume that another elementary sig-

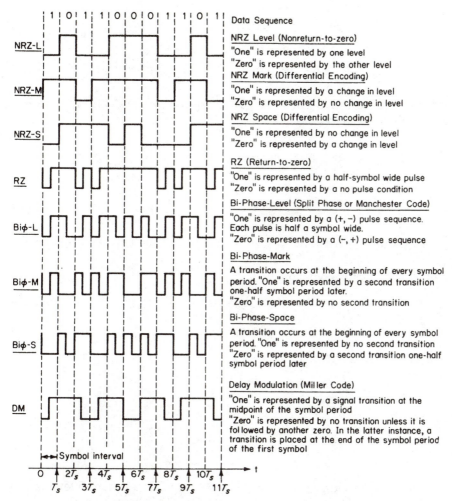

Figure 11.2   Various binary PCM waveforms. (*William C. Lindsey and Marvin K. Simon,* Telecommunication Systems Engineering, © *1973, p. 11. Reprinted by permission of Prentice-Hall, Inc., Englewood Cliffs, N.J.*)

nal is transmitted in a signaling interval that is independent of signals transmitted in previous signaling intervals. Under these conditions, the transition matrix can be expressed as follows [1]:

$$[P] = \begin{bmatrix} p & 1-p \\ p & 1-p \end{bmatrix} \qquad (11.4)$$

The relationship between the transition probability and the power spectrum can be expressed as follows [2]:

$$S(f) = \frac{1}{T_s^2} \sum_{n=-\infty}^{\infty} \left\{ \left| pS_1\left(\frac{n}{T_s}\right) + (1-p)S_2\left(\frac{n}{T_s}\right) \right|^2 \delta\left(f - \frac{n}{T_s}\right) \right\}$$

$$+ \frac{1}{T} p(1-p) |S_1(f) - S_2(f)|^2 \tag{11.5}$$

where $S_1(f)$ and $S_2(f)$ are the power spectra of $s_1(t)$ and $s_2(t)$, respectively, defined as follows:

$$S_1(f) = \int_0^T s_1(t)e^{-j2\pi ft}\, dt \tag{11.6}$$

$$S_2(f) = \int_0^T s_2(t)e^{-j2\pi ft}\, dt \tag{11.7}$$

and $s_1(t)$ and $s_2(t)$ are the waveforms representing 0 and 1, respectively.

For NRZ baseband signaling

$$s_1(t) = A \qquad 0 \leq t \leq T_s \tag{11.8}$$

$$s_2(t) = -s_1(t) \tag{11.9}$$

where $s_1(t)$ and $s_2(t)$ are rectangular pulses of width T_s. Their Fourier transforms are:

$$S_1(f) = -S_2(f) = AT_s \exp(-j2\pi fT_s) \frac{\sin \pi fT_s}{\pi fT_s}$$

$$= AT_s \exp(-jx) \frac{\sin x}{x} \tag{11.10}$$

where $x = \pi fT_s$. Substituting Eq. (11.10) into Eq. (11.5), and letting the power of the rectangular pulse become $E_s = A^2 T_s$, yields:

$$\frac{S(f)}{E_s} = \frac{1}{T_s}(1-2p)^2\, \delta(f) + 4p(1-p)\frac{\sin^2 x}{x} \tag{11.11}$$

and when $p = \frac{1}{2}$, Eq. (11.11) becomes:

$$\frac{S(f)}{E_s} = \frac{\sin^2 \pi fT_s}{\pi fT_s} \tag{11.12}$$

The function for Eq. (11.12) is plotted in Fig. 11.3.

Assuming that the signaling rate $f_s = 10$ kHz, then $T_s = 1/f_s = 100$ μs. For a voiceband, where the upper frequency is $f = 3000$ Hz, then, $T_sf = 0.3$. Thus the energy of the NRZ waveform lies in the voiceband.

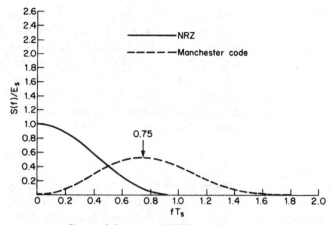

Figure 11.3   Spectral densities of NRZ and Manchester-code modulation waveforms.

There is no way to separate the NRZ signaling waveforms from the voice signals.

For Manchester-code-waveform signaling   The expression for Manchester-code-waveform signaling is:

$$s_1(t) = A \qquad 0 \le t \le \frac{T_s}{2}$$

$$s_2(t) = -A \qquad \frac{T_s}{2} \le t \le T_s \qquad (11.13)$$

$$s_1(t) = -s_2(t)$$

Substituting the Fourier transforms of Eq. (11.13) into Eq. (11.5) yields the following:

$$\frac{S(f)}{E_s} = \frac{1}{T_s} (1 - 2p)^2 \sum_{n=-\infty}^{\infty} \left(\frac{2}{n\pi}\right)^2 \delta\left(f - \frac{n}{T_s}\right)$$

$$+ 4p(1-p) \left[\frac{\sin^4 \pi f T_s / 2}{\pi f T_s / 2}\right] \qquad (11.14)$$

For $p = \frac{1}{2}$, Eq. (11.14) becomes:

$$\frac{S(f)}{E_s} = \frac{\sin^4 \pi f T_s / 2}{\pi f T_s / 2} \qquad (11.15)$$

The function for Eq. (11.15) is also shown in Fig. 11.3. Note that most of the energy of the Manchester-coded waveform lies between 0.4 and 1.2 on the $T_s f$ scale. For a signaling rate of $f_s = 10$ kHz, the energy is concentrated within the 4000- to 12,000-H range, which is above the normal range of mobile-radio voice transmissions. Hence, it is preferable to use Manchester-coded signaling rather than NRZ signaling to facilitate easy separation of the analog voice and the digital control signaling information. In a fully digitized voice communication system, both the voice and control signaling components are digital, and therefore the criteria for improving the quality of the signaling responses are dependent upon the implementation of techniques for compressing the bandwidth.

Waveform shaping

The bandwidth of a transmitted data stream can be compressed by shaping the waveform prior to transmission. The waveform-shaping technique also reduces intersymbol interference. There are several different pulse shapes that have been found to be effective; they are described as follows:

1. Squared-waveform pulse [see Fig. 11.4(1)]:

$$s(t) = \begin{cases} A & -\dfrac{T}{2} \le t \le \dfrac{T}{2} \\ 0 & \text{otherwise} \end{cases} \tag{11.16}$$

2. Triangular-waveform pulse [see Fig. 11.4(2)]:

$$s(t) = \begin{cases} \left(1 - \dfrac{2t}{T}\right)A & 0 \le t \le \dfrac{T}{2} \\ 0 & \text{otherwise} \end{cases} \tag{11.17}$$

3. Cosine-waveform pulse [see Fig. 11.4(3)]:

$$s(t) = \begin{cases} A \cos \dfrac{\pi t}{T} & -\dfrac{T}{2} \le t \le \dfrac{T}{2} \\ 0 & \text{otherwise} \end{cases} \tag{11.18}$$

4. Raised-cosine-waveform pulse [see Fig. 11.4(4)]:

$$s(t) = \begin{cases} \dfrac{A}{2}\left(1 + \cos \dfrac{2\pi t}{T}\right) & \dfrac{-T}{2} \le t \le \dfrac{T}{2} \\ 0 & |t| > \dfrac{T}{2} \end{cases} \tag{11.19}$$

5. Exponential-waveform pulse [see Fig. 11.4(5)]:

$$s(t) = \begin{cases} 0 & t < 0 \\ Ae^{-t/\tau} & t \ge 0 \end{cases} \tag{11.20}$$

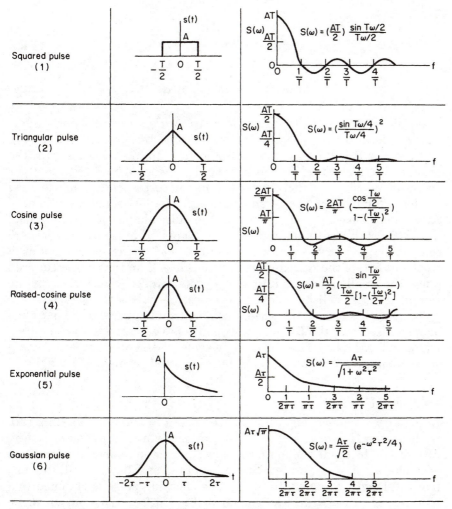

Figure 11.4   Effect of waveform shape on power spectral density.

6. Gaussian-waveform pulse [see Fig. 11.4(6)]:

$$s(t) = Ae^{-(t/\tau)^2} \tag{11.21}$$

Assume that the pulse $s(t)$ can be represented by the Fourier series:

$$s(t) = \frac{1}{\pi} \int_0^\infty S(\omega) \cos(\omega t + \Phi(\omega))\, d\omega$$

$$= \frac{1}{\pi} \int_0^\infty \sqrt{a^2(\omega) + b^2(\omega)} \cos\left[\omega t + \tan^{-1}\frac{b(\omega)}{a(\omega)}\right] d\omega \tag{11.22}$$

where

$$a(\omega) = \int_{-\infty}^{\infty} s(t) \cos \omega t \, dt \tag{11.23}$$

$$b(\omega) = \int_{-\infty}^{\infty} s(t) \sin \omega t \, dt \tag{11.24}$$

Then the spectral density functions $S(\omega) = \sqrt{a^2(\omega) + b^2(\omega)}$ of each different waveform pulse $s(t)$ can be obtained from Eqs. (11.23) and (11.24), and the result will be as shown in Fig. 11.4. In most applications, the shapes of $S(\omega)$ are used for waveform pulses denoted by $S(t)$. Then, their spectrum density functions are the shapes of corresponding $s(t)$ denoted by $s(\omega)$. Both of them can be shown in Fig. 11.4 by simply changing the parameters $T\omega \leftrightarrow 4\pi t/T$. Among all the pulse shapes of Fig. 11.4, note that $S(t)$ for the raised-cosine pulse and squared pulse have their zero component at intervals that are a multiple of $T/2$. Since the amount of ripple for $S(t)$ of the raised-cosine pulse is less than that of the squared pulse, the small amount of intersymbol-interference jitter, caused by the sampling clock, will be less for raised-cosine pulses than for squared pulses. For this reason, the raised-cosine pulse shape $S(t)$ is the most frequently used and the frequency response is very similar to $s(\omega)$, illustrated in Fig. 11.4(4). The duobinary waveform described in the following subsection is a good example of the use of raised-cosine pulse shaping.

Duobinary-waveform signaling [4]

Duobinary-waveform signaling is used when the binary signaling rate is above the Nyquist rate. The Nyquist rate states that the signaling rate should be $2f_1$ at a given frequency f_1. The duobinary waveform has good characteristics and can be implemented as follows. First, the normal full-length pulse width of the raised-cosine pulse is changed to a half length pulse width, as shown in Fig. 11.5(a). The raised-cosine waveform characteristics of Fig. 11.4(4) are modified by interchanging the time response and frequency response, as follows:

$$s(t) = \frac{\sin \pi t/T}{\pi t/T} \frac{\cos \pi t/T}{1 - 4t^2/T^2} \tag{11.25}$$

The spectral density can be expressed:

$$S(\omega) = \begin{cases} T & 0 \le \omega \le \dfrac{\pi}{T} \\[2ex] \dfrac{T}{2}\left\{1 - \sin\left[\dfrac{T}{2}\left(\omega - \dfrac{\pi}{T}\right)\right]\right\} & \dfrac{\pi}{T}(1 - \alpha) \le \omega \le \dfrac{\pi}{T}(1 + \alpha) \end{cases} \tag{11.26}$$

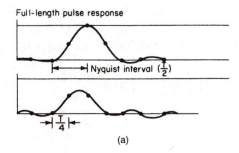

Nyquist interval ($\frac{T}{2}$)

(a)

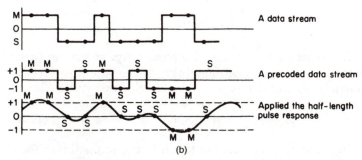

A data stream

A precoded data stream

Applied the half-length pulse response

(b)

Figure 11.5 Characteristics of duobinary precoded waveforms: (*a*) illustration of Nyquist principle; (*b*) relationships of a duobinary half-length pulse response to a full-length pulse response. (Note that in the data stream, $M = 1$, $a_i = 0$; $S = -1$, $a_i = 1$.)

Second, the bits of the input data stream $\{a_k\}$ are precoded to $\{b_k\}$ as follows:

$$b_k = a_k \oplus b_{k-1} \tag{11.27}$$

where the sign $\oplus$ represents a modulo-2 summation, which means that if a_k is 0, $b_k = b_{k-1}$, and if a_k is 1, then $b_k = -b_{k-1}$. Last, the half-length pulses are inserted and responsed into the data stream as three-level-coded pulses, with marks (M) occurring at either the +1 or −1 level and spaces (S) occurring at the 0 level, as shown in Fig. 11.5(*b*). Since the rate is double, the bit-error rate must be investigated. If there is no noise, the levels are 1, 0, and −1, and the probabilities of receiving these are ¼, ½, and ¼, respectively. Therefore the probability of error is ¾ times that of a two-level signal expressed in Eq. (2.118):

$$P_e = \tfrac{3}{2}P(x > \tfrac{1}{2}) \tag{11.28}$$

The average power output for a duobinary filter can be approximated by assuming that the filter is a simple cosine type. The response of a cosine filter in terms of its transfer function can be expressed:

$$S(\omega) = \begin{cases} 2T \cos\left(\dfrac{T}{2}\,\omega\right) & |\omega| \le \dfrac{\pi}{T} & (11.29) \\[2ex] 0 & |\omega| > \dfrac{\pi}{T} & (11.30) \end{cases}$$

and

$$s(t) = \frac{4}{\pi}\left(\frac{\cos \pi t/T}{1 - 4t^2/T^2}\right)$$

The relationship between a precoded duobinary signal that is transmitted at one end of a communications link and the signal received at the other end can be expressed:

$$S_T(\omega) = S_R(\omega) = [S(\omega)]^{1/2} \tag{11.31}$$

The transmitting power, after passing through the transmitter filter, can be expressed:

$$P_s = P_d \frac{1}{2\pi} \int_{-\pi/T}^{\pi/T} |S_T(\omega)|^2 \, d\omega$$

$$= P_d \frac{1}{2\pi} \int_{-\pi/T}^{\pi/T} \left[2T \cos\left(\frac{T}{2}\,\omega\right)\right]^2 d\omega = \frac{4}{\pi} P_d \tag{11.32}$$

where P_d is the average power of the duobinary waveform before transmitter filtering. Equation (11.32) shows a transmitting power gain factor of $4/\pi$. Therefore, at the receiving end the duobinary-waveform power would be reduced by the same factor of $4/\pi$, assuming that the same type of filtering is used in the receiver. However, the noise at the output of the receiver filter would be increased by a factor of $4/\pi$ over the input receiver noise N_0:

$$N = \frac{4}{\pi} N_0 \tag{11.33}$$

Then the signal-to-noise ratio at the output of the receiver filter becomes:

$$\frac{P_d}{N} = \frac{(\pi/4)P_r}{(4/\pi)N_0} = \left(\frac{\pi}{4}\right)^2 \gamma \tag{11.34}$$

The probability of error can be determined by using the same procedure described in Chap. 8, as derived from Eq. (8.70) and expressed as follows:

$$P_e(x > \tfrac{1}{2}) = P_e\left(y > \frac{1}{2}\sqrt{\frac{P_d}{N}}\right)$$

$$= \frac{1}{2}\,\mathrm{erfc}\left(\frac{\pi}{4}\sqrt{\gamma}\right) \tag{11.35}$$

Equation (11.28) becomes:

$$P_e = \frac{3}{4}\,\mathrm{erfc}\left(\frac{\pi}{4}\sqrt{\gamma}\right) \tag{11.36}$$

For a binary AM signal, assuming that ideal filtering is used (see Sec. 8.3 of Chap. 8), the probability of error can be expressed:

$$P'_e = \tfrac{1}{2}\,\mathrm{erfc}\,(\sqrt{\gamma}) \tag{11.37}$$

In evaluating the desirability of the precoded duobinary signaling option, the erfc argument is the dominant factor. Comparing the erfc of Eq. (11.36) with that of Eq. (11.37) shows that the performance of duobinary waveforms is poorer by a factor of $\pi/4$, or 2.1 dB. This is the price that must be paid in order to achieve a binary signaling rate that is higher than the Nyquist rate.

11.3 Bit and Frame Synchronization

Receiving digital transmissions requires that a digital clock signal be synchronized to the received data stream [3]. If the clock output signal experiences time jitter or frequency drift, the bit- and word-error rate will be increased. This type of problem can be eliminated by using improved bit- and frame-synchronization techniques, which are described in the following subsection.

Bit synchronization

The input digital signal to a bit synchronizer can be characterized as follows:

$$y(t) = r(t)\sum_{k=0}^{K} s(t; a_k, \varepsilon_1) + n(t) \tag{11.38}$$

where a_k is the polarity (± 1) of the kth transmitted bit $s(t; a_k, \varepsilon_1)$, random epoch ε_1 is assumed to be constant for KT seconds, $n(t)$ is Gaussian noise, and $s(t; a_k, \varepsilon_1)$ can be expressed in terms of waveshape $P_s(t)$ as follows:

$$s(t; a_k, \varepsilon_1) = a_k P_s[t - (k - 1)T - \varepsilon_1] \tag{11.39}$$

Assume that the undistorted bit-stream pattern $x(t)$ is:

$$x(t) = \sum_{k=0}^{K} s(t; a_k, \varepsilon_2) \tag{11.40}$$

where ε_2 is the error due to the instability of the clock at the receiving end. The cross-correlation of $x(t)$ and $y(t)$ is:

$$R_{xy}(\tau) = \int_0^{KT} x(t - \tau)y(t)\, dt \tag{11.41}$$

Equation (11.41) must be maximum, or the estimator δ, expressed:

$$\delta = |R_{xy}(\tau) - \langle x^2(t)\rangle| \tag{11.42}$$

will be minimum at $\tau = \tau_m$. τ_m is the time difference, to which the receiver clock must be adjusted, as shown in Fig. 11.6.

The value of $R_{xy}(\tau_m)$ always decreases as the number of errors in the bit-stream pattern increases. The error rate tends to increase as a function of Rayleigh fading within the mobile-radio environment. The commonly used bit-synchronization pattern is a series of K symbols consisting of alternate 0s and 1s. Depending upon the severity of fading, the values of K are determined so that δ in Eq. (11.42) meets the minimum value required.

Example 11.2   Assume that a received bit-synchronization pattern $y(t)$ consists of a sequence of alternating +1s and −1s, as in +1, −1, +1, −1, +1, −1, . . . , and that these bit symbols correspond to logic 1 and 0 voltage levels, respectively. How can this pattern be used to time the clock synchronization at the receiver?

solution   For purposes of explanation, assume that the sequence of clock timing pulses generated within the receiver is expressed by the term $x(t)$. Then, the timing relationship between $y(t)$, the received bit-synchronization pattern, and $x(t)$,

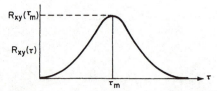

Figure 11.6   Cross-correlation in time difference between incoming-signal pulses and clock pulses.

the local receiver-generated bit-synchronization pattern, can be as illustrated in Fig. E11.2.1. The correlation between $x(t)$, the local receiver-generated bit-synchronization pattern, and $y(t)$, the received bit-synchronization pattern, is expressed:

$$R_{xy}(\tau) = \int_0^{kT} x(t - \tau)y(t)\, dt \qquad \text{(E11.2.1)}$$

and the autocorrelation function $R_{xy}(\tau)$ then becomes:

$$R_{xy}(\tau) = \frac{[T - \tau]}{T} \qquad \text{(E11.2.2)}$$

The autocorrelation is based on the waveform parameters shown in Fig. E11.2.2, and the function of Eq. (E11.2.2) is illustrated in Fig. E11.2.3. If $R_{xy}(\tau) = 0.5$, from Eq. (E11.2.2), the bit-synchronization clock pulses have slipped in time to one-half the pulse-duration interval, expressed as $\tau_m = T/2$. A general expression for finding the value of τ_m is obtained from Eq. (E11.2.2):

$$\tau_m = T[1 - R_{xy}(\tau_m)] \qquad \text{(E11.2.3)}$$

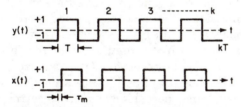

Figure E11.2.1 Relationship between bit-synchronization patterns $y(t)$ and $x(t)$.

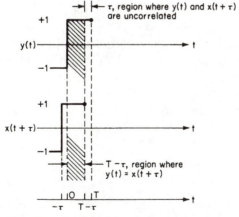

Figure E11.2.2 Waveform parameters for determining autocorrelation function $R_{xy}(\tau)$.

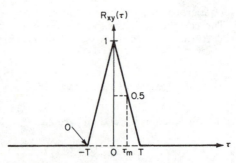

Figure E11.2.3 Plot of autocorrelation function $R_{xy}(\tau)$.

Frame synchronization

A data bit stream contains data arranged in a series of words, with a prescribed number of words making up a block and a given number of blocks constituting a frame. Each data frame is identified by one or more bits that are inserted at the beginning or ending of each frame or, in certain schemes, at both the beginning and ending of each frame. A commonly used frame code, known as the Barker code, is often used for frame synchronization. The sync-frame code word can be recognized during reception by a specific bit sequence consisting of N-bit segments, which are compared with an identical sync-frame code word stored within the receiver data processing circuits. The sync-frame coding information can be thought of as a finite number k of symbols having the unique property that their sidelobes have a correlation of $R(k) = 1/N$. The correlation function $R(k)$ of a specific bit sequence, separated by k bits from the sequence, can be expressed:

$$R(k) = \sum_{i=1}^{N-k} x_i x_{i+k} \tag{11.43}$$

where x_i ($i = 1, \ldots, N$) has the values listed in Table 11.1. The autocorrelation function $R(k)$ for values of $N = 7$ and $N = 11$ is shown in Fig. 11.7.

TABLE 11.1   Barker Sync-Frame Coding for *N*-Bit Segments

N	Code
1	+
2	++ or +−
3	++−
4	+++− and ++−+
5	+++−+
7	+++−−+−
11	+++−−−+−−+−
13	+++++−−++−+−+

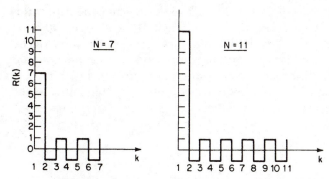

Figure 11.7 Plot of the autocorrelation function $R(k)$ for $N = 7$ and $N = 11$.

The Barker code [5] has excellent autocorrelation properties that make it ideal for sync-frame coding applications; however, in transmission through a multipath-fading environment, synchronization is dependent upon the correlation coefficient to remain above a prescribed threshold level.

Example 11.3 An 11-b Barker sync-frame coding sequence is transmitted at a rate of 16 kb/s, at a frequency of 900 MHz. If the receiving mobile unit is traveling at a speed of 30 km/h, what is the average duration of fades, and how often will synchronization errors occur at levels of -15 dB with respect to the rms value of the received signal?

solution The average duration of fades $\bar{t}(-15$ dB) can be determined from Eq. (6.99), as follows:

$$\bar{t}(-15 \text{ dB}) = 0.00288 \text{ s} \tag{E11.3.1}$$

The level-crossing rate for $n(-15$ dB) can be determined from Eq. (6.65), as follows:

$$n(-15 \text{ dB}) = 10.76 \text{ s}^{-1} \tag{E11.3.2}$$

On the basis of the results of Eqs. (E11.3.1) and (E11.3.2), there will be an average of 11 fades, each lasting for a period of 0.00288 s, and approximately 46 b will be lost during each fade. Under these conditions, it will be necessary to repeat the 11-b Barker sync-frame coding sequence at least five times to ensure synchronization of the received signal. This requirement is based on the following calculation:

$$\frac{11 + 46 \text{ bits}}{11} \approx 5 \text{ times} \tag{E11.3.3}$$

Signaling formats used in mobile-radio transmission are designed to reduce the probability of error. The autocorrelation of a signal envelope for a single mobile-radio channel operating at 850 MHz is 0.8λ

(~0.93 ft). At mobile speeds of $V = 60$ mi/h (88 ft/s), the time interval is (0.93 ft)/(88 ft/s) ≈ 10 ms. If the signaling rate is 10 kb/s, then 100 b can be transmitted in 10 ms of time. Under these conditions, the first bit and the 101st bit are not correlated, since they are separated far enough in time (in different frames) and therefore do not have the same fading characteristics. To reduce the probability of error, the following signaling format can be used:

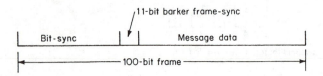

An error-correcting code and repetition techniques can be incorporated into the frame word to improve the reliability of error-free reception. These types of framing techniques are described in Chap. 13.

11.4   Syllabic Companding

Communication channels are designed to accommodate different human voices as well as message data from different sources. In order to ensure adequate performance across a wide range of signaling applications, the mobile-radio system design must provide sufficient signal power to overcome fading, random noise, and other types of interference encountered in a typical communications link. In practice, the signal power is proportionate with the message power, and excessive power can cause excessive interference in other systems. One method that can be used to overcome this interference problem is to compress the message power range prior to transmission, with a complementary expansion of the message power range at the time of reception. For voice transmissions, the techniques of compression and expansion are implemented at a syllabic rate. Hence, the technique is referred to as syllabic companding. Syllabic companding can be used to reduce noise and cross talk in a voice transmission system.

A simplified block diagram for a compandor is shown in Fig. 11.8. The conventional syllabic compandor consists of a compressor and an expandor. The compressor is used to compress the voice signal prior to modulation of the carrier for transmission. The compressor circuit contains a variable-loss device VL1, which increases signal attenuation as the unidirectional control current i_c increases.

The control current is developed by rectifying a portion of the output signal from VL1 and feeding it back as a control signal for VL1. The expandor circuit contains a similar variable-loss device, VL2, which performs the complementary function of VL1, increasing the control

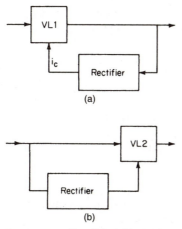

Figure 11.8 Simplified block diagram for a compandor: (*a*) compressor (at the transmitter); (*b*) expandor (at the receiver).

current i_c to VL2 as input-signal attenuation decreases. The control current for VL2 is developed by rectifying a portion of the input signal and feeding it forward as a control signal for VL2. The expandor circuit is used only during signal reception.

The unidirectional control current i_c is designed to follow variations in signal level occurring at a syllabic rate, rather than the individual variations in the speech waveform.

To better understand the principle of compression, let σ_x represent the rms amplitude of a particular message waveform, $\hat{\sigma}_x$ the maximum amplitude of the message waveform, and σ_y the compressed output-message waveform. Then, the compressor characteristic $g(x)$ can be expressed in the following relationship:

$$y = g(x) = g\left(\frac{\sigma_x}{\hat{\sigma}_x}\right) \tag{11.44}$$

and the expandor characteristic $h(y)$ can be expressed in a similar manner:

$$x = h(y) = h\left(\frac{\sigma_y}{\hat{\sigma}_y}\right) \tag{11.45}$$

For purposes of explanation, when the companding characteristics are expressed as [6]

$$y = g(x) = \frac{\sinh^{-1} \mu x}{\sinh^{-1} \mu} \tag{11.46}$$

and where μ is the compression factor. When $\mu = 0$, the equation shows that no compression takes place. Then

$$y = g(x) = 1 \tag{11.47}$$

is the case where

$$\sigma_x = \sigma_y \tag{11.48}$$

The expandor characteristic $h(y)$ can be obtained from Eq. (11.46) and expressed as follows:

$$x = h(y) = \frac{1}{\mu} \sinh (y \sinh^{-1} \mu) \tag{11.49}$$

To better understand the principle of the expandor, let the rms noise amplitude at the expandor input be represented by σ_n, where $n = \sigma_n/\hat{\sigma}_x$, and where the output $n_0 = \sigma_{n_0}/\hat{\sigma}_x$. Then, the following relationship is applicable:

$$x + n_0 = \frac{1}{\mu} \sinh [(y + n) \sinh^{-1} \mu] \tag{11.50}$$

n can also represent the total additive noise from the phase terms described in Sec. 8.2 of Chap. 8:

$$n = n_{\mathrm{rfm}} + n_{\mathrm{cn}} + n_{\mathrm{sn}} + n_{\mathrm{gn}}$$

For $y \gg n$, as would ordinarily be the case, the following approximation is applicable:

$$x + n_0 \approx x + n \frac{\sinh^{-1} \mu}{\mu} \cosh (\sinh^{-1} \mu x) \tag{11.51}$$

The noise reduction factor η, in decibels, as a function of the ratio of x in decibels, for values of $\mu = 50$ and $\mu = 500$, expressed as $\eta = 10 \log n/n_0$, is shown in Fig. 11.9.

When there is crosstalk interference between similar systems, the companding requirements become more complex and may require special consideration [7–9]. However, most of the syllabic compandors currently in use are of the 2:1 companding type, where the ratio between input and output signals is approximately 2:1 for compression and expansion. Figure 11.10 shows a typical input-output signal relationship for a 2:1 compandor, where the compressor has a slope of 1/2 and the expandor has a slope of 2/1.

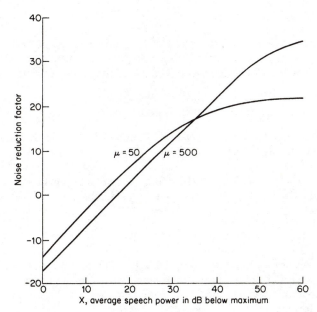

Figure 11.9 Plot of noise-reduction factors.

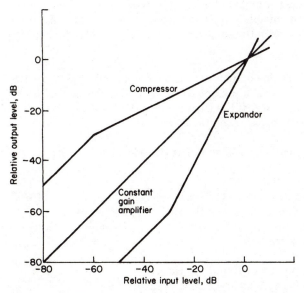

Figure 11.10 Typical input-output relationship in a 2:1 compandor.

Problem Exercises

1.   Given the parameters of $P_R = -61.7$ dBm at the 1-mi intercept point and $\gamma = 38.4$ dB per decade, use Eq. (11.2) to find the received power at a mobile location 16 mi from the base-station transmitter. Calculate the signal-to-noise ratio from the noise floor:

$$N = kT \text{ (dBm)} + B \text{ (dB)} + \text{receiver front-end noise}$$

where $kT = -174$ dBm at a room temperature of 17°C, bandwidth B of the IF filter is 30 kHz, and receiver front-end noise is 7 dB.

2.   On the basis of the Barker sync-frame coding for $N = 13$ b of Table 11.1 in Sec. 11.3, find the correlation function $R(k)$ of x_i with its sidelobes, where k is the number of bits shifted.

$$x_i = +++++--++-+-+$$

3.   If a mobile-radio channel has a bandwidth of 25 kHz and a maximum signaling rate of 100 kb/s, what is the maximum distance at which the signal can be received without degrading performance? The condition given in Example 11.1 can be used as a reference.

4.   A mobile unit traveling at a speed of 50 km/h receives a signal with a data rate of 30 kb/s at an operating frequency of 450 MHz. Based on the rms level of the signal, what is the average number of bits that can be received during a signal fade? If the speed of the mobile unit increases to 70 km/h, what signaling rate is required in order to obtain the same average number of bits during each signal fade?

5.   Given an FSK mobile-radio channel with a bandwidth of 25 kHz, a time-delay spread of 0.05 μs for the medium, and a cnr of 15 dB, what is the recommended maximum signaling rate? If the time delay spread is 3 μs, what effect will this have on the maximum signaling rate?

6.   In Fig. 11.5(*b*), the "space" symbols all fall at the 0 level. Can this property be used to detect errors?

7.   A bit-sync pattern has a duration of 62.5 μs and a correlation value of –0.3, after passing through a comparator. Give the rationale for having a negative correlation value, describe the amount of time slip, and explain how the synchronization error can be corrected.

8.   Apply Eqs. (11.49) and (11.51) to find the ratio of average message power to average noise power, where the average speech power is $\mu = 50$.

9.   What is the recommended Barker sync-frame coding for a data rate of 5 kb/s at a frequency of 1 GHz? Assume that the average speed of the receiving mobile unit is 50 km/h. Refer to Table 11.1.

10.   Prove that 70 percent of the signal power in a Manchester-coded signal is within a power spectral density band of $1/T_s$, and that 95 percent of the signal power is within a band of $2/T_s$.

References

1. J. M. Wazencraft and I. M. Jacobs, *Principles of Communication Engineering,* Wiley, New York, 1965, p. 33.
2. R. C. Titsworth and L. R. Welch, "Power Spectra of Signals Modulated by Random and Pseudorandom Sequences," Tech. Rep. no. 32-140, Jet Propulsion Laboratory, Pasadena, Calif., October 1961.
3. W. C. Lindsay and M. K. Simon, *Telecommunication System Engineering,* Prentice-Hall, Englewood Cliffs, N.J., 1973, p. 11.
4. A. Lender, "The Duobinary Technique for High-Speed Data Transmission," *IEEE Trans. on Communication and Electronics,* vol. 82, May 1963, pp. 214–218.
5. R. H. Barker, "Group Synchronization of Binary Digital Systems," in W. Jackson (ed.), *Communication Theory,* Academic Press, New York, 1953, pp. 273–287.
6. E. D. Sunde, *Communications Systems Engineering Theory,* Wiley, New York, 1969, pp. 69–70.
7. R. O. Carter, "Theory of Syllabic Compandors," *Proc. IEEE,* vol. 3, no. 3, March 1964, pp. 503–513.
8. R. Toumani, "A High Performance Syllabic Compandor," U.S. Patent 3,919,654, Nov. 11, 1975.
9. R. C. Mathes and S. B. Wright, "The Compandor—An Aid against Static in Radio Telephony," *Bell System Technical Journal,* vol. 7, July 1934, pp. 315–332.

Chapter
12

Interference Problems

12.1 Effects of Interference

A primary design objective for any commercial or military mobile-radio system is to conserve the available spectrum by reusing allocated frequency channels in areas that are geographically located as close to each other as possible. The limitation in distance for reusing frequency channels can be determined by the amount of cochannel interference. The separation between adjacent channels and the assignment of frequency channels within specified geographic areas is limited by the parameters for avoidance of adjacent-channel interference. To achieve a satisfactory frequency channel-assignment plan it is necessary to fully understand the effects of cochannel and adjacent-channel interference on mobile-radio reception. The following paragraphs discuss cochannel interference and are followed by a discussion of adjacent-channel interference for different situations involving FM reception. At the present time, most mobile-radio systems in use are of the FM type.

Three other types of interference will also be described in this chapter. The problems are near-to-far ratio, intermodulation, and intersymbol interference. The near-to-far ratio interference problem is inherent in the nature of mobile communications; e.g., at times the mobile unit may be too close to an undesired transmitter and too far away from the desired transmitter. The intermodulation (IM) interference problem is usually experienced in multichannel, frequency-division multiplex applications. Finally, the problem of intersymbol interference because of digital transmission is also described.

12.2   Cochannel Interference

Assuming that $s(t)$ is the desired signal and $i(t)$ is the interfering signal, the relationship between them can be expressed by the following equations:

$$s(t) = \exp j(\omega_c t + \phi(t)) \tag{12.1}$$

and
$$i(t) = r \exp j(\omega_c t + \phi_i(t) + \phi_0) \tag{12.2}$$

where $\phi(t)$ and $\phi_i(t)$ are the phase-modulation components of the desired and interfering signals, respectively, ϕ_0 is the phase difference between $s(t)$ and $i(t)$, and r is the ratio of the amplitude of the interfering signal to that of the desired signal. With these relationships established, the composite signal $v_c(t)$, the input to the FM detector, can be derived by adding $s(t)$ and $i(t)$:

$$
\begin{aligned}
v_c(t) &= s(t) + i(t) \\
&= \exp j[\omega_c t + \phi(t)] \,\{1 + r \exp j[\phi_i(t) - \phi(t) + \phi_0]\} \\
&= \exp j[\omega_c t + \phi(t)]\{R(t) \exp j\delta(t)\} \tag{12.3}
\end{aligned}
$$

The phase angle $\delta(t)$ in Eq. (12.3) can be determined as follows. Let

$$e^x = 1 + r \exp j[\phi_i(t) - \phi(t) + \phi_0] \tag{12.4}$$

and
$$x = x_r + j\delta(t) \tag{12.5}$$

then, from Eq. (12.4), the following is obtained:

$$x = \ln \{1 + r \exp j[\phi_i(t) - \phi(t) + \phi_0]\} \tag{12.6}$$

and from Eq. (12.5), $\delta(t)$ expresses the imaginary part of x:

$$\delta(t) = \operatorname{Im}(x) = \operatorname{Im}(\ln \{1 + r \exp j[\phi_i(t) - \phi(t) + \phi_0]\}) \tag{12.7}$$

Note that r^2 is the interference-to-carrier ratio.
Since

$$\ln(1 + y) = \sum_{k=1}^{\infty} (-1)^{k+1} \frac{1}{k} y^k \tag{12.8}$$

then Eq. (12.7) becomes:

$$
\begin{aligned}
\delta(t) &= \operatorname{Im}\left\{ \sum_{k=1}^{\infty} (-1)^{k+1} \frac{1}{k} r^k \exp j\, k[\phi_i(t) - \phi(t) + \phi_0] \right\} \\
&= \sum_{k=1}^{\infty} (-1)^{k+1} \frac{1}{k} r^k \sin k[\phi_i(t) - \phi(t) + \phi_0] \tag{12.9}
\end{aligned}
$$

When Eq. (12.9) is substituted into Eq. (12.3) and it is assumed $v_d(t)$ is the signal output after limiting, then Eq. (12.3) changes to:

$$v_d(t) = A \exp j[\omega_c t + \phi(t) + \delta(t)]$$ (12.10)

where A is of constant amplitude and $\delta(t)$ is as expressed in Eq. (12.9). $\delta(t)$ represents the phase interference.

Nonfading environment

To estimate the spectrum of $\delta(t)$ in a nonfading environment, it is necessary to obtain the correlation by averaging the random phase angle ϕ_0 from Eq. (12.9). Assuming that ϕ_0 is uniformly and independently distributed from 0 to 2π, then the correlation $R_\delta(\tau)$ of $\delta(t)$ is [1]:

$$R_\delta(\tau) = \lim_{T \to \infty} \frac{1}{2T} \int_{-T}^{T} E[\delta(t)\delta(t+\tau)]_{\phi_0} dt$$

$$= \frac{1}{2} \sum_{k=1}^{\infty} \frac{r^{2k}}{k^2} \lim_{T \to \infty} \frac{1}{T} \int_{-T}^{T} \cos k[\phi_i(t+\tau) - \phi(t+\tau) - (\phi_i(t) - \phi(t))] \, dt$$

$$= \frac{1}{2} \sum_{k=1}^{\infty} \frac{r^{2k}}{k^2} \exp \left[-R_{\Delta_k}(0) + R_{\Delta_k}(\tau)\right]$$ (12.11)

where $R_{\Delta_k}(t)$ is the covariance function of $k[\phi_i(t) - \phi(t)]$. In practice, it has been found that band-limited random Gaussian noise has a rectangular power spectrum and statistically simulates wideband composite speech signals [1].

Since

$$\Delta_k = k[\phi_i(t) - \phi(t)]$$ (12.12)

the spectral density $S_{\Delta_k}(f)$ can be expressed as follows:

$$S_{\Delta_k}(f) = \begin{cases} R_{\Delta_k}(0) & |f| < W \\ 0 & \text{otherwise} \end{cases}$$ (12.13)

and the covariance function $R_{\Delta_k}(\tau)$ of Δ_k is given by:

$$R_{\Delta_k}(\tau) = \int_{-\infty}^{\infty} S_{\Delta_k}(f) e^{j\omega\tau} \, df = R_{\Delta_k}(0) \frac{\sin 2\pi W\tau}{2\pi W\tau}$$ (12.14)

The functions of Eqs. (12.13) and (12.14) are plotted in Fig. 12.1.

Given Eq. (12.14), $R_\delta(\tau)$ of Eq. (12.11) can be determined. The spectrum $S_\delta(f)$ of $\delta(t)$ can be written as follows [2]:

$$S_\delta(f) = \int_{-\infty}^{\infty} R_\delta(\tau) \exp(-j2\pi f\tau) \, d\tau$$

$$= \frac{1}{2} \sum_{k=1}^{\infty} \frac{r^{2k}}{k^2} S_{vk}(f)$$ (12.15)

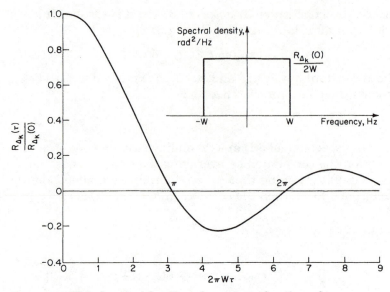

Figure 12.1 Plot of spectral density and covariance functions of Δ_k.

where

$$S_{vk}(f) = \int_{-\infty}^{\infty} \exp\,[-R_{\Delta_k}(0) + R_{\Delta_k}(\tau)]\,\exp\,[-j2\pi f\tau]\,d\tau \qquad (12.16)$$

Substituting Eq. (12.14) into Eq. (12.16) yields the following:

$$S_{vk}(f) = \frac{1}{2\pi W} \int_{-\infty}^{\infty} \exp\left[-R_{\Delta_k}(0)\left(1 - \frac{\sin p}{p}\right)\right] \exp\,(-j\lambda p)\,dp \qquad (12.17)$$

where

$$\lambda = \frac{f}{W} \qquad (12.18)$$

and

$$p = 2\pi W\tau$$

Equation (12.17) is difficult to solve by using analytic methods. The curve for $R_{\Delta_k}(0) = 6$ rad^2, obtained numerically [2], is shown in Fig. 12.2; the small amount of discontinuity in the spectrum at $f/W = 1$ is apparent.

The modulation in a Gaussian stochastic process is only a rough approximation of the statistics for a single voice signal. This statistical difference mainly affects the tails of the IF FM spectrum and does not have any appreciable effect on the analysis of the cochannel baseband

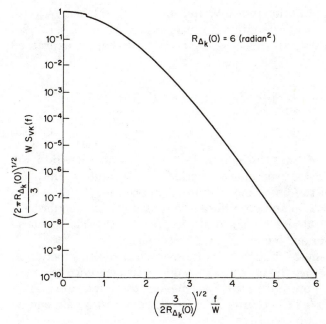

Figure 12.2   Plot of $R_{\Delta_k}(0) = 6$ rad^2. *(Copyright 1969, American Telephone and Telegraph Company. Reprinted with permission.)*

interference. Assuming that $\phi(t)$ and $\phi_i(t)$ are independent stationary Gaussian stochastic processes with rectangular two-sided power spectra ($-W$ to $+W$ Hz), then the spectrum of $S_\phi(f)$ is expressed:

$$S_\phi(f) = \begin{cases} \dfrac{\Phi^2}{2W} & \text{rad}^2/\text{Hz} & |f| \le W \\[2mm] 0 & & \text{otherwise} \end{cases} \tag{12.19}$$

The mean square modulation index Φ^2 is defined as follows:

$$\Phi^2 = E[\phi^2(t)] = \int_{-\infty}^{\infty} S_\phi(f)\, df \tag{12.20}$$

The autocorrelation of $\phi(t)$ is

$$R_\phi(\tau) = \Phi^2 \frac{\sin 2\pi W\tau}{2\pi W\tau} \tag{12.21}$$

Similarly, the spectrum of $S_{\phi_i}(f)$, of the interference, is:

$$S_{\phi_i}(f) = \begin{cases} \dfrac{\Phi_i^2}{2W} & \text{rad}^2/\text{Hz} & |f| \le W \\[2mm] 0 & & \text{otherwise} \end{cases} \tag{12.22}$$

The mean square modulation index Φ_i^2 and the autocorrelation $R_{\phi_i}(\tau)$ of interference signal $\phi_i(t)$ are similar to those expressed in Eqs. (12.20) and (12.21), respectively. In this case the term $R_{\Delta_k}(0)$ can be expressed in terms of Φ^2 and Φ_i^2:

$$R_{\Delta_k}(0) = \lim_{T \to \infty} \frac{1}{2T} \int_{-T}^{T} k^2(\phi^2(t) + \phi_i^2(t))\, dt = k^2[R_\phi(0) + R_{\phi_i}(0)]$$

$$= k^2(\Phi^2 + \Phi_i^2) \tag{12.23}$$

Substituting $R_{\Delta_k}(0)$ of Eq. (12.23) into Eq. (12.17) and then into Eq. (12.15) causes Eq. (12.15) to become a function of r^2, Φ^2, and Φ_i^2. The spectral density $S_\delta(f)$ of the baseband cochannel interference is plotted in Fig. 12.3 for $\Phi = \Phi_i = 2.55$ and for different values of r^2, the interference-to-carrier ratio (icr) at the intermediate frequency (IF). The value of $S_\delta(f)$ decreases either as f increases or as r decreases.

Figure 12.3 also shows that $S_\delta(f)$ contains a Dirac delta function at the frequency $f = 0$. Since the maximum baseband interference occurs at the lowest frequency present in the system—i.e., the minimum signal-to-interference ratio (s.i.r.) occurs at $f = 0$—the minimum s.i.r. can be expressed:

$$\text{s.i.r.} = \left(\frac{S}{I}\right)_{\min} = \left(\frac{\Phi^2}{2W}\right)\Big/ S_\delta(0) = \frac{\Phi^2}{2WS_\delta(0)} \tag{12.24}$$

where $S_\delta(f)$ is as shown in Eq. (12.15). The analytical solution for Eq. (12.15) is complicated and therefore difficult to obtain. The less compli-

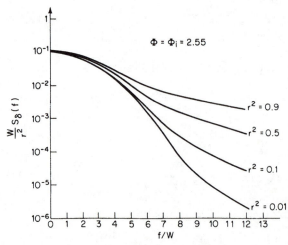

Figure 12.3   Spectral density of baseband cochannel interference in a nonfading environment.

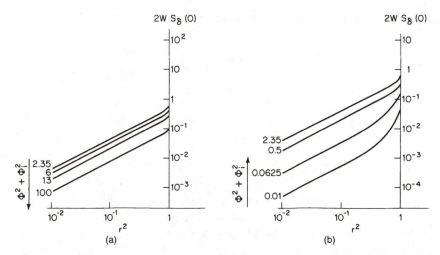

Figure 12.4 Plot of power of baseband cochannel interference under nonfading conditions, $r^2 < 1$. *(From Ref. 3.)*

cated numerical calculation [3] of Eq. (12.15) is plotted in Fig. 12.4 for the function $r^2 < 1$, which is the value most frequently encountered. When r is small in value, the slopes of the curves are constant, as shown in Fig. 12.4. Once the value of $S_\delta(0)$ for a given r^2, Φ^2, and Φ_i^2 is known, then $(S/I)_{\min}$ of Eq. (12.24) can be obtained.

Example 12.1 Assume that there is cochannel interference under nonfading conditions and that the values of $\Phi^2 = \Phi_i^2 = 3$ and $r^2 = 0.1$ are given. Then, the following value can be obtained from Fig. 12.4:

$$2WS_\delta(0) = 0.032$$

Thus,

$$\left(\frac{S}{I}\right)_{\min} = \frac{3}{0.032} = 93.75 \approx 19.7 \text{ dB} \tag{E12.1.1}$$

This means that when the cochannel interference is 10 dB below the carrier level, the baseband signal-to-interference ratio (s.i.r.) can still attain a value of 20 dB.

An approximation of interference power, $2WS_\delta(0)$, can be determined by examining the curves in Fig. 12.4 to verify the following relationships:

$$2WS_\delta(0) = \begin{cases} r^2\sqrt{\dfrac{3}{2\pi(\Phi^2 + \Phi_i^2)}} & \Phi^2 + \Phi_i^2 > 2.35, r^2 < 1 \\[3mm] r^2\left(\dfrac{\Phi_i^2 + \Phi^2}{2}\right) & \Phi^2 + \Phi_i^2 < 2.35, r^2 < 1 \end{cases} \tag{12.25}$$

The $(S/I)_{\min}$ of Eq. (12.24) then becomes:

$$\left(\frac{S}{I}\right)_{\min} = \begin{cases} \dfrac{\Phi^3}{r^2 \sqrt{\dfrac{3}{2\pi[1+(\Phi_i/\Phi)^2]}}} & \Phi^2 + \Phi_i^2 > 2.35,\, r^2 < 1 \\[4ex] \left[\dfrac{r^2}{2}\left(1 + \dfrac{\Phi_i^2}{\Phi^2}\right)\right]^{-1} & \Phi^2 + \Phi_i^2 < 2.35,\, r^2 < 1 \\[4ex] \left[1 + \dfrac{\Phi_i^2}{\Phi^2}\right]^{-1} & r^2 > 1 \end{cases} \tag{12.26}$$

Equation (12.26) shows that the output signal-to-interference ratio (s.i.r.) at the baseband level improves as a function of the cube of the modulation index Φ of the signal itself when the signal is greater than the cochannel interference and where $\Phi^2 + \Phi_i^2 > 2.35$. The signal-to-interference ratio at the baseband level is equal to $\Phi^2/(\Phi_i^2 + \Phi^2)$, when the interference is greater than the signal.

This analysis can be applied even when the baseband signal is passed through linear networks, such as preemphasis or deemphasis networks, before modulation [1].

Rayleigh-fading environment

In a Rayleigh-fading environment, the quasi-static approximation described in Sec. 8.2 of Chap. 8 can be used to determine the average s.i.r. at the baseband level as a function of the average IF interference-to-carrier ratio (icr). The icr at the IF is

$$r^2 = \frac{i^2(t)}{s^2(t)} \tag{12.27}$$

With both the signal $s(t)$ and the interferer $i(t)$ undergoing independent Rayleigh fading, the probability density function (pdf) $p_{r^2}(x)$ can be expressed as follows [4]:

$$p_{r^2}(x) = \frac{\Gamma}{(1 + \Gamma x)^2} \tag{12.28}$$

Note The pdf of $p_{r^2}(x)$ can be derived from Prob. 6 of Chap. 2.

Equation (12.28) is referred to as the F distribution for two degrees of freedom, and Γ is defined:

$$\frac{1}{\Gamma} = E[r^2] = \int_0^\infty r^2 p_{r^2}(x)\, dx \tag{12.29}$$

where Γ is the average cir, $\overline{\text{cir}}$.

Now, the cochannel interfering power $E[I]$ at the baseband level (evaluated at $f = 0$, the minimum s.i.r. in a Rayleigh-fading case) can be expressed by two terms, as follows:

$$E[I] = E[2WS_\delta(0)] + E[N_s] \tag{12.30}$$

The term $E[2WS_\delta(0)]$ is the interference averaged over r from 0 to ∞, where $2WS_\delta(0)$ is found from Eq. (12.25). Then, for the case of $\Phi^2 + \Phi_i^2 > 2.35$,

$$E[2WS_\delta(0)] = \int_0^\infty p_{r^2}(x) 2WS_\delta(0)\, dx$$

$$= \int_0^1 \frac{\Gamma x}{(1 + \Gamma x)^2} \sqrt{\frac{3}{2\pi(\Phi^2 + \Phi_i^2)}}\, dx + \int_1^\infty \frac{\Gamma}{(1 + \Gamma x)^2} \left(\frac{\Phi^2 + \Phi_i^2}{2} \right) dx$$

$$= \sqrt{\frac{3}{2\pi(\Phi^2 + \Phi_i^2)}} \left(\frac{1}{\Gamma} \ln(1 + \Gamma) - \frac{1}{1 + \Gamma} \right) + \frac{(\Phi^2 + \Phi_i^2)}{2(1 + \Gamma)} \tag{12.31}$$

N_s in Eq. (12.30) is the power of the signal-suppression noise $n_s(t)$ due to Rayleigh fading, as previously described in Eq. (8.52), and $n_s(t)$ can be expressed as follows:

$$n_s(t) = \phi(t) - E[\phi(t)] \tag{12.32}$$

where $\phi(t)$ is the baseband output signal for a nonfading case and $E[\phi(t)]$ is the signal output for a Rayleigh-fading case. In the latter instance, the following expression applies:

$$E[\phi(t)] = \int \phi(t) p_{r^2}(x)\, dx = \begin{cases} \phi(t) \int_0^1 p_{r^2}(x)\, dx = \dfrac{\Gamma \phi(t)}{1 + \Gamma} & r^2 < 1 \\[4mm] \phi(t) \int_1^\infty p_{r^2}(x)\, dx = \dfrac{\phi(t)}{1 + \Gamma} & r^2 > 1 \end{cases} \tag{12.33}$$

Substituting Eq. (12.33) into Eq. (12.32) yields:

$$n_s(t) = \phi(t) - E[\phi(t)] = \begin{cases} \dfrac{1}{1 + \Gamma} \phi(t) & r^2 < 1 \\[4mm] \dfrac{\Gamma}{1 + \Gamma} \phi(t) & r^2 > 1 \end{cases} \tag{12.34}$$

The time average of $N_s^2(t)$ of Eq. (12.34) is

$$N_s = \langle n_s^2(t) \rangle = \begin{cases} \dfrac{1}{(1 + \Gamma)^2} \langle \phi^2(t) \rangle = \dfrac{1}{(1 + \Gamma)^2} \Phi^2 & r^2 < 1 \\[4mm] \dfrac{\Gamma^2}{(1 + \Gamma)^2} \langle \phi^2(t) \rangle = \dfrac{\Gamma^2}{(1 + \Gamma)^2} \Phi^2 & r^2 > 1 \end{cases} \tag{12.35}$$

Then the average power $E[N_s]$ of signal-suppression noise over Rayleigh fading is

$$E[N_s] = \int_0^\infty p_{r^2}(x)N_s\,dx = \int_0^1 \frac{\Phi^2}{(1+\Gamma)^2}\frac{\Gamma}{(1+\Gamma x)^2}\,dx$$

$$+ \int_1^\infty \frac{\Gamma^2\Phi^2}{(1+\Gamma)^2}\frac{\Gamma}{(1+\Gamma x)^2}\,dx = \frac{\Gamma}{(1+\Gamma)^2}\Phi^2 \qquad (12.36)$$

Substituting Eq. (12.31) and Eq. (12.36) into Eq. (12.30) gives the average power of interference $E[I]$. The average output signal power is

$$E[\Phi^2] = \left\langle\left(\int_0^1 \phi(t)p_{r^2}(x)\,dx\right)^2\right\rangle = \frac{\Gamma^2}{(1+\Gamma)^2}\langle\phi^2(t)\rangle = \frac{\Gamma^2}{(1+\Gamma)^2}\Phi^2 \quad (12.37)$$

where $\Phi^2 = \langle\phi^2(t)\rangle$ and the average s.i.r., $\overline{\text{s.i.r.}}$, is:

$$\overline{\text{s.i.r.}} = \frac{E[\Phi^2]}{E[I]} = \frac{\Gamma^2\Phi^2/(1+\Gamma)^2}{\sqrt{3/(2\pi(\Phi^2+\Phi_i^2))}\,(1/\Gamma\,\ln\,(1+\Gamma) - 1/(1+\Gamma)) + Q}$$

$$(12.38)$$

where

$$Q = (\Phi^2 + \Phi_i^2)/(2(1+\Gamma)) + (\Gamma/(1+\Gamma)^2)\,\Phi^2$$

The signal-to-interference ratio $(\overline{\text{s.i.r.}})$ of Eq. (12.38) is plotted in Fig. 12.5, assuming $\Phi = \Phi_i$ and $r^2 < 1$.

When $\Phi^2 > 5$ rad^2, the $\overline{\text{cnr}}$, Γ, versus $\overline{\text{s.i.r.}}$ is independent of Φ^2. When $\Gamma \gg 1$ and $\Phi_i^2 = \Phi^2$, Eq. (12.38) becomes:

$$\overline{\text{s.i.r.}} = \frac{\Gamma\Phi^3}{(1/2)\sqrt{3/\pi}\,\ln\,\Gamma + 2\Phi^3} \qquad (12.39)$$

Comparing the terms $\Phi^2 + \Phi_i^2 > 2.35$, $r^2 < 1$, and $\Phi^2 = \Phi_i$ in Eq. (12.26) with the terms in Eq. (12.39) shows that the improvement of s.i.r. due to the cube of the modulation index is lost with Rayleigh fading.

As seen from Fig. 12.5, the improvement of SIR by increasing the modulation index Φ from 1 to 5 is small. It means that increasing the modulation index to reduce the cochannel interference plus fading is not a good approach. Therefore, spatial separation is used in cellular systems to reduce the cochannel interference.

Example 12.2   For values of $\Gamma = 33$ and $\Phi_i^2 = \Phi^2 = 9$ rad^2, the baseband s.i.r. due to cochannel interference in a Rayleigh fading environment is:

$$\overline{\text{s.i.r.}} = \frac{33(27)}{(1/2)\sqrt{3/\pi}\,\ln\,33 + 2\times 27} = \frac{891}{1.67 + 54}$$

$$= 16 \approx 12\text{ dB} \qquad (E12.2.2)$$

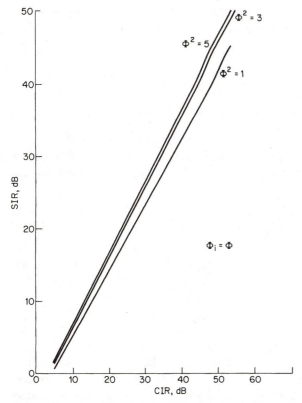

Figure 12.5   Plot of baseband s.i.r. in a case of cochannel
interference plus fading.

12.3   Adjacent-Channel Interference

The method used in Sec. 12.2 for cochannel interference calculations
can be extended to adjacent-channel interference calculations. Adjacent-
channel interference can be classified as either "in-band" or "out-of-
band" adjacent-channel interference. The term "in-band" is applied
when the center of the interfering signal bandwidth falls within the
bandwidth of the desired signal. The term "out-of-band" is applied
when the center of the interfering signal bandwidth falls outside the
bandwidth of the desired signal. The following discussion of adjacent-
channel interference mainly concentrates on in-band adjacent-channel
interference, since this type of interference always has a pronounced
effect on the desired signaling channel, whereas out-of-band signals
are less of a problem. To simplify the discussion, the same model used
to describe cochannel interference, where $r < 1$, is used to describe in-
band adjacent-channel interference. For cases where $r > 1$, the interfering signal becomes dominant and should be treated as Gaussian

noise, which is discussed later in the chapter. The following analysis assumes that $r < 1$ and the desired signal is

$$s(t) = \exp j[\omega_c t + \phi(t)] \tag{12.40}$$

and the adjacent-channel interfering signal is

$$i(t) = r \exp j[(\omega_c + \omega_a)t + \phi_i(t) + \phi_0] \tag{12.41}$$

where $(1/2\pi)(\omega_c + \omega_a)$ is the frequency of the adjacent channel.

The same approach as that used in Eqs. (12.3) through (12.9) shows that the phase interference $\delta(t)$ is similar to that obtained in Eq. (12.9):

$$\delta(t) = \sum_{k=1}^{\infty} (-1)^{k+1} \frac{1}{k} r^k \sin k[\omega_a t + \phi_i(t) - \phi(t) + \phi_0] \tag{12.42}$$

The adjacent-channel interference in two different environments, nonfading and Rayleigh, is discussed in the following subsections.

Nonfading environment

In a nonfading environment, it can be assumed that the interference-to-carrier ratio r^2 is a constant and ϕ_0 in Eq. (12.42), the phase difference between $s(t)$ and $i(t)$, is uniformly and independently distributed. Then the correlation $R_\delta(\tau)$ of $\delta(t)$ can be obtained in a manner similar to that expressed in Eq. (12.11) [1]:

$$R_\delta(\tau) = \lim_{T \to \infty} \frac{1}{2T} \int_{-T}^{T} E[\delta(t)\, \delta(t + \tau)]\, dt$$

$$= \frac{1}{2} \sum_{k=1}^{\infty} \frac{r^{2k}}{k^2} \exp[-R_{\Delta_k}(0) + R_{\Delta_k}(\tau)] \cos \omega_a \tau \tag{12.43}$$

and the power density $S_\delta(f)$ of $\delta(t)$ is similar to that obtained in Eq. (12.15):

$$S_\delta(f) = \int_{-\infty}^{\infty} R_\delta(\tau) \exp j(-2\pi f \tau)\, d\tau$$

$$= \frac{1}{4} \sum_{k=1}^{\infty} \frac{r^{2k}}{k^2} [S_{vk}(f - kf_a) + S_{vk}(f + kf_a)] \tag{12.44}$$

where $S_{vk}(f)$ is as expressed in Eq. (12.17). The power density function of Eq. (12.44) is plotted in Fig. 12.6, for $\omega_a = 2\pi \times 5W$, $\Phi^2 = 9$, and $\Phi_i^2 = 4$.

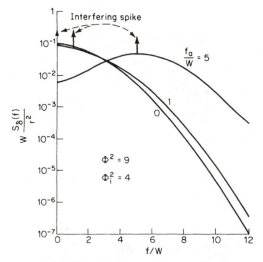

Figure 12.6 Plot of power density for adjacent-channel interference with different adjacent frequencies. *(Copyright 1969, American Telephone and Telegraph Company. Reprinted with permission.)*

The same baseband signal-to-interference ratio (s.i.r.) expressed in Eq. (12.24) can be used to find the approximate adjacent-channel interference for the in-band condition:

$$\text{s.i.r.} = \frac{\Phi^2}{2WS_\delta(0)} \qquad r^2 < 1 \tag{12.45}$$

where $S_\delta(f)$ is as shown in Eq. (12.44). The signal-to-interference ratio of Eq. (12.45) is plotted in Fig. 12.7 as a function of rms phase-modulation indices Φ, for $\Phi_i = 6$ rad and $\Phi_i = \Phi$.

In comparing the cochannel interference power $f_a/w = 0$ with adjacent interference power $f_a/w = 5$, it should be noted that the baseband interference due to an adjacent-channel interferer is smaller than that resulting from a cochannel interferer of equal strength (same value of r) at the IF, and that the adjacent-channel interference is mainly concentrated at the higher baseband frequencies.

For the case of $r > 1$, the above analysis is not valid, and therefore the adjacent-channel interferer must be treated as Gaussian noise. In this case, the noise band is centered away from the carrier by $f_c - f_a$, where f_a is the offset frequency. The case of $r > 1$ is the more complicated to handle, but the interference is not as critical as for $r < 1$, which has been discussed in the preceding paragraphs. The signal-to-interference ratio for the condition $r > 1$ at the baseband level can be derived by using the techniques previously described in Chap. 8.

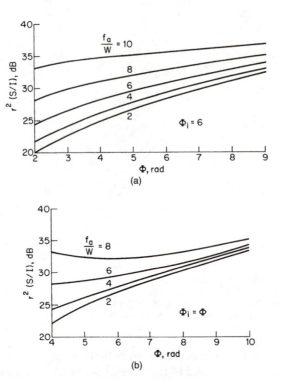

Figure 12.7 Plot of adjacent-channel signal-to-interference ratio (s.i.r.) as a function of rms phase-modulation indices. *(Copyright 1969, American Telephone and Telegraph Company. Reprinted with permission.)*

Rayleigh-fading case

In a Rayleigh-fading environment, the quasi-static approximation described in Sec. 8.2 of Chap. 8 can again be applied to determine the average s.i.r. at the baseband level as a function of the average IF cir. In the mobile-radio environment, the desired signal and the adjacent-channel signal may be partially correlated with their fades. Then the probability exists that $r_2 \geq \alpha r_1$, where r_1 and r_2 are the two envelopes of the desired and interfering signals, respectively; and that probability can be obtained from the joint density function of Eq. (10.31), assuming that $E[r_1^2] = E[r_2^2] = 2\sigma^2$ and that α is a constant:

$$
\begin{aligned}
P(r_2 \geq \alpha r_1) &= \int_0^\infty dr_1 \int_{\alpha r_1}^\infty p(r_1, r_2)\, dr_2 \\
&= \int_0^\infty dr_1 \int_{\alpha r_1}^\infty r_1 r_2 \exp\left[-\frac{r_1^2 + r_2^2}{2\sigma^2(1 - \rho_r)}\right] I_0\left[\frac{r_1 r_2}{\sigma^2}\frac{\sqrt{\rho_r}}{(1 - \rho_r)}\right] dr_2 \\
&= \frac{1}{2} + \frac{1}{2}\frac{1 - \alpha^2}{\sqrt{(1 + \alpha^2)^2 - 4\rho_r\,\alpha^2}}
\end{aligned}
\tag{12.46}
$$

where ρ_r is the correlation coefficient between r_1 and r_2. Then the probability density function $p_r(y)$ of $r = r_2/r_1$ can be obtained as follows:

$$p_r(y) = -\frac{d}{d\alpha} P\left(\frac{r_2}{r_1} \geq \alpha\right)\bigg|_{y=\alpha} \tag{12.47}$$

Let $R = \sqrt{G}r$, where G is the power gain at the IF filter output for the desired signal relative to the adjacent-channel interferer. Then

$$p_{R^2}(x) = p_r(y)\frac{1}{2yG}\bigg|_{y=\sqrt{x/G}}$$

$$= \frac{(1-\rho_r)\left(1+\dfrac{x}{G}\right)}{G\left[\left(1+\dfrac{x}{G}\right)^2 - 4\rho_r\dfrac{x}{G}\right]^{3/2}} \tag{12.48}$$

where ρ_r can be found from Eq. (6.126), with $\tau = 0$:

$$\rho_r = \frac{1}{1 + (\Delta\omega)^2 \Delta^2} \tag{12.49}$$

and where $\Delta\omega/2\pi$ is the difference in frequency between the desired signal and the interferer. The term Δ is the time-delay spread, which will vary in value depending on the different types of mobile environments. ρ_r decreases proportionately as either Δ or $\Delta\omega$ increases. As ρ_r decreases, the adjacent-channel interference also decreases. The same procedure used to find the cochannel baseband signal-to-noise ratio for the pdf expressed in Eq. (12.28) can also be used to find the baseband signal-to-noise ratio due to an adjacent-channel interferer in a fading environment, by substituting the pdf of Eq. (12.46) in place of Eq. (12.28). The baseband signal-to-interference ratio then becomes a function of G, Δ, $\Delta\omega$, and Φ_i; and the calculation can be obtained by straightforward numerical integration methods.

The problem of adjacent-channel interference can be significantly reduced by developing a frequency-allocation plan that ensures an adequate frequency separation between adjacent channels. Also, the use of high-gain bandpass filters in the receivers will assist in reducing adjacent-channel interference. One of the few advantages that results from time-delay spread within the mobile-radio environment is that the delay spread tends to decorrelate the adjacent-channel frequency toward the desired channel, and thus reduces adjacent-channel interference. As a final consideration, when adjacent-channel interference is compared with cochannel interference at the same level of interfering power, the effects of the adjacent-channel interference are always less.

Out-of-band adjacent-channel interference

In the out-of-band case, only the tail of the adjacent-channel signal enters the FM demodulator from the IF filter. Considerable phase-to-amplitude conversion causes the adjacent-channel signal to appear noiselike at the output of the IF filter. The noiselike adjacent-channel signal emerges from the IF filter as Gaussian noise. The techniques discussed in Sec. 8.2 of Chap. 8 for frequency demodulation in the presence of noise can be used for the out-of-band analysis. The significant difference in this case is that the noise band is not centered with respect to the carrier. For this reason, the expression previously given in Eq. (12.30) for cochannel interfering power is not valid for out-of-band interference.

The techniques previously used to obtain the baseband average adjacent-channel interference with Rayleigh fading could be used [5], but the solution is more complicated and the effect on out-of-band interference is less pronounced. For these reasons, it is not recommended. Further information based on additional studies is available in published literature [1–3].

12.4   Near-End to Far-End Ratio Interference

Near-end to far-end ratio interference occurs when the distance between a mobile receiver and the base-station transmitter becomes critical with respect to another mobile transmission that is close enough to override the desired base-station signal. This situation usually occurs when a mobile unit is relatively far from its desired base-station transmitter at a distance of d_1, but close enough to its undesired nearby mobile transmitter at a distance d_2 such that $d_1 > d_2$. Under such conditions, if the two transmitters are assumed to be transmitting simultaneously at the same power and frequency, then the signals received by the mobile unit from the desired source will be masked by the signals received from the undesired source. This same type of interference can occur at the base station when signals are received simultaneously from two mobile units that are at unequal distances (one close, the other far) from the base station. In this case, the closer mobile transmission at distance d_2 may interfere with the distant mobile transmission at distance d_1. The power difference due to the path loss between the receiving location and the two divergently located transmitters is called the "near-end to far-end ratio interference" and can be expressed:

$$\text{Near-end to far-end ratio} = \frac{\text{path loss } d_1}{\text{path loss } d_2} \qquad (12.50)$$

To obtain a quick approximation of the near-end to far-end ratio (in decibels), a propagation-loss slope of 40 dB per decade can be used for a typical mobile-radio environment. Then the following expression is valid:

$$\text{Near-end to far-end ratio (dB)} = 40 \log (\text{distance ratio}) \quad (12.51)$$

Example 12.3 Consider the case where a base station is simultaneously serving two mobile units. One mobile unit is 0.1 km from the base station and the other is 15 km distant. Under these conditions, the near-end to far-end ratio can be approximated by first determining the distance ratio: $R_d = 15/0.1 = 150$. The distance ratio is then applied to the propagation-loss slope for the mobile-radio environment (40 dB per decade):

$$\text{Near-end to far-end ratio} = 40 \log (\text{distance ratio})$$

$$= 40 \log 150 = 87 \text{ dB} \qquad (E12.3.1)$$

If the required signal-to-interference ratio (s.i.r.) is 15 dB, then the adjacent-channel frequency separation must be far enough apart to provide a minimum of 102 dB (87 dB + 15 dB) of signal isolation between the two adjacent channels.

As an alternate solution, the required signal isolation can be obtained through the use of time-division duplex coding, or through the use of spread-spectrum schemes. However, these schemes require equalization of the transmitting power, inversely proportional to the distance to the receiving location. This would require automatic control logic within the mobile-radio and base-station transceivers. These techniques can be effective in overcoming problems in performance that are related to near-end to far-end ratio interference in the mobile-radio environment.

12.5 Intermodulation (IM) Interference

There are three major sources that can cause intermodulation interference problems: (1) AM-PM conversion through amplification, (2) antenna mismatching, and (3) interaction between colocated transmitting antennas. The causes and remedies for these three problem areas are discussed in the following subsections.

AM-PM conversion

When an IF signal passes through a nonlinear power amplifier, intermodulation of the third order is produced as a result of the AM to PM conversion process. The calculation of third-order intermodulation can be found in published textbooks [5]. The intermodulation introduced by a nonlinear amplifier can be eliminated by using either one of the following two methods: using a feed-forward amplifier [6, 7], or using linear amplification with nonlinear components (LINC) [8, 9].

Feed-forward amplification Figure 12.8 illustrates the block diagram for a basic feed-forward amplification circuit. Without the feed-forward circuit, the nonlinearity of the amplifier will generate undesired odd-

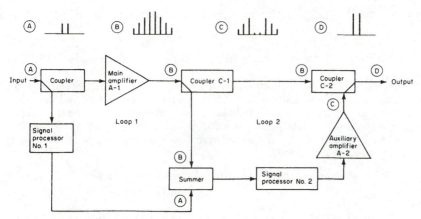

Figure 12.8 Block diagram of a basic feed-forward circuit.

order intermodulation. It is of interest to know that even-order IM generally falls outside the band and therefore can be filtered out. The basic feed-forward circuit contains a direct signal path and an auxiliary loop that generates an error-correction signal.

In the feed-forward circuit of Fig. 12.8, multiple input signals are coupled to main amplifier A-1 and to signal processor 1. The main signal is amplified by main amplifier A-1 and passed by couplers C-1 and C-2 to the output at point D. A portion of the amplified signal at point B is coupled to a summing network, where it is summed with the reshaped and retimed output of signal processor 1. The summed signal is adjusted in phase and amplitude by signal processor 2. After amplification of the error-correction output from signal processor 2, the output of auxiliary amplifier A-2 is used to null out the error portion of the main amplifier signal. The resultant output-signal spectrum at point D is a linearly amplified replica of the input-signal spectrum at point A. This feed-forward amplification technique can be used for wideband signal applications, where 35 dB of internal suppression can be achieved. Even better IM suppression can be achieved by using additional stages of feed-forward amplification; however, the signal-to-intermodulation ratio (S/I_{mo}) is better for saturated class A amplifiers than it is for class C amplifiers. Also, the feed-forward amplification technique does not fully utilize power efficiency.

Linear amplification with nonlinear components (LINC) The LINC technique for producing linear amplification with nonlinear components was proposed by Cox [8, 9]. In general, a bandpass carrier frequency with equal amplitude and angle modulation cannot be applied to a nonlinear amplifier without becoming distorted. However, if the time-

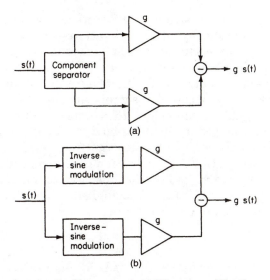

Figure 12.9 Two examples of linear amplification with nonlinear components (LINC): (*a*) component-separation method; (*b*) inverse-sine modulation method.

varying amplitude modulation of the carrier can be translated into a correlated time-varying angle modulation of the carrier, then the signal can be passed through an amplifier with severe nonlinear characteristics without serious impairment of waveform fidelity.

Figure 12.9 illustrates two versions of the LINC technique, one using inverse-sine modulation and the other using signal component separation. In Fig. 12.9(*a*), the input signal $s(t)$ can be represented:

$$s(t) = P \cos (\omega t + \theta) \sin \psi(t) \qquad (12.52)$$

where $\psi(t)$ is the information term and P and θ are the amplitude and phase of the carrier at a frequency ω_0.

In the LINC scheme, all of the amplitude and phase information in the original signal is converted into conjugate phase modulation of two component signals:

$$s(t) = \frac{P}{2} \cos (\omega_0 t + \theta) \sin \psi(t) + \frac{P}{2} \sin (\omega_0 t + \theta) \cos \psi(t)$$

$$+ \frac{P}{2} \cos (\omega_0 t + \theta) \sin \psi(t) - \frac{P}{2} \sin (\omega_0 t + \theta) \cos \psi(t)$$

$$= \frac{P}{2} \sin (\omega_0 t + \theta + \psi(t)) - \frac{P}{2} \sin (\omega_0 t + \theta - \psi(t)) \qquad (12.53)$$

When the component signals are subtracted after separate amplification, the resultant signal is an exact replica of the original band-pass signal. Since each component signal is constant in amplitude, the two signals can be amplified in highly nonlinear devices prior to component-signal subtraction without generating distortion products. For multiple-carrier-frequency signals at the LINC input, the complex input signal is equivalent to a single-frequency carrier signal with a complex amplitude-modulation component and a complex angle-modulation component about the carrier frequency.

A critical step in LINC amplification is the generation of signals proportional to the inverse sine of the envelopes of the signals to be amplified. This requirement can be accomplished as follows. Let the input signal be expressed in another form:

$$s(t) = e(t) \cos (\omega_0 t + \theta) \tag{12.54}$$

where $e(t)$ is the information that is being sent. Since Eqs. (12.52) and (12.54) are equal, the phase $\psi(t)$ in Eq. (12.52) should be:

$$\psi(t) = \sin^{-1} \frac{e(t)}{P} \tag{12.55}$$

The information term $e(t)$ is now converted from an envelope form to a phase form. The inverse-sine relationship represented in Eq. (12.55) is essentially a function of the hardware, and there are two ways that it can be realized. One way is to generate the two conjugate components of Eq. (12.53) by using the component-separator method of Cox [9], as shown in Fig. 12.9(a). The component separator can generate the two conjugate phase-modulation components expressed in Eq. (12.53) and pass each signal component through its corresponding nonlinear amplifier.

The other way is to use the inverse-sine modulation method of Fig. 12.9(b) to generate two conjugate phase components, as was suggested by Rustako and Yeh [10].

In summary, the significant advantages of the LINC technique over the feed-forward amplification technique are (1) that the inherent distortions (IM_3) are independent of the nonlinearity of the amplifier and therefore a class C amplifier with inherently higher efficiency can be used; (2) that a two-tone third-order IM_3 level, 50 dB below the carrier level, is achievable in the forward transmission path of a single LINC amplifier; and (3) that the LINC amplifier can be operated over a wider band of 6 to 12 percent of the total bandwidth. However, the LINC amplifier, unlike the feed-forward amplification techniques, cannot be cascaded or looped for further suppression on IM_3 levels, since the distortion will be generated by the cascading multiple LINC.

Antenna terminal mismatch

When the antennas and the antenna terminal are mismatched, a small portion of the transmitting power is reflected back into the power amplifier. If the power amplifier is not perfectly linear, additional intermodulation interference will be generated. To correct this situation, an improved matching device is needed to minimize the reflected power and thus eliminate the additional intermodulation interference caused by mismatching between the power amplifier and the antenna feed.

Colocated transmitting antennas

When two transmitting antennas are placed close together, the signal of one antenna becomes a source of interference for the other system. Signals from an adjacent antenna transmission are strong enough to be picked up at the input to the power amplifier of the desired signal transmitter, which results in intermodulation interference. One method that can be used to reduce the effects of this type of intermodulation interference is to install an interference-cancellation system, such as the one shown in Fig. 12.10.

In Fig. 12.10, the interfering signal $a_1 \exp j\phi_1$ becomes $a_1' \exp j\phi_1'$ upon arrival at the antenna of the desired signal. The change in amplitude is the effect of signal loss within the transmission medium, whereas

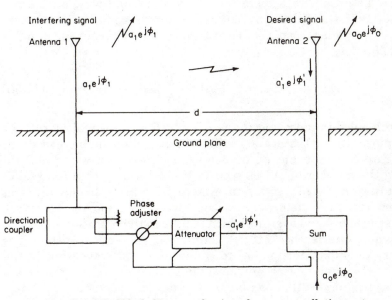

Figure 12.10 Simplified block diagram of an interference-cancellation system.

the change in phase is the result of propagation delay. The magnitude of the changes in signal amplitude and phase is dependent upon the separation distance d between the two colocated antennas. The cancellation circuit provides an internal path for the interfering signal that forms an error-cancellation loop between the antenna for the interfering signal (antenna 1), and the antenna for the desired signal (antenna 2). The objective of the interference-cancellation circuit is to cancel the interfering signal components as they enter antenna 2, prior to reaching the power-amplifier stage, where intermodulation can occur. To accomplish this action, a portion of the interfering signal is directionally coupled to a phase- and amplitude-adjusting circuit and then combined with the interfering signal picked up by antenna 2. If the amount of phase shift and attenuation imparted to the cancellation signal is correct, then the signal from the cancellation circuit will cancel out the interfering signal picked up by antenna 2. Additional details on interference-cancellation systems are available in published books [11].

12.6   Intersymbol Interference

If a digital transmission system is linear and distortionless over all frequencies, then the system has a bandwidth that is theoretically infinite. However, in practical terms, every digital transmission system has a finite bandwidth with a certain varying amount of frequency-response distortion. In a distortionless system, one with an infinite bandwidth, a basic pulse $s(t)$ would suffer no degradation as a consequence of frequency response. For instance, let $x(t)$ represent a pulse train of N pulses:

$$x(t) = \sum_{k=1}^{N} a_k \, s\left(t - \frac{k}{R}\right) \qquad (12.56)$$

where the pulse-to-pulse spacing is $1/R$ and R is the signaling rate for binary pulses $a_k = \pm 1$. Since $s(t)$ suffers no distortion, an increased signaling rate can be achieved by decreasing the pulse width and increasing the number of pulses transmitted within a given interval of time. However, when a practical system with finite bandwidth and frequency-response distortions is considered, the individual pulses tend to spread out and overlap, which results in the condition known as "intersymbol interference" (ISI). Therefore, in a practical system it is necessary to employ techniques to shape the pulse waveforms of the transmitted signal, as was previously described in Sec. 11.2 of Chap. 11. Other techniques that can be used to minimize intersymbol interference due to overlapping of pulses, while at the same time increasing the signaling rate, are dependent upon the use of equalizers [12] or special digital coding circuits [13].

Another consideration that is independent of the signaling rate and transmitting frequency is the effects of time-delay spread Δ due to the mobile-radio environment. Time-delay spread is affected by bandwidth limitations, and by multipath signal reflections that arrive at the receiving antenna at different times. Since it is impractical to selectively choose the mobile-radio environment, the improvements that can be obtained through more conventional techniques must be evaluated.

When the signaling rate is relatively low, as when

$$\frac{1}{R} \gg \Delta \qquad (12.57)$$

then the effects of time-delay spread have a negligible effect on received pulses. For instance, if the delay spread in a typical suburban area is 0.5 μs and the signaling rate is 16 kb/s, then $1/R = 6.25$ μs, which is much greater than the delay spread. In this situation, no intersymbol interference (ISI) exists. When the signaling rate increases, the ISI increases; the ISI can be greatly reduced by using signal waveform-shaping techniques, and by using an equalizer. There are many reference publications that deal with this type of problem [5, 14, 15, 16].

Figure 12.11 is a conventional "eye" diagram that can be used to evaluate the causes of intersymbol interference in a mobile-radio environment. The effects of ISI on signal degradation can be classified as amplitude and timing degradation, corresponding to the vertical and horizontal displacements shown in Fig. 12.11. Signal amplitude de-

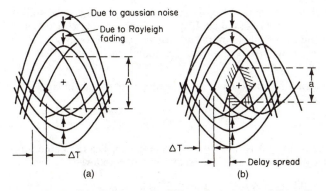

Figure 12.11 Eye diagrams for ISI analysis in a mobile-radio environment: (a) case 1, where delay-spread effects are negligible; (b) case 2, where delay-spread effects are significant.

grading is caused by the summed effects of Gaussian noise, limited bandwidth distortions, variations in local oscillator stability, decision-threshold uncertainties, and Rayleigh fading. Signal timing degradations are caused by limited frequency-response distortions, time slippage due to clock error, time jitter due to clock instability, static misalignment of timing signals, and time-delay spread in the mobile-radio medium. The effects of Gaussian noise and Rayleigh fading on signal amplitude were described in Secs. 8.2 and 8.3 of Chap. 8. The effects of timing distortion due to limited frequency response are described by Sunde [5], Bennett [14], Ho [15], and Taub [16].

Degradation effects of time-delay spread on signal timing

The following analysis of the effects of time-delay spread on signal timing is based on the assumption that the probability of error due to Gaussian noise and/or Rayleigh fading is the same as it would be for the case where there is no delay spread. Therefore, in the presence of time-delay spread, the carrier-to-noise ratio must be increased by

$$\Delta \frac{S}{N} = 20 \log \frac{A}{a} \tag{12.58}$$

where A is the vertical opening of the eye diagram in the no-delay-spread condition, case 1, Fig. 12.11(a). A represents the signal amplitude degraded either by Gaussian noise alone or by Gaussian noise and Rayleigh fading combined. The term a is the vertical opening of the eye diagram in the presence of significant delay spread, case 2, Fig. 12.11(b). Under this condition, Eq. (12.58) can be calculated by using the method applied in the following example.

Example 12.4   A noncoherent binary FSK signal transmitted in a typical suburban area has a time-delay spread of 0.5 μs. If the signaling rate is 200 kb/s and the waveform is a cosine shape due to amplitude distortion in a Rayleigh-fading mobile-radio environment, what is the amount of signal degradation that will result from the time-delay spread?

solution   From Eq. (11.19), the expression for a cosine-pulse waveform is:

$$s(t) = A \cos \frac{\pi t}{T} \tag{E12.4.1}$$

where a decision is made at $t = 0$. At a delay spread of 0.5 μs, and for a signaling rate of $R = 1/T = 200$ kb/s, the value of a can be expressed as follows:

$$a = s(0.5 \ \mu s) = A \cos (\pi \times 0.5 \times 10^{-6} \times 2 \times 10^5)$$
$$= A \cos 0.1\pi = A \times 0.95 \tag{E12.4.2}$$

Then, by substituting Eq. (E12.4.2) into Eq. (12.58), the expression for the effects of time-delay spread becomes

$$\Delta \frac{S}{N} = 0.44 \text{ dB} \qquad\qquad (E12.4.3)$$

The signal degradation of 0.44 dB gives an approximation of the relatively small magnitude of signal degradation that can result from time-delay spread in a typical suburban mobile-radio environment. The value of 0.44 dB in Eq. (E12.4.3) is an additional signal degradation which degrades slightly on the performance curves shown in Fig. 8.14 of Chap. 8. In other words, it has only a slight effect on the probability of error in a Rayleigh-fading environment. In Chap. 13, the irreducible error rate resulting from time-delay spread will be discussed.

The results obtained in Example 12.4 can also be applied to systems that use coherent FSK, DPSK, and PSK signaling techniques, since all of these experience the same type of signal degradation as a consequence of time-delay spread within the transmission medium. However, the results in terms of the difference in the amount of signal degradation will vary from system to system, because of the different values of time-delay spread and the different waveform shapes. The total probability of error for the different signaling systems was shown in Fig. 8.14 of Chap. 8.

A few points that are worth remembering are (1) that waveform shaping can be effective in reducing ISI when the signaling rate is low, as expressed in Eq. (12.57); (2) that shaping the waveform at the transmitting end does little to reduce ISI at the receiving end when the signaling rate is high and the time-delay spread is severe; and (3) that the solution of using an equalizer to reduce ISI in a Gaussian-noise environment can actually contribute to degraded signal performance in a multipath Rayleigh-fading environment, depending on the degree of time-delay spread involved [12].

Problem Exercises

1. Given the following parameters for cochannel interference in a nonfading environment, what is the baseband signal-to-interference ratio (s.i.r.)? The modulation index for the desired signal is $\Phi = 2$ rad, and the interferer-to-signal ratio is $r^2 = 0.5$, with a modulation index of $\Phi_i = \sqrt{6}$ rad for the interfering signal.

2. Given the same parameters as in Prob. 1 for cochannel interference but in a fading environment, what is the baseband s.i.r.?

3. Given the same parameters as in Prob. 1 but for adjacent-channel interference in a nonfading environment, and given a center frequency of $f_a = 8W$ away from the carrier, what is the baseband s.i.r.?

4. Given the same parameters as in Prob. 1 but for adjacent-channel interference in a fading environment, and using the procedure of Sec. 12.2 for finding the s.i.r. for cochannel interference, find the s.i.r. for $f_a = 5W$.

5. Consider a situation where a base-station receiver capable of providing 100 dB of isolation between channels is receiving a signal from a mobile unit 1 km away. What is the minimum distance that a second mobile unit can transmit its signal without causing interference with the signal from the near-end mobile unit?

6. What are the considerations for using a component-separator technique to obtain linear amplification with nonlinear components? [Reference Fig. 12.9(a).]

7. What are the arguments for using an inverse-sine modulation technique to obtain linear amplification with nonlinear components? [Reference Fig. 12.9(b).]

8. Calculate the differences in amplitude and phase between the signal at point A and the signal at point B over the radio communications link based on the parameters given in Fig. P12.8.

9. Assuming a DPSK cosine signal waveform, a signaling rate of 200 kb/s, and a time-delay spread of 3 ms, what amount of signal degradation will occur, and what rms threshold level should be stipulated to maintain the probability of error on the order of 10^{-3}?

10. Apply the same procedure used in Example 12.4 to calculate the signal degradation for a PSK signal due to time-delay spread where the signaling rate is 300 kb/s. Compare the two results for PSK and FSK signaling and explain the advantages and disadvantages of the PSK signaling scheme.

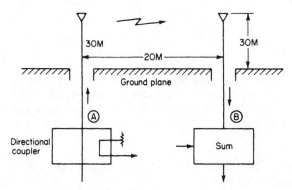

Figure P12.8   Model for Prob. 8.

References

1. V. K. Prabhu and L. H. Enloe, "Interchannel Interference Considerations in Angle-Modulated Systems," *Bell System Technical Journal,* vol. 48, September 1969, pp. 2333–2358.
2. V. K. Prabhu and H. E. Rowe, "Spectral Density Bounds of a PM Wave," *Bell System Technical Journal,* vol. 48, March 1969, pp. 789–811.
3. W. C. Jakes, *Microwave Mobile Communications,* Wiley, New York, 1974, p. 180.
4. J. S. Bendat and A. G. Piersol, *Random Data Analysis and Measurement Procedures,* Wiley-Interscience, New York, 1971, p. 107.
5. E. D. Sunde, *Communication System Engineering Theory,* Wiley, New York, 1969, chap. 8, pp. 422–454.
6. H. Seidel, "A Microwave Feed-forward Experiment," *Bell System Technical Journal,* vol. 50, no. 9, November 1971, pp. 2879–2916.
7. H. Seidel, H. R. Beurrier, and A. N. Friedman, "Error-Controlled High Power Linear Amplifiers at VHF," *Bell System Technical Journal,* vol. 47, no. 5, May–June 1968, pp. 651–722.
8. D. C. Cox, "Linear Amplification with Non-linear Components," *IEEE Trans. Comm.,* vol. 22, December 1974, pp. 1942–1945.
9. D. C. Cox and R. P. Leck, "Component Signal Separation and Recombination for Linear Amplification with Non-linear Components," *IEEE Trans. Comm.,* vol. 23, November 1975, pp. 1281–1287.
10. A. J. Rustako and Y. S. Yeh, "A Wideband Phase Feedback Inverse-Sine Phase Modulator with Application towards a LINC Amplification," *IEEE Trans. Comm.,* vol. 24, October 1976, pp. 1139–1143.
11. W. A. Sauter, *Interference Cancellation System,* Interference Technology Engineers Master, R&B Enterprises, Plymouth Meeting, Pa., 1979, p. 181.
12. R. W. Lucky, J. Salz, and E. J. Weldon, Jr., *Principles of Data Communication,* McGraw-Hill, New York, 1968, pp. 128–165.
13. A. J. Viterbi and J. K. Omura, *Principles of Digital Communication and Coding,* McGraw-Hill, New York, 1979, p. 284.
14. W. R. Bennett and J. R. Davey, *Data Transmission,* McGraw-Hill, New York, 1965, chap. 7.
15. E. Y. Ho and Y. S. Yeh, "A New Approach for Evaluating Error Probability in the Presence of Intersymbol Interference and Additive Gaussian Noise," *Bell System Technical Journal,* vol. 49, November 1970, pp. 2249–2266.
16. H. Taub and D. L. Schilling, *Principles of Communication Systems,* McGraw-Hill, New York, 1971, chap. 7.

Signal-Error Analysis versus System Performance

13.1 System and Signaling Criteria

Performance criteria

In Chap. 11, the signaling or message-stream format, including the different types of bit syncs and word syncs, was discussed. In this chapter, the methods for analyzing the word-error rate of a message are discussed. The word-error rate is a major parameter for measuring the quality of signaling performance and can also be used for comparing the performance of different systems. The word-error rate is, to a large degree, dependent upon the bit-error rate; and the bit-error rate is directly affected by the signaling rate, which is determined by the channel capacity, as shown in Eq. (11.1). In a typical mobile-radio case, the curve for predicting the received power in dBm versus distance can easily be plotted, as was illustrated in Fig. 3.13 of Chap. 3. Given a transmission path of 8 mi, the average receiving power at this distance, as shown in Fig. 3.13, is –96.5 dBm. If a noise level of –121 dBm is assumed, based on a bandwidth of 30 kHz and a front-end noise of 8 dB, then, by applying Eq. (11.1),* the following is obtained:

$$C = B \log_2 \left(1 + \frac{S_c}{N}\right) = B \times 8.17 = 245 \text{ kb/s} \tag{13.1}$$

where C is defined as the theoretical maximum rate at which the channel can be expected to supply reliable information to the terminal sub-

---

* See the footnote to the text following Eq. (11.1).

scriber at the receiving destination. In actual practice, this rate cannot be achieved. In order to obtain an increase in reliability, it is necessary to lower the bit rate. In addition, for bit rates of the order of 10 kb/s, it is usually necessary to provide coding redundancy in order to obtain reliability. The amount of coding redundancy required depends on the efficiency of signal detection and error correction applicable to a fading channel.

Testing criteria

Before the bit-error rate and word-error rate of the mobile-radio signals can be examined, a set of test standards must be developed that will set the criteria for comparisons between one system and another. Without standard testing criteria, system analysts may inadvertently select certain road locations and environmental conditions that will produce the desired results and meet the system requirement, while under less optimal conditions the test results may prove to be marginal or unsatisfactory. To ensure repeatable desired results over a wide range of locations and environmental conditions, subjective test criteria must be established prior to the start of performance test and evaluation. Three accepted methods can be used to determine the test standards:

1. Select a unique geographical route that has been carefully evaluated against the specified test parameters and one that can be used effectively to test the different systems under consideration. The difficulty is that one geographical route may not be adequate in satisfying all of the test parameters. In this case, a set of routes with varying properties may have to be used.

2. Use the same type of simulators and measurement techniques to test the performance of the different systems. There is a great selection of simulators offered by different manufacturers, and the use of a standard portable simulator may not be feasible. The selection of simulators and measuring devices should be given careful consideration.

3. Establish standard analytical techniques based on statistical data. The route of tested roads should be marked off in measured distances from the base station, and signal strength should be averaged from measured and recorded data taken at intervals of 20 to 40 wavelengths along the route. A cumulative probability distribution (cpd) curve should be plotted to determine the long-term-fading characteristics along the transmission path. The cpd curve always follows a lognormal distribution, as was previously discussed in Chap. 3. The mean of the cpd curve is a reference point that can be

set by adjusting the level of transmitting power to the dBm level specified by the evaluator; however, the variance of the cpd curve will vary from one route to another and from one area to another. Since it is difficult to maintain a specified cpd reference level from area to area, a quantitative analysis of the cpd is as follows. The cpd curve for lognormal fading can be expressed:

$$P(x \leq X) = \frac{1}{\sqrt{2\pi}} \int_{-\infty}^{(X-m)/\sigma} \exp\left(-\frac{x^2}{2}\right) dx \tag{13.2}$$

where X is expressed in decibels above the mean m, which is assumed to be 0 dB referenced to a given specified level. Figure 13.1 shows the cpd curves for two different areas. The variances of the two cpd's can be obtained by finding the levels at a given value of cpd, such as $P(x \leq X) = 90\%$. Then the following expression can be obtained from published mathematical tables:

$$\frac{X-m}{\sigma} = \frac{X}{\sigma} = 1.29$$

and

$$X' = 15 \text{ dB}; \qquad \sigma_A = 11.6 \text{ dB} \qquad \text{from curve } P_A$$

$$X = 10 \text{ dB}; \qquad \sigma_B = 7.7 \text{ dB} \qquad \text{from curve } P_B$$

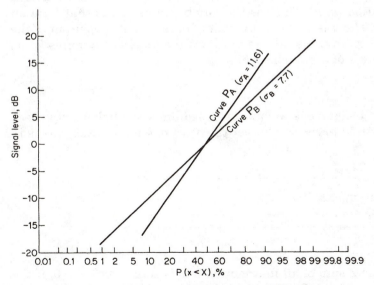

Figure 13.1 Cpd curves of lognormal fading for two different areas.

Based on the above findings, the performance for the P_B area is much better than that for the P_A area, using the same evaluation criteria for both areas.

Equivalent performance criteria

The equivalent performance for two systems measured in two different areas can be compared as follows:

Let P_{eA} and P_{eB} denote the bit-error rates for the measured data curves P_A and P_B of Fig. 13.1. Then, the equivalent performance P'_{eB}, as though it were obtained from the P_A area, can be expressed:

$$P'_{eB} = P_{eB} \frac{P_A}{P_B} \qquad \text{(at a level } X \text{ below its mean)} \qquad (13.3)$$

Therefore, if the values for P_{eA} and P_A are specified by the evaluator, then the bit-error rate P_{eB} of any other system measured in any other area can be converted to its equivalent P'_{eB} from the above equation. Then P'_{eB} can be used to compare with P_{eA}. If $P'_{eB} \leq P_{eA}$ at a specified signal level can be achieved, then the system B being tested is qualified. Equation (13.3) can also be used to compare the system performance of two systems measured in two different areas. If $P'_{eB} \leq P'_{eA}$, the system performance of system B is better.

13.2   Linear Block Code

The linear block code is a class of parity check code. It is characterized by the notation (n, k). The k is the number of data bits that form an input block. The n is the total number of data bits at the output of the encoder. The ratio k/n is called the rate of the code. The n minus k bits are the number of parity check bits.

The single-parity check code

To transmit a signal over a low-noise medium, the single-parity check code can be used to check whether the block of k information bits is in error.

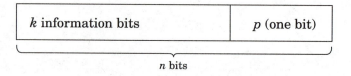

k information bits	p (one bit)

n bits

If the modulo-2 sum of all information bits is zero (even), $p = 0$. If the modulo-2 sum of all information is one (odd), $p = 1$. The rate of the code is $(n - 1)/n$ and is the highest rate. It can be considered the most efficient code that does not need to increase the bandwidth.

Parity-check matrix

When a block of n information bits $\{x_m\}$ is transmitted in a high noise medium, the parity-check matrix is used to generate a number of parity-check bits and form a coded word. The coded word $[C]$ can be expressed as

$$[C] = [x_m][G] \tag{13.4}$$

where

$$[x_m] \text{ is the information bits matrix}$$

$[G]$ is the generation matrix that contains the parity checks. It can be expressed as

$$[G] = [I \quad P] \tag{13.5}$$

where I is the identity matrix and P is the parity matrix.

For example, we let the generation matrix $[G]$ be

$$[G] = [I \quad P] = \begin{bmatrix} 1000 & 101 \\ 0100 & 111 \\ 0010 & 110 \\ 0001 & 001 \end{bmatrix} \tag{13.6}$$

The modulo-2 additions are

$$1 + 1 = 0,\ 1 + 0 = 1,\ 0 + 0 = 0,\ 0 + 1 = 1$$

Let $[x_m] = 1001$; then,

$$[C] = [x_m][G] = [1001]\begin{bmatrix} 1000 & 1 & 01 \\ 0100 & 1 & 11 \\ 0010 & 1 & 10 \\ 0001 & 0 & 01 \end{bmatrix} = [1001 \quad C_5 \quad C_6 \quad C_7] \tag{13.7}$$

To find C_5, we take

$$C_5 = [1001]\,[1110] = [1 \cdot 1 + 0 \cdot 1 + 0 \cdot 1 + 1 \cdot 0] = 1 + 0 + 0 + 0 = 1$$

Use the same way to find C_6 and C_7 as follows

$$C_6 = [1001]\,[0111] = 0 + 0 + 0 + 1 = 1$$

$$C_7 = [1001]\,[1101] = 1 + 0 + 0 + 1 = 0$$

Therefore, the information bits (1001) become a coded word

$$[C] = 1001110$$

Coding gain

Coding gain is the comparison of the dB difference between the coded and uncoded bits at a given bit error probability value P_B. The bit error probability for coherent BPSK with two various (n, k) codes over a Gaussian channel is shown in Fig. 13.2. For example, the coding gain is 1.2 dB for the (24,12)-coded word and 2 dB for the (127,92)-coded word at $P_B = 10^{-4}$. The rate of code for the (24,12) code is 0.5, and the parity-check bits are 12. The rate of code for the (127,92) code is 0.72, and the parity-check bits are 35.

Therefore, in the Gaussian channel, the low rate of code does not necessarily provide the higher coding gains.

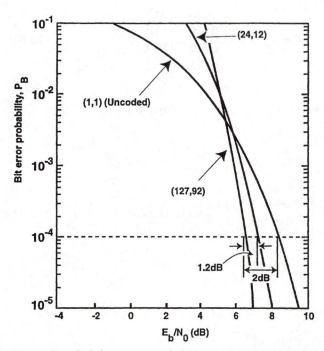

Figure 13.2   Coded versus uncoded bit error performance for coherent PSK with two codes.

Coding strength

Hamming distance. The error patterns in the decoded bits are related to the Hamming distance between two code vectors $\underline{U}$ and $\underline{V}$. The Hamming distance is denoted by $d(\underline{U},\underline{V})$, measured by the different bits between two code vectors. There are six different bits marked by "x".

$$\underline{U} = 1\ 1\ 0\ 1\ 0\ 0\ 1\ 0\ 1\ 0$$
$$\underline{V} = 0\ 1\ 1\ 1\ 1\ 0\ 0\ 1\ 0\ 0$$
$$\text{x}\quad\text{x}\quad\text{x}\quad\ \text{x}\,\text{x}\,\text{x}$$
$$d(\underline{U},\underline{V}) = 6$$

The larger the number of $d(\underline{U},\underline{V})$, the less the chance of having errors.

Hamming weight. The weight $W(\underline{U})$ is defined as nonzero bits of $\underline{U}$.

$$W(\underline{U}) = 5$$
$$W(\underline{V}) = 5$$
$$W(\underline{U} + \underline{V}) = W(1010101110) = 6$$

Therefore the weight of $\underline{U} + \underline{V}$ is equal to the distance of $\underline{U}$ and $\underline{V}$.

$$W(\underline{U} + \underline{V}) = d(\underline{U},\underline{V}) \tag{13.8}$$

Since calculating the weight is simpler than finding the distance, the vector $W(\underline{U} + \underline{V})$ is usually used to find $d(\underline{U},\underline{V})$.

Error detection and error correction

The optimal decoder strategy can be expressed in terms of the maximum likelihood algorithm. By deciding in favor of $\underline{U}_i$ when the received code is $\underline{r}$,

$$P\,(\underline{r}\,|\,\underline{U}_i) = \max\,\{P\,(\underline{r}\,|\,\underline{U}_j)\} \qquad \text{for } j = 1 \text{ to } N \tag{13.9}$$

where N is all N code vectors. Also, we may find the minimum distance $d(\underline{r}, \underline{U}_i)$ as

$$d\,(\underline{r}\,|\,\underline{U}_i) = \min\,\{d(\underline{r},\underline{U}_j)\} \qquad \text{for } j = 1 \text{ to } N \tag{13.10}$$

The minimum distance of the code is denoted $d_{\min}$, which is the shortest link in a chain toward the $\underline{U}_j$.

In general, the error-detecting capability e of a code is defined as

$$e = d_{\min} - 1 \tag{13.11}$$

and the error-correcting capability t of a code is

$$t = \frac{d_{\min} - 1}{2} \tag{13.12}$$

Therefore, in a block code with the minimum $d_{\min}$, we can detect $d_{\min} - 1$ errors and correct one-half of them.

Erasure correction

If a symbol cannot be decided in a receiver, instead of guessing a state, allow the bit to erase, but retain the location information. If the code has minimum distance $d_{\min}$, then any pattern of α errors and β erasures can be corrected simultaneously with the condition:

$$d_{\min} \geq 2\alpha + \beta + 1 \tag{13.13}$$

Error detection and correction algorithm

Let us assume that the information $m(X)$ with a multiple of the generator polynomial $g(X)$ creates a code word $U(X)$:

$$U(X) = m(X)g(X) \tag{13.14}$$

$U(X)$ is sent and received by $Y(X)$, which is the corrupted version of $U(X)$,

$$Y(X) = U(X) + e(X) \tag{13.15}$$

The decoder tests the received code word $Y(X)$ by dividing $g(X)$, the remainder $S(X)$ is created as

$$\frac{Y(X)}{g(X)} = q(X) + \frac{S(X)}{g(X)}$$

or

$$Y(X) = q(X)g(X) + S(X) \tag{13.16}$$

If $S(X) = 0$, then $q(X)g(X) = U(X)$, and $q(X) = m(X)$.

If $S(X)$ is not equal to zero, then $Y(X)$ is corrupted. By subtracting Eq. (13.15) with Eq. (13.16) and applying the relationship of Eq. (13.14), we achieve

$$e(X) = q(X)g(X) - U(X) + S(X)$$
$$= [q(X) - m(X)]g(X) + S(X) \tag{13.17}$$

The syndrome $S(X)$ of the received polynomial $\underline{Y}(X)$ contains the information needed for correction of the errors.

There are many well-known block codes listed in Table 13.1. Details on this topic can be found in books on coding theory [1, 2].

Cyclic redundancy check (CRC)

The cyclic redundancy check (CRC) is used to detect the error only. For example, the generator polynomial $g(X)$ of a 7-bit CRC can be expressed as

$$g(X) = 1 + X + X^2 + X^3 + X^4 + X^5 + 0 \cdot X^6 + X^7 \qquad (13.18)$$

An input polynomial $a(X)$ with m bits is

$$a(X) = \sum_{k=0}^{m} B_k X^k \qquad (13.19)$$

where B_k is the bit sequence.

The polynomial $b(X)$ is the remainder of the division of $a(X)$ and $g(X)$ obtained from

$$\frac{a(X)X^7}{g(X)} = q(X) + \frac{b(X)}{g(X)} \qquad (13.20)$$

TABLE 13.1   Commonly Used Coding Schemes with Error-Correcting Capability

Coding schemes	Number of errors capable of correcting (t)	Remarks
Hamming code (L, k)	1	$L - k = m$ where $L = 2^m - 1$ and $k = 2^m - m - 1$.
Golay code (23, 12)	3	The code word length and the information bits are fixed.
Reed-Muller code (L, k)	$(L + 1) \cdot 2^{-Q} - 1$	Q is the number of steps of a majority logic decoder.
BCH code (L, k)	$\dfrac{L - k}{\log_2 (L + 1)} \leq t \leq \dfrac{d - 1}{2}$	d is the minimum Hamming distance.
Reed-Solomon code (L, k)	$\dfrac{L - k}{2}$	A nonbinary BCH code.
Fire code (L, k)	$\dfrac{k + 1 - \log_2 (L + 1)}{2}$	Only detecting and correcting single burst of t errors.
Convolutional code (L, k)	$\dfrac{L - k}{1 + k/L}$	For burst correcting [1] with large L.

where $q(X)$ is the quotient of the division that is discarded for transmission. The remainder $b(X)$ can be generated from Eq. (13.19) and Eq. (13.20).

$$b(X) = \sum_{k=1}^{7} C_k X^{k-1} \tag{13.21}$$

The seven bits of $C_1 \cdots G_1$ will be sent with the message bits of $\{B_k\}$. The message sent out will be $\{B_k\}$ plus $\{C_k\}$. The CRC is performed at the receiving end. The received message $a'(X)$ will be divided by the same $g(X)$ stored at the receiver.

$$\frac{a'(X)X^7}{g(X)} = q'(X) + \frac{b'(X)}{g(X)} \tag{13.22}$$

Then the new generated remaining bits b'_k can be obtained as

$$b'(X) = \sum_{k=1}^{7} C'_k X^{k-1} \tag{13.23}$$

Comparing $\{C_k\}$ from the received signal and $\{C'_k\}$ from the generated bits, if

$$\{C_k\} = \{C'_k\}$$

then the message has zero error. If $\{C_k\} \neq \{C'_k\}$, then either the message has errors or the $\{C_k\}$ is in error due to the transmission. This section only demonstrates a 7-bit CRC application. The number of CRC bits varies depending on the particular needs. The popular ones are 7, 16, and 32.

13.3   Convolutional Code

A convolutional code is described by three integers, n, k, and K. K is the constraint length to control the redundancy of the message bits, while k is a number of M-ary bits ($M = 2^k$). For the input of binary bits, $k = 1$. For the input of 4-ary bits, $k = 2$. kK is the number of a shift register. n is the coded bits at the output of the shift register. The ratio k/n is the rate of the code. For the input of binary bits, $k = 1$, the kK-stage shift register can be referred to simply as a K-stage register. The encoder is then called a $1/n$ encoder. The convolutional encoder with constraint length K and rate k/n is shown in Fig. 13.3.

Convolutional encoder

To illustrate the mechanism of the convolutional encoder, we take an encoder of rate $\frac{1}{2}$, $K = 3$, and $k = 1$, as shown in Fig. 13.4. Since $k = 1$, the

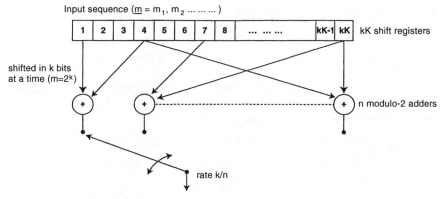

Figure 13.3   Convolutional encoder.

input bits m_i are binary. The generator polynomial $g(X)$ is expressed

$$g(X) = \sum_{i=1}^{K} a_i x^i \qquad (13.24)$$

where $a_i = 0,1$. The generator polynomial $g_1(X)$ for the left adder (point a) is

$$g_1(X) = 1 + X + X^2 \qquad (13.25)$$

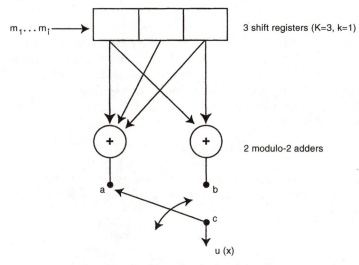

Figure 13.4   A 3-register, 2-adder convolutional encoder.

and for the right adder (point b) is

$$g_2(X) = 1 + X^2$$

The output sequence is found at point c as

$$\underline{U}(X) = \underline{m}(X)g_1(X) \text{ alternated with } \underline{m}(X)g_2(X) \qquad (13.26)$$

First, let the message vector $m = 011$, then $\underline{m}(X) = X + X^2$, and

$$\underline{m}(X)g_1(X) = (X + X^2)(1 + X + X^2) = X + X^2 = X^2 + X^3 + X^4$$
$$= X + X^4 \qquad \text{(after modulo-2 addition)} \qquad (13.27)$$

$$\underline{m}(X)g_2(X) = (X + X^2)(1 + X^2) = X + X^2 + X^3 + X^4 \qquad (13.28)$$

From Eq. (13.27) and (13.28),

$$\underline{m}(X)g_1(X) = 0 + 1 \cdot X + 0 \cdot X^2 + 0 \cdot X^3 + 1 \cdot X^4$$

$$\underline{m}(X)g_2(X) = 0 + 1 \cdot X + 1 \cdot X^2 + 1 \cdot X^3 + 1 \cdot X^4$$

The output $\underline{U}(X)$ can be found from Eq. (13.26) as

$$\underline{U}(X) = \{\underline{m}(X)g_1(X), \underline{m}(X)g_2(X)\}$$
$$= (0,0) + (1,1)X + (0,1)X^2 + (0,1)X^3 + (1,1)X^4$$

The vector $\underline{U}$ is then

$$U = 00\ 11\ 01\ 01\ 11$$

The tree diagram

It is very simple to code the input message to the output of the encoder by using the tree diagram. Every encoder has its own tree diagram. The output of the encoder shown in Fig. 13.4 can have a tree representation, as shown in Fig. 13.5. Suppose that the message $m = 10\ 110\cdots$ enters the encoder with the bit on the far left entering first. Bit 1 goes downward. Bit 0 goes upward. Then, the heavy trace indicates the paths, and the coded word is

$$U = 1110000101$$

The tree diagram is expanded upward and downward as the time passes. It is very hard to store the tree diagram in memory, even though it is easy to use.

Trellis diagram

The trellis diagram has a repetitive structure, and its size is more manageable than the size of the tree diagram. We can demonstrate a new

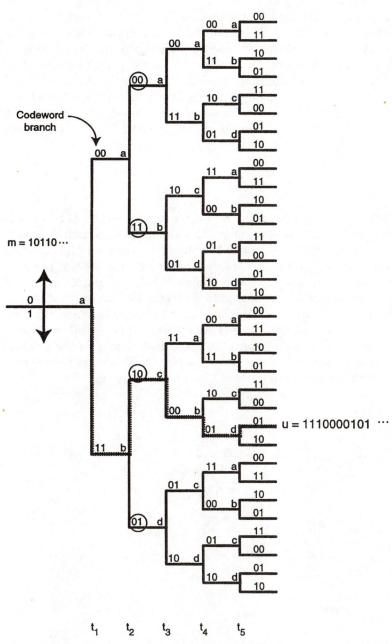

Figure 13.5   Tree representation of encoder (rate ½ $K = 3$).

trellis diagram for the output of the encoder in Fig. 13.4. In Fig. 13.5, we circle four states (00, 11, 10, 01) at time t_2 and show that every state can split into two new states. Following the diagram trace, we convert the tree diagram from Fig. 13.5 to the trellis diagram shown in Fig. 13.6. In the trellis diagram, there are solid lines for the input bit 0 and the dotted line for the input bit 1. We realize that the diagram between time t_3 and t_4 repeats thereafter as the shaded region indicated in Fig. 13.6. Then, $m = 10110$ will be coded by $U = 11\ 10\ 00\ 0101$ at the output of the encoder.

The Veterbi convolutional decoding algorithm

The Veterbi decoding algorithm was discovered in 1967 [11]. It performs maximum likelihood decoding. It calculates a measure of similarity between the received signal Y_i and all the trellis paths entering each state at time t_i. Remove all the candidates that are not possible based on the maximum likelihood choice. When two paths enter the same state, the one with the best path metrics (the sum of the distance of all branches) along the path is chosen. This path is called the surviving path. This selection of surviving paths is done for all the states and makes decisions to eliminate the least likely paths in early calculation stages to reduce the decoding complexity. The material regarding the Veterbi decoder can be found in Refs. 13 and 14.

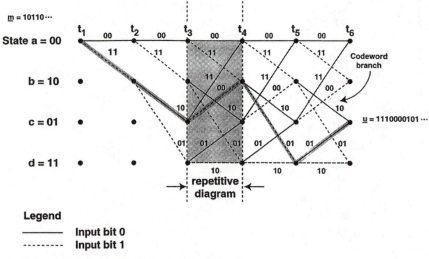

Figure 13.6 Encoder trellis diagram (rate $\frac{1}{2}K = 3$).

13.4   Trellis Coded Modulation (TCM) [12, 13, 14]

TCM is a modulation that combines coding and modulation to achieve a great coding gain without sacrificing bandwidth efficiency. The modulations use a rate $k/(k+p)$ where k is the information bits and $k+p$ are the coded bits. The convolutional encoder can map the coded bits onto signal points $\{x_k\}$ through a technique called mapping by set partitioning.

Trellis-coded modulation uses a multilevel/phase-signaling set that signals constellations with multiple amplitudes and multiple phases. The signaling sets combine with a state-oriented trellis coding scheme.

The TCM achieves coding gain without any bandwidth expansion but at the expense of decoder complexity. The coding gain can be observed from block codes or trellis codes. Due to the advantage of the Viterbi decoding algorithm, trellis decoding becomes attractive. TCM uses coding to gain redundancy but increases the signal alphabet to achieve multilevel/phase signaling, or $M = 2^n$ (n coded bits), so that the channel bandwidth is not increased.

The error performance of an uncoded M-ary PAM or PSK or QAM depends on the distance between the closest pair of signal points. To keep a constant average power, the minimum distance between points decreases as the number of points increases, which also creates the redundancy of the transmitted signal. This is the concept of TCM. TCM does not decrease the bandwidth efficiency or increase the transmitted power, and it does not increase the outage from error.

Trellis coded modulation can be implemented with a conventional encoder where k current bits and $k(K-1)$ prior bits are used to produce a convolutional code with a number of parity bits p. The $n = k + p$ coded bits require 2^n binary channel symbols for transmission. The encoding increases the signal size from 2^k to 2^{k+p}. In Fig. 13.7(a), an uncoded 4-ary (2^2) PAM (Pulse Amplitude Modulation) has the transmitted power P_{t_1} and transmission rate R_1. Its $k = 2$ and $p = 0$. The uncoded PAM is coded to a TCM by taking $k = 2$ and $p = 1$, which is a rate of $k/(k+p) = 2/3$. The TCM is the same as a 2/3 coded 8-ary PAM with the same power P_{t_1} and the same rate R_1.

Figure 13.7(b) shows an uncoded 4-ary PSK ($k = 2, p = 0$) to a TCM by taking $k = 2$ and $p = 1$, which is a rate of $k/(k+p) = 2/3$. This new TCM is the same as a 2/3-coded 8-ary PSK with the same power P_{t_2} and the same rate R_2 as the uncoded 4-ary PSK.

Figure 13.7(c) shows an uncoded 16 QAM ($k = 4, p = 0$) to a TCM by taking $k = 4$ and $p = 1$, which is a rate of $k/(k+p) = 4/5$. This new TCM is the same as a 4/5-coded 32-ary QAM with the same power P_{t_3} and the same rate R_3 as the uncoded 16-ary QAM.

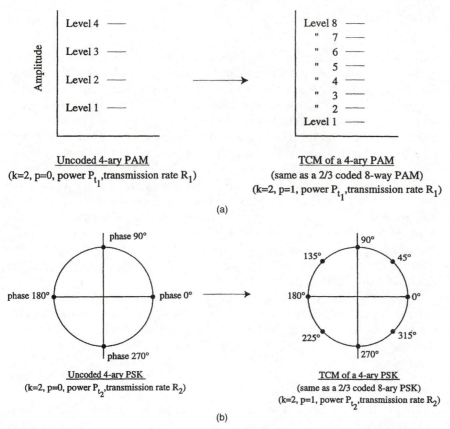

Figure 13.7 Increase of signal set size $(k + p)$ for TCM: (*a*) TCM of a 4-ary PAM; (*b*) TCM of a 4-ary PSK; (*c*) a 16 QAM is coded to TCM.

Although the increase in signal set size provides coding redundancy, the increase in the number of signals does not result in increasing bandwidth because the transmission rate remains the same.

Coding gain for trellis coding

The average signal power P_{av} is computed as

$$P_{av} = \frac{\sum_{1}^{M} d_i^2}{M} \qquad (13.29)$$

where d_i is the Euclidean (true) distance of the ith signal from the center of the space, and M is the number of the code word symbols. We may find the average power of a 4-PAM signal set and an 8-PAM set to be the same, as shown in Fig. 13.8.

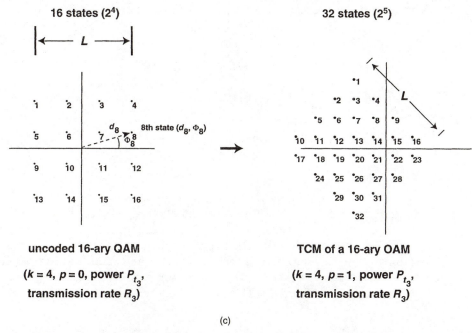

16 states (2⁴)

32 states (2⁵)

uncoded 16-ary QAM

($k = 4$, $p = 0$, power P_{t_3},
transmission rate R_3)

TCM of a 16-ary OAM

($k = 4$, $p = 1$, power P_{t_3},
transmission rate R_3)

(c)

Figure 13.7 (*Continued*) Increase of signal set size ($k + p$) for TCM: (*c*) TCM of a 16-ary QAM.

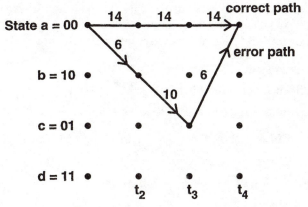

Figure 13.8 Minimum distance error event for rate 2/3 convolutional code.

The average power P_{av} of an 8-ary PAM can be obtained from Eq. (13.29).

$$P_{av} = \frac{\sum di^2}{M} = \frac{672}{8} = 84$$

The average power P_{av} of a 4-ary PAM can have the same power as the 8-ary PAM

$$P_{av} = \frac{2[d^2 + (3d)^2]}{4} = 84$$

$$d = 4.1$$

The signal set of a 4-ary PAM has the same average power of an 8-ary PAM.

In general, the P_{av} values for the two cases do not need to be the same.

We may show the error event for a 2/3 rate 8-ary PAM and compare to the uncoded 4-ary PAM by using the coding gain as a measurement.

The coding gain G from the TCM can be described as

$$G = \frac{(d_{min}^2/P_{av})_{TCM}}{(d_{min}^2/P_{av})_{uncoded}} \tag{13.30}$$

Suppose that the transmitted sequence is 000, which corresponds to an all-zero (state 00) path. The error event diverges from the all-zero path to state 10 first, then state 01, then returns to state 00 as shown in Fig. 13.8 for a TCM signal. This TCM is the same as a 2/3-coded 8-ary PAM where $K = 3$. The minimum distance d_{min} is calculated as follows:

The error distance from state 00 to state 10 is $14 - 6 = 8$.

The error distance from state 00 to state 01 is $14 - 10 = 4$.

The error distance from state 01 to state 00 is $14 - 6 = 8$.

The error path of the TCM of a 2/3-rate 8-ary PAM shown in Fig. 13.8 results in its d_{min} as

$$(d_{min}^2)_8 = (14 - 6)^2 + (14 - 10)^2 + (14 - 6)^2 = 64 + 16 + 64 = 144$$

For uncoded 4-PAM, the d_{min}^2 value for each error event is

$$(d_{min}^2)_4 = (12.3 - 4.1)^2 = 67.2$$

The coding gain G can be obtained from Eq. (13.30) as

$$G = \frac{144}{67.2} = 2.14 \qquad \text{or} \qquad 3.31 \text{ dB}$$

Therefore, even for the simple register $K = 3$, a noticeable amount of coding gain is shown without increasing the bandwidth and the average power for a TCM signal.

13.5   Types of Signaling Errors

In mobile communications, there are two types of errors which may be observed. One is the unrecognizable word. The other is the falsely recognizable word. Both of these kinds of errors can be reduced by adding additional coding to the basic coded word. The probability of receiving falsely recognizable words is usually very small when compared with the probability of receiving unrecognizable words. It should be noted that when the mobile-radio system receives an unrecognizable word, it does not take any action in response to the unrecognizable word. Under this condition, even though a word is unrecognizable, the system will not malfunction. However, if the system receives a falsely recognizable word, it will act in response to the falsely recognizable word, and the system may malfunction with a certain degree of probability. Because of this, it is necessary to keep the rate of falsely recognizable words (the false-alarm rate) as low as possible. The false-alarm rate can be calculated in terms of probability of error p of each bit based on false recognition. If two code words have a length of L bits and are different from each other by t bits, then the false-alarm rate can be expressed:

$$P_f = \text{false-alarm rate} = p^t(1 - p)^{L-t} \qquad (13.31)$$

The word-error rate calculation considers only the probability that a word is in error, which means that if one or more bits in a given word are in error, then the entire word is considered in error. The word-error rate can be found by the following expression:

$$P_w = \text{word-error rate} = 1 - (1 - p)^L \qquad (13.32)$$

If a word of L bits length is coded by the insertion of g bits, then the new word of $(L + g)$ bits has the capability for detection and correction of t errors, depending upon the coding scheme that is used and the number g of bits that have been inserted. Table 13.1 lists some of the commonly used coding schemes [1, 2], together with their error-correction capabilities. The code-word-error rate of a code word consisting of N bits, with t errors corrected, can be expressed:

$$P_{\text{cw}} = 1 - \sum_{k=0}^{t} C_k^N p^k (1 - p)^{N-k} \qquad (13.33)$$

where

$$C_k^N = \frac{N!}{(N-k)!k!} \tag{13.34}$$

It can readily be seen that the word-error rate P_w is always greater than the false-alarm rate P_f. Also, P_w is always greater than the code-word-error rate; however, a required false-alarm rate of 10^{-6} is usually specified. Reduction of the false-alarm rate can become a problem in analyzing the various code patterns, and unfortunately, the solutions for such problems are beyond the scope of this book. Therefore, analyses of system performance are based either on word-error rate or on the coded-word-error rate, exclusively.

Example 13.1   If there are two code words, 1010 and 1100, and the bit-error rate is 10^{-2}, then false recognition can occur when the two middle bits of the first code word are in error and mistakenly identified as though the second code word were sent. Under these conditions, the false-alarm rate is

$$P_f = p^2(1-p)^2 = 9.8 \times 10^{-5} \tag{E13.1.1}$$

and the word-error rate is

$$P_w = 1 - (1-p)^4 = 3.94 \times 10^{-2} \tag{E13.1.2}$$

Example 13.2   As indicated in Table 13.1, the Golay code (23, 12) has the capability for correcting up to three errors. If the Golay code is used and the bit-error rate is 10^{-1}, the word-error rate P_{cw} can be found from Eq. (13.33) as follows:

$$P_{cw} = 1 - \sum_{k=0}^{3} C_k^{23} p^k (1-p)^{23-k} = 0.1927 \tag{E13.2.1}$$

If no coding is added to the word, then the word length is 12 b instead of 23 b and the word-error rate can be obtained from Eq. (13.32) as follows:

$$P_w = 1 - (1-p)^{12} = 0.7175 \tag{E13.2.2}$$

It is apparent that P_{cw} is approximately 3.7 times less than P_w; however, the throughput for Golay-coded words is $12/23 \approx 50$ percent. This means that when the signaling rate is fixed, only half of the received bits of a Golay-coded word are information bits, whereas for an uncoded word all of the received bits are information bits.

Example 13.3   In an FM digital transmission scheme, most mobile receivers use a discriminator for detection of FM voice transmissions. Therefore, it is reasonable to assume that some form of direct digital noncoherent FSK carrier modulation is used to transmit the data. This type of modulation can then be demodulated by the same type of discriminator used for voice, although with different baseband filtering, as discussed in Chap. 11. With a Manchester-coded waveform, the energy is typically concentrated around $fT_s = 0.75$ (see Fig. 11.3). For the values of 10 kHz and $T_s = 100$ μs, then $f = 8$ kHz. According to this type of analysis, the filter for a digital FSK signal should be centered at 8 kHz.

13.6   Bit-Error Analysis Based on the
M-Branch Maximal-Ratio Combiner

To simplify bit-error analysis, and to minimize problems associated with the error function, it is best to employ a signaling scheme such as DPSK, which has a relatively simple exponential expression, as was discussed in Sec. 8.3 of Chap. 8. Also, the bit-error rate for a DPSK signal received over a fading channel is essentially the same as the bit-error rate for a coherent FSK received signal, at a given signal level. For noncoherent FSK signaling, the same signaling rate will cause approximately 3 dB of degradation in signal level, as shown in Fig. 8.14 of Chap. 8. By knowing the bit-error rate for a DPSK system with *M*-branch maximal-ratio combining, the bit-error rate for other types of signaling schemes can be predicted with a fair degree of certainty. For these reasons, an *M*-branch DPSK diversity-combiner system is the model of choice for analyzing the error rate for mobile-radio transmission in a multipath-fading environment. In studying the bit-error rate for a signal in an *M*-branch DPSK diversity-combining system, it is assumed that all signal branches are mutually independent. During analysis, the error rate for the DPSK *M*-branch diversity system will be compared with different types of correlated signals, where the correlation coefficients typically vary from 0 to 1. The analysis will also examine the word-error rate for *M*-branch DPSK diversity combining.

The average bit-error rate for a DPSK signal from an *M*-branch maximal-ratio combiner is described in the following paragraphs. The main advantage in using this type of combiner is that the cumulative probability distribution can easily be obtained by using analytical methods.

The actual difference in the cumulative probability distribution between an equal-gain type of combiner and the maximal-ratio combiner is about 1 dB, making it possible to estimate the bit-error rate for one system from the solution obtained for the other system. Figures 10.8 and 10.10 of Chap. 10 show a comparison between an equal-gain combiner and a maximal-ratio combiner. The average bit-error rate for a DFSK system with a maximal-ratio combiner operating in a mobile fading environment is

$$\langle P_e \rangle = \int_0^\infty P_e p_M(\gamma)\, d\gamma = \int_0^\infty \frac{1}{2}\, e^{-\gamma} p_M(\gamma)\, d\gamma \qquad (13.35)$$

where $p_M(\gamma)$ is as shown in Eq. (10.61):

$$p_M(\gamma) = \frac{1}{(M-1)!}\, \frac{\gamma^{M-1}}{\Gamma^M}\, \exp\left(-\frac{\gamma}{\Gamma}\right) \qquad (13.36)$$

where Γ is the average carrier-to-noise ratio for a single channel and P_e for the DPSK case is shown in Eq. (8.82). Then Eq. (13.35) becomes:

$$\langle P_e \rangle = \frac{1}{2} \left(\frac{1}{\Gamma + 1} \right)^M \tag{13.37}$$

The average bit-error rate for an M-branch maximal-ratio-combined DPSK signal is plotted in Fig. 13.9. The word-error rate depends on the basic detection scheme and on the rapidity of the fading as described in the following section.

13.7   Effects of Fading on Word-Error Rate

The average word-error rate (also called frame-error rate) depends on the rapidity of the fading, which is a function of mobile speed V. When V changes, the rapidity of the fading changes, and the average word error changes accordingly. An analysis that uses V as a variable is complicated and very difficult to solve. However, an appropriate solution can

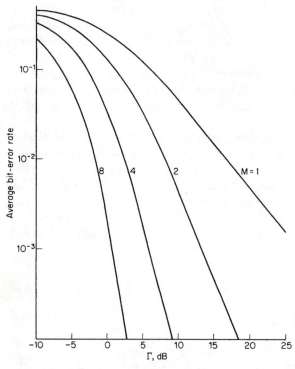

Figure 13.9 Average bit-error rate for M-branch maximal-ratio-combined DPSK signaling.

be obtained for the two extreme conditions: the fast-fading condition and the slow-fading condition [3].

Fast-fading condition

In the fast-fading condition ($V \to \infty$), the fading is assumed to be fast enough that the cnr's of two adjacent bits are independent and therefore the bit errors of individual bits are mutually independent. Equation (13.37) can be used to calculate the error probability for each bit. Given a word length of N bits, the probability that m bits will be in error can be obtained through binomial expansion:

$$p(N, m) = \binom{N}{m}(1 - \langle P_e \rangle)^{N-m} \langle P_e \rangle^m \qquad (13.38)$$

where

$$\binom{N}{m} = \frac{N!}{m!(N-m)!} \qquad (13.39)$$

The difference between Eqs. (13.31) and (13.38) is that Eq. (13.31) considers only one particular possibility, whereas Eq. (13.38) considers (N m) possibilities. If no error-correcting codes are used, then one or more errors in a word of N bits will cause the word to be in error. The word-error rate for a noncorrecting code can be expressed:

$$P_{ew} \text{ (no correcting code)} = 1 - p(N, 0) \qquad (13.40)$$

The calculation of word-error rate, using Eq. (13.40) for a 22-b word, is shown in Fig. 13.10 for a single channel, 2-branch and 4-branch combiners operating under fast-fading conditions. If correcting codes are used, the word-error rate decreases as the number of error corrections, t, increases. The word-error rate for an error-correcting code can be expressed:

$$P_{cw} \text{ (}t \text{ error correcting)} = 1 - p(N, 0) - \sum_{m=1}^{t} p(N, m) \qquad (13.41)$$

where $p(N, m)$ is a function of $\langle P_e \rangle$.

Slow-fading condition

For a slow-fading condition, the fading is assumed to be such that the carrier-to-noise ratio throughout the length of the word does not change significantly. For example, at a carrier frequency of 850 MHz and a vehicle speed of 60 mi/h, the average time the carrier spends at 10 dB or more below its mean strength is about 2 ms, as shown in

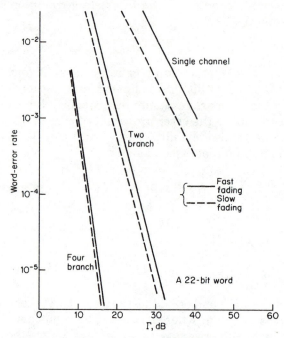

Figure 13.10   Word-error rate during fast and slow fading, for an M-branch maximal-ratio DPSK signal.

Fig. 6.7 of Chap. 6. For a signaling rate of 10 kb/s, approximately 20 b will be in the fade. At slower vehicular speeds, the number of bits in the fade will be even higher. For simplicity, let p replace P_e in Eq. (8.82):

$$p = P_e = \frac{1}{2} e^{-\gamma} \tag{13.42}$$

In the slow-fading case, it can be assumed that a word-error rate of all bits in a word can be treated as equal to a single bit-error rate, as expressed in Eq. (13.42). The probability that exactly m bits will be in error in a word of N bits is

$$p(N, m) = \binom{N}{m}(1 - p)^{N-m}p^m$$

$$= \binom{N}{m}\left[\sum_{k=0}^{N-m} \binom{N-m}{k}(-p)^k \right]p^m \tag{13.43}$$

The average word-error rate can be obtained by averaging $p(N, m)$ over γ in Eq. (13.43); the calculation is as follows:

$$\langle p(N, m) \rangle_\gamma = \int_0^\infty p(N, m) p_M(\gamma)\, d\gamma$$

$$= \binom{N}{m} \sum_{k=0}^{N-m} \binom{N-m}{k} (-1)^k \left(\frac{1}{2}\right)^{m+k} \frac{1}{[(m+k)\Gamma + 1]^M} \quad (13.44)$$

In the noncorrecting error-code case, the probability that the word is in error can be expressed:

$$P_{\text{ew}} \text{ (no correcting code)} = \sum_{m=1}^{N} \langle p(N, m) \rangle_\gamma = 1 - \langle p(N, 0) \rangle_\gamma \quad (13.45)$$

The calculation for the word-error rate, using Eq. (13.45), for a 22-b word is shown in Fig. 13.10 for single-channel, 2- and 4-branch combiners operating under slow-fading conditions. When a correcting code is used, the word-error rate decreases as the number of error-correction codes, t, increases. The expression of Eq. (13.45) then becomes:

$$P_{\text{ew}} \text{ (t error corrections)} = 1 - \sum_{m=0}^{t} \langle p(N, m) \rangle_\gamma \quad (13.46)$$

There is a significantly large difference between word-error rates for a single channel under fast- and slow-fading conditions. However, this difference diminishes as the number of branches increases, as is shown in Fig. 13.10.

13.8 Error Reduction Based on Majority-Voting Processes

To decrease the word-error rate even more, it is necessary to employ the techniques of word repetition. The word-repetition format for sending words repeatedly requires that an algorithm be developed for detecting the repeated words at the time of reception. Assuming that each word is repeated 5 times during transmission, the 5 repeats of the received message-bit stream must be aligned bit by bit. A 3-out-of-5 majority-voting process is then used to determine each valid message bit. The resulting majority-voted message words then constitute the improved message stream, on which additional single-bit error correction is used to further improve the chances for obtaining error-free content in the message stream. This error-correction technique can be applied to the message format under either fast-fading or slow-fading conditions.

Fast-fading condition

Under fast-fading conditions, and on the assumption that no correlation exists between any two repeated message bits, the improved bit-error rate, P_e', after a 3-out-of-5 majority vote, can be expressed:

$$\langle P'_e \rangle = \sum_{k=3}^{5} \binom{5}{k} \langle P_e \rangle^k (1 - \langle P_e \rangle)^{5-k} \qquad (13.47)$$

where P_e can be found in Eq. (13.37). Each bit can be considered an individual event after a 3-out-of-5 majority vote. The average improved bit-error rate of Eq. (13.47) is plotted in Fig. 13.11(a), for $M = 1$ (that is, $\rho_r = 1$), and $M = 2$, with different values of correlation coefficient ρ_r. The improved bit error rates of different majority votes are shown in Fig. 13.11(b) for comparison.

Given a word length of N bits, the probability that 1 or more bits will be in error can be expressed similarly to Eq. (13.40):

$$P_{\text{ew}} \text{ (no error correcting)} = 1 - (1 - \langle P'_e \rangle)^N \qquad (13.48)$$

The error probability expressed in Eq. (13.48) is plotted in Fig. 13.12, for $N = 40$. Given a word length of N bits, the probability that more than t bits are in error can be obtained as follows:

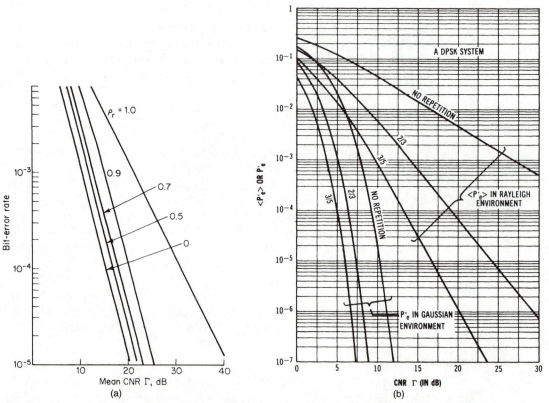

Figure 13.11 Bit-error rate (a) after 3-out-of-5 majority vote, for 2-branch DPSK signaling, (b) after different majority vote arrangements for a single branch.

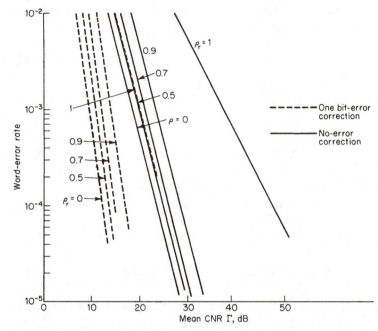

Figure 13.12 Word-error rate for a 40-b improved word message, for 2-branch DPSK signaling in a fast-fading case.

P_{ew} (t error correcting)

$$= 1 - (1 - \langle P'_e \rangle)^N - \sum_{m=1}^{t} \binom{N}{m}(1 - \langle P'_e \rangle)^{N-m}\langle P'_e \rangle^m \qquad (13.49)$$

The error probability expressed in Eq. (13.49) is also plotted in Fig. 13.12 for $t = 1$ and $N = 40$.

Slow-fading condition

Under slow-fading conditions, the improved message stream, after 3-out-of-5 majority voting, is used to find the word-error rate. It is assumed that this same 3-out-of-5 majority vote on the bits in a word repeated 5 times has been taking place as if all of the bits were under the same fading conditions. The word-error rate is then obtained by averaging the cnr, γ, in a Rayleigh-fading environment. The result is the average word-error rate for a slow-fading condition.

The improved bit-error rate, P'_e, after a 3-out-of-5 majority vote, is given by:

$$P'_e = \sum_{k=3}^{5} \binom{5}{k} P_e^k (1 - P_e)^{5-k} \qquad (13.50)$$

where P_e is as shown in Eq. (13.42). Then, the word-error rate for exactly m errors in an N-bit word is:

$$p'(N, m) = \binom{N}{m}(1 - P'_e)^{N-m}P'^m_e \tag{13.51}$$

The average improved bit-error rate for exactly m errors can be expressed:

$$\langle p'(N, m)\rangle_\gamma = \int_0^\infty p'(N, m)p_M(\gamma)\, d\gamma \tag{13.52}$$

The word-error rate when no error correction is used is:

$$P_{\text{ew}} = 1 - \langle p'(N, 0)\rangle \tag{13.53}$$

The word-error rate when t error corrections are used is:

$$P_{\text{ew}} = 1 - \langle p'(N, 0)\rangle - \sum_{m=1}^{t} \langle p'(N, m)\rangle \tag{13.54}$$

The word-error probabilities based on the results of Eqs. (13.49) and (13.54) are plotted in Fig. 13.13.

Comparing differences between the word-error rates for fast fading and slow fading (cnr $\Gamma = 15$ dB) shows a difference greater than three orders of magnitude for a word length greater than 32 b. Usually the mobile speed is less than 15 mi/h when the traffic is jammed. Therefore, the word-error rate curves derived for slow-fading conditions are more suitable for error analysis.

13.9   Bit-Error Analysis Based on Correlated Signals

If a signal that is received from one branch is correlated with a signal from another branch, then the performance of the combiner becomes degraded as the correlation coefficient increases. This situation can be studied for two cases, fast fading and slow fading.

Fast-fading condition

The probability density function $p_2(\gamma)$ of a combined signal from a 2-branch maximal-ratio combiner is shown in Eq. (10.71). In the fast-fading condition, substituting $p_2(\gamma)$ into Eq. (13.35) yields the following:

$$\langle P_e\rangle = \int_0^\infty \frac{1}{2} e^{-\gamma}\, \frac{1}{2|\rho|\Gamma}\, (e^{-\gamma/(1+|\rho|)\Gamma} - e^{-\gamma/(1-|\rho|)\Gamma})\, d\gamma$$

$$= \frac{1}{2[\Gamma^2(1 - |\rho|) + 2\Gamma + 1]} \tag{13.55}$$

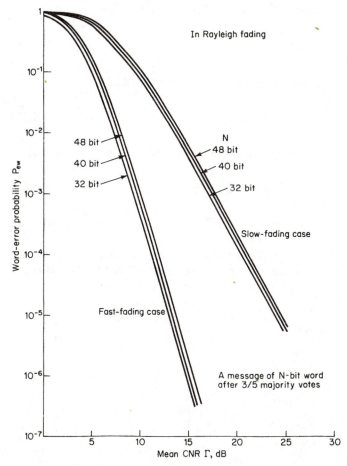

Figure 13.13   Word-error probability for an improved word message in a slow-fading case.

where ρ is the correlation coefficient of two complex Gaussian fadings received from two branches and $|\rho|$ is the magnitude of ρ. The correlation $|\rho|$ can be expressed:

$$|\rho| = \frac{E[X_1 X_2]}{E[X_1^2]} = \frac{E[Y_1 Y_2]}{E[Y_1^2]} \tag{13.56}$$

where $E[X_1 X_2]$ is as shown in Eq. (6.108) and $E[Y_1 Y_2] = E[X_1 X_2]$. Note that $r_1 = \sqrt{X_1^2 + Y_1^2}$ and $r_2 = \sqrt{X_2^2 + Y_2^2}$ are the terms for two Rayleigh envelopes. Equation (6.107) indicates that the correlation coefficient ρ_r between two Rayleigh envelopes r_1 and r_2 is:

$$\rho_r = |\rho|^2 \tag{13.57}$$

Inserting Eq. (13.57) into Eq. (13.55) gives the following:

$$\langle P_e \rangle = \frac{1}{2[\Gamma^2(1 - \sqrt{\rho_r}) + 2\Gamma + 1]} \tag{13.58}$$

When $\rho_r = 1$, the two signals are totally correlated, and

$$\langle P_e \rangle = \frac{1}{2[2\Gamma + 1]} \tag{13.59}$$

When $\rho_r = 0$, the two signals are completely uncorrelated, and thus:

$$\langle P_e \rangle = \frac{1}{2(\Gamma + 1)^2} \tag{13.60}$$

The function for Eq. (13.58), with different values of ρ_r, is plotted in Fig. 13.14.

Calculation of the word-error rate for a system employing word repetition during transmission can easily be performed in the following

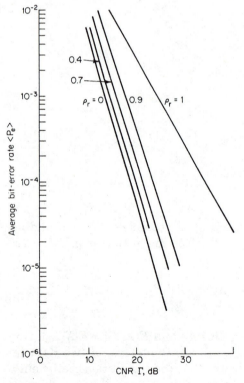

Figure 13.14 Average bit-error rate for a 2-branch combined correlated signal.

manner. Assume that a word-repetition rate of 5 times is used, and that a bit-by-bit, 3-out-of-5 majority-voting process is applied in the receiving process to determine the validity of each message bit, as previously described in Sec. 13.8. It is therefore necessary to first substitute Eq. (13.58) into Eq. (13.47) to obtain the improved bit-error rate $\langle P'_e \rangle$. Then, by substituting $\langle P'_e \rangle$ into Eq. (13.49), the word-error rate can be found. The word-error rates for a 40-b word without error correction and with 1-b error correction are both shown in Fig. 13.12, for different values of correlation coefficient ρ.

Slow-fading condition

For the slow-fading condition, the same procedure described in Sec. 13.8 can again be used, except that this time, p_2 as shown in Eq. (13.52) will be replaced by $p(\gamma)$, as expressed in Eq. (10.71). When $\rho_r = 1$ (corresponding to a single channel), the word-error rate, after 3-out-of-5 majority voting, should agree with the curves shown in Fig. 13.13 for the slow-fading case.

13.10   Irreducible Error Rate Due to Random FM

DPSK system

For a DPSK System, Eq. (13.37) is the average bit-error rate assuming that time delay T shown in Fig. 8.13 is negligible while the received signal is in a fading environment. When the time delay T is not negligible, the two complex Gaussian signals z_1 and z_2, separated by T, will have a correlation coefficient equivalent to that expressed in Eq. (6.108) in Chap. 6, which can be expressed:

$$\rho = \rho_{z_1 z_2^*}(T) = \frac{E[z_1 z_2^*]}{E[z_1 z_1^*]} = J_0(\beta V T) = |\rho|$$

$$= J_0\left(2\pi \frac{V}{\lambda} T\right) = J_0(2\pi f_m T) \leq 1 \qquad (13.61)$$

where z_1 and z_2 in a fading environment have the same expression as shown in Eq. (8.81) with $t + T$ replacing t for z_2.

The pdf of v, $v = \mathrm{Re}\ [z_1 z_2^*]$, given that a mark M is transmitted for $v < 0$ is [4]:

$$p(v \mid M) = \frac{1}{2(\Gamma^2 + 1)\sigma_n^2} \exp\left[\frac{v/\sigma_n^2}{\Gamma(1 - |\rho|) + 1}\right] \qquad v < 0$$

where $2\sigma_n^2$ is the noise power and $|\rho|$ is given in Eq. (13.61). The average error probability can be found by integrating $p(v\,|\,M)$ over v from $-\infty$ to 0, as

$$\langle P_e \rangle = \int_{-\infty}^{0} p(v\,|\,M)\,dv = \frac{1 + \Gamma(1 - |\rho|)}{2(\Gamma + 1)} \tag{13.62}$$

The term $\langle P_e \rangle$ can be separated into two parts, as follows:

$$\langle P_e \rangle = \frac{1}{2(\Gamma + 1)} + \frac{\Gamma(1 - |\rho|)}{2(\Gamma + 1)} \tag{13.63}$$

The first part is the same as for a single channel, as previously stated in Eq. (13.37) for $M = 1$. The second part is an additional error caused by the correlation coefficient $|\rho|$. When $|\rho| = 1$, the second part is diminished. When the average cnr of a single branch becomes very large, i.e., when $\Gamma \gg 1$, then the error rate of the second part becomes the irreducible error rate due to random FM and can be expressed:

$$\langle P_e \rangle_{\text{rfm}} = \frac{\Gamma(1 - |\rho|)}{2(\Gamma + 1)} \rightarrow \frac{1 - |\rho|}{2} \qquad \Gamma \gg 1 \tag{13.64}$$

where $|\rho|$ can be approximated by

$$|\rho| = J_0(2\pi f_m T) \approx 1 - \left(\frac{2\pi f_m T}{2}\right)^2 \tag{13.65}$$

when the pulse duration T becomes small.

Substituting Eq. (13.65) into Eq. (13.64) yields the following:

$$\langle P_e \rangle_{\text{rfm}} \approx \frac{1}{2} (\pi f_m T)^2 \tag{13.66}$$

Digital FM (FSK) system

For digital FM systems, the irreducible error due to random FM can be expressed:

$$\langle P_e \rangle_{\text{rfm}} = \langle P(\dot{\psi}_{\text{rfm}} \leq -2\pi f_d) \rangle \tag{13.67}$$

where f_d is the peak deviation of the FM system, or where $2f_d$ is the separation between two keying frequencies. For the average cnr, $\Gamma \gg 1$ implies that the error is due only to $\dot{\psi}_{\text{rfm}} \leq -2\pi f_d$. The probability density function of $\dot{\psi}_{\text{rfm}}$, $p(\dot{\psi}_{\text{rfm}})$, is as previously shown in Eq. (7.68). Thus, averaging Eq. (13.67) over γ gives:

$$\langle P_e \rangle_{\text{rfm}} = \frac{1}{2} \left[1 - \sqrt{2}\, \frac{f_d}{f_m} \left(1 + 2\, \frac{f_d^2}{f_m^2} \right)^{-1/2} \right] \tag{13.68}$$

where $f_m = V/\lambda$. For the case of $f_d \gg f_m$, Eq. (13.68) reduces to

$$\langle P_e \rangle_{\text{rfm}} \approx \frac{1}{8} \left(\frac{f_m}{f_d} \right)^2 \tag{13.69}$$

Example 13.4   On the basis of the following given parameters, find the irreducible error rate due to random FM, in both an FSK and a DPSK system. The receiving mobile unit is traveling at 40 km/h and the carrier frequency is 800 MHz. FSK frequency deviation is 12 kHz. The DPSK signaling rate is $1/T = 6$ kHz.

solution   Given $f_d = 12$ kHz and $f_m = V/\lambda = 29.6$ Hz, the FSK irreducible error rate from Eq. (13.69) becomes:

$$\langle P_e \rangle_{\text{rfm}} = \frac{1}{8} \left(\frac{29.6}{12,000} \right) = 3 \times 10^{-4} \qquad \text{FSK system} \tag{E13.4.1}$$

Given $T = 1/(6 \text{ kHz}) = 1.66 \times 10^{-4}$ s and $f_m = 29.6$ Hz, the DPSK irreducible error rate from Eq. (13.66) becomes:

$$\langle P_e \rangle_{\text{rfm}} = \frac{1}{2} (\pi \times 29.6 \times 1.66 \times 10^{-4})^2 = 1.19 \times 10^{-4} \tag{E13.4.2}$$

Note that the irreducible error rate due to random FM for the DPSK system is lower than that for the FSK system.

13.11   Irreducible Error Rate Due to Frequency-Selective Fading

As the time-delay spread increases and therefore is no longer negligible, the irreducible error rate due to frequency-selective fading becomes more noticeable [5]. In Chap. 1, Sec. 1.3, the effects of frequency-selective fading as a result of the multipath-propagation phenomenon were discussed. Bello and Nelin [6–8] have calculated the error probability due to frequency-selective fading and Gaussian noise for square-pulse FSK and DPSK systems with diversity combining. Bailey and Lindenlaub [9] have extended Bello and Nelin's results to include DPSK systems with fully raised-cosine signaling pulse shaping for both Gaussian time-delay-spread function and rectangular time-delay-spread function. The irreducible error rate is an error rate which will not be reduced by further increasing the transmitting power. The irreducible error rate can be obtained by letting the average cnr become infinite in the expression of error probability and solving for error probability. The following paragraphs describe Bello and Nelin's work, with examples of their techniques for obtaining the irreducible error rate as demonstrated by Bailey and Lindenlaub [9].

If a received signal $y(t)$ at the input of a receiver can be represented in the frequency domain by

$$Y(f) = S(f) + N(f) \tag{13.70}$$

where $S(f)$ is the desired signal spectrum and $N(f)$ is the noise spectrum, then it can be assumed that the signal has arrived via a frequency-flat and time-flat channel. Frequency-flat fading usually occurs when the bandwidth of the digital pulse is considerably less than the correlation bandwidth of the channel. Time-flat fading usually occurs when the digital pulse width is considerably less than the correlation time of the fading.

In the case of frequency-selective fading, the received-signal spectrum can be expressed:

$$Y(f) = S(f)T(f) + N(f) \tag{13.71}$$

where $T(f)$ is a random transfer function assumed to be a complex-valued Gaussian process attributable to the medium, since the complex Gaussian characteristics of the channel have been taken for granted. For flat-frequency fading where $T(f) = 1$, Eqs. (13.70) and (13.71) are identical. The autocorrelation function of $T(f)$ is often referred to as the "frequency-correlation function" $R(\Omega)$ and can be defined:

$$R(\Omega) = \langle T^*(f)T(f + \Omega)\rangle \tag{13.72}$$

Equation (13.72) can also be used to determine the error probabilities resulting from frequency-selective fading in certain types of systems which are discussed in the following sections. The irreducible error rate can be deduced from those results.

13.12   Error Probability for Incoherent Matched-Filter Receivers

The block diagram for an incoherent matched-filter receiver with its functional equivalent is shown in Fig. 13.15. The term $y_k(t)$ represents the complex envelope of the kth branch of the signal received by an

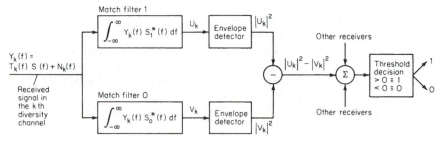

Figure 13.15   System of incoherent-matched-filter receiver.

M-branch diversity system. The terms $s_0(t)$ and $s_1(t)$ represent the complex envelope for the corresponding spectra of the mark ($S_0(f)$) and space ($S_1(f)$) waveforms, respectively, in a noise-free time interval, $0 \leq t \leq T$. The outputs of the two matched filters are given by:

$$U_k = \int_0^T y_k(t)s_1^*(t) \, dt = \int_{-\infty}^{\infty} Y_k(f)S_1^*(f) \, df \tag{13.73}$$

$$V_k = \int_0^T y_k(t)s_0^*(t) \, dt = \int_{-\infty}^{\infty} Y_k(f)S_0^*(f) \, df \tag{13.74}$$

where

$$Y_k(f) = S(f)T_k(f) + N_k(f) \tag{13.75}$$

The random variable q, the combined input of M receivers to the "threshold decision" block of Figure 13.15, is expressed by:

$$q = \sum_{k=1}^{M} (|U_k|^2 - |V_k|^2) \tag{13.76}$$

where U_k and V_k are normal complex random variables; i.e., a linear operation in a Gaussian network always produces a Gaussian result. The probability density function $W(q)$ of q is shown in Ref. 5, and the desired error probabilities $P_r(q > 0)$ and $P_r(q < 0)$ of the system can be expressed:

$$P_r(q > 0) = \int_0^{\infty} W(q) \, dq$$

$$= \left(\frac{1+\rho}{2+\rho}\right)^M \sum_{m=0}^{M-1} \frac{1}{(2+\rho)^m} C_m^{M-1+m} \tag{13.77}$$

$$P_r(q < 0) = \int_{-\infty}^{0} W(q) \, dq$$

$$= \frac{1}{(2+\rho)^M} \sum_{m=0}^{M-1} \left(\frac{1+\rho}{2+\rho}\right)^m C_m^{M-1+m} \tag{13.78}$$

where M is the number of diversity branches and ρ is defined:

$$\rho = \frac{2(m_{11} - m_{00})}{\sqrt{(m_{11} - m_{00})^2 + 4(m_{11}m_{00} - |m_{10}|^2)} - (m_{11} - m_{00})} \tag{13.79}$$

In this definition of ρ,

$$m_{11} = \langle |U_k|^2 \rangle$$

$$m_{10} = \langle U_k^* V_k \rangle \qquad (13.80)$$

$$m_{00} = \langle |V_k|^2 \rangle$$

and $\qquad C_m^n = \dfrac{n!}{m!(n-m)!} \qquad (13.81)$

Note that ρ is a function of snr γ. The procedures for finding the values of m_{00}, m_{10}, and m_{11} are described in greater detail in Bello and Nelin [6]. The two error probabilities of major concern are:

$$p_0 = P_r[q > 0 \mid s(t) = s_0(t); 0 < t < T] \qquad (13.82)$$

$$p_1 = P_r[q < 0 \mid s(t) = s_1(t); 0 < t < T] \qquad (13.83)$$

Because of the frequency-selective behavior of the channel, inter-symbol interference is introduced in Eqs. (13.82) and (13.83). In actual practice, as frequency-selective fading is gradually introduced into an otherwise flat-fading channel, only values of $s(t)$ for time instants adjacent to the interval $0 \le t \le T$ are effective in producing intersymbol interference. Intersymbol interference can be considered an interaction between adjacent pulses. For example, to find the probability p_0 for a 1 to be printed when a 0 has been transmitted, it is first necessary to calculate the probabilities for a 1 to be printed when 000, 101, 100, and 001 have been transmitted. That is,

$$p_0 = \frac{1}{4} \{p_{000} + p_{101} + p_{100} + p_{001}\} \qquad (13.84)$$

Similarly

$$p_1 = \frac{1}{4} \{p_{111} + p_{010} + p_{011} + p_{110}\} \qquad (13.85)$$

where

$$p_{a0c} = P_r[q > 0 \mid s(t) = s_{a0c}(t)] \qquad (13.86)$$

$$p_{a1c} = p_r[q < 0 \mid s(t) = s_{a1c}(t)] \qquad (13.87)$$

Equations (13.86) and (13.87) can be obtained from Eqs. (13.77) and (13.78), respectively. Examples for calculating p_0 and p_1 follow.

Example 13.5 The following method can be used to find the irreducible error rate for an M-branched noncoherent FSK signal. Bello and Nelin [8] calculated the irreducible error probability for continuous-phase FSK signaling with frequency separation between mark and space signals of $1/T$, where T equals the square-pulse width. In this case, a Gaussian random channel is assumed. Since the symmetry of the FSK signaling is satisfactory, it is only necessary to evaluate p_1. The values of ρ that are involved when the snr $\gamma \to \infty$ are:

$$\rho_{111} = \infty \tag{E13.5.1}$$

$$\rho_{010} = \frac{\pi^2}{12d^4}\left(1 - \frac{32}{3}\frac{d^3}{\pi\sqrt{\pi}}\right) \tag{E13.5.2}$$

$$\rho_{011} = \rho_{110} = \frac{\pi^2}{5d^4}\left(1 - \frac{16}{3}\frac{d^3}{\pi\sqrt{\pi}}\right) \tag{E13.5.3}$$

where $d = 1/B_cT$. Figure E13.5 shows the curve that results from substituting Eqs. (E13.5.1) through (E13.5.3) into Eq. (13.78), then applying the result to Eq. (13.87),

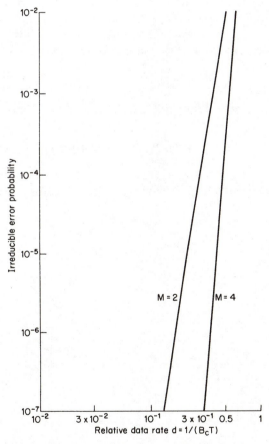

Figure E13.5   Irreducible error probability as a function of relative data rate for FSK. *(From Ref. 6.)*

and then substituting that result into Eq. (13.85). The curve is for a postdetection, maximal-ratio-combined signal. For delay spread $\Delta = 5$ μs, $B_c = 1/2\pi\Delta = 32$ kHz, and assuming that $1/T = 16$ kHz, then $d = 1/B_cT = 0.5$. The irreducible error rate therefore becomes 8×10^{-3} for a 2-branch and 8×10^{-4} for a 4-branch signal.

13.13   Error Probability for Differentially Coherent Receivers

The block diagram for a differentially coherent matched-filter receiver is shown in Fig. 13.16.

Assume that only $s_1(t)$ and $s_0(t)$ are transmitted. In a differentially coherent receiver, the transmitted bit is encoded to indicate either a change or a lack of change in successive adjacent pulses. In this context, the transmission of a pair encoded $s_1(t) + s_0(t + T)$ or $s_0(t) + s_1(t + T)$ denotes a 0, whereas the transmission of a pair encoded $s_1(t) + s_1(t + T)$ or $s_0(t) + s_0(t + T)$ denotes a 1. The term $y_k(t)$ represents the complex envelope of the kth branch of the signal received by a diversity system. The complex envelope of the sampled undelayed output of the matched filter receiver is:

$$V_k = \int_0^T y(t)[s_1^*(t) - s_0^*(t)] \, dt$$

$$= \int_{-\infty}^{\infty} [S(f)T_k(f) + N_k(f)][S_1^*(f) - S_0^*(f)] \, df \qquad (13.88)$$

The sampled output of the delayed matched-filter receiver is:

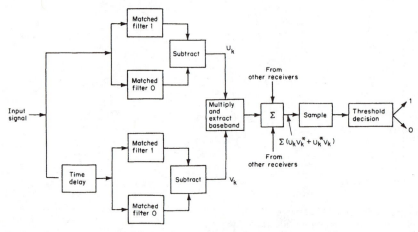

Figure 13.16   System of differentially coherent matched-filter receiver.

$$U_k = \int_0^T y_k(t - T)[s_1^*(t) - s_0^*(t)] \, dt$$

$$= \int_{-\infty}^{\infty} e^{-j2\pi fT}[S(f)T_k(f) + N_k(f)][S_1^*(f) - S_0^*(f)] \, df \qquad (13.89)$$

The random-variable input to the "threshold decision" block of Fig. 13.16 is expressed:

$$q = \sum_{k=1}^{M} [U_k^* V_k + U_k V_k^*] \qquad (13.90)$$

The procedure for obtaining $P_r(q < 0)$ and $P_r(q > 0)$ is the same as that followed in Eqs. (13.77) and (13.78), except that the parameter ρ is expressed:

$$\rho = \frac{2(m_{10} + m_{10}^*)^2}{(m_{10} + m_{10}^*)^2 + 4(m_{11}m_{00} - |m_{10}|^2) - (m_{10} + m_{10}^*)} \qquad (13.91)$$

All values of m are as defined in Eq. (13.80), except that for this application, $m_{10}^* = \langle U_k V_k^* \rangle$ and the two error probabilities of major concern, with no inter-symbol-interference effect, are:

$$p_0 = \frac{1}{2} P_r[q > 0 \mid s(t) = s_1(t + T) + s_0(t); -T < t < T]$$

$$+ \frac{1}{2} P_r[q > 0 \mid s(t) = s_0(t + T) + s_1(t); -T < t < T] \qquad (13.92)$$

and

$$p_1 = \frac{1}{2} P_r[q < 0 \mid s(t) = s_0(t + T) + s_0(t); -T < t < T]$$

$$+ \frac{1}{2} P_r[q < 0 \mid s(t) = s_1(t + T) + s_1(t); -T < t < T] \qquad (13.93)$$

In determining the effects of intersymbol interference, only the pulses adjacent to the interval $-T < t < T$ are considered, since the degree of frequency-selective fading is small. The probability p_0 that a 1 is printed when a transition 0 is transmitted, with the effect of intersymbol interference, is:

$$p_0 = \frac{1}{8} \sum_{a=0}^{1} \sum_{b=0}^{1} \sum_{c=0}^{1} p_{ab\bar{b}c} \qquad (13.94)$$

where

$$p_{ab\bar{b}c} = P_r[q > 0 \mid s(t) = s_{ab\bar{b}c}(t)] \qquad (13.95)$$

and where $s_{ab\bar{b}c}$ is equal to that portion of $s(t)$ corresponding to the transmission of the bit sequence $ab\bar{b}c$, assuming that in the absence of selective fading the bit sequence $b\bar{b}$ is actually received during the $-T < t < T$ interval. The term $\bar{b}$ is the binary complement of b; that is, $\bar{1} = 0$ and $\bar{0} = 1$.

For purposes of analysis, the probability that a 0 is printed when no transition is transmitted, as a result of intersymbol interference, is given by:

$$p_1 = \frac{1}{8} \sum_{a=0}^{1} \sum_{b=0}^{1} \sum_{c=0}^{1} p_{abbc} \qquad (13.96)$$

where

$$p_{abbc} = P_r[q < 0 \mid s(t) = s_{abbc}(t)] \qquad (13.97)$$

and where $s_{abbc}(t)$ has an interpretation which is obviously analogous to $s_{ab\bar{b}c}(t)$.

Example 13.6 The procedure for finding the irreducible error rate for an M-branch DPSK signal, as discussed in the preceding section, can be summarized as follows.

Given the pulse shape $S(f)$ and the transfer function $T(f)$, the values of V_k and U_k can be found. Once V_k and U_k are known, the signal q for each branch can be derived as shown in Eq. (13.90). When all branch values of q_k are combined, the total sum of the signal branches q can be used to make a determination of the irreducible error rate. The probability functions $P_r(q > 0)$ and $P_r(q < 0)$, are shown in Eqs. (13.77) and (13.78), respectively, as a function of ρ, expressed in Eq. (13.91). The conditional probabilities for different sequences of combinations can be derived from $P_r(q < 0)$ and $P_r(q > 0)$. Finally, the error probabilities resulting from intersymbol interference, p_1 and p_0, in terms of conditional error probabilities, are shown in Eqs. (13.96) and (13.94), respectively. Bello and Nelin [8] have shown that the total system-error probability in terms of the conditional error probabilities for an M-branch DPSK system is:

$$P_e = \frac{1}{8} [p_{0101} + p_{1100} + 2p_{0100} + p_{0110} + p_{1111} + 2p_{0111}] \qquad (E13.6.1)$$

In Eq. (E13.6.1) the cnr γ becomes infinite in determining the irreducible error rate.

Figure 11.4 of Chap. 11 illustrates the spectra for rectangular and raised-cosine pulses. If these two pulse shapes $S(f)$ are used in the signaling scheme, the transfer function $T(f)$ due to the mobile-radio medium can be represented by the frequency-correlation function shown in Eq. (13.72). The two frequency-correlation functions can be assumed to be Gaussian and sinc functions, expressed:

$$R_1(f) = 2\sigma^2 \exp\left(-\frac{4f^2}{B_c}\right) \qquad \text{for a GF channel} \qquad (E13.6.2)$$

and $\qquad R_2(f) = 2\sigma^2 T_m \, \text{sinc} \, 2fT_m \qquad \text{for an SF channel} \qquad (E13.6.3)$

where B_c is the coherence bandwidth, σ^2 is the average power, and the scaling factor between the two frequency correlation functions is

$$\frac{T_m}{T} = \frac{0.7}{B_c T} \tag{E13.6.4}$$

The following three cases have been considered [10]:

1. Square-pulse signaling over a Gaussian-function (GF) channel
2. Square-pulse signaling over a sinc-function (SF) channel
3. Raised-cosine-spectrum signaling over a sinc-function (SF) channel

The irreducible error probabilities for the three different types of channel-signal combinations calculated by Lindenlaub and Bailey [10] are plotted in Fig. E13.6.

In the case of no diversity, $M = 1$, the raised-cosine pulse and the SF channel are assumed, and the time-delay spread is given as 0.5 μs. The value of d can be calculated as follows:

$$d = \frac{T_m}{T} = R_b T_m = R_b \times 5 \times 10^{-7} \tag{13.98}$$

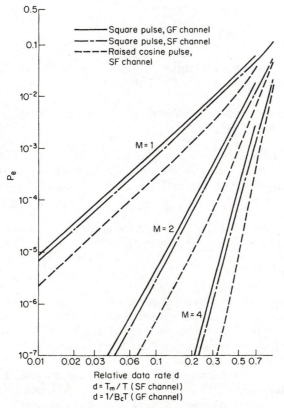

Figure E13.6 Comparison of irreducible-error probabilities due to frequency-selective fadings for the three channel-signal combinations investigated. (From Ref. 10.)

In the case where a bit-error rate (ber) of 10^{-4} is required, then the maximum bit-rate R_b is at $d = 0.06$, expressed

$$R_b \leq \frac{0.06}{5 \times 10^{-7}} = 12 \text{ kHz} \tag{13.99}$$

Figure E13.6 provides three examples of (1) how the use of diversity branches helps to increase the maximum bit-rate, (2) how different types of signaling affect the irreducible error rate, rather than different kinds of channels, and (3) how different types of signals affect performance. In conclusion, it appears that there is an optimum signal that should be used for frequency-selective channels, and the signal of choice is the raised-cosine pulse.

Comparing the plot of $d = 0.1$ with that of $M = 2$, from Fig. E13.5 for the FSK case and from Fig. E13.6 for the DPSK case, it can be shown that for a given bit rate, an FSK system can tolerate a higher transmission bit-error rate than can a DPSK system in a frequency-selective medium.

Problem Exercises

1. What are the basic differences in obtaining the word-error rate between the fast-fading case and a slow-fading case?

2. Does the mobile speed affect the bit-error rate over a given distance, or over a given period of time?

3. Assuming that the bit-error rate is 10^{-3}, what would the word-error rate be for a 22-b word transmitted through a Rayleigh-fading environment after a 2-out-of-3 majority-vote algorithm has been applied? A mobile speed of 15 mi/h is assumed.

4. Assuming that the bit-error rate is 10^{-2}, what would be the false-alarm rate for a 30-b code word transmitted through a nonfading environment at a Hamming distance of $d = 6$? What would the false-alarm rate be for a fading environment?

5. Assuming that a 20-b code word has a bit-error rate of 10^{-1} in a nonfading environment, and allowing correction of three bit errors, what would the average word-error rate be for a single channel and for a 2-branch diversity-combined DPSK signal, respectively, in a slow-fading environment?

6. If a 2-branch DPSK signal with a correlation of 0.7 between branches is used to transmit a 30-b word, what would be the word-error rate at a threshold of 10 dB below mean power in a fast-fading environment?

7. If a DPSK signal with a 32 kb/s signaling rate is transmitted by a mobile unit traveling 60 km/h at a frequency of 600 MHz, what is the irreducible error rate due to random FM? If an FSK signaling scheme is used with 16-kHz separation between the two keying frequencies, what would be the irreducible error rate? Which signal, DPSK or FSK, has the higher irreducible error rate?

8. A 2-branch FSK signal is received in a suburban area with a time-delay spread of 0.5 µs. If the signaling rate is 30 kb/s, what is the irreducible error rate?

9. A 2-branch DPSK signal is received in an urban area with a time-delay spread of 3 µs. If the signaling rate is 20 kb/s and a raised-cosine signal pulse and sinc-function channel are used, what is the irreducible error rate?

10. Name the parameters that affect the irreducible error rate for random FM, but which are not affected by frequency-selective fading.

References

1. W. W. Peterson and E. J. Weldon, Jr., *Error-Correcting Codes,* MIT Press, Cambridge, Mass., 1972, chap. 5 and p. 115.
2. Shu Lin, *An Introduction to Error-Correcting Codes,* Prentice-Hall, Englewood Cliffs, N.J., 1970, chaps. 5–6.
3. R. E. Langseth and Y. S. Yeh, "Some Results on Digital Signaling over the Mobile Radio Channel," *Microwave Mobile Symp.,* Boulder, Colo., 1973.
4. H. B. Voelker, "Phase-Shift Keying in Fading Channels," *Proc. IEE,* vol. 107, pt. B, January 1960, p. 31.
5. E. D. Sunde, "Digital Troposcatter Modulation and Transmission Theory," *Bell System Technical Journal,* vol. 43, January 1964, pp. 143–214.
6. P. A. Bello and B. D. Nelin, "The Effect of Frequency Selective Fading on the Binary Error Probabilities of Incoherent and Differentially Coherent Matched Filter Receivers," *IEEE Trans. Comm. Sys.,* vol. 11, June 1963, pp. 170–186.
7. P. A. Bello and B. D. Nelin, "Predetection Diversity Combining with Selectivity Fading Channels," *IEEE Trans. Comm. Sys.,* vol. 10, March 1962, pp. 32–42. Correction in vol. 10, December 1962, p. 466.
8. P. A. Bello and B. D. Nelin, "The Influence of Fading Spectrum on the Binary Error Probabilities of Incoherent and Differentially Coherent Matched Filter Receivers," *IEEE Trans. Comm. Sys.,* vol. 10, June 1962, pp. 160–168.
9. C. C. Bailey and J. C. Lindenlaub, "Further Results Concerning the Effect of Frequency Selective Fading on Differentially Coherent Matched Filter Receivers," *IEEE Trans. Comm.,* vol. 16, October 1968, pp. 749–751.
10. J. C. Lindenlaub and C. C. Bailey, "Digital Communication Systems Subject to Frequency Selective Fading," Purdue University School of Elec. Eng. Tech. Rep. TR-EE67-17, Lafayette, Ind., November 1967.
11. A. J. Viterbi, "Error Bounds for Convolutional Codes and An Asymptotically Optimum Decoding Algorithm," *IEEE Trans. Inf. Theory,* vol. IT-13, April 1967, pp. 260–269.
12. G. Underboeck, "Trellis-Coded Modulation with Redundant Signal Sets, Part I. Introduction, Part II. State of the Art," *IEEE Communication Magazine,* vol. 25, no. 2, February 1987, pp. 5–21.
13. D. Divsalar, M. K. Simon, and J. H. Yuen, "Trellis Coding with Asymmetric Modulation," *IEEE Trans. Commun.,* vol. Com 35, no. 2, February 1987.
14. Bernard Sklar, *Digital Communications, Fundamentals and Applications,* Section 7.10, Prentice Hall, New Jersey, 1988.

Chapter 4

Voice-Quality Analysis versus System Performance

14.1 Mobile-Telephone Voice Characteristics

The mobile-telephone voice signal is essentially a sequence of audible sounds varying at a syllabic rate, with occasional random pauses of silence. This time-varying waveform associated with human speech (200 to 3000 Hz) is not easy to characterize, since the speaker's pauses between phrases and sentences result in a concentration of speech energy in "talk spurts" of about 1-s average duration, separated by random gaps of varying lengths. Thus, the resulting speech signal consists of randomly spaced bursts of energy of random duration. As a consequence, accurate means for analyzing the characteristics of the speech signal are very difficult to define. The following paragraphs will identify and describe certain characteristics of speech waves that can be used to measure the voice-quality performance of a mobile radiotelephone system.

The primary characteristics of the speech wave are (1) the distribution of average speech power, (2) the instantaneous amplitude probability distribution of speech power, and (3) the power spectrum.

Distribution of average speech power

The distribution of average speech power among talkers closely follows the lognormal law [1]. Let S be the average speech power of a particular talker, S_r some arbitrary reference power, and $v = S/S_r$. Then the speech volume V can be expressed:

$$V = 10 \log \frac{S}{S_r}$$

$$= 10 \log v \qquad \text{dB} \qquad (14.1)$$

The probability distribution for V less than V_1, based on Eq. (2.45), can be expressed as:

$$P(V < V_1) = \int_{-\infty}^{V_1} \frac{1}{\sqrt{2\pi}\sigma_V} \exp\left(-\frac{(V - \langle V \rangle)^2}{2\sigma_V^2}\right) dV$$

$$= \frac{1}{2}\left[1 + \text{erf}\left(\frac{V_1 - \langle V \rangle}{\sqrt{2}\sigma_V}\right)\right] \tag{14.2}$$

where

$$\langle V \rangle = E[V] \tag{14.3}$$

$$\sigma_V = \{E[V^2] - (E[V])^2\}^{1/2} \tag{14.4}$$

The quality V is the power that is normally referred to as speech volume, and $\langle V \rangle$ is the power that is normally referred to as average speech volume. Also, $\bar{v} = \langle 10^{V/10} \rangle$ is the expression for the average speech power related to $\langle V \rangle$ and σ_V, and can be written as:

$$\bar{v}_{\text{dB}} = 10 \log \bar{v} = \langle V \rangle + 0.115\sigma_V^2 \tag{14.5}$$

Equation (14.5) indicates that the average speech volume $\langle V \rangle$ is 0.115 σ_V^2 dB lower than the volume $\bar{v}_{\text{dB}}$ corresponding to the average speech power $\bar{v}$. The average speech volume V_0 per talker can be measured as follows. When the reference power S_r is 1 mW, the volumes are in volume units (VU), and V_0 can be expressed as:

$$V_0 = \langle V \rangle + 10 \log \tau \qquad \text{VU} \tag{14.6}$$

where τ is the percentage of time that the talker's activity threshold is 20 dB below the average power. Typically, the activity τ of a continuous talker is found to be about 72.5 percent. Thus, Eq. (14.6) becomes:

$$V_0 = \langle V \rangle - 1.4 \qquad \text{VU} \tag{14.7}$$

The average speech-volume power, V_0, is often referred to as the long-term average volume of the talker. The distribution for talker levels less than V_0 is shown in Fig. 14.1.

Example 14.1 To analyze the relationship between the average speech power and the speech volume, expressed in volume units (VU), let S represent the speech power measured in watts or milliwatts. Then, the speech volume can be measured and expressed as follows:

$$V = 10 \log \frac{S}{1 \text{ mW}} \qquad \text{VU or dBm} \tag{E14.1.1}$$

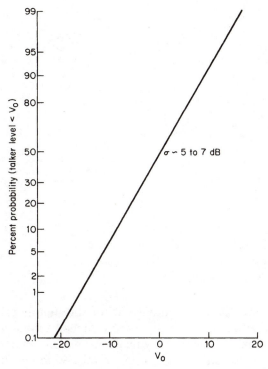

Figure 14.1 Speech-volume distribution of talker levels.

and the average speech volume is

$$\langle V \rangle = \left\langle 10 \log \frac{S}{1 \text{ mW}} \right\rangle \tag{E14.1.2}$$

The long-term average speech volume of a talker, allowing for pauses, can be expressed from Eq. (14.7):

$$V_0 = \langle V \rangle - 1.4 = E \left[10 \log \frac{S}{1.37 \text{ mW}} \right] \quad \text{VU or dBm} \tag{E14.1.3}$$

and the average speech power $\overline{V}_{\text{dB}}$ is shown in Eq. (14.5) as

$$\overline{V}_{\text{dB}} = 10 \log E \left[\frac{S}{1 \text{ mW}} \right] = \langle V \rangle + 0.115 \sigma_V^2 \quad \text{dB} \tag{E14.1.4}$$

where the value of σ_V is as expressed in Eq. (14.4).

Instantaneous amplitude probability

The instantaneous amplitude probability is virtually independent of the average speech power $\overline{V}_{\text{dB}}$ but depends on the type of microphone

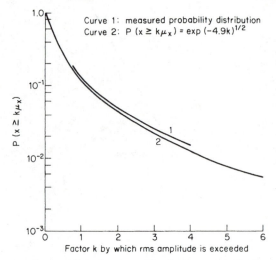

Figure 14.2 Plot of rms amplitude for a typical telephone voice signal. *(From Ref. 2.)*

and the point of measurement within the mobile-telephone circuit. It also depends on whether or not the measurement relates to natural telephone conversation. A typical curve has been obtained for telephone conversation using a commercial telephone set and is shown in Fig. 14.2. The curve plots the following approximation and is valid for $k \leq 4$:

$$P(x \geq k\mu_x) = \exp(-4.9k)^{1/2} \tag{14.8}$$

where μ_x is the rms amplitude [2].

Example 14.2 Find the probability that the instantaneous speech amplitude remains above its rms value. For this case, $k = 1$ and:

$$P(x \geq \mu_x) = 0.1$$

Note that 90 percent of the instantaneous amplitudes are below μ_x, as shown in Fig. 14.2.

Power spectrum of speech

The normalized acoustic power spectrum of speech is shown in Fig. 14.3. It is independent of the average power, and most of the power is confined to the frequency band below 1 kHz, whereas the nominal bandwidth of voice channels is about 3 kHz [2].

Knowing the voice characteristics is essential in controlling the voice quality of a mobile-radio receiver, whether the voice is in digital form

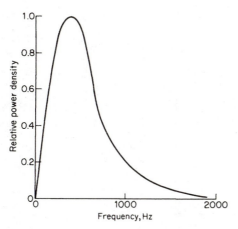

Figure 14.3   Relative power density for the acoustic-power spectrum of speech. *(From Ref. 2.)*

or in analog form. Usually the requirement for controlling the bit-error rate of digital voice signals is less than that for controlling the quality of analog voice signals; therefore, if the requirements of the analog voice signals are met, then the digital requirements are also met. However, the degree of distortion in digital form will be determined by the type of vocoder and the voice processor. For analog voice signals, the compandor is used, and the degree of distortion is determined by the type of compandor used.

14.2   Parameters for Determining Grade of System

The term "grade of system $(G(R))$" is used to describe the performance parameters of acceptability for a voice channel. The term $G(R)$ can be applied to a given system parameter and usually combines such factors as customer opinion and the distribution of system performance within the range of system parameters that have been specified.

Grade of system, $G(R)$, is defined (similar to one shown in Ref. 3) as follows:

$$G(R) = \int_{-\infty}^{\infty} P(x\,|\,R)f(x)\,dx \qquad (14.9)$$

where x is the selected parameter, which could be noise, volume, bandwidth, or any other parameter of interest. R is the opinion category (such as poor, fair, good, excellent, etc.) and $f(x)$ is the probability density function of obtaining that parameter. $P(x\,|\,R)$ is the percentage of

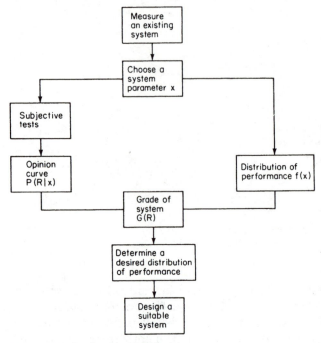

Figure 14.4 Grade-of-system decision process.

users choosing an identical level of a given parameter based on a given opinion category obtained through subjective tests, and $f(x)$ characterizes overall system performance for a given parameter. Usually, an adequate $f(x)$ can be determined by knowing $G(R)$ and $P(x|R)$.

Figure 14.4 illustrates the decision process for determining the grade of system for a given parameter of interest. To calculate the grade of system for a selected parameter, assume that $f(x)$ and $P(x|R)$ are normally distributed with x in decibel levels, with respective means of F_0 and P_0 and standard deviations of σ_F and σ_P. Then, by substituting these terms into Eq. (14.9), the grade of system $G(R)$ for an opinion category R is obtained. For example, let the term $Z(x)$ be the resulting distribution of satisfaction, as in:

$$Z(x) = x_F - x_P$$

Therefore, $Z(x)$ is also a normally distributed parameter expressed in decibels with a mean of:

$$Z_0 = F_0 - P_0 \tag{14.10}$$

and a standard deviation of:

$$\sigma_Z = \sqrt{\sigma_F^2 + \sigma_P^2} \tag{14.11}$$

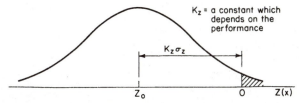

Figure 14.5   Normal probability density function of $Z(x)$.

The normal probability density function of $Z(x)$ is illustrated in Fig. 14.5.

For example, let the parameter be the system noise. As the system noise increases, the number of unfavorable opinions from customers increases. By analysis, it is obvious that when $Z(x)$ is less than 0, the noise level is less than that which is needed to ensure good or better service. When $Z(x) = 0$ the function $f(x)$ is an optimum design. Thus, it is only when $Z(x)$ is greater than 0 that the noise level becomes too great to ensure good or better service. Therefore, the following relationship holds true:

$$G(R) = P(Z(x) \leq 0) = \text{grade of system that is good or better}$$

$$= \frac{1}{\sqrt{2\pi}\sigma_Z} \int_{-\infty}^{0} \exp\left(-(Z - Z_0)^2/2\sigma_Z^2\right) dZ \qquad (14.12)$$

This integral function can be found in published mathematical tables for given normal values of Z_0 and σ_Z. In general, the fixed-telephone system goal is to provide a grade of system that is 95 percent good or better, i.e., with less than 5 percent in the fair category and a negligible number in the poor category. The mobile-telephone system goal is to match the grade of system to that of the fixed-telephone system.

Example 14.3   It is desirable to test the voice quality of a mobile-radio system, on the basis of subjective tests from a large group of listeners. Sixty percent of listeners say good voice quality is present when the noise level is at 5 dBm or below; the standard deviation at that signal level based on all listeners' opinions is 3 dB. What should the adequate value of F_0 be if σ_F of the system is 4 dB?

solution   The values $P_0 = 5$, $\sigma_P = 3$, $\sigma_F = 4$, and $P = 60$ percent are given. Then by finding values for Z_0 and σ_Z from mathematical tables, the following can be derived. Let

$$P(r \geq R) = 60\% \qquad (E14.3.1)$$

then $Z_0/\sigma_Z = -0.26$ is obtained. σ_Z can be obtained from Eq. (14.11).

$$\sigma_Z = \sqrt{3^2 + 4^2} = 5$$

Then $Z_0 = -0.26 \times \sigma_Z = -1.3$. The mean value F_0 can be found from Eq. (14.10) as

$$F_0 = Z_0 + P_0 = -1.3 + 5 = 3.7 \text{ dBm}$$

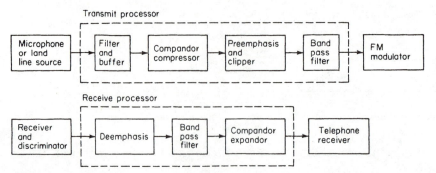

Figure 14.6 Block diagram of a typical transmit and receive processor in an FM mobile radio.

14.3 Voice Processing in an FM System

In a typical mobile-radio FM system, there are normally two processors, a transmit processor and a receiving processor, as shown in Fig. 14.6. In the transmit processor, the compandor-compressor and preemphasis functions must be evaluated. In the receive processor, the compandor-expandor and deemphasis functions must be evaluated. In the following paragraphs, preemphasis and deemphasis are discussed first, followed by a discussion of compandor compressor and expandor functions.

Preemphasis and deemphasis

As shown in Eq. (8.30), the Gaussian noise-power density spectrum at the base-band output of an FM radio system is approximately parabolic in shape in the voice-frequency region under large-cnr conditions. Therefore, if the power spectrum of the carrier signal remains a constant, P_0, over the entire band, then the noise components will contaminate the signal at the high-frequency end of the band during reception.

The effects of the noise spectrum on signal components are illustrated in Fig. 14.7. The Gaussian noise can be expressed as follows:

$$S_n(f) = N_0 \left[1 + \left(\frac{f}{f_0} \right)^2 \right] \tag{14.13}$$

where N_0 is the power density of a flat noise spectrum at the low end of f_B and f_0 is the frequency at which the flat noise spectrum and the FM triangular noise spectrum cross, as shown in Fig. 14.7. Then the preemphasis circuit must be designed to filter and shape the signal-power density spectrum to match the curvature of the noise-power density spectrum. This relationship can be expressed as follows:

$$S_s(f) = P_0 \left[1 + \left(\frac{f}{f_0} \right)^2 \right] \tag{14.14}$$

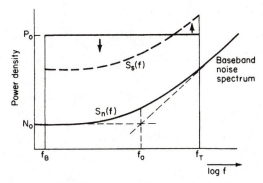

Figure 14.7   Effects of noise on signal components during reception.

This type of shaping effectively increases the total transmitted signal power over the original flat-spectrum total power, $P_0 f_T$, by the ratio

$$r = \frac{1}{f_T} \int_0^{f_T} \left[1 + \left(\frac{f}{f_0} \right)^2 \right] df = 1 + \frac{1}{3} \left(\frac{f_T}{f_0} \right)^2 \qquad (14.15)$$

The preemphasis function $S_{\mathrm{pre}}(f)$ in decibels is then

$$S_{\mathrm{pre}}(f) = 10 \log \frac{S_s(f)}{r} \qquad \mathrm{dB} \qquad (14.16)$$

The deemphasis filter is used to obtain the reverse of this operation during reception:

$$S_{\mathrm{de}}(f) = \frac{1}{S_{\mathrm{pre}}(f)} \qquad (14.17)$$

By the use of these techniques, the message can be preemphasized before modulation when the noise is absent, and deemphasized when the noise level is relative to the message after demodulation, in order to suppress the noise.

Example 14.4   What effect does noise have on a preemphasized signal?

Since the preemphasis filter shapes the signal spectrum to match the parabola of the noise spectrum, the signal-to-noise ratio will be a constant value across the entire voiceband and can be expressed:

$$\mathrm{snr} = \frac{P_0}{r N_0} \qquad (E14.4.1)$$

where r is the power ratio defined in Eq. (14.15).

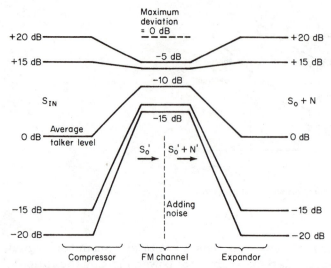

Figure 14.8   Audio levels in an FM system with 4:1 companding.

Compandor: compressor and expandor

Figure 14.8 shows the relative audio levels in a mobile-radio FM system with a 4:1 compandor [4, 5]. If a mobile talker has an audio level that is 20 dB above the level of an average talker, then the compandor will compress the signal to 5 dB below the maximum frequency deviation at 12 kHz. Then the −5-dB level with respect to the maximum deviation is equivalent to a frequency deviation at 3.8 kHz. After reception, the receiving compandor expands the signal to the original +20-dB audio level. When speech is not present, the noise is drastically reduced by companding, as shown in Fig. 14.9. This action is referred to as "receiver quieting."

Example 14.5   The characteristics of compandor performance contribute to receiver quieting. Analysis of the audio levels of Figs. 14.8 and 14.9 shows that when $S_i = 0$ dB, the compressor compresses S_i to $S'_o = -10$ dB. If an additive noise N' is introduced by the FM channel, say 5 dB below S'_o, $N' = -15$ dB. Then, during each pause in speaking, only the noise exists. The noise level is dropped from

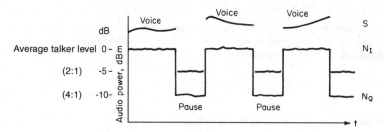

Figure 14.9   Audio output after companding.

−15 dB to −20 dB after passing through an expandor at the receiving site. This drop in noise level during a nontalk condition is referred to as receiver quieting.

Using a 4:1 compandor or a 2:1 compandor does not make any performance improvement when the ambient noise (including the interference from other systems) is low. But when the noise level is high, then, as we see from Fig. 14.9, the 4:1 compandor can make a better quieting effect than the 2:1 compandor.

14.4 Subjective Analysis of FM System Performance

The performance of a system parameter, such as noise, volume, bandwidth, or any other, can be obtained by holding all other parameters at their typical or normal values and changing only the one under investigation.

The mobile-radio FM signal-to-noise performance for a system like that shown in Fig. 14.6 is shown in Fig. 14.10 for different speeds of the mobile unit: $V = 0$, $V = 35$ mi/h, and $V = 50$ mi/h. Also, two curves are illustrated for two different combining techniques at $V = 50$ mi/h. At $V = 50$ mi/h, a cnr of 20 dB corresponds to an snr of 36 dB for a single channel with no fading. However, with fading this same 20-dB cnr corresponds to an snr of 31 dB for a 2-branch equal-gain combiner, an snr of 22 dB for a 2-branch switched combiner, and an snr of 17 dB for a single channel.

The system shown in Fig. 14.6 can be evaluated in two parts. First, an evaluation can be made on the basis of whether or not preemphasis

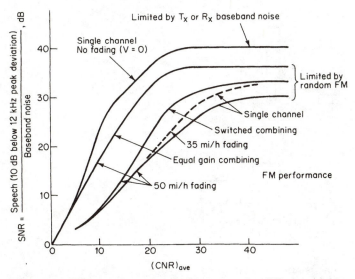

Figure 14.10   FM signal-to-noise performance using different combining techniques.

and deemphasis are used, excluding the compandor function. Second, an evaluation can be made on the assumption that preemphasis and deemphasis are included, on the basis of whether or not the compandor functions are used.

Preemphasis and deemphasis without companding

There are two methods that can be used to perform a subjective evaluation of preemphasis and deemphasis where companding is not used: (1) comparison based on a cnr reference [6] and (2) comparison based on circuit merit [7]. In making the comparison based on cnr reference, the tester disables the preemphasis and deemphasis circuits of the reference system and measures the system gain. The same system gain is then measured on the system under test with the preemphasis and deemphasis circuits enabled. These two system-gain measurements are compared and the difference is recorded. This gain difference is called the "processing gain." Figure 14.11 shows several histograms of processing gain where preemphasis and deemphasis were used for comparison against the reference system. With a Doppler frequency of 50 Hz and a bandwidth of 30 kHz, the median processing gain increases as cnr increases. The median values of processing gain versus the cnr are plotted in Fig. 14.12.

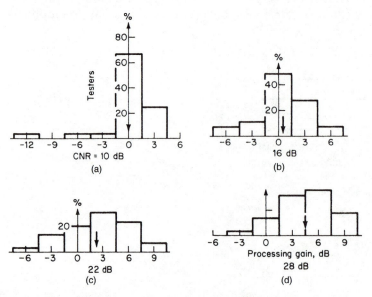

Figure 14.11 Histograms of processing gain using preemphasis and deemphasis [Doppler frequency = 50 Hz (V = 40 mi/h) and bandwidth = 30 kHz].

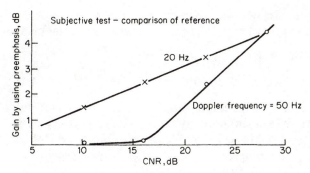

Figure 14.12   Processing gain vs. cnr.

In making the comparison based on circuit merit determined by the listeners, the circuit merit is defined in levels, as follows [7]:

CM5—Speech perfectly understandable, negligible noise (excellent).

CM4—Speech easily understandable, some noise (good).

CM3—Speech understandable with a slight effort; occasional repetitions needed for conversation clarification (fair).

CM2—Speech understandable only with considerable effort; frequent repetitions needed for intelligible conversation (poor).

CM1—Speech not understandable (unusable).

Among the five group levels, only two are significant for comparison purposes: CM3 or better and CM4 or better, which are compared in Fig. 14.13.

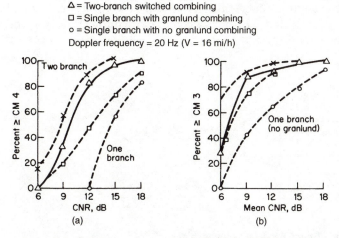

Figure 14.13   System-performance comparisons based on circuit merit CM4 vs. CM3.

The following criteria can be used to state goals for cnr's that will achieve a circuit merit that will satisfy 90 percent or more of the users:

$$\overline{cnr} = 12 \text{ dB} \qquad \text{for } 90\% \geq \text{CM3}$$

$$\overline{cnr} = 18 \text{ dB} \qquad \text{for } 90\% \geq \text{CM4} \qquad (14.18)$$

for a single branch with Granlund combining. The CM's are purely subjective ratings; for a 12-dB cnr, 90 percent said "CM3 or better," for an 18-dB cnr, 90 percent said "CM4 or better." The statistics from CM are used to obtain the MOS (mean opinion score), which is used to evaluate the speech coders or system quality.

Preemphasis and deemphasis with and without companding

In performing a subjective evaluation of the compandor function, it is always assumed that preemphasis and deemphasis are used in the system, as in Fig. 14.6. The first requirement is to establish an equivalent frequency deviation for each talker, based on the test configuration illustrated in Fig. 14.14.

In Fig. 14.14, when the switch is set to position A, the VU level of the speaker is recorded on the VU meter. When the switch is set to position B, the attenuator is adjusted to obtain a corresponding level on the VU meter for a 1-kHz modulating frequency. The phase of the modulated signal at the FM output is expressed as:

$$\phi_m(t) = \beta' \cos \omega_m t = \beta' \cos (2\pi \times 100t) \qquad (14.19)$$

where β' is the equivalent deviation for a 1-kHz modulation frequency. When a speaker talks louder, β' increases. By measuring this response, the equivalent deviation for each talker, based on the reference level of the 1-kHz signal, is obtained. This β' is used to determine the snr at the baseband level for each talker.

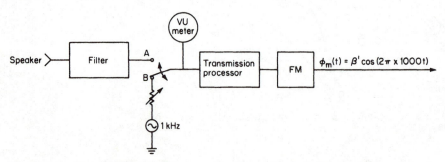

Figure 14.14   Test configuration for measuring frequency deviation per talker.

Subjective evaluation [7] consists of comparisons of circuit merit, with and without compandors, based on the data plotted in Fig. 14.15. Here, curve 1*N* represents the subjective test results [7] for a system with no companding and no combining. By comparison, for curve 4*D* for a system with 4:1 companding and equal gain combining, a cnr of 12 dB is obtained, which is a quality of performance equal to CM4 or better for 95 percent of the users at the velocity of 35 mph (56 kmph).

When the mobile speed increases, all the curves shown in Fig. 14.15 shift toward the right and the voice quality is degraded. We may use the same criterion that the voice quality is acceptable if 75 percent of the people say "good" or "excellent." The similar curve shown in Fig. 14.15 indicates that the *C/N* or *C/I* is around 18 dB for the speed of 60 mph (96 kmph). The value of *C/I* = 18 dB is based on the voice quality at a mobile speed at 60 mph. For designing a system, we are considering the worst case. Therefore, *C/I* = 18 dB at the boundary of each cell is the value we use to find the *D/R* ratio shown in Fig. I.2 of Introduction.

Example 14.6   The performance of voice quality due to vehicular speeds is demonstrated in this example. The curves plotted in Fig. 14.13(*a*) are the result of subjective tests performed on the system shown in Fig. 14.6, but without the compandor. The comparison is based on different methods of combining and a vehicular speed of *V* = 16 mi/h. The curves in Fig. 14.15 are also the result of subjective tests based on companding, with and without equal-gain combining, and a vehicular speed of *V* = 35 mi/h. The comparison between these two figures shows that voice quality improves at lower vehicular speeds, and with companding.

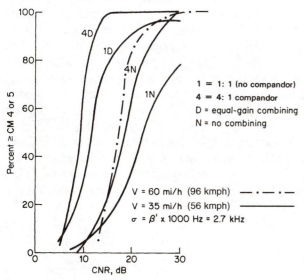

Figure 14.15 System-performance comparisons based on circuit merit CM4 or CM5.

The performance of an FM mobile-radio system can be improved through the use of techniques incorporated into the functional design of the equipment. These techniques have been identified and described as preemphasis during transmission, deemphasis during reception, compandor compression and expansion, and equal-gain combining. Comparisons based on subjective evaluation of test results show favorable increases and improvements in circuit merit, approaching CM3 or better, when one or more of these improvement techniques are used.

14.5   Digital Processing of Voice Signals

For a digital mobile-radio system, the analog voice must be converted into a digital form before it can be transmitted. Though the operation of an analog-to-digital converter is simple enough in principle, the distortion and granulation noise are of major concern. There are many different methods for converting analog voice signals to digital voice signals, such as linear pulse-code modulation (PCM), differential pulse-code modulation (DPCM), delta modulation (DM), adaptive delta modulation (ADM), linear predictive coding (LPC), and other methods. Among these different methods, the primary asset of the DM method is that it is extremely simple in hardware design, though the required bit-transmission rate is usually much higher than that of PCM for a given quality. Because of its simple hardware design, the DM technique is most attractive for mobile-radio systems. ADM is also an attractive option, since it can be used to improve voice quality by varying the step size as a function of signal behavior. For signals rapidly increasing in level, the step size is increased to avoid slow-rate limiting. For small signals, the step size is correspondingly decreased to minimize granulation noise [8–10]. The LPC method is an alternative to the representation of vocal-tract information by spectral smoothing, which is based on estimating the parameters of a vocal-tract model. LPC is developed and well matched to the current state of the art in micro-processors and other digital technology [11, 12]. In digital mobile-radio systems, the two digital voice schemes, ADM and LPC, are more attractive than the others. After analog voice is converted to digital form, the analysis of the system performance on voice can be based only on the bit-error rate, which can be treated the same as signaling was in Chaps. 8 and 13.

Problem Exercises

1.   If the average speech volume is 10 mW and an activity of 90 percent characteristic of a continuous fast talker is assumed, what is the long-term average speech volume V_0 and what is the probability that the talker's level is below V_0?

2. Predict the threshold level in terms of rms value when an instantaneous speech amplitude is above the threshold level 1 percent of the time.

3. Use a normal distribution model to solve the problem of monthly service of renting a mobile phone. If 50 percent of the customers prefer $60 per month, the standard deviation of the customers' opinion is $30. What price would please 95 percent of the customers?

4. Assume the voiceband is from 200 Hz to 3000 Hz and the frequency of flat noise is 500 Hz. What is the power ratio r after the voice is preemphasized, and what is the improvement in snr?

5. A subjective test is given for the distance range of a newly designed communication system. It uses the received signal level as a parameter and the customers judge the system based on its voice quality at a given signal level. The result is that 70 percent of the customers say the system is good or excellent at a mean received power of 0 dBm with $\sigma_P = 4$ dB due to other customers. The system mean received power based on an accepted voice quality specification is 1 dBm. What is the distribution of satisfaction, $Z(x)$, and what will be the system characteristics function $f(x)$ of the received power x?

References

1. H. B. Bolbrook and J. T. Dixon, "Load Rating Theory of Multi-channel Amplifiers," *Bell System Technical Journal,* vol. 18, October 1939, p. 624.
2. E. D. Sunde, *Communication System Engineering Theory,* Wiley, New York, 1969, pp. 65–66.
3. Bell Telephone Laboratories, *Transmission System for Communications,* 4th ed., Western Electric Company, Inc., 1970, p. 45.
4. G. A. Arredondo and J. I. Smith, "Voice and Digital Transmission in a Mobile Radio Channel at 850 MHz," *IEEE Veh. Tech. Conf.,* Washington, D.C., Mar. 24–26, 1977, pp. 74–79.
5. J. C. Feggeler, "Experiments with Companding in Mobile Telephone," *NTC Conf. Proc.,* 1977, pp. 27.5–27.10.
6. R. E. Langseth, "Some Results on Subjective Testing of FM Transmission through Multipath Fading Channels," Microwave Mobile Symp., Boulder, Colo., 1974.
7. S. W. Halpern, "Techniques for Estimating Subjective Opinion in High-Capacity Mobile Radio," *Microwave Mobile Symp.,* Boulder, Colo., 1976.
8. R. Steele, *Delta Modulation System,* Wiley, New York, 1975.
9. L. Zetterberg and J. Uddenfeldt, "Adaptive Delta Modulation with Delayed Decision," *IEEE Trans. Comm.,* vol. 22, 1974, pp. 1195–1198.
10. S. Song, J. Garodmck, and D. Schilling, "A Variable Step Size Robust Delta Modulator," *IEEE Trans. Comm.,* vol. 19, 1971, pp. 1033–1044.
11. J. D. Markel and A. H. Gray, Jr., *Linear Prediction of Speech,* Springer-Verlag, New York, 1976.
12. J. Makhoul, "Linear Prediction: A Tutorial Review," *Proc. IEEE,* vol. 63, September 1975, pp. 561–580.

Multiple-Access (MA) Schemes

15.1 Introduction

Multiaccess schemes are used to provide resources for establishing calls. Since spectrum is a scarce resource and can be directly related to capacity, efficiency enables the spectrum to handle a large number of calls. In order to make a call, either a physical channel served by a circuit switch or a virtual channel served by a packet switch is needed. A call can be defined as a voice call or a data transmission, a short-message call or a long-message call. Therefore, the evaluation of the capacity for each multiple-access scheme is very different. The calculation of capacity should be stated as, "The calculation of channels (or calls) in a cell served at the same time (simultaneously) by one multiple-access scheme."

For example, one paging channel can serve 80,000 tone-only pagers in one hour. But at any instant, only one pager can be served. We still consider the paging channel one channel per cell or per base station. Based on this definition, the spectrum used for physical or virtual channels makes no difference in finding radio capacity from different multiple-access schemes.

Multiple-access (MA) schemes on physical channels

To access physical channels, there are five major multiple-access schemes. FDMA (frequency division MA) serves the calls with different frequency channels. TDMA (time division MA) serves the calls with different time slots. CDMA (code division MA) serves the calls with different code sequences. PDMA (polarization division MA) serves the calls with different polarizations. SDMA (space division MA) serves the calls by spot beam antennas. In SDMA, the cells serve a different area covered by corresponding spot beams and can be shared by the same frequency—the frequency reuse concept. With the exception of PDMA (see Sec. 5.6), all other MA schemes can be applied to mobile communication. Those MA schemes use the following dimensions:

Frequency, Time, Code, Space

From the purpose of evaluating spectrum efficiency, we have to calculate radio capacity of each MA scheme in two kinds of environments for mobile radio communications: the noise-limited and the interference-limited. TDMA must convert its time-slot channels to the equivalent FDMA channels before calculating its capacity.

15.2   The Noise-Limited Environment

In this environment, a frequency-reuse scheme is not applied in the deployment of the system. We can then prove that FDMA, TDMA, or CDMA provide roughly the same number of traffic channels with the same voice quality and thus have the same spectrum efficiency. To estimate the number of channels from a spectrum band of 1.23 MHz, we should base the estimate on a required $(C/I)_s$ as a reference level and then, compare the number of channels served from different MA schemes.

The $(C/I)_s$ level can be found from a subjective test of the voice quality. In Sec. 14.4, we found that, in order to achieve a toll quality of telephone voice, the C/N or C/I equals 18 dB for a channel bandwidth of 30 kHz in an FM (analog) system. We can have a total of 41 FM channels in 1.23 MHz.

In digital systems, we may divide the $C/I > 1$ systems and $C/I < 1$ systems into two kinds.

$C/I > 1$ systems

The FDMA or TDMA systems belong to $C/I > 1$ systems. In this kind of digital systems, the voice quality is based on the North American TDMA system [1], which is a 30-kHz channel, 3 time slots, and $C/I = 18$ dB. Also, we may derive the relation for converting the old channel bandwidth B_c to the new channel bandwidth B_c' when the old (C/I) is changed to the new $(C/I)'$ as follows.

The noise-limited environment can be also called the adjacent-channel-interference-only environment. In this environment, no co-channel interference exists. However, we can assume that the total number of channels cannot be handled at a single site due to adjacent channel interference. *Adjacent channel interference* is a common term.

Strictly speaking, it means non-cochannel interference. Adjacent channels usually have to be served in different cells. Then the adjacent channel interference I_a at the cell boundary (at cell radius R) relates to both the neighboring channels (not necessarily the adjacent channels) at the distance R of the same cell and the adjacent channels at the distance L from a different cell as

$$I_a = \sum_{i=1}^{p-1} k_i R^{-4} + \sum_{j=1}^{q} k_j L^{-4} \tag{15.1}$$

where k_i and k_j are the constants, $p-1$ is the number of the neighboring channels, and q is the number of adjacent channels. $M = p + q$ is the total number of channels. In general, the first terrain in Eq. (15.1) is small and can be neglected. Then,

$$I_a \approx \sum_{j=1}^{q} k_j L^{-4} = \overline{k} L^{-4} \tag{15.2}$$

where k is a constant. In this situation, L is always equal or greater than R. The carrier-to-interference C/I_a at the cell boundary is

$$\frac{C}{I_a} = \frac{R^{-4}}{\overline{k} L^{-4}} = \frac{\left(\dfrac{L}{R}\right)^4}{\overline{k}} \tag{15.3}$$

To calculate the number of cells K_a where the total number of channels are assigned under the frequency non-reuse condition, we can use the formula based on the hexagon cell configuration as follows:

$$K_a = \frac{(L/R)^2}{3} \tag{15.4}$$

The number of channels per cell in a non-reuse condition is

$$m = \frac{M}{K_a} = \frac{B_T/B_c}{\dfrac{(L/R)^2}{3}} \tag{15.5}$$

where $M = p + q$ is the total number of channels, B_T is the total spectrum bandwidth, and B_c is the channel bandwidth.

Then, substituting Eq. (15.3) into Eq. (15.5) yields

$$m = \frac{B_T/B_c}{\sqrt{\dfrac{\overline{k}}{9}\left(\dfrac{C}{I_a}\right)}} \qquad \text{number of channels/cell} \tag{15.6}$$

From Eq. (15.6), we can obtain a new bandwidth B'_c with its corresponding $(C/I_a)'$ to achieve the same number of channels per cell. The relationship becomes

$$\left(\frac{B'_c}{B_c}\right)^2 = \frac{\left(\dfrac{C}{I_a}\right)}{\left(\dfrac{C}{I_a}\right)'} \qquad (15.7)$$

Equation (15.7) is the relationship between B_c and C/I_a.

Now we may find the equivalent channel bandwidths of TDMA systems such as GSM [2] and PDC systems [3] from Eq. (15.7) based on $C/I = 18$ dB as shown in Table 15.1.

The calculation for the total number of channels for four systems is based on the total number of NA-TDMA channels. We find that the same number of channels provided from each of the four systems (using different multiple-access schemes) ranges from 100–120 channels, as shown in Table 15.1. Because there are many other concerns, such as voice quality, that may not be closely related to C/I, no accurate calculation can be taken. Nevertheless, the conclusion can still be drawn that, in the noise-limited environment, the three TDMA systems almost provide the same number of traffic channels.

$C/I < 1$ systems

The CDMA systems belong to $C/I < 1$ systems. The voice quality for CDMA is based on E_b/I_0, where E_b is the energy/bit and I_a is the interference per hertz. E_b/I_0 in the noise-limited environment is around 5 dB, where the voice quality is the same as that of NA-TDMA at its $C/I = 18$ dB. The C/I can be expressed as

$$\frac{C}{I} = \frac{E_b/I_0}{(B/R_b)} \qquad (15.8)$$

TABLE 15.1   Number of Channels from Spectrum Band of 1.23 MHz

Equation no.	System	Channel bandwidth	Time slots	Equivalent bandwidth B_{eq}	C/I	No. of channels (1.23 MHz)
15.6	NA-TDMA	30 kHz	3	10 kHz	18 dB	123
15.6	GSM	200 kHz	8	25 kHz	11 dB	
15.7				11.3 kHz	18 dB	109
15.6	PDC	25 kHz	3	8.3 kHz	20 dB	
15.6				10.4 kHz	11 dB	118
15.12	CDMA	1.23 MHz	N/A	1.23 MHz	N/A	124.5

Let $E_b/I_0 = 5$ dB in the noise-limited environment. Also, in the IS-95 CDMA system [4], the bandwidth B is 1.23 MHz, and the transmission rate R_b is 9.6 kHz. Then, the value of C/I can be obtained from Eq. (15.8).

$$\frac{C}{I} = \frac{3.16}{128} = \frac{1}{40.5} \tag{15.9}$$

C/I can also be evaluated at the boundary of a single cell or sector for CDMA. Suppose that one of M traffic channels is the desired channel. It follows that the rest of the M-1 channels are interference to the desired channel from the same cell site. Then,

$$\frac{C}{I} = \frac{R^{-4}}{(M-1)R^{-4}} = \frac{1}{M-1} \tag{15.10}$$

and

$$M = \frac{I}{C} + 1 \tag{15.11}$$

Since we are using three-sector cells for all multiple-access schemes, the radio capacity M_t of a CDMA cell is

$$M_t = 3\left(\frac{I}{C} + 1\right) = 3(41.5) = 124.5 \text{ channels} \tag{15.12}$$

Equation (15.12) is also shown in Table 15.1. In examining the number of channels derived from all the multiple-access schemes, we may conclude that, with a given spectrum band and voice quality, the same number of traffic channels are provided, regardless of different multiple-access schemes.

15.3   The Interference-Limited Environment

The frequency reuse is applied in the system to increase the capacity. The interferences, cochannel, and adjacent channel interferences prevail in this environment. The radio capacity is limited from the interferences. In this environment, we may show that CDMA is superior to FDMA or TDMA in capacity enhancement.

Capacity in an interference-limited environment [5]

All the multiple-access schemes can be applied with a means of multiplexing, such as FDM, TDM, CDM, or SDM where F stands for frequency, T for time, C for code, DM for division multiplexing, and S for space. In order to have large radio capacity, a certain multiple-access scheme applied to a certain multiplexing scheme is chosen. The radio capacity of an SDM system uses any kind of multiple-access scheme and can be expressed as

$$\overline{m} = \frac{M}{K} \qquad \text{channels/unit area}/B_T \qquad (15.13)$$

where M is the total number of channels in a given spectrum B_T and K is a frequency reuse factor usually defined by both the radius of unit area R and the separation between two cochannel areas D as

$$K = k\left(\frac{\pi D^2}{\pi R^2}\right) = k\,\frac{D^2}{R^2} \qquad (15.14)$$

In Eq. (15.14), k is a constant; $k = 1/3$ in a hexagon-cell configuration. The unit area with radius R indicated in Eq. (15.14) can be a cell, a sector, or a footprint. In cellular systems, the unit area is a cell.

The radio capacity calculated in a high-capacity cellular or PCS (personal communication service) system can be expressed in the following multiple-access schemes.

C/I > 1 systems—FDMA/SDM systems and TDMA/SDM schemes

SDM uses the spatial separation to provide frequency reuse. The AMPS system is a FDMA/SDM system, and both NA-TDMA and GSM system are TDMA/SDMs. In these systems, the radio capacity m is

$$m = \frac{M}{K} \qquad \text{channels/cell} \qquad (15.15)$$

where K is the frequency reuse factor

$$K = \frac{(D/R)^2}{3} \qquad (15.16)$$

D and R are shown in Fig. I.2 of the introduction. In Fig. 15.1, the six interferers, indicated by "1" at the first tier, circle around the desired cell (center "1"). The calculation of the number of interference N around the pth tier is as follows:

$$N = \frac{2\pi D \cdot p}{D} = 2\pi p \approx 6p \qquad (15.17)$$

Equation (15.17) is plotted in Fig. 15.1. Since the interference coming from the second-tier interferers (12 cells) is very weak and can be neglected, the first-tier interferers are considered a worst-case inter-

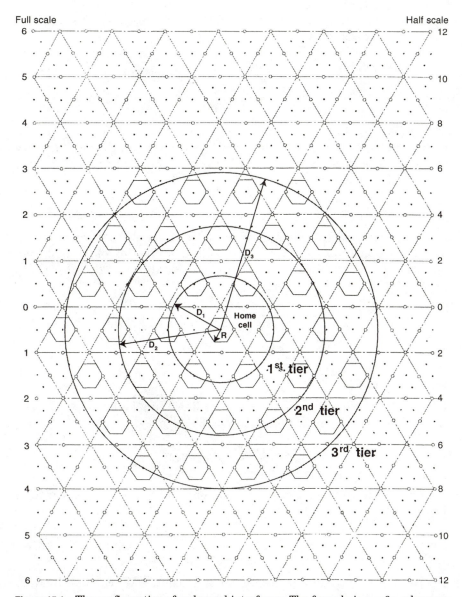

Figure 15.1   The configuration of cochannel interferers. The formula is $n = 6p$, where $n =$ the number of interfaces and $p =$ the pth tier.

ference model. Based on the six-interferer model and the fourth-power propagation-path loss law, the C/I (carrier-to-interferer ratio) at the desired cell can be found approximately by Lee [5] as

$$\frac{C}{I} = \frac{R^{-4}}{6 \cdot D^{-4}} = \frac{(D/R)^4}{6} \tag{15.18}$$

Substituting Eq. (15.18) into Eq. (15.16), then inserting the result in Eq. (15.15), the capacity m becomes

$$m = \frac{M}{K} = \frac{M}{\sqrt{\frac{2}{3}\left(\frac{C}{I}\right)_s}} \qquad \text{channels/cell} \qquad (15.19)$$

M in FDMA is the total frequency channels and in TDMA is the total time slot channels. $(C/I)_s$ is the required C/I based on the voice quality or the error performance for data. M can be expressed as

$$M = \frac{B_T}{B_c} \qquad (15.20)$$

B_T is the total spectrum band and B_c is the channel bandwidth. Therefore, the capacity m shown in Eq. (15.19) is a function of two parameters B_c and C/I. Equation (15.19) can be normalized by $B_T = 1$ MHz in Eq. (15.13). Then,

$$\overline{m} = \frac{1}{B_c\sqrt{\frac{2}{3}\left(\frac{C}{I}\right)}} \qquad \text{channels/cell/1 MHz} \qquad (15.21)$$

C/I < 1 systems—CDMA/SDM schemes

In CDMA, all the cells use the same radio channels (i.e., $D = 2R$). The frequency reuse factor K from Eq. (15.16) becomes

$$K = \frac{(D/R)^2}{3} = \frac{4}{3} = 1.33$$

Therefore, the radio capacity equation of CDMA can be obtained from Eq. (15.13) as

$$m = \frac{M}{1.33} \qquad (15.22)$$

where in CDMA, the total channel M is unknown. We can obtain M by calculating C/I of a mobile unit in a CDMA system illustrated in Fig. 15.2. Assuming all M traffic channels in a radio channel have equal power while receiving, one M channel is the desired signal, and the rest $(M - 1)$ of the channels are interference from the same site,

$$C/I = \frac{C}{I_s + I_a} = \frac{C}{C(M - 1) + I_a} = \frac{1}{(M - 1) + \dfrac{I_a}{C}} \qquad (15.23)$$

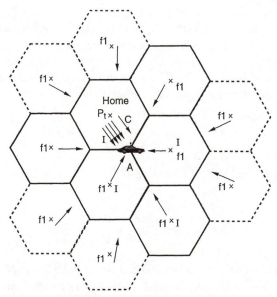

Figure 15.2 CDMA system and its interference.
(Processing-Gain Intelligent-Cell System)

where I_s is the self interference, I_a is the adjacent cell interference, and
$I = I_s + I_a$. The value of M can be obtained from Eq. (15.23) as

$$M = \frac{I - I_a}{C} + 1 \tag{15.24}$$

and Eq. (15.22) becomes

$$m = \frac{1}{1.33}\left[\frac{I - I_a}{C} + 1\right] \quad \text{channel/cell} \tag{15.25}$$

For the case of $I_a = 0$, the condition is only for a single cell or for multiple cells without adjacent-cell interference. The adjacent-cell interference reduction can be achieved by using intelligent microcells or a smart antenna arrangement. When $I_a = 0$, the number of channels per cell reaches the maximum and is called *pole capacity*.

$$m_p = \frac{1}{1.33}\left[\frac{I}{C} + 1\right] \quad \text{channel/cell} \tag{15.26}$$

In CDMA, the voice activity cycle from human behavior is about 40 percent, which means that there is only 40 percent usage for any traffic channel. This voice activity cycle is very beneficial to CDMA because all the traffic channels share one radio channel. Also, using the sectorial cell can increase radio capacity for CDMA, since three sectors can be treated like three cells. All the sectors can use the same radio channels. The sectors use directional antennae, so there is no interference among them. The capacity reduction due to cell overlap is taken care of by $K = 1.33$. The pole capacity of a single three-sector cell is therefore derived from Eq. (15.26) as

$$m = \frac{3}{1.33} \left[\frac{I}{0.4C} + 1 \right] \tag{15.27}$$

The relation between C/I and E_b/I_0 can be found in Eq. (15.8) as

$$\frac{C}{I} = \frac{\dfrac{E_b}{I_0}}{\text{P.G.}} \tag{15.28}$$

where P.G. stands for processing gain and P.G. $= B/R$.

If E_b/I_0 and B/R are known, then C/I is known. In an IS-95 CDMA system, if

$$\frac{E_b}{I_0} = 7 \text{ dB} \qquad \text{and} \qquad \frac{B}{R} = 128$$

then $C/I = -14$ dB or 0.039. Substituting $C/I = 0.039$ into Eq. (15.27), the maximum pole capacity m_p becomes

$$m_p = \frac{3}{1.33} \left[\frac{25.6}{0.4} + 1 \right] = 146 \text{ channels/cells}$$

15.4   Comparison of Radio Capacity Between FDMA, TDMA, and CDMA

In $C/I < 1$ systems—FDMA and TDMA systems

The radio capacity m_1 can be expressed as

$$m = \frac{M_1}{K_1} \qquad \text{channels/cell} \tag{15.29}$$

where the total number of channels M_1 is known, but the frequency reuse factor K_1 is a variable related to C/I:

$$K_1 = \sqrt{\frac{3}{2} \left(\frac{C}{I} \right)} = \frac{(D/R)^2}{3} \tag{15.30}$$

Furthermore, $(C/I)_s$ is related to D/R. In a frequency reuse (SDM) system, the general formula is

$$m_1 = \frac{M_1}{\sqrt{\frac{3}{2}\left(\frac{C}{I}\right)}} = \frac{M_1}{\frac{(D/R)^2}{3}} \tag{15.31}$$

where D/R is based on C/I, based on the accepted voice quality or error performance. In FDMA or TDMA, C/I is always greater than one.

In $C/I < 1$ systems—CDMA systems

The radio capacity m_2 is also expressed as

$$m_2 = \frac{M_2}{K_2} \qquad \text{channels/cell} \tag{15.32}$$

where, in this system, the frequency reuse factor K_2 is known since $D = 2R$ and

$$K_2 = \frac{1}{3}\left(\frac{D}{R}\right)^2 = 1.33 \tag{15.33}$$

But, the total number of traffic channel M_2 is a variable related to C/I, shown in Eq. (15.24) for an omni-cell system, is

$$M_2 = \frac{I - I_a}{C} + 1 = \frac{I - I_a}{\left(\frac{E_b/I_0}{\text{P.G.}}\right)} \cdot I + 1 = \frac{\text{P.G.}}{E_b/I_0} \cdot \left(1 - \frac{I_a}{I}\right) + 1 \tag{15.34}$$

The total traffic channel M_3 for a three-sector-cell system is

$$m_3 = \frac{M_3}{K_2} = \frac{3M_2}{K_2} = \frac{\left\{\frac{3 \cdot \text{P.G.}}{E_b/I_0}\left(1 - \frac{I_a}{I}\right) + 3\right\}}{1.33} \tag{15.35}$$

Therefore, the radio capacity in CDMA shown in Eq. (15.32) or Eq. (15.35) indicates that K_2 is known, but M_2 and M_3 depend on C/I.

Capacity of CDMA is greater

In FDMA or TDMA, $C/I \approx (E_b/I_0)_T$, where $(E_b/I_0)_T$ is the E_b/I_0 of TDMA. Then, Eq. (15.31) becomes

$$m_1 = \frac{M_1}{\sqrt{\frac{3}{2}\left(\frac{E_b}{I_0}\right)_T}} \tag{15.36}$$

In CDMA, Eq. (15.32) becomes

$$m_2 = \frac{\dfrac{\text{P.G.}}{E_b/I_0}\left(1 - \dfrac{I_a}{I}\right) + 1}{1.33} \qquad \text{for an omni-cell} \qquad (15.37)$$

From Eq. (15.35),

$$m_3 = 3m_2 \qquad \text{for a three-sector cell} \qquad (15.38)$$

In a noise-limited environment, we mentioned that the number of traffic channels provided by each of the multiple-access schemes is the same. The total number of channels M_1 and M_3 from TDMA and CDMA, respectively, are the same.

$$M_1 = M_3 \qquad (15.39)$$

If we let the adjacent cell interference $I_a = 0$ in Eq. (15.37), then

$$M_1 = \frac{\text{P.G.}}{E_b/I_0} + 1 \qquad (15.40)$$

and Eq. (15.36) becomes

$$m_1 = \frac{\dfrac{\text{P.G.}}{E_b/I_0} + 1}{\sqrt{\dfrac{3}{2}\left(\dfrac{E_b}{I_0}\right)_T}} = \frac{\dfrac{\text{P.G.}}{E_b/I_0} + 1}{K_1} \qquad (15.41)$$

Comparing Eq. (15.37) with Eq. (15.41), we find that the CDMA capacity is degraded by taking a portion of P.G. to reduce I_a, but TDMA capacity is degraded by implementing the frequency reuse, which is K_1 or

$$\sqrt{\frac{3}{2}\left(\frac{E_b}{I_0}\right)_T}$$

The condition that $m_3 > m_1$ can be obtained by using Eq. (15.38) and Eq. (15.41):

$$\frac{m_3}{m_1} = \frac{3\,\dfrac{\text{P.G.}}{E_b/I_0}\left(1 - \dfrac{I_a}{I}\right) + 3}{\left(\dfrac{\text{P.G.}}{E_b/I_0} + 1\right)} \frac{K_1}{1.33} > 1 \qquad (15.42)$$

Assume that

$$\frac{\text{P.G.}}{E_b/I_0} \gg 1 \qquad (15.43)$$

Then

$$\frac{I - I_a}{I} = \frac{I_s}{I} > \frac{1.33}{K_1} \tag{15.44}$$

Where I_s is the self interference or intracell interference for CDMA. The best performance for TDMA or FDMA is a $K_1 = 3$ system (which applies the intelligence in the system), then

$$I_s > \left(\frac{1.33}{3}\right) \cdot I = 44.3\% \cdot I$$

or

$$55.7\% \cdot I > I_a$$

This means that the capacity of CDMA is always higher than that of TDMA or FDMA as long as the adjacent interference is not more than 55.7 percent of the total interference, which usually does not occur.

15.5   Processing Gain (P.G.)

The processing gain (P.G.) is from the spread-spread (SS) modulation. The processing gain can be used to defeat enemy jamming (shown in Chap. 18) or any intentional or unintentional interference. Also, in CDMA communication systems, the processing gain is used to increase system capacity. However, the total processing gain (P.G.) is a definite value:

$$(P.G.)_t = (P.G.)_1 + (P.G.)_2 + (P.G.)_3 + (P.G.)_4 \tag{15.45}$$

where $(P.G.)_1$ is used for capacity, $(P.G.)_2$ is used to defeat jamming, $(P.G.)_3$ is used to defeat unintentional interference, and $(P.G.)_4$ maybe used to compensate for the contamination of an imperfect power-control effect. In commercial systems where there is no intentional interference or jamming, then $(P.G.)_2 = 0$. If we can reduce the unintentional interference, such as adjacent cell or sector interference by using intelligent microcells or small antennae, then we can save the use of $(P.G.)_3$; $(P.G.)_3 = 0$. If the power control performance is much improved, then $(P.G.)_4$ can be saved; $(P.G.)_4 = 0$, and then

$$(P.G.)_t = (P.G.)_1 \tag{15.46}$$

This is the strategy to obtain the highest capacity in CDMA systems. The next section deals with $(P.G.)_t$ used for generating capacity. We simply change the notation $(P.G.)_t$ to P.G.

Processing gain is not a real gain, just as diversity gain is not a real gain. The gain becomes effective only if interference exists.

For a single-cell CDMA system, only the interference within the cell (*self-interference*) is considered.

$$\frac{C}{I} = \frac{E_b/I_0}{B/R} = \frac{E_b/I_0}{\text{P.G.}} = \frac{1}{M-1} \tag{15.47}$$

The number of traffic channels M is

$$M = \frac{\text{P.G.}}{E_b/I_0} + 1 \tag{15.48}$$

For a multiple-cell CDMA system, the adjacent cell interference I_a is added.

$$\frac{C}{I} = \frac{1}{(M'-1) + I_a/C} = \frac{E_b/I_0}{\text{P.G.}} \tag{15.49}$$

The number of traffic channels M' is

$$M' = \frac{\text{P.G.}}{E_b/I_0} - \frac{I_a}{C} + 1 = \frac{\text{P.G.} - (I_a/C)(E_b/I_0)}{E_b/I_0} + 1 = \frac{(\text{P.G.})'}{E_b/I_0} + 1 \tag{15.50}$$

Now, to reduce P.G. due to any additional interference, the new (P.G.)' becomes

$$(\text{P.G.})' = \text{P.G.} - (I_a/C)(E_b/I_0) \tag{15.51}$$

where comparing Eq. (15.48) with Eq. (15.50), $(\text{P.G.})_3$ in Eq. (15.45) becomes

$$(\text{P.G.})_3 = (I_a/C)(E_b/I_0) \tag{15.52}$$

15.6   Ambient Noise in Cellular and PCS Bands and Its Impact on the CDMA System Capacity and Coverage [6]

Introduction

Industry focus has shifted to digital cellular technologies in order to expand cellular capacity and effectively build a PCS network to complement or compete with cellular communications services. Due to its natural attributes [7], the code division multiple access (CDMA) airlink interface is becoming the technology of choice for many cellular carriers and potential PCS providers. A well-known fact of PCS networks is that the RF signal-path loss at the PCS band (1.8–1.9 GHz) is 9 to 10 dB higher than that of the cellular band (0.8–0.9 GHz). However, it is a lesser known fact that the ambient noise level at the PCS band is 2–7 dB lower than that at the cellular band. This section will address the impact of ambient noise on CDMA system capacity and coverage. We will first develop a set of formulae that relate the ambient noise level with the CDMA capacity and coverage. Then we will compare the CDMA system capacity and coverage at the cellular band with that at the PCS band. The gains from CDMA system capacity and coverage at the PCS band due to the weaker ambient noise level is addressed.

Theory

A CDMA cellular system may consist of many CDMA cell sites. We will concentrate on one particular cell called the *serving cell,* which is surrounded by the other cells in the system [7]. The receiver P_b at the serving cell is

$$P_b = \eta_n N_0 B + P_s + \sum_i P_i \qquad (15.53)$$

where η_n is the total noise figure that includes ambient noise and receiver noise that exceeds the thermal noise level. η_n is dependent on RF carrier frequency (see Fig. 15.3) [8]. N_0 is the thermal noise density.

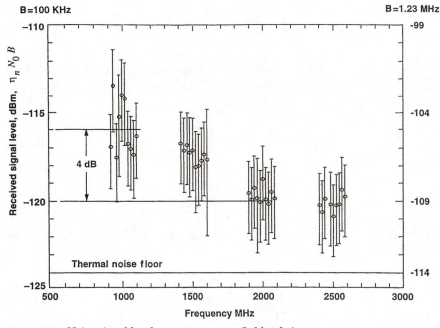

Figure 15.3 Noise signal level measurements—Oakland sites.

B is the receiver bandwidth. The P_i summation is the total serving-cell receiving CDMA power from all other cells in the system. P_s is the total received CDMA power from all users within the serving cell that can be expressed as

$$P_s = M v P_r \qquad (15.54)$$

where M is the number of users within the serving cell, v is the channel (voice) activity, and P_r is the single mobile signal level at the base receiver. Assuming the CDMA system uses a $k = 1$ cell reuse pattern, we define the frequency reuse efficiency q_f [9] as

$$q_f = \frac{P_s}{P_s + \sum_i P_i} \qquad (15.55)$$

Now we have

$$P_b = \eta_n N_0 B + \frac{M v P_r}{q_f} \qquad (15.56)$$

The defined receiver power rises over the thermal-noise floor factor Z [10] as

$$Z = \frac{P_b}{N_0 B} = \eta_n + \frac{M v P_r}{q_f N_0 B} \qquad (15.57)$$

Through Z, we can easily establish a relationship between the theory and measurement results. Recall the CDMA system energy per bit-to-interference power per Hz ratio, E_b/I_0,

$$E_b = \frac{P_r}{R_b} \qquad (15.58)$$

$$I_0 = \eta_n N_0 + \frac{v P_r}{B} \left(\frac{M}{q_f} - 1 \right) \qquad (15.59)$$

where R_b is the signal bit rate. Combining Eqs. (15.57), (15.58) and (15.59), we have,

$$\frac{\left(\dfrac{E_b}{I_0} \right) v}{\left(\dfrac{B}{R_b} \right)} = \frac{\dfrac{(Z - \eta_n) q_f}{M}}{Z - \dfrac{(Z - \eta_n) q_f}{M}} \qquad (15.60)$$

Expressing the pole capacity m_p of a CDMA carrier in another way as

$$m_p = \left[1 + \frac{\left(\dfrac{B}{R_b}\right)}{\left(\dfrac{E_b}{I_0}\right)\nu} \right]\Bigg/ K \qquad (15.61)$$

For simplicity, let $K = 1$ in Eq. (15.61); then the number of users within the serving cell M $(= m_p)$ becomes

$$M = m_p q_f \left(1 - \frac{\eta_n}{Z} \right) \qquad (15.62)$$

Defining CDMA cell loading X as

$$X = \frac{M}{m_p q_f} \qquad (15.63)$$

we have

$$Z = \frac{\eta_n}{1 - X} \qquad (15.64)$$

Equation (15.62) is the CDMA capacity formula, and Eq. (15.64) relates to the CDMA system coverage.

Calculation results

We assume the total cellular-band noise is 9 dB and the total PCS-band noise is 5 dB as shown in the measurement data (see Fig. 15.3). Assume that

$$B = 1.23 \text{ MHz}$$
$$R_b = 9.6 \text{ kbps}$$
$$\frac{E_b}{I_0} = 7 \text{ dB}$$
$$N_0 = -174 \text{ dBm/Hz}$$
$$q_f = 0.7$$
$$\nu = 0.45$$

The CDMA capacity of Eq. (15.62) is plotted in Fig. 15.4. The x-axis of Fig. 15.4 is Z, which is the noise rising over the thermal noise level. At the same Z level, Fig. 15.4 shows that the PCS system has a capacity advantage due to a lower ambient noise level.

On CDMA coverage or cell size calculation, we use a reverse link budget, since IS-95's CDMA system is limited by the reverse link. Lee's

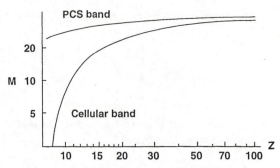

Figure 15.4   Ambient noise impact on CDMA capacity.

model from Chap. 4 is used. Lee's area-to-area model expresses the median path loss L_p at a distance d as shown in Eq. (3.47).

$$L_p = P_r/P_t = L_0 + \gamma \log d + \alpha_0 \tag{15.65}$$

where L_p is the ratio of the received power P_r to the transmit power P_t. L_0 is the median path loss at a range of 1 km, γ is the slope of the path-loss curve, d is the distance in km from 1 km, and α_0 is the adjustment factor. We assume:

$$L_0 + \alpha_0 = 156 - 20 \log h_1 - 10 \log h_2 \tag{15.66}$$

where h_2 is the mobile antenna height and h_1 is the base-station antenna height. The cell radius r can be found by equating the path loss at r with a maximum allowed path loss $L_{\max}$, namely

$$L_p + F_M = L_{\max} \tag{15.67}$$

$$L_{\max} = P_t + G_2 + G_1 - \frac{E_b}{N_0} + \frac{B}{R_b} - (kTB) - Z \tag{15.68}$$

where F_M = fade margin added to ensure 90% coverage
   P_t = maximum transmission power of a CDMA mobile phone
   G_2 = net mobile antenna gain
   G_1 = net base-station antenna gain
   k = Boltzman constant ($k = 1.3807 \times 10^{-23}$ J/K)
   T = absolute temperature ($T = 300$ K)

In calculating the CDMA system coverage from Eq. (15.64), we assume that

   P_t = 23 dBm
   G_2 = 2 dB
   G_1 = 10 dB

$$F_M = 4.5 \text{ dB}$$
$$h_2 = 2 \text{ m}$$
$$h_1 = 30 \text{ m}$$
$$\gamma = 35 \text{ dB/decade}$$
$$X = 0.7$$

The system coverage plot is shown in Fig. 15.5. The x-axis of Fig. 15.5 is X, which is the load of the CDMA system. With the same load or capacity, the PCS cell size remains smaller than that of the cellular due to the higher RF path loss. In other words, the cellular cells have better coverage than the PCS cells. However, since the ambient noise is also lower on the PCS band, the net effect on the link budget is 5 dB, not 9 dB, although the difference of the path loss between the PCS and the cellular system is 9 dB. The PCS cell radius is about 70 percent of that of the cellular-band cell radius.

Conclusion

The CDMA system capacity and coverage results are compared in our calculation. Due to the lower ambient noise level at the PCS band, these results have a profound impact on the deployment strategy of the PCS system. The additional power required for PCS over cellular is based on the difference between the two RF path losses and the two ambient noise levels. Since the path loss of the PCS system is 9 dB higher and the ambient noise level is 4 dB lower, the ERP of the PCS system needs only a 5-dB increase to be adequate. At the same E_b/I_0 level, the CDMA system capacity of PCS is 20 to 35 percent higher than that of cellular due to a lower ambient noise level in the PCS band. This capacity increase meets the requirements of PCS in order to serve vast consumer markets that require large traffic capacity.

On the coverage side, if both PCS and cellular CDMA systems operate at the same capacity, the PCS cell size is one-half that of cellular. Without the 4-dB-lower ambient noise level of the PCS band, the PCS

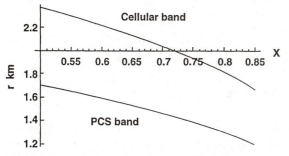

Figure 15.5 Ambient noise impact on CDMA coverage.

cell size would be only one quarter that of cellular. Therefore, the net effect of lower ambient noise doubles the PCS cell size, thus reducing the PCS deployment costs by half.

15.7 Limits of Interference Cancellation for CDMA Systems [11]

When we calculate the performance limits of error-correcting coding using the Shannon's channel capacity formula, the coding is a part of the optimum modulation process. For a CDMA-DS system, each user's data rate R_b is much less than the total spread spectrum bandwidth B. The situation is more favorable in the interference cancellation when all the users commonly share the same bandwidth. In theory, the multiple users from a single cell can have an error-free reception in an additive white-noise environment (AWGN), as long as the total rate (which is summed over the individual rates of all users) does not exceed Shannon capacity. The Shannon capacity is based on a given summed power of all users and at an interference level within the bandwidth. In the mobile radio environment, the Shannon channel capacity formula needs to be modified as described in Chap. 17 (Sec. 17.1). However, the Rayleigh fading can be reduced to Gaussian noise by applying M-branch diversity reception at the receiver while $M \to \infty$. Therefore, the Shannon channel capacity can also be treated as an upper limit for mobile radio channels.

If there are M users, each transmitting at a rate R_i ($i = 1, M$) corresponding to carrier-to-interference density ratio C_i/I_0 over a spread bandwidth B, the total rate R_T becomes

$$R_T = \sum_{i=1}^{M} R_i \tag{15.69}$$

The total carrier-to-interference density rate is

$$\frac{C_T}{N_0} = \sum_{i=1}^{M} \frac{C_i}{N_0} \tag{15.70}$$

where $N_0 = N/B$ is the noise power in watts per bandwidth in Hz.

The Shannon's channel capacity for AWGN channel is

$$R_T < B \log_2 \left(1 + \frac{C_T}{N_0 B}\right) \tag{15.71}$$

Equation (15.71) is reached by employing successive interference cancellation of each user's demodulated and decoding signal at a common cell site.

If all $R_i = R_b$, $R_T = MR_b$, the maximum number of users is obtained from Eq. (15.71):

$$M < \frac{B}{R_b} \log_2 \left(1 + \frac{C_T}{N}\right) \qquad (15.72)$$

where N is the total noise power, $N = N_0 B$. Equation (15.72) is applied for a single-cell system. Let $\eta = N/C_T$; then, Eq. (15.72) becomes

$$M < \frac{B}{R_b} \log_2 \left(1 + \frac{1}{\eta}\right) \qquad (15.73)$$

For a multicell system, the interference cancellation is performed for the users within each individual cell. However, the interference generated by the users of other cell sites cannot be canceled. Thus, a component of background noise is introduced. The relationship among the interference from other cells is not the same as the interference from a single cell. This is why the interference cannot be canceled out and degrade the capacity. The self-interference I_s is expressed in Eq. (15.23) as

$$I_s = (M - 1) \cdot C \approx M \cdot C = C_T \qquad (15.74)$$

and the parameter η' is expressed as a ratio between the adjacent interference and self-interference:

$$\eta' = \frac{I_a}{I_s} = \frac{I_a}{C_T} \qquad (15.75)$$

If so, the C/I in Eq. (15.49) for a multicell CDMA system can be expressed in a different form, as

$$\frac{C}{I} = \frac{C_T}{M'(I_a + I_s)} = \frac{1}{M'(1 + \eta')} = \frac{E_b/I_0}{B/R_b} \qquad (15.76)$$

where $I = I_0 B$ and M' is the number of users in a multicell system. M' can be obtained from Eq. (15.76) as

$$M' = \frac{B/R_b}{(E_b/I_0)(1 + \eta')} \qquad (15.77)$$

where E_b/I_0 has a lower bound in an AWGN channel as

$$E_b/N_0 > \ln 2 \qquad (15.78)$$

When the interference I_0 is approaching N_0, then η' approaches η.

Applying Eq. (15.78) into Eq. (15.77) for the lower bound of the E_b/I_0 case, we derive

$$M' < \frac{B/R_b}{(1 + \eta) \ln 2} \tag{15.79}$$

Let G be the ratio between the number of users in the single-cell system of Eq. (15.73) and the number of users in the multiple-cell system of Eq. (15.79).

$$G = \frac{M}{M'} = \frac{\text{limit on } M \text{ with perfect cancellation (single cell)}}{\text{limit on } M \text{ without perfect cancellation (multiple cells)}}$$

$$= (1 + \eta) \ln \left(1 + \frac{1}{\eta} \right) \tag{15.80}$$

Equations (15.73), (15.79), and (15.80) are plotted in Fig. 15.6. In the figure, we observe two cases, one with and one without perfect interference cancellation. When $C_T/N_0 B$ is high (i.e., η is small), the number of users in both cases increases, and the ratio G is more than two. When $C_T/N_0 B$ is weaker, the two curves show that the number of users in

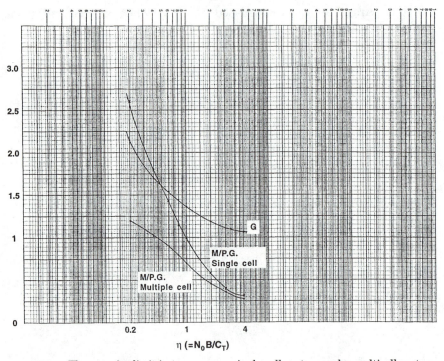

$$\eta \ (=N_0 B/C_T)$$

Figure 15.6 The capacity limit in two cases: a single-cell system and a multicell system.

both cases decreases and its ratio G approaches to one. This means that the number of users in these two cases is nearly the same.

15.8 Multiple-Access Schemes on Virtual Channels

The following MA schemes increase the virtual channels. Every one of the five multiple-access schemes mentioned above can be used to work on the following virtual channels in order to increase capacity further.

ALOHA

ALOHA [13] is a random access scheme. The users do not pay attention to what other users are doing when they attempt to transmit data packets. When packet A is beginning to transmit at time t_0, the vulnerable period for packet A to collide with packets is two times the packet interval, as shown in Fig. 15.7(a). Let's assume that the traffic load L is

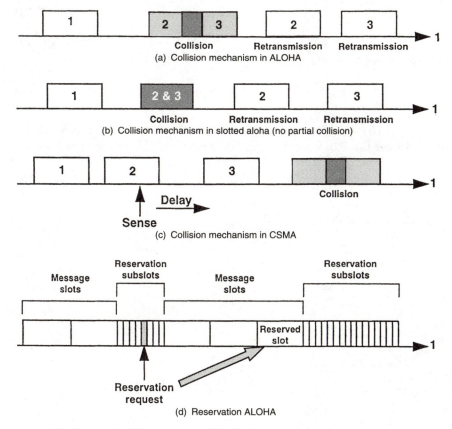

Figure 15.7   Protocols of four versions of ALOHA.

the number of packets that arrive in one packet time. If so, the traffic load becomes $2L$ arriving in two packet times. The probability that n packets arrive in two packet times following a Poisson distribution is

$$p(n) = \frac{L^n e^{-2L}}{n!} \qquad (15.81)$$

The probability P_0 that a packet is successfully received without colliding is calculated by letting $n = 0$ in Eq. (15.81).

$$P_0 = P(0) = e^{-2L} \qquad (15.82)$$

P_0 also can be interpreted as the probability that no packet can arrive within the two packet times. The throughput S cannot have one Erlang (i.e., continuous transmission for one hour) without having collisions. We may calculate throughput S with a traffic load L (number of packets arriving in one packet time) as follows:

$$S = L \cdot P_0 = L \cdot e^{-2L} \qquad (15.83)$$

Equation (15.83) is plotted in Fig. 15.8. We may see that the maximum throughput is obtained by the derivative of Eq. (15.83) with L

$$\frac{dS}{dL} = L \cdot e^{-2L} \cdot (-2) + e^{-2L} = 0 \qquad (15.84)$$

Equation (15.84) indicates that the maximum throughput occurs at the offered load $L = 0.5$. Substituting $L = 0.5$ in Eq. (15.83), we obtain

$$S_{\max} = \frac{1}{2} e^{-1} = 0.184$$

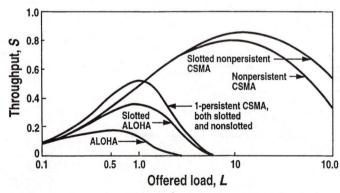

Figure 15.8 Comparison of throughput versus traffic load for ALOHA and CSMA protocols. Normalized propagation delay is $a = 0.01$. *(From Ref. 12.)*

Slotted ALOHA

Slotted ALOHA [13] increases the efficiency of the ALOHA protocol. The transmission time is divided into time slots (packet size). All the users are synchronized to these time slots. Thus, the packet-time interval of vulnerability to having a collision for any packet is reduced from two packet times, in conventional ALOHA, to one packet time. This means that the two packets can collide only in one packet time, as shown in Fig. 15.7(*b*). In slotted ALOHA, the probability of no collision is now

$$P_0 = e^{-L} \tag{15.85}$$

and the throughput S becomes

$$S = L \cdot P_0 = L \cdot e^{-L} \tag{15.86}$$

The maximum S_{max} is obtained by

$$\frac{dS}{dL} = e^{-L} - L \cdot e^{-L} = 0 \tag{15.87}$$

or

$$L = 1$$

Substituting $L = 1$ in Eq. (15.86) yields the maximum throughput S_{max}, shown in Fig. 15.8 as

$$S_{max} = 1 \cdot e^{-1} = 0.368$$

CSMA

CSMA (carrier-sense multiple access) [14] is sometimes called the listen-before-talking protocol. The collision mechanism in CSMA is shown in Fig. 15.7(*c*).

There are two kinds of CSMA: nonpersistent CSMA and 1-persistent CSMA. In the nonpersistent CSMA, after sensing the busy condition, the user waits a randomly selected amount of time before sensing again. For the unslotted nonpersistent CSMA, the throughput S is the number of successfully received packets per packet time T_p, and L is the offered traffic load (i.e., number of packets per packet time). In the CSMA, the propagation delay τ is a parameter that can be specified by a, where a is

$$a = \frac{\tau}{T_p} \tag{15.88}$$

The throughput of two nonpersistent CSMAs are [12]

$$S = \frac{Le^{-aL}}{L(1 + 2a) + e^{-aL}} \qquad \text{(unslotted)} \qquad (15.89)$$

$$S = \frac{aLe^{-aL}}{1 - e^{-aL} + a} \qquad \text{(slotted)} \qquad (15.90)$$

Both Eqs. (15.89) and (15.90) are plotted in Fig. 15.8 for $a = 0.01$. The throughput is much higher for large L. When L is small ($L < 1$), the throughput of nonpersistent CSMA is lower.

In a 1-persistent CSMA, each user first listens. If the channel is busy, the user continuously listens, and when the channel becomes free, the user then sends the packet immediately. The term *1-persistent* means the probability of transmission is unity. If a collision does occur, there are no acknowledged returns, and the users will wait for a random time interval and repeat the process.

The throughput of 2 1-persistent CSMAs is

1. For unslotted 1-persistent CSMA

$$S = \frac{L \, [1 + L + aL(1 + L + aL/2)]e^{-L(1 + 2a)}}{L(1 + 2a) - (1 - e^{-aL}) + (1 + aL) \, e^{-L(1 + a)}} \qquad (15.91)$$

2. For slotted 1-persistent CSMA

$$S = \frac{L \, [1 + a - e^{-aL}] \, e^{-L(1 + a)}}{(1 + a) \, (1 - e^{-aL}) + ae^{-L(1 + a)}} \qquad (15.92)$$

Equations (15.91) and (15.92) are plotted in Fig. 15.8 for the case of $a = 0.01$. Since a is a small value, Eqs. (15.91) and (15.92) are alike as shown in Fig. 15.8.

Reservation ALOHA

Reservation ALOHA [15] is a controlled random access method. It is a combination of Slotted-ALOHA and TDM protocols, with the transmission time to each protocol being varied in response to the traffic demand. The protocol is shown in Fig. 15.7(d). As you can see in this figure, the fixed-length frame is longer than the longest propagation delay in the network. Each frame is divided into equal-length slots. Some slots are message slots, and some are reserved slots. The reserved slots can further subdivide into short reservation subslots. In this reservation ALOHA, there are two modes: unreserved and reserved. The unreserved mode is only composed of reservation subslots, not message slots. The user sends a request for transmission in the reservation subslots and then waits for a slot (or multiple) assignment and the time to send from the base station. At this time, the system switches to the reserved mode. In the reserved mode, one slot in each frame is assigned a reservation subslot, and the remaining slots are available for message slots. The number of subslots should be small enough to keep the transmission overhead low but large enough to handle the number of reservation requests. A reasonable design guideline is to provide three reservation subslots per message slot.

In the reservation ALOHA scheme, there is no centralized control function in the network. The control is handled by each base station in the network.

PRMA

PRMA (packet reservation multiple access) [16] is a derivative of reservation ALOHA that combines TDMA and slotted ALOHA. The advantage is that it can utilize the voice activity detector (VAD) to detect the human voice activity cycle (35–40 percent talking time one way) and increase capacity of radio channels by increasing the virtual channels. PRMA may perform poorly when the frequency-reuse factor K reduces.

PRMA is developed for centralized network operation over short-range radio channels. The transmission data is organized by frames. Each frame contains a fixed number of time slots. The user terminal identifies each slot, either reserved or available, from the base station at the end of the feedback message. The reserved slot can only be used by the user. An available slot can be used by any users. There are two types of information to send: periodic and random. Speech packets are sent periodically. Data packets are sent randomly if the packets are short or periodically if the data streams are long. One bit in the packet header indicates the type of information. During the transmission, the user owns the reserved slot in each of all succeeding frames. The base station will indicate the slots available as soon as the user finishes sending. To send a packet, two requirements need to be met. The time slots must be available, and permission to send must be received from the base station. The maximum holding time T_{max} is determined by a delay constraint on speech communication that is a quarter of a second. The user terminal can keep sending the packets in the reserved slots. When packets of speech are sent over the air and the time exceeds T_{max}, a timeout occurs, and the communication link is terminated. To send data, the packet can be stored and sent over the reserved slots and not be affected by the T_{max}.

In designing a PRMA system, the maximum number of packets per channel for conversation under an accepted packet-dropping probability (i.e., is less than 1 percent). The speech code rate is 32 kbps, and the length of the header is 64 bits in each packet. The simulation shows some encouragement in this multiple access.

PRMA++

PRMA++ (packet reservation multiple access++) is an enhanced version of PRMA where slot allocation and frame-based structures in the original protocol can be dynamically changed for better channel management. A centralized algorithm makes a transformation easily achievable. However, the rigid TDMA transmission scheme is still kept.

BTMA

BTMA (busy-tone multiple access) [13] is a technique to solve the hidden terminal problem. For example, two terminals can be within the range of each other but are interfered with by the third terminal. In this case, the system spectrum bandwidth is divided into two kinds of channels: message channels and busy-tone channels. When a user terminal senses a signal on the message channel, it turns on its busy-tone channel. As a user terminal detects that another user is on the message channel, the terminal sounds alarms on the busy-tone channel to inform other users. This technique is used in the military where mobile units must stay in communication with each other all the time.

CSMA/CD

CSMA/CD (CSMA with collision detection) [13] is a variant technique of the *listen-while-talking* scheme. It can be in nonpersistent or 1-persistent of CSMA. In this scheme, no user terminal begins to transmit when it senses that the channel is busy. The CSMA/CD provides for detection of a collision and stops transmission as soon as it senses one.

DSMA

DSMA (data-sense multiple access) [13] is used in AMPS (advanced mobile phone service), CDPD (cellular digital packet data), and ARDIS (advanced radio data information service). DSMA is a full-duplex wireless data communication network. The forward channel (base to mobile) transmits a busy-idle bit to each data frame informing the user terminals to send when the busy-idle bit becomes idle. Then, the base station will change to a busy bit as soon as the user picks up.

ISMA

ISMA (idle signal multiple access) [17] is an improved version of CSMA. It is a more flexible packet access technique.

RAMA

RAMA (resource auction multiple access) [18] is a fast resource assignment and dynamic channel-allocation technique based on deterministic contention resolution.

CDPA

CDPA (capture division packet access) is based on the packet-switching technique and uses the same channels in all cells. An ALOHA retransmission scheme is used to solve contentions among transmissions of different cells. Within a cell, CDPA uses a hybrid reservation-polling (HRP) mechanism that can easily integrate different types of traffic due to the short propagation delay. The CDPA allows the use of connectionless communication between the mobile unit and the base station. CDPA is integrated with a coding procedure and provides constant throughput by increasing the transmission rate to respond to the changes in propagation conditions and interference. CDPA uses the same frequency channel set in each cell, so that cochannel collisions in the system prevail. However, when the cochannel collision occurs, based on the location of the user terminals relative to their base station as well as to the cochannel interference, the survived link can be found or dropped using the received C/I for each link. If the received C/I exceeds the threshold, the link survives. Also, the received C/I values are different from the forward link and reverse link. Therefore, when a collision occurs, in certain conditions all the links can survive, or all the links can be dropped, or some survive and some are dropped. A simulation can be generated using the ALOHA protocol. The system has two-tier interferers or 18 interferers. The following assumptions are made: The message packet size follows the truncated Gaussian distribution varied from 40 bytes to 256 bytes. The C/I threshold is 7 dB. The propagation-path loss slope is 40 dB/dec. The propagation log-normal fading is 0 dB or 8 dB with a correlation of 0.5 between two cell sites. The transmission rate is 9.6 kbps. The capacity plot is shown in Fig. 15.9. The percentage of message loss increases as the number of messages/cell/hour increases.

Let the acceptance criteria for the system performance be "less than 1 percent of the message delivered unsuccessfully." Then, from the figure, the number of messages is reduced from 3960 to 2160 as the log-normal fading increases from 0 dB to 8 dB. In this simulation, no adjacent channel interference and Rayleigh fading degradation are considered.

Conclusion

Many kinds of multiple access have been mentioned to increase the number of virtual channels. The packet data transmission for voice and data can be treated differently.

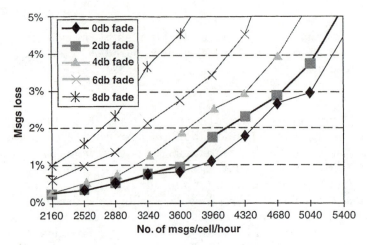

Figure 15.9 Capacity vs. System loading for CDPA system (a simulation result).

Using packet transmission for voice

Over an FDMA or TDMA channel, the packet data transmission can send a piece of voice data by a packet from a sender to one destination and send a different piece of voice data by another packet from the second sender to another destination using the same physical channel. Then, if the voice-activity cycle is 40 percent, each FDMA or TDMA channel is virtually two conservation channels.

Over a CDMA channel, the packet data transmission does not help in increasing radio capacity. The nature of a wideband CDMA channel can take advantage of the voice-activity cycle of human conversation. Therefore, there is no need to transmit packet data in CDMA channels for increasing the voice capacity.

Using packet transmission for data

For long-length data transmission. Packet transmission using virtual channels does not increase capacity better than continuous transmission using physical channels.

For short-length data transmission. One physical channel can send to many different destinations as long as each packet that is transmitted contains an address in the header. Thus, the radio capacity in all kinds of multiple-access channels, including CDMA channels, increases.

15.9 Mobile Satellite Systems

Mobile satellite systems use the same multiple-access schemes as the terrestrial mobile systems. In late 1989, the United Kingdom had issued three licenses for the Personal Communication Network (PCN) in the 1800-MHz band. Then, in the early 1990s, the United States started to define the Personal Communication Service (PCS) in both terrestrial mobile and satellite mobile communication. The Mobile Satellite Systems (MSS) use the same multiple-access schemes; some use TDMA or FDMA; others use CDMA. In this chapter [21, 22], we expose the new development of MSS systems. Mobile Satellite Service (MSS) disclosed new space mobile systems, providing three types of mobile communication service from nongeostationary orbits.

1. Short data transmission is used for very small and inexpensive satellites, sometimes called little LEO (low-earth orbit). They are the space equivalent of paging systems and are only in service for a limited time of the day.

2. Voice and data-communication satellites in orbit above the earth are either MEO (midEarth orbit) or big LEO satellites.

3. Wideband data transmission is the extension of the global information initiative (GII). These systems offer computer-to-computer links and videoconferencing with a data rate of 1.544 Mbps or higher.

Therefore, the MSS (mobile satellite service) provides three distinct satellite technologies:

1. GEO (geosynchronous earth orbit)
2. MEO (midearth orbit)
3. LEO (low-earth orbit)

The orbits used by the three MSS systems are shown in Fig. 15.10. The allocated frequency spectrum for MSS systems is shown in Fig. 15.11. Specifications are listed in Table 15.2.

We can understand the ability and inability of each technology to offer mobile and handheld services as follows:

1. *GEO.* Long distance from a GEO satellite to a subscriber unit needs a large power requirement and causes delays in conversation. GEO systems include Omni TRACS for data-only service. INMARSAT, a global consortium of PTTs (post and telephone and telegraph) from major countries, provide service for voice, fax, and data to maritime, airplane, and mobile users. AMSC provides MSS services within the United States.

2. *MEO.* MEO satellite communication has advantages and disadvantages compared to both GEO and LEO. The MEO satellites include GPS (global positioning satellites) owned by the military and civilian

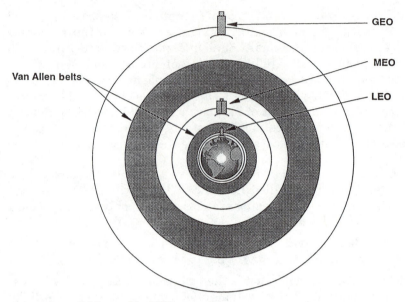

Figure 15.10   Orbits used by MSS systems.

users. The GPS system has 24 satellites that can provide the location and the altitude from a GPS receiver. New proposed MEO satellites are the TRW Odyssey and INMARSAT P.

3. *LEO.*   LEO satellites can further be divided into two kinds: little LEO and big LEO.

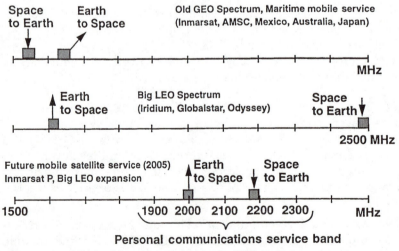

Figure 15.11   Allocated frequency spectrum for MSS systems.

TABLE 15.2 Specification of Three Types of Mobile Satellite Communications

Type of satellite	Altitude of satellite	Average number of satellites needed for global coverage	Band of operation	Subscriber unit power requirement	Delay	Life span
GEO	35,000 km	3	1.6–2.5 GHz	Large power	¼ sec	15 years
MEO	5000–10,000 km	10	1.6–2.5 GHz	Medium power	16.5 ms–33 ms	10 years
LEO	500–1500 km	Little < 20	Little = 150 MHz	Small power	2.6 ms–8 ms	5 years
		Big < 100	Big = 1.6–2.5 GHz			

- Little LEO allocates frequencies below 1 GHz. It uses frequency-scanning technology to send packets and create a time delay that cannot adequately support voice communication. Thus, it will provide data-only services. Satellite systems such as Orbcomm, Starsye, and VITA (Volunteers in Technical Assistance) are little LEO systems.

- Big LEOs use frequencies above 1 GHz and provide integrated voice, data paging, fax, and geopositioning services. The allocated frequencies are 1.6 GHz uplink and 2.4 to 2.5 GHz downlink. The big LEO systems are Aries, Globalstar, and Iridium. Teledesic is also a big LEO, but it is a nonmobile system that proposes high-speed data and video communication. Teledesic belongs to the GII family and provides wideband data transmission.

Comparison of GEO, MED, and LEO systems

The system comparisons of two GEOs (AMSC and Spaceway), two MEOs (Odyssey and Inmarsat P), one little LEO (OrbComm), and five Big LEOs (Globalstar, Iridium, Teledesic, Aries, and Ellipso) are shown in Table 15.3. Some data is not available, as indicated. The most complete data concentrates on the four systems: Globalstar [23], Inmarsat P [21], Odyssey [24], and Iridium [25], all of which are shown in the table.

Description of LEO operation

In satellite mobile environment, the line-of-sight condition prevails. Thus, the propagation-path loss follows 20 dB/dec instead of 40 dB/dec. The MSS power P_M can be saved by one-half dB value compared to the power P_r needed in mobile radio environment for the same distance, or it can be expressed in linear scale as

$$P_M = \sqrt{P_r} \qquad (15.93)$$

Besides, the signal arrival will not suffer from multipath fading unless it is at a low elevation angle. Therefore, the signal suffers by Gaussian noise or Rician fading [26].

TABLE 15.3 Global Mobile Satellite System Comparison

	Globalstar Big Leo	Inmarsat P-21 ICO, MEO	Odyssey MEO	Iridium Big Leo	Teledesic Big Leo	AMSC GEO	OrbComm Little Leo	Spaceway GEO	Aries Big Leo	Ellipso* Big Leo
Altitude (km)	1400	10,400	10,400	900	700	35,000	750	35,000	2000	(7846 × 520), (7846)
Number of satellites	48	12	12	66	840	2	8	8	46	(8), (6)
Number of spares	8	3	2	12	80		8	8		3
BOL mass (kg)	425	1244	1250	700						
Power (watt)	823	3760	3050	1200						
Signal modulation	CDMA	TDMA/FDMA	CDMA	TDMA	TDMA	FDMA	TDMA	TDMA	FDMA/CDMA	CDMA
Antenna size (m)	1	>3	3	>2						
Beam number	16	61	61	48						61
Onboard processing	No	Yes	No	Yes	Yes	Yes	on-board memory			
Intersatellite links	No	No	No	Yes	Yes	Yes	No			
Capacity (crt)	65,000	24,000	27,600	56,000						
System life	7.5	10	10	5	5	15	5		7.5	7.5
Communications link Uplink	1610–1626.5 MHz	2.0 GHz	1600 MHz	1610–1626.5 MHz	20 GHz	1.6 GHz	137.5 MHz	20 GHz	1610–1626.5 MHz	1610–1626.5 MHz
Downlink	2483.5–2500 MHz	2.2 GHz	2500 MHz	2483.5–2500 MHz	30 GHz	1.5 GHz	150 MHz	30 GHz	2483.5–2500 MHz	2483.5–2500 MHz
Feed link Uplink	5091–5250 MHz	5 GHz	20 GHz	19.3–19.6 GHz	20 GHz	13 GHz		20 GHz	15.45–15.65 GHz	15.45–15.65 GHz
Downlink	6875–7055 MHz	15 GHz	30 GHz	29.1–29.3 GHz	30 GHz	11 GHz		30 GHz	6825–7025 MHz	6725–7025 MHz
Data rate	9.6 kbps		4.8 kbps	9.6 kbps	1.5 Mbps		2.4 kbps			
No. of orbits	8	2	3	6	21	N/A	4	1	5, equator	(2), (1)
No. of satellites/orbit	6	6	4	11	40	N/A	2	8	7, 11	(4), (6)
Inclination	52°	45°	50°	90°	98.2°	0°	70°	0°		116.5°
One revolution time	114 minutes									
Ground stations	90	8–12	8	20	16		4	2		
Elevation angle	10°	16°	22°	8.2°	40°					
Convolutional code	$n = \frac{1}{2}, K = 9$		$n = \frac{3}{4}, K = 7$	$n = \frac{3}{4}, K = 7$						
Required BER	$<10^{-3}$		$<10^{-2}$	$<10^{-2}$						

*(elliptical orbit) and (circular orbit)

The Rician pdf represents a direct wave plus reflected waves

$$p(r) = 2\,\frac{r}{\overline{r^2}}\,\exp\left(-\frac{r^2 + a^2}{\overline{r^2}}\right) I_0\left(\frac{r}{\sqrt{\overline{r^2}/2}} \cdot \frac{a}{\sqrt{\overline{r^2}/2}}\right) \qquad (15.94)$$

where r is the envelope of the Rician fading signal, $\overline{r^2}$ is the average of the fading signal, a is the amplitude of a direct wave, and $I_0(\cdot)$ is the modified Bessel function of zero order, which can be expressed as

$$I_0(z) = \sum_{n=0}^{\infty} \frac{z^{2n}}{2^{2n} n! n!} \qquad (15.95)$$

The Rician cumulative probability distribution (CPD) is obtained by integrating Eq. (15.94):

$$P(r \le R) = \int_0^{R_0} r_0 \exp\left(-\frac{r_0^2 + a_0^2}{2}\right) I_0\,(a_0 r_0)\,dr_0 \qquad (15.96)$$

where $r_0 = \dfrac{r}{\sqrt{\overline{r^2}/2}}$

$a_0 = \dfrac{a}{\sqrt{\overline{r^2}/2}}$

$R_0 = \dfrac{R}{\sqrt{\overline{r^2}/2}}$

The Rician CPD with different values of a_0 is shown in Fig. 15.12. The direct wave is stronger than the reflected waves by 7 dB ($a_0 = 5$). The slope shown in the figure is very steep, and the dynamic range of fading is around 4–5 dB. Figure 15.12 can be used to determine how much power can be saved in the mobile satellite system.

Now we can take the Globalstar frequency plan (see Fig. 15.13) to explain LEO operation. The 16 beams for each satellite are illustrated in the figure for both communication links, uplink (L band) and downlink (S band).

Every LEO satellite covers roughly 2 percent of the Earth's surface area. The total surface area of the Earth is 510,000,000 sq. km (200,000,000 sq. miles). Therefore, each LEO satellite covers 10,000,000 sq. km. Globalstar has 16 antenna beams. Each beam may cover 625,000 sq. km. This means that the footprint radius r_e is

$$r_e = \sqrt{\frac{A}{\pi}} = 446 \text{ km (or 278 miles)} \qquad (15.97)$$

Since the footprint area is so large, it can be treated as a huge cell. According to the radio capacity from Eq. (15.13), we realize the number

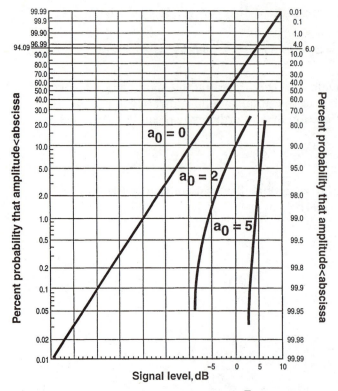

Figure 15.12 CPD of Ricean distribution ($a_0/\sqrt{2}$ = amplitude of direct path/rms value).

of channels per cell is used to calculate the capacity of LEO. The number of channels per cell can be calculated as follows: the total capacity number 65,000 channels is found in Table 15.3. Globalstar has 48 satellites with 16 beams per satellite. The number of channels per cell (beam) is

$$\frac{65{,}000}{48 \times 16} \approx 85 \text{ channels/cell}$$

Since the cell is so large, the channel per sq. km becomes

$$\frac{85 \text{ channels}}{625{,}000 \text{ sq. km}} = 0.00014 \text{ channels/sq. km}$$

The above number just indicates that LEO cannot be a high-capacity system due to the large footprint. Therefore, all LEOs claim to be complementary and enhanced systems for terrestrial mobile systems.

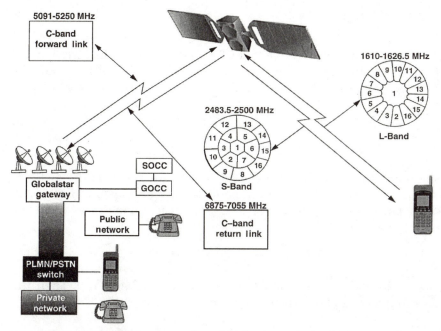

Figure 15.13 Globalstar frequency plan. GOCC = ground operations control center; PLMN = public land mobile network; PSTN = public switched telephone network; SOCC = satellite operations control centers.

Operation concept of LEO

The speed of a LEO satellite circling around the earth is 114 minutes (see Table 15.3). The total circumference of the earth is 40,000 km (or 25,000 miles), and the radius of the earth is 3183 km. Also, the altitude of a Globalstar satellite above the earth is 1400 km. Therefore, the time for the mobile unit to move out of a LEO footprint (covered by one beam) is based on the information of Eq. (15.97) as

$$446 \times \frac{114}{4\pi(3183 + 1400)} = 0.88 \text{ minutes}$$

Each footprint (beam) will move out every 0.88 minute. In this condition, the cells are moving so fast that the ground mobile unit moves can be treated as stationary. The handoffs from cell to cell will be handled by the gateways in the Globalstar system. This is because the Globalstar satellites are transponders. An average three-minute conversation may experience three handoffs, which is more frequent than that of terrestrial mobile systems.

15.10   Nature of Cochannel Interference in a SDMA System

The configuration of cochannel interferers is shown in Fig. 15.1. Assume that the transmit power at every cell site is identical in Eq. (15.98). The C/I received by a mobile unit within the home cell can be expressed as:

$$\frac{C}{I} = \frac{r^{-4}}{\sum\limits_{i=1}^{N} d_j^{-4}} \tag{15.98}$$

where r is the distance from the home-cell site and d_j is the distance from the interference site j to the mobile unit, as shown in Fig. 15.14. The separation D_i is the distance between the home cell site and the interfering site j. Take

$$d_i = \{D_j^2 + r^2 - 2D_j r \cos (\alpha_m - \alpha_j)\} \tag{15.99}$$

where the polar coordination of the mobile unit is (r, α_m). Then the interference I can be derived from Eq. (15.98):

$$I = C \cdot \frac{\sum\limits_{1}^{N} d_j^{-4}}{r^{-4}} \tag{15.100}$$

Now the nature of I within the SDMA cells is studied as follows. The SDMA system is a frequency reuse system. Let the cells in the SDMA form an ideal hexagon layout with flat-earth-propagating path loss slopes 20 dB/dec, 30 dB/dec, and 40 dB/dec. The three-tier interfering cells ($6 + 12 + 18 = 36$ cells) are used. From Eq. (15.100), the value C is assumed to be -75 dBm and the frequency reuse factor $K = 7$ is used, then

$$r \leq R = \frac{D_1}{4.6} \quad \text{or} \quad \frac{r}{R} \leq \frac{D_1}{4.6R} \tag{15.101}$$

where R is the cell radius, and D_1 is the distance between two cochannel cells in the first tier.

The simulation is to pick randomly a mobile-unit location within a cell and find its I from Eq. (15.100). There are 1000 simulated I samples for each propagating path loss slope generated. The distribution of all the I samples are shown in Fig. 15.15. Figure 15.15(a) shows the interference I causes by a path loss of 20 dB/dec. The $C_R/I = 5$ dB at the cell boundary, and the variation of I within the cell is about 0.2 dB. Figure 15.15(b) shows the interference I caused by a path loss of

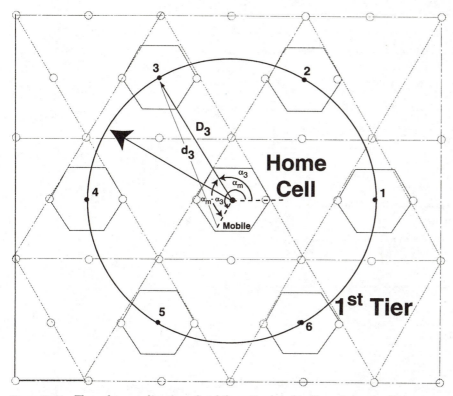

Figure 15.14 The polar coordination of mobile unit related to its cochannel cells.

30 dB/dec. The $C_R/I = 11$ dB at the cell boundary, and the variation of I built into the cell is about 0.5 dB/dec.

Figure 15.15(c) shows the interference I caused by a path loss of 40 dB/dec. The $C_R/I = 18$ dB at the cell boundary, and the variation of I within the cell is about 0.9 dB.

From Fig. 15.15, we may draw the following conclusions:

1. The cochannel interference within the cell is apparently constant.

2. The interference level I obtained from the flat earth propagation is the maximum level. If the interference I' is generated from a nonflat earth, or if the interference I'' is generated from the shadow condition, then I' and I'' are always less than I.

$$I \geq I' \text{ or } I'' \tag{15.102}$$

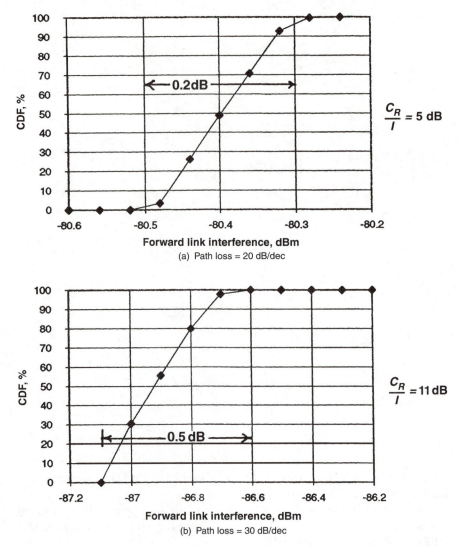

Figure 15.15 Distribution of forward link interference under the assumptions: (1) ideal hexagon layout; (2) flat-earth propagation with path loss slopes 20 dB/dec shown in a, 30 dB/dec in b, and 40 dB/dec in c; (3) seven-cell reuse, three tiers of cells; (4) –75 dBm of received signal level at cell boundary. The interference levels are collected at the center cell.

15.11   The Requirement of Alien-Interference Tolerance

Often more than one wireless communication system shares the same spectral band allocated by the same or different authorities. The primary system service would be very concerned regarding alien-interference

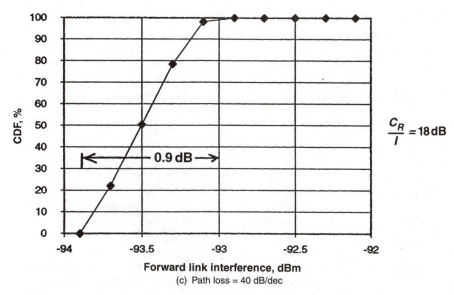

Figure 15.15   *(Continued).*

from the secondary system service. For example, the PDC (personal digital cellular) authority in Japan is concerned that satellite mobile communication service operates in a portion of the PDC spectrum band. In this section, we are able to calculate the tolerant interference of a primary system service that is either cellular or PCS.

For the purpose of illustration, we will use PDC as a primary service. PDC is a digital cellular system using TDMA. At the 1.5-GHz band, the outbound frequency band is 1477–1501 MHz, and the inbound frequency band is 1429–1453 MHz. The carrier spacing is 50 kHz, and each carrier has six time slots. This system uses the frequency reuse scheme. Thus, the cochannel interference is of great concern in this system. The required C/I is 13 dB for the static case and 16 dB for the fading case.

Designing a wireless communication system

The worst-case scenario   A wireless communication system is generally designed based on a worst-case scenario. The required C/I is determined by the BER and is based on the voice quality. In most systems, such as PDC, the required C/I is based on the BER $\leq 10^{-2}$.

In order to keep BER $\leq 10^{-2}$ for voice quality, the required C/I varies with different vehicle speeds. At the PDC specification indicated, the required $C/I \geq 16$ dB needs to be used for the maximum fading frequency $f_D = 40$ Hz.

The $f_D = 40$ Hz is equivalent to the mobile speed $V = 28.8$ km/h at 1.5 GHz and $V = 54$ km/h at 800 MHz.

The tolerant interference requirement of an alien interference The tolerant interference requirement of an alien interference may be set as low as 10 percent of the total interference caused by the system itself.

The noise figures The noise figures obtained from ambient noise are different when received at the base and received at the mobile.

The ambient noise level at the mobile is higher than at the base. This is because the base station antenna is placed high above the ground and away from the street curb. Therefore, the ambient noise received by the base station is roughly 2 dB less. At the base station, the noise figure due to ambient noise is 8 dB at 800 MHz and 5 dB at 1.5 MHz. At the mobile site, the additional 2 dB is added.

Assumption of the natural propagation parameters The propagation loss is 40 dB/dec, and the standard deviation of the long-term shadowing is 8 dB.

Based on the requirement of system design The requirement of a system design in general is to find a specific level X such that the received signal strength level x is equal or greater than the specific level X within a cell at a probability of 90 percent [27].

$$P_c \, (x \geq X) = 90 \text{ percent} \tag{15.103}$$

The notation P_c is the probability of $(x \leq X)$ within a cell.

When a specific level X is set at the cell boundary, the received signal strength x has a 50 percent chance of being above X and 50 percent chance of being below X; that is,

$$P_R \, (x \geq X) = 50 \text{ percent} \tag{15.104}$$

The notation P_R is the probability of $(x \geq X)$ at the cell boundary of a cell radius R (see Fig. 15.16). In this case, the probability of x being greater than X within the cell can be found from Eq. (4.62). The answer is 77.5 percent (see Fig. 15.17) [28].

$$P_c \, (x \geq X) = 77.5 \text{ percent} \tag{15.105}$$

Equation (15.105) does not meet the requirement of Eq. (15.103).

Since, in most systems, the designing criterion is based on Eq. (15.103), we need to calculate a new level Y such that

$$P_R \, (x \geq Y) = 50 \text{ percent} \tag{15.106}$$

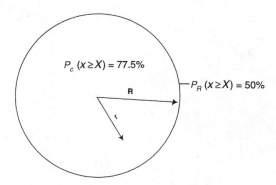

Figure 15.16   Relationship between P_c and P_R.

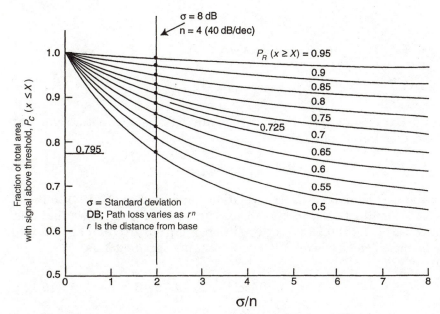

Figure 15.17   Fraction of total area with signal above threshold X [28].

at the cell boundary to satisfy the level X in Eq. (15.103) as illustrated in Fig. 15.18.

Calculation of the specific level X

The level X such that $P_c (x \geq X) = 90$ percent can be obtained from Fig. 15.16 by first obtaining the probability of x being greater than X at the cell boundary of a cell radius r as

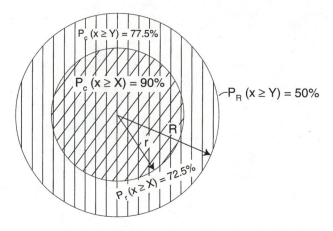

$$Y = X + 40 \log \frac{R}{r} \ \text{(in dB)}$$

Figure 15.18   The relationship between X and Y.

$$P_r \ (x \geq X) = 72.5 \ \text{percent} \qquad (15.107)$$

at the cell boundary r, and the probability of x being greater than Y at the cell boundary of a cell radius R as

$$P_R \ (x \leq Y) = 50 \ \text{percent} \qquad (15.108)$$

at the cell boundary R. The relation between Eq. (15.103) and Eq. (15.107) is illustrated in Fig. 15.18. Then, from Fig. 4.16, we find that the ratio of r/R is 0.76 for the condition of Eq. (15.107) and Eq. (15.108) to be met. The new level Y is found from the specified X as

$$Y = X + 40 \log \frac{R}{r} = X + 4.76 \ \text{dB} \qquad (15.109)$$

Obtaining the tolerant interference level

The calculation of the tolerant interference level should be based on the system design requirement, not the radio specifications. For example, the required $C/I = 16$ dB is a radio specification. In the system design requirement, we have to set up a percentage that the required $C/I = 16$-dB level will be kept in a cell. The interference I is usually uniform within a cell based on the six-interferer model shown in Fig. 15.14 and described in Sec. 15.10. The simulation result shows in Fig. 15.15 that the interference is nearly a constant within the home cell area. Therefore, we are only considering the received signal strength level C. In this section, C is represented by x in Eq. (15.103) and Eq. (15.108).

Let us consider that the PDC outbound frequency band is commonly shared with a secondary system service and calculate the tolerant interference of PDC mobile units.

$$\text{Thermal noise } (N_{th} \text{ in a 21-kHz band}) = -130.6 \text{ dBm} \quad (15.110)$$

$$\text{Noise figure (NF)} = 7\text{-dB (received at the mobile)} \quad (15.111)$$

$$\text{Ambient noise } (N) = N_{th} + \text{NF} = -123.6 \text{ dBm}$$
$$= 4.36 \times 10^{-13} \text{ mw} \quad (15.112)$$

The cochannel interference level is about 6 to 7 dB above the N level.

$$I + N = -117 \text{ dBm} \quad (15.113)$$

$$\frac{\text{Required } C}{I + N} = 16 \text{ dB} \quad (15.114)$$

Minimum receiving level

$$= -101 \text{ dBm (the level } X \text{ within the cell)} \quad (15.115)$$

According to Eq. (15.109), the level Y at the cell boundary is

$$Y = -101 \text{ dBm} + 4.76 \text{ dB} = -96.24 \text{ dBm} \quad (15.116)$$

and can meet the system design requirement of Eq. (15.103). The required $C/(I + N) = 16$ dB is still held. Since the level C becomes -96.24 dBm, the new $N + I'$ level is

$$N + I' = -96.24 - 16 = -112.24 \text{ dBm (mw/21 kHz)}$$
$$= -149.44 \text{ dBw (w/4 kHz)} \quad (15.117)$$

The level I' is the total interference allowance, which can be obtained by substituting N from Eq. (15.112) into $N + I'$ of Eq. (15.117) as

$$I' = -122.56 \text{ dBm} \quad (15.118)$$

The tolerant interference level for an alien-interference is 10 percent of I', which is

$$10\% \, I' = -122.56 \text{ dBm} = -159.76 \text{ dBw (w/4 kHz)} \quad (15.119)$$

**The maximum interference level
of an alien system can be tolerant**

In order to obtain the tolerant alien-interference level at the PDC mobile, we must take care of the antenna gain at the PDC mobile unit.

The PDC mobile antenna is assumed to be a dipole antenna with its gain of 2 dBi; that is, $G = 1.58$. The wavelength of 1.5 GHz is 0.2 m. The aperture of a dipole antenna is therefore

$$A_e = \frac{G \cdot \lambda^2}{4\pi} = \frac{1.58 \times (0.2)^2}{4\pi} = 0.00503\text{m}^2 \ (=) -22.9 \text{ dB (m}^2) \quad (15.120)$$

The maximum interference level z for the PDC mobile units is

$$z \ [\text{w/(m}^2 \cdot 4 \text{ kHz)}] - 22.9 \text{ dB (m}^2) < -159.76 \text{ dBw (w/4 kHz)} \quad (15.121)$$

or

$$z < -159.76 + 22.9 = -136.86 \text{ dBw } [\text{w/(m}^2 \cdot 4 \text{ kHz)}] \quad (15.122)$$

Problem Exercises

1.   What is the fundamental difference in the radio capacity formula $m = M/K$ between TDMA and CDMA?

2.   What is the processing gain, and how do you use it wisely to increase CDMA capacity?

3.   What is the pole capacity of CDMA if the bandwidth B is 2.5 MHz, the data rate is 28.8 kbps, and $E_b/I_0 = 7$ dB?

4.   If a CDMA channel bandwidth B is 1.25 MHz, its data rate R_b changes from 9.6 kbps to 14.4 kbps for voice quality improvement. What is the change in radio capacity?

5.   Three companies are developing TEMA systems based on a stated voice quality. Company A's system specifies channel bandwidth $B_c = 18$ kHz and $C/I \geq 17$ dB. Company B's system specifics are $B_c = 25$ kHz and $C/I \geq 14$ dB. Company C's system specifics are $B_c = 15$ kHz and $C/I \geq 20$ dB. Among these three systems, which one has the highest radio capacity?

6.   The limit of interference cancellation for multiple-cell CDMA systems is found in Eq. (15.79). But Eq. (15.79) does not tell us how to achieve it. Can you find how to achieve interference cancellation?

7.   The cellular radio capacity is large due to the frequency reuse scheme. However, it generates cochannel interference, which is very damaging to the cellular system. The D/R ratio is a parameter to reduce cochannel interference. In the packet data transmission, CDPA is currently being used. It may use narrowband channels, but the same channel is in every cell. How does it work to avoid cochannel interference?

8.   Why do we count the number of physical channels, not the number of virtual channels to measure radio capacity?

9. Why do the mobile satellite systems provide low radio capacity?

10. Why do the LEO systems become good candidates for mobile satellite systems, and what are the drawbacks?

References

1. Cellular System, IS 136.1, "800 MHz TDMA Cellular—Radio Interface—Mobile Station—Base Station Compatibility—Digital Control Channel," Telecommunications Industry Association, Committee TR 45-3, EIA Engineering Department, December 1994.
2. "European Digital Cellular Telecommunications System (Phase 2): General Description of a GSM Public Land Mobile Network," GSM 01-12, ETSI, France, October 1993.
3. "PDC—Digital Cellular Telecommunication System, RCR STF-27A Version," Research & Development Center for Radio System (RCR), Nippon Ericsson K. K., January 1992.
4. Cellular System, IS 95, "Dual-Mode Mobile Station-Base Station Wideband Spread Spectrum Compatibility Standard," PN 3118, EIA Engineering Department, December 1992.
5. W. C. Y. Lee, *Mobile Cellular Telecommunication Systems,* McGraw-Hill, 1995, p. 405.
6. Y. E. Lu and W. C. Y. Lee, "Ambient Noise in Cellular and PCS Bands and Its Impact on the CDMA System Capacity and Coverage," 1995 ICC Conference record, Seattle, Washington, June 18–22, 1995, pp. 708–712.
7. William C. Y. Lee, "Overview of Cellular CDMA," *IEEE Trans. Commun.,* May 1991, pp. 291–302.
8. Telesis Technologies Laboratory, Pacific Telesis, "Experimental License Progress Report," August 1991.
9. K. Gilhousen, I. Jacobs, R. Padovani, A. Viterbi, L. Weaver, and C. Wheatley, "On the Capacity of a Cellular CDMA System," *IEEE Transactions on Vehicular Technology,* vol. 40, May 1991, pp. 303–312.
10. QUALCOMM Inc., "CDMA Capacity 2.1 Test Report," August 1993.
11. A. J. Viterbi, "Performance Limits of Error-Correcting Coding in Multicellular CDMA Systems with and without Interference Cancellation," in S. G. Glisic and P. A. Leppanen (eds.), *Code Division Multiple Access Communications,* Kluwer Academic Publishers, 1995, pp. 47–52.
12. J. L. Hamond and P. J. P. O'Reilly, *Performance Analysis of Local Computer Networks,* Addison-Wesley, Reading, Massachusetts, 1986.
13. K. Pahlavan and A. H. Levesque, *Wireless Information Networks,* John Wiley & Sons, 1995.
14. L. Kleinrock and F. A. Tobagi, "Carrier Sense Multiple Access for Packet Switched Radio Channels," *Proc. IEEE ICC '74,* June 1974, pp. 21B1–21B7.
15. W. Crowther et al., "A System for Broadcast Communication: Reservation-ALOHA," *Proc. 6th Hawaii Int'l. Conf. Sys. Sci.,* January 1973, pp. 596–603.
16. D. J. Goodman, "Cellular Packet Communications," *IEEE Commun.,* vol. COM-38, August 1990, pp. 1272–1280.
17. G. Wu, K. Mukumoto, and A. Fukuda, "Performance Evaluation of Reserved Idle Signal Multiple Access Scheme for Wireless Communication," *IEEE Trans. Vehicular Tech.,* vol. VT-43, August 1994, pp. 653–658.
18. N. Amitay and L. J. Greenstein, "Resource Auction Multiple Access (RAMA) in the Cellular Environment," *IEEE Trans. Vehicular Tech.,* vol. 43, Nov. 1994, pp. 1101–1111.
19. F. Borgonovo, M. Zorzi, L. Fratta, V. Trecordi, and G. Branchi, "Capture-Division Packet Access for Wireless Personal Communications," *IEEE Jour. on Selected Areas in Comm.,* vol. 14, May 1996, pp. 609–622.
20. F. Borgonovo, L. Fratta, P. diMilano, M. Zorzi, and A. Acampora, "Capture-Division Packet Access: A New Cellular Access Architecture for Future PCNs," *IEEE Comm. Mag.,* September 1996, pp. 154–162.

21. R. Rush, "The Market and Proposed Systems for Satellite Communications," *Applied Microwave & Wireless,* fall 1995, pp. 10–34.

22. G. M. Comparetto and Neal D. Hulkower, "Personal Satcom," *Mobile Europe,* January 1995, pp. 20–25.

23. R. A. Wiedeman, "The Globalstar Mobile Satellite System for Worldwide Personal Communication," Proc. Third International Mobile Satellite Conference, Pasadena, June 16–18, 1993, pp. 291–296.

24. C. Spitzer, "Odyssey Personal Communication Satellite System," *Proc. Third International Mobile Satellite Conference,* Pasadena, June 16–18, 1993, pp. 291–296.

25. J. E. Hatlelid and L. Casey, "The Iridium System: Personal Communications Any-Time, Anyplace," *Proc. Third International Mobile Satellite Conference,* Pasadena, June 16–18, 1993, pp. 285–290.

26. W. C. Y. Lee, *Mobile Communication Design Fundamentals,* 2d ed., John Wiley & Sons, 1993, p. 30.

27. W. C. Y. Lee, *Mobile Cellular Telecommunications, Analog and Digital Systems,* McGraw-Hill, 1995, p. 10.

28. W. C. Jakes, *Microwave Mobile Communications,* Wiley, 1974, p. 127.

Clarification of the Concepts of Sensitive Topics

16.1   The Motivation for Writing This Chapter

After studying the first 15 chapters, readers should understand the fundamental concept of the mobile radio field. This chapter will try to give readers further in-depth clarification of several topics that are not generally treated in books on this subject.

It is very important to understand that this chapter will also help readers find new problems and do their own research. As an example: many people can read and appreciate novels; but after reading many novels, only a few can write their own. Many stimulating and interesting topics will help readers dig for more unsolved problems.

Since the 1960s, many engineers and scientists had joined the new mobile radio field. But at that time, not all of them understood the mobile radio environment. As they began to learn about the environment, they found it different from other communication environments such as satellite, microwave links, ionospherical propagation, and troposcattering. Based on their knowledge at that time, they selected and applied the proper technologies and were successful. Some of the engineers simply adapted existing equipment used for other environment applications such as wireline or fixed-to-fixed radios. For example, some modified the data modem used for the wireline in cellular by simply adding a scheme to protect the data stream from the handoff process. They totally neglected the multipath dispersive fading phenomenon. Of course, that kind of modem did not perform as expected. Therefore, it is important to clarify first the concept and then apply proper technologies in the mobile radio environment.

16.2   Research Direction in the 1960s

In the 1960s, research projects that pursued the feasibility of mobile radio communication systems had at least two phases: (1) understanding the nature of multipath fading and (2) searching for techniques to reduce signal fades.

The mobile radio environment

The conditions for forming the mobile radio environment

1. The antenna height h_2 of the mobile unit is close to the ground, as shown in Fig. 16.1.
2. The mobile unit travels at a variable speed.

Causes for the environment

1. The ground-reflected wave cannot be ignored in reception when the antenna height is close to the ground.
2. Signal reception is effected by the natural terrain contours.
3. Human-made structures, like buildings, are situated along the propagation path from the base station to the mobile unit and create multipath waves if the antenna height of the mobile unit is lower than the surrounding structure heights. Only those reflected multipath waves, due to the structures surrounding the mobile units, are collected at the mobile unit.
4. Human-made noise also affects signal reception because the mobile antenna height is closer to the building structures, and the mobile antenna easily receives the noise.

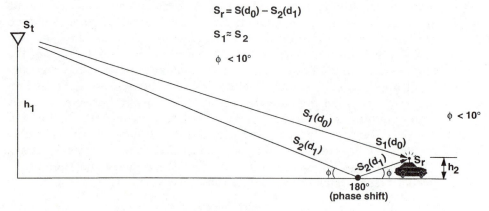

Figure 16.1   Mobile radio reception at the mobile unit.

Signal impacted by the mobile radio environment

Impact factors and phenomena

Excessive path loss. Excessive path loss can be explained by a two-wave model showing the direct and ground-reflected waves added together at the mobile receiver. Because of the reflection, coefficiency is always close to −1 due to the small incident angles of the reflected wave. The resulting signal, based on the sum of two waves, becomes very weak, and the power path loss is roughly $1/d^4$ (40 dB/dec), as shown in Eq. (3.44). Two wave models are only used to understand the path loss mechanism, not to explain the multipath-fading phenomenon.

Long-term fading. Due to the nature of the terrain, the average signal is strong when the mobile unit is at the high spot and weak when at the low spot. This average signal is called local mean. The local mean is a random variable. Its statistics follow the log-normal distribution.

Antenna height. When the antenna height at the base increases, reception at the mobile is stronger.

Area-dependent path loss. The multipath wave is created after the transmit wave reflects from human-made structures along the propagation path. These waves head in different directions, so the energy carried along a certain path becomes much weaker. The degree of signal weakness depends on the size and height of the human-made structures in cities and towns. This area-dependent path loss factor must be measured because the structure of each city is different. Those multipath waves reflected by the structures surrounding the mobile unit and then received at the mobile unit do not contribute any additional path loss to the reception.

Signal strength. Multipath waves are generated by the surrounding structures. Thus, the summed-up signal received can be strong in one area but weak in another. The mobile unit can be situated in a signal null spot. The raising or lowering of the signal received at the mobile unit depends on the location of the standstill mobile.

Signal fading. The medium becomes a moving medium when the mobile unit is moving. The signal received varies up and down. It is called *signal fading*. Its statistical nature is called *Rayleigh fading*. The fading frequency is a function of mobile speed (i.e., $f_d = V/\lambda$). When the speed is high, fading frequency (fading rate) is also high (i.e., above 3 kmph). This is called *fast fading*. When the speed is low (i.e., below 3 kmph), the fading frequency is low. We call it *slow fading*. The 3 kmph is based on the human walking speed.

Dispersive medium. Due to the path length of each multipath wave created along the path from the base station to the mobile unit, the time arrival of each multipath wave is different. If we send a data symbol at the base, the mobile unit receives the same data symbol many times, like echoes. The time used for collecting all of the echoes is called *time delay spread,* Δ. The time delay spread is measured in microseconds, and the spread is wider in an urban area than in a suburban one. If the time interval (T) between two symbols (where $T = 1/R$ and R is the transmission rate) sent from a transmitter is larger than the time delay spread of the medium, then no intersymbol interference will occur at the reception. Otherwise, an equalizer is used to eliminate the delayed echoes. It is mentioned in Sec. 16.6 that this delay spread can be also reduced with a diversity combining technique.

The rough-terrain criterion. The rough-terrain criterion is expressed in Eq. (3.29) as:

$$H_R = \frac{\lambda d}{8(h_1 + h_2)} \tag{16.1}$$

where h_1 and h_2 are the base antenna height and the mobile antenna, respectively. Rough terrain criteria H_R will increase as the distance increases. For example, the same terrain height H can be treated as a rough terrain ($H > H_{R_1}$) at a short distance but as smooth terrain ($H < H_{R_2}$) when the distance is longer, as shown in Eq. (16.1). Also, the value of H_R is smaller when the antenna height h_1 is higher at the base, as shown in Eq. (16.1).

Causes of signal reception

1. Signal coverage is much smaller due to 40 dB/dec.

2. Voice quality is degraded.

3. The data stream has burst errors instead of random errors due to multipath fading. The mobile speeds are always changing, and the duration of signal fades change in real time accordingly. These unpredictable changes make it difficult for interleaving and coding schemes in reducing burst errors effectively.

4. Excessive propagation path loss (40 dB/dec) forms a better interference reduction environment. Thus, the excessive propagation path loss nature helps in achieving a high-capacity frequency-reuse system that is applied to cellular and PCS.

16.3   AWGN Channels

Transmission of digital information over the AWGN (additive white Gaussian noise) channels most often applies to wireline and fixed-to-fixed radio channels.

A process $x(t)$ is called white noise if its spectral density $S_n(f)$ is a constant over all frequencies. White noise is not a real physical process. However, thermal noise has almost a constant spectral density over a wide spectrum as white noise. Its spectral density is

$$S_n(f) \approx \frac{kT}{2} \qquad 0 \leq f < f_1 \qquad (16.2)$$

$$S_n(f_1) = (0.9)\frac{kT}{2} \qquad \text{at } f_1 = 2 \times 10^{12} \text{ Hz} \qquad (16.3)$$

where $kT = -174$ dBm/Hz at 17°C. $kT = N_0$ is the thermal noise power/Hz. $S_n(f)$ is referred to as a two-sided power-spectral density. The autocorrelation of white noise is an inverse Fourier transformation of $S_n(f)$, an impulse function.

$$R_n(\tau) = F^{-1}\left(\frac{N_0}{2}\right) = \frac{N_0}{2}\,\delta(\tau) \qquad (16.4)$$

or

$$R_n(\tau) = \frac{N_0}{2} \qquad \text{at } \tau = 0 \qquad (16.5)$$

This means that the noise has no correlation if $\tau \neq 0$. This property is used for generating pseudorandom noise (PN) for synchronization or for spread spectrum.

The Gaussian process

Most random variables such as age or grade distribution in a class have the Gaussian property.

In the real environment, thermal noise or any multipath fading signal has the same nature as noise that performs Gaussian density function.

Noise has a complex Gaussian process

$$n(t) = n_r(t) + jn_s(t) \qquad (16.6)$$

and the amplitude of $n(t) = \{n_r^2(t) + n_s^2(t)\}^{1/2}$ is a Rayleigh. Their spectral density $S_n(f)$ shown in Eq. (16.2) is a constant (white process) over a bandpass filter B. The thermal noise power over B is

$$P_n = \int_{-B/2}^{B/2} S_n(f)\,df = \int_{-B/2}^{B/2} \frac{kT}{2}\,df = kTB \qquad (16.7)$$

In the multipath fading signal, spectral density is no longer a white process, but the complex Gaussian process still holds true for the signal. The fading signal spectral density is different depending on the mobile speed and the position of the directional antenna related to the heading of the mobile shown in Fig. E7.1.2.

16.4   Mobile Radio Fading Model

Many textbooks published recently give credit to sources that may be incorrect.

I would like to give you a bit of history. In 1964, J. D. Pierce was the executive director of Bell Lab Research Division 13. Pierce initially suggested that the energy-density diversity expression

$$V_{\mathrm{III}}^2 = |E_z|^2 + |H_x|^2 + |H_y|^2 \tag{16.8}$$

could be used to reduce signal fading, as indicated in Sec. 9.5. He asked E. N. Gilbert to study mobile radio energy-density reception. At the same time, I was assigned to study the implementation of the new concept. Six months later, Gilbert came up with a statistical model. Gilbert used his statistical model and derived the first order of statistical properties for E_z, H_x, and H_y. His paper was published in *BSTJ* (*Bell System Technical Journal*) in October 1965. Earlier in 1964, only Gilbert had the knowledge to form the statistical model for the newly developed mobile radio field at Bell Labs. In 1964, R. H. Clarke and I were just joining the Labs, where I learned of the model from Gilbert's internal memo in 1964 and the classic S. O. Rice paper. I first derived the formula from the model for the level crossing rates and average durations of fades for E_z, H_x, and H_y that appeared in Sec. 6.5 and Sec. 6.6 of this book. My paper was finished in late 1965, but it was published in *BSTJ* (*Bell System Technical Journal*) in February 1967. The cause of the delay was my inexperience with writing.

R. H. Clarke published his paper in July-August 1968. He used Gilbert's statistical model and included my material in his paper. Clarke never claimed Gilbert's model as his. He just did not give proper credit to Gilbert.

In 1974, Jake's book was published. He included Gilbert's statistical model and the formula and curves of level-crossing-rate and average duration of fades from my paper in his Chap. 1. Unfortunately, he forgot to quote properly Gilbert's or my paper in the Chap. 1 references. Lee's work of the measurement of electric and magnetic fields for studying the value of an energy density antenna [$\omega = \frac{1}{2}(\varepsilon E^2 + \mu H^2)$] was published in *BSTS,* September 1967. The power spectrum density of mobile radio signals was published in IEEE transactions on vehicular technology in 1967 [1]. Except for the last reference, these sources can be found from the references in Chap. 6.

Gilbert and I concentrated on the problem of energy reception, where the E field was just one of three components. The theory for energy reception was developed at that time. I wish that new authors would read Gilbert's paper in order to appreciate his contribution to mobile communications field. Indeed, it is time to name the statistical mobile radio model, "Gilbert's model."

16.5   Effect of the Height of an Antenna

1. *Effect on the propagation loss due to the mobile station antenna height.*   Because the height of the mobile unit antenna is very close to the ground, the ground-reflected wave cannot be ignored. Also, the reflected coefficient is close to -1 based on the small incident angle θ. The angle θ is a function of the base station antenna height h_1, the mobile antenna height h_2, and the distance as shown in Fig. 3.5 or Fig. 16.1.

$$\frac{h_1 + h_2}{r} = \tan \theta \approx \theta \tag{16.9}$$

Fortunately, the two waves, direct and reflected, do not cancel each other because of the finite difference in their phases, βd_0 and βd_1, as shown in Fig. 16.1. Since $d_1 > d_0$ always, the resultant signal is always weak but never null. As a result, the path loss of the fourth-power rule, or 40 dB/dec, is observed.

2. *Effect on the antenna height gain due to the base station antenna height.*   The antenna height gain ΔG_x is measured based on the effective antenna height, not the actual antenna height. But the reference in calculating gain is based on the actual antenna height. At the location x, the gain ΔG_x can be expressed as

$$\Delta G_x = 20 \log \frac{h_e(x)}{h_a} \quad \text{in dB} \tag{16.10}$$

The ΔG_x always varies as the location x changes. If $h_e(x) < h_a$, then ΔG_x is a negative value in dB.

3. *Terrain roughness due to both base station and mobile station antenna height.*   The terrain roughness affects the signal reception when the mobile antenna height is close to the ground (see Fig. 3.8). When the mobile antenna height is low, the criterion for terrain roughness can be affected by base station antenna height h_1 as shown in Eq. (16.1). Increase the antenna height; the criterion of terrain roughness height is reduced. In other words, the actual terrain, thus, is increasing its chance to exceed the criterion of height of terrain roughness H_R to be claimed as rough ground surface. Therefore, the terrain has been claimed to be a smooth ground at a certain antenna height, and when the antenna height increases, the same terrain is claimed to be a rough ground surface.

4. *Effect on the angle spread of signal arrival due to the base station antenna height.*   The angle spread BW is described in Sec. 6.9. The angle BW will become wider as the base station antenna height is lower. This phenomenon will be described in Sec. 16.18.

16.6   The Difference between Path Loss Prediction and Local Mean Prediction

Generation of path loss curve

Path loss prediction is an averaging process based on path loss over radio path. Since the mobile station is moving along a mobile path (a road), to generate a predictive curve for path loss, we need to collect many data points from the different mobile paths and plot the data points corresponding to the radio-path distance from the base station. From the plot, we can average many data points at the same radio-path distance x_1 to obtain an average value at x_1. Actually, the average process for generating a path loss curve is based on a best fit of all the points along the x-axis of the radio path. However, the eyeball-fit curves are normally used. In Fig. 16.2, we show that the eyeball-fit curves are insensitive to measured data. The four data-fitting curves all represent the data well. This is the reason that the Okumara-Hata path loss formula can generally represent the data fairly well. Besides, it is easy to use.

Standard deviation of path loss

The standard deviation of the data points along the path loss curve depends on the number of sample points in a confined area. From Okumura's data, based on the data measured in Tokyo, the standard deviation is 6 dB. From suburban data in the United States based on many different geographical areas, the standard deviation is 8 dB. The value of the standard deviation is proportional to the size of the area and the number of different areas where the measured data is collected. The larger the size of the area and the larger the number of different areas, the standard deviation is bigger. The standard deviation is totally due to the variation of the terrain configuration and the human-made structure.

Local mean prediction

A local mean is generated along the street designated as the mobile path. The local mean is obtained along only one mobile path, *not* averaged on all of them. The predicted local mean is then compared with the measured one along a particular path. Therefore, the predicted path loss curve obtained from an average process cannot be used for predicting the local means.

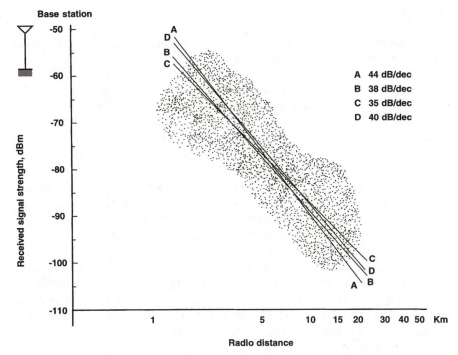

Figure 16.2   Insensitivity of curve fitting in path loss curves.

Actually, the design of a cellular or PCS system is always based on the local mean prediction. The Lee model is a local mean prediction model and is used to save manpower and cost during the startup system stage. Also, it is used for troubleshooting and improvement of capacity during the stages of system operation and expansion.

Although the Okumura-Hata formula generally represents path loss prediction, the Lee model can predict the local means more accurately. Therefore, there are comments that the Okumura model is good based on the data generally fitted to the path loss prediction curve, and there are comments that the Lee model is good based on the data generally fitted to the local mean prediction. Of course, deploying a cellular system or studying the handoff performance should *always* be based on the local mean prediction.

The effect of human-made structures

We can first either collect enough measured data of signal strength in any given city by covering all possible terrain contours or selected high and low spots. Although the standard deviation from all data points retains the terrain-contour variation, the generated path loss curve through the averaging process washes out the terrain contour variation [2]. The explanation of the washout process while generating the path loss curve is shown in Fig. 16.3. The signal strengths are stronger at the higher spots and weaker at the lower spots as shown in the figure. As a result, the averaging signal strength at a given radio distance corresponds to the signal strength as if it is received on a flat ground. The path loss curve now represents the character of the city as if it were situated on flat ground. The explanation is described in Ref. 2. We call this curve the area-specified path loss curve. The curve is described by both the one-mile (or one-kilometer) intercept and the slope of path loss to characterize the structures of the given city. The explanation of the impact on these two parameters is shown in Fig. 16.4. In Fig. 16.4(a), the base-station antenna is in the urban area. The signal strength is attenuated fast due to the surrounding buildings. Therefore, the one-mile intercept is low and the slope is flat. In Fig. 16.4(b), the base-station antenna is in the suburban area. The signal strength is generally strong due to clear surroundings. In this case, the one-mile intercept is high and the slope is steep due to the strong attenuation in the far-out urban area. In Fig. 16.4(c), the base station is situated in an evenly distributed building environment. The one-mile intercept point is up or down depending on the density of buildings S, but the slope will be kept as 40 dB/dec.

In general, the one-mile intercept value is more sensitive and impacted than the slope value. This means that a 1-dB error in the one-mile intercept affects the prediction more than a 1-dB error in the slope value.

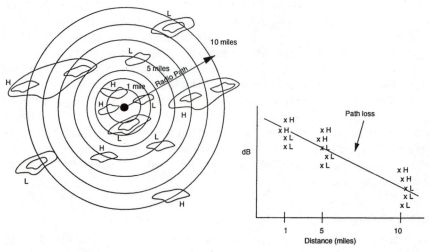

Figure 16.3 Explanation of the wash-out process of terrain contour variation while generating a path loss curve.

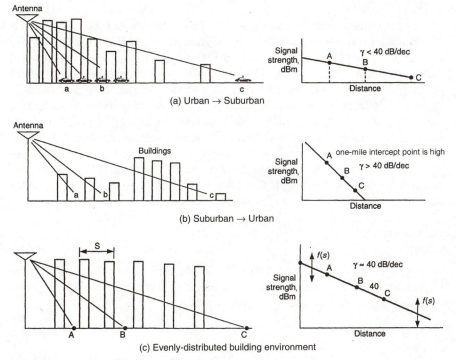

Figure 16.4 Explanation of the path loss phenomenon.

16.7 Combined Signal Diversity

Diversity is a means of generating two or more uncorrelated or less correlated fading signals at the reception. Diversity can only be used in the multipath-fading environment. In a line-of-sight environment, it is impossible to generate two uncorrelated fading signals because, in a line-of-sight condition, no multipath waves can be generated. The signal is not distorted. Diversity is not needed. Diversity combining is used by summing two or more uncorrelated fading signals to reduce the signal fading caused by the multipath waves during reception. The fading is reduced by many different combining techniques, as described in Chap. 10.

The diversity can reduce the signal fading, reduce the time-delay spread (Sec. 10.12), increase the coverage and so on. However, the clarification of using diversity should be stated as follows.

Diversity gain

Diversity gain is not a real gain. To find a diversity gain, first find each threshold of diversity signal and nondiversity signal so that 90 percent or 95 percent of each of the signals is above its corresponding threshold. Then, compare the difference of the two threshold levels between a diversity signal and a nondiversity signal in dB. If the threshold level of a signal at 90 percent is −10 dB (10 dB below the average power) before the diversity, and the level of a diversity-combined signal at 90 percent is −6 dB after the diversity (as shown in Fig. 10.8 for comparing a single-branch and a two-branch diversity), the diversity gain is 4 dB. The diversity gain is different based on the specified percentage of a signal above the threshold level. When a high percentage of signal above a threshold is used, more diversity gain can be observed. At the level of a signal at 50 percent, the diversity gain is small. But we are not interested at this level. Usually, the threshold above which is 90 percent of the signal strength, is used to calculate the diversity gain.

**Diversity does not improve *C/I*
in postdetection combining**

We have to emphasize that diversity combining can reduce fading and the delay spread but does *not* reduce the *C/I* of a signal. When three branch signals are added together in power,

$$\frac{\sum_{1}^{3} C_i}{\sum_{1}^{3} I_i} = \frac{C}{I} \tag{16.11}$$

If all C_i and all I_i are identical, then

$$\frac{3C_i}{3I_i} = \frac{C_i}{I_i} = \frac{C}{I} \tag{16.12}$$

**Diversity does improve *C/I*
in predetection combining**

Take the equal gain combining, as expressed in Eq. (10.73):

$$\frac{\left(\sum R_i\right)^2}{\sum I_i} = \frac{C}{I}$$

If all R_i^2 are equal and all I_i are equal, then

$$\frac{C}{I} = \frac{\left(\sum_{1}^{M} R_i\right)^2}{\sum_{1}^{M} I_i} = M\frac{R_i^2}{I_i} \tag{16.13}$$

Predetection combining provides the improvement of 3 dB in a 2-branch ($M = 2$) diversity combining, as shown in Eq. (16.13).

**Diversity reception is different
from antenna-array reception**

Diversity is not combined at the RF stage, but antenna array is. For example, when a signal $S_o(t)$ is transmitted and received by two RF signals after going through the mobile radio environment, the two signals received are

$$S_1(t) = R_1\, e^{j\phi_1}\, S_o(t) \tag{16.14}$$

$$S_2(t) = R_2\, e^{j\phi_2}\, S_o(t) \tag{16.15}$$

where R_1 and R_2 are the Raleigh fading envelopes and ϕ_1 and ϕ_2 are random phases. Diversity combining is combining $R_1 S_o(t)$ from Eq. (16.14) and $R_2 S_o(t)$ from Eq. (16.15). Antenna-array combining is combining $S_1(t)$ from Eq. (16.14) and $S_2(t)$ from Eq. (16.15), involving the phase terms ϕ_1 and ϕ_2. The phase term ϕ is related to the wave or the resultant wave arrival in different directions in space:

$$\phi = \beta d \cos \theta \tag{16.16}$$

where d is the spacing between two antennae, and θ is the angle arrival of the wave.

Since diversity combining does not involve the phase, diversity combining is independent from the arrival of the waves. The antenna-array combining $[S_1(t) + S_2(t)]$ has a directivity related to the combining.

Diversity combining is used to reduce signal fading and time delay spread. Antenna array is used to change the antenna pattern for direction finding or interference reduction.

**Diversity from the transmit
or at the reception**

Usually, diversity is used at the reception for an FDD (frequency division duplexing) system for two reasons: (1) no additional signal transmitted power needs to be generated—it is a power saver; and (2) in the meantime, no additional interference is created to other users.

However, there is an advantage to using diversity at the transmit end for TDD (time division duplexing) systems. Diversity for transmitting and receiving signals can be engineered at the base stations and used for both terminals. The arrangement is illustrated in Fig. 16.5.

In Fig. 16.5, the mobile unit has only a regular transmitter and receiver, but it can still benefit from diversity reception. Take the two received signals at the base station, as expressed in Eq. (16.14) and Eq. (16.15), where the difference in two phases is

$$\Delta\phi = \phi_1 - \phi_2 \tag{16.17}$$

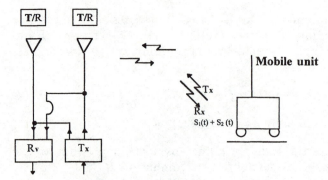

Figure 16.5   Diversity scheme for the TDD system.

Now, since TDD uses the same frequency at both the transmitting and receiving ends, the phase difference is the same at the mobile unit. Therefore, at the base station, the two transmitted signals can be arranged as

$$S_{t_1}(t) = e^{-j(\phi_1 - \phi_2)} S_o(t)$$
$$S_{t_2}(t) = S_o(t)$$

$$(16.18)$$

Then, when two signals arrive at the mobile unit,

$$S_1(t) = R_1 e^{j\phi_1} [e^{-j(\phi_1 - \phi_2)} S_o(t)] = R_1 e^{j\phi_2} S_o(t)$$
$$S_2(t) = R_2 e^{j\phi_2} S_o(t)$$

$$(16.19)$$

The single antenna receiving the two signals at the RF stage at the mobile is

$$S_1(t) + S_2(t) = (R_1 + R_2)e^{j\phi_2} S_o(t) \qquad (16.20)$$

The signal received at the mobile unit is a diversity combined signal. The common phase ϕ_2 can be factored out. Figure 16.5 shows that a diversity advantage is gained at both the base station and the mobile unit but is only engineered at the base station.

16.8   Antenna Height Gain and Diversity Gain

Antenna height gain Δ_{AHG} can be expressed as

$$\Delta_{\mathrm{AHG}} = 20 \log\left(\frac{h_1'}{h_1}\right) \qquad (16.22)$$

where h' is the new effective antenna height. If $h'_1 < h_1$, then Δ_{AHG} is a minus in dB.

Diversity gain Δ_{DG} is the measured difference in dB between the diversity signal and nondiversity signal at 90 percent signal level. Diversity gain is inverse proportional to the correlation coefficient ρ, where ρ is proportional to the base station antenna height h_1:

$$\rho_{h_1} = f\left(\frac{h_1}{d}\right)$$

This is shown in Fig. 9.4.

It is assumed that the antenna spacing d is a given. Thus, if $h'_1 < h_1$, $\rho'_{h_1} < \rho_{h_1}$, then the diversity gain shown in Fig. 10.9 is based on different values of ρ, which is from h_1/d.

$$\Delta_{\text{DG}}(h'_1) - \Delta_{\text{DG}}(h_1) = +\,\text{dB} \tag{16.23}$$

To obtain the proper diversity gain, there are two necessary conditions:

1. The difference in average signal strengths of two fading signals S_1 and S_2 must be less than 3dB. That is,

$$\Delta_{S_1 S_2} = |\,\overline{S}_1 - \overline{S}_2\,| = \left|\,20 \log \frac{h'_{S_1}}{h'_{S_2}}\,\right| \le 3\,\text{dB} \tag{16.24}$$

where $\overline{S}_1$ and $\overline{S}_2$ are average strengths.

2. The requirements of the correlation coefficient ρ as shown in Eq. (9.5a) as

$$\rho_{S_1 S_2} < 0.7 \tag{16.25}$$

Combining these two necessary conditions becomes a sufficient condition.

Now, examine Eqs. (16.22) and (16.23). We may conclude that

$$\Delta_{\text{AHG}} \propto \frac{1}{\Delta_{\text{DG}}} \tag{16.26}$$

When h'_1 increases, Δ_{AHG} increases but Δ_{DG} decreases. From Eq. (16.24), we point out that the condition $h'_{S_1} = h'_{S_2}$ is the optimum condition. The horizontal separation can achieve this condition. Therefore, for diversity gain, horizontal separation is better than vertical separation. However, when the frequency is at 1.8 GHz or higher, the wavelength becomes much smaller. Therefore, the vertical separation can be used since the necessary condition in Eqs. (16.24) and (16.25) are met.

Vertical separation versus horizontal separation

At base stations. The horizontal separation between two omnidirectional antennae has a drawback. A 20° sector on each side of $\alpha = 90°$ presents the nondiversity gain region as shown in Fig. 16.6(*a*). On the other hand, the vertical separation has some merit, as long as the two required conditions are met. The two vertically mounted omnidirectional antennae can cover all 360°, as shown in Fig. 16.6(*b*). Also, the vertical separation is much easier to install on the antenna mast.

At mobile units. The horizontal antenna separation at the mobile unit does not create the nondiversity-gain region. The reason is that the signal that arrives at the mobile unit is from 360° horizontally, unlike the signal that arrives at the base station. (See Fig. 6.11.) Therefore, with a 0.5-λ horizontal antenna separation at the mobile unit, the requirements of Eqs. (16.24) and (16.25) are met. The diversity gain is achieved. We also find that the two horizontally separated antennas at the mobile unit should line up with the direction of motion (front to back) in order to achieve the maximum diversity gain [3]. The vertical separation between the omnidirectional mobile antennas can also be implemented as shown in Fig. 16.7(*a*).

Per Ref. 4, the correlation between the two received signals is

$$\rho\left(\frac{d}{\lambda}, \theta\right) = \frac{\sin\,[(\pi d/\lambda)\,\sin\,\theta]}{(\pi d/\lambda)\,\sin\,\theta} \tag{16.27}$$

Equation (16.27) is plotted in Fig. 16.7(*b*) with the measured data.

From the measurement data, we found that, in most cases, the wave arrived at the mobile unit is at an elevation angle of 30° or more above the horizontal. Therefore, when the vertical antenna separation $d = 1.5\,\lambda$ at the angle of wave arrival $\theta \geq 30°$, the correlation $\rho \leq 0.4$ is met with Eq. (16.25). Also, at the mobile unit, the height gain of the antenna is 3 dB/oct. The two antennae are vertically separated by 1.5 λ at 850 MHz, or 0.52 m. Assuming that the lower antenna height is 3 m, then the height gain is

$$\Delta G = 10\,\log\,\frac{3.52\text{ m}}{3\text{ m}} = 0.69\text{ dB}$$

which is met with Eq. (16.24). Therefore, the vertical antenna separation can be used at the mobile unit.

Details on the vertical separation of two base-station antennae and two mobile antennae can be found in Refs. 4 and 5.

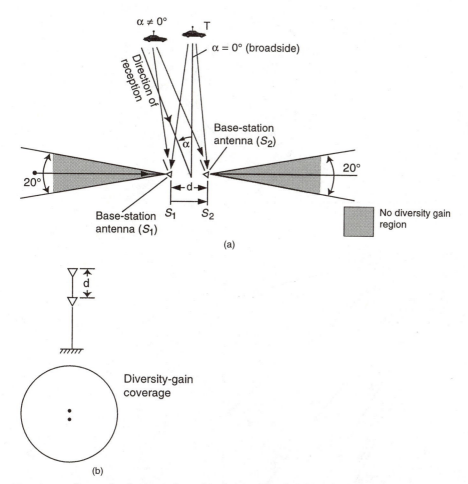

Figure 16.6   Comparing horizontal spacing: h = antenna height in wavelength; d = antenna spacing in wavelength. (a) Regions for no diversity gain from two omnidirectional antennae separated horizontally (applying at 850 MHz or lower); (b) diversity gain covers all 360° from two omnidirectional antennae separated vertically (applying at 1.8 GHz or higher).

16.9   Time Delay Spread Δ

When the transmission rate R of a data stream is high and the time interval $1/R$ exceeds the time delay spread Δ, then the ISI (intersymbol interference) affects the received signal regardless of whether the mobile unit is moving or at a standstill. Thus, the phenomenon of time delay spread is not like the multipath fading phenomenon that occurs only while the mobile unit is in motion.

When the mobile speed $V = 0$, the pulse shape of time delay spreads of two adjacent symbols are the same. It is easy to use the training symbol sequence implemented in the equalizer to reduce ISI effectively.

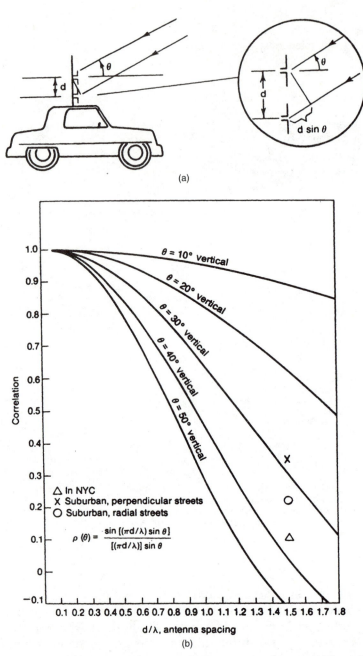

Figure 16.7 Two vertically spaced antennae mounted on a mobile unit: (*a*) Vertical separation between two mobile antennae; (*b*) correlation vs. antennae spacing.

When the speed $V \neq 0$, especially when the mobile speed is very high, the pulse shapes of time delay spreads of two adjacent symbols are not the same, and the training symbol sequence may not be effectively used in the equalizer. The equalizer is described in Sec. 2.15.

**Time delay spread does not affect
the analog voice signal**

If the voice frequency f_a is from 300 to 3000 Hz, then $1/f_a$ is from 3 to 30 milliseconds. This is much longer than the value of time delay spread that is usually less than 100 µs. Therefore, the human ear cannot notice the effect. But transmitting the data over the analog system when the transmission rate is high, causes the time delay spread to affect the data signal at the reception.

**Less effect on FDMA and CDMA
than on TDMA**

The time delay spread is very effective to the transmission rate of the data, either on analog or digital. For this reason, we must examine the transmission rate for each multipath access scheme. The FDMA channel sends the transmission rate per channel R, and the TDMA channel sends the equivalent transmission rate per slot channel R_1. If $R = R_1$, then the total transmission rate of an eight-time-slot TDMA is $8R$. Over the air link, the transmission rate R of FDMA is much less than $8R$ of TDMA. The time delay spread always affects less on the lower transmission rate as FDMA than the higher transmission rate as TDMA.

In CDMA, although the chip rate is high (1.23 Mcps), the bit rate is 9.6 or 14.4 kbps. This means that each information bit has spread in 128 chips for a data rate of 9.6 kbps and 88 chips for 14.4 kbps using spread spectrum modulations. Therefore, each bit carries a great deal of redundancy. The spread spectrum modulation helps overcome the effect of time delay spread by collecting the redundancy chip signal through a correlator at the receiving end. Although only partial energy from the chip signal for each information bit is received, each information bit is sufficiently recovered, and no equalizer is needed.

**Time delay spread can be reduced
by diversity scheme**

In Sec. 10.12, we demonstrated that the time delay spread can be reduced by the M-branch diversity, as shown in Eq. (10.159).

$$\Delta_d = \frac{\Delta}{M} \qquad (16.28)$$

The new Δ_d is reduced by the factor of M. The large number of M branches reduces Δ_d drastically, but the cost of the M-branch ($M > 2$) diversity receiver is very high. Usually, $M = 2$ is the most cost-effective number to be used to compare the cost and benefit of using it.

16.10   Noise Figure (NF) Issues

Noise figure (NF) has many definitions. In general, one is used for ambient noise, one is used for receiver noise, and one is used for the antenna noise.

Ambient noise

We measure the ambient noise received from the antenna over the thermal noise. The equivalent circuit of the antenna is shown in Fig. 16.8(a). The available thermal noise power of a resistor R_L, where complex impedance $Z_A = Z_L^*$, is

$$P_a = \frac{V_s^2}{4R_L} = \frac{4kTB \cdot R_L}{4R_L} = kTB \qquad (16.29)$$

Z_L^* is the complexed conjugate of the load impedance, Z_L is the load impedance, $Z_L = R_L + JX_L$. Z_A is the equivalent impedance of antenna; $Z_A = R_A + jX_A$.

The available thermal noise power does not depend on the resistance of R_L, as shown in Eq. (16.29).

NF is used to measure the ambient noise P_{ao} received from the environment above the thermal noise P_a, or

$$F = \frac{P_{ao}}{P_a} = \frac{Q \cdot (kTB)}{kTB} = Q$$

Receiver noise

S_i and N_i are the inputs of signal and noise, respectively, as shown in Fig. 16.8(b). S_0 and N_0 are the outputs of the final stage; S_1 and S_2 and N_1 and N_2 are the signal and noise of intermediate stages; and n_1 and n_2 are the internal noise of amplifiers G_1 and G_2, respectively.

The noise figure F_1 at first stage is

$$F_1 = \frac{\frac{S_i}{N_i}}{\frac{S_1}{N_1}} = \frac{\frac{S_i}{N_i}}{\frac{G_1 S_i}{G_1 N_i + n_1}} = \frac{\frac{S_i}{N_i}}{\frac{S_i}{N_i + \frac{n_1}{G_1}}} = \frac{N_i + \frac{n_1}{G_1}}{N_i} = 1 + \frac{n_1}{G_1 N_i} \qquad (16.30)$$

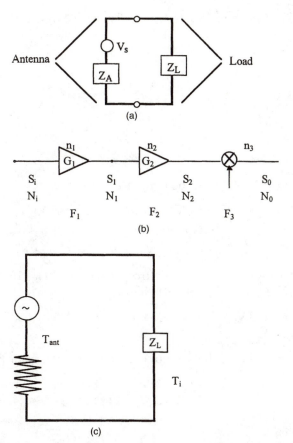

Figure 16.8 Equivalent circuits of (*a*) antenna load, (*b*) receiver, and (*c*) antenna temperature.

The value of n_1 is obtained from Eq. (16.30) as

$$n_1 = G_1 N_i (F_1 - 1) \tag{16.31}$$

Let $S_2' = G_2 S_i$ and $N_2' = G_2 N_i + n_2$. The noise figure F_2 at the second stage defined as

$$F_2 = \frac{\dfrac{S_i}{N_i}}{\dfrac{S_2'}{N_2'}} = \frac{\dfrac{S_i}{N_i}}{\dfrac{G_2 S_i}{G_2 N_i + n_2}} = 1 + \frac{n_2}{G_2 N_i} \tag{16.32}$$

$$n_2 = G_2 N_i (F_2 - 1) \tag{16.33}$$

Let $S_3' = G_3 S_i$ and $N_3' = G_3 N_i + n_3$. The noise figure F_3 at the third stage is

$$F_3 = \frac{\dfrac{S_i}{N_i}}{\dfrac{S_3'}{N_3'}} = 1 + \frac{n_3}{G_3 N_i} \tag{16.34}$$

and

$$n_3 = G_3 N_i (F_3 - 1) \tag{16.35}$$

The output noise can be expressed as

$$N_0 = G_3 N_2 + n_3 = G_3 (G_2 N_1 + n_2) + n_3$$
$$= G_3 G_2 (G_1 N_i + n_1) + G_3 n_2 + n_3$$
$$= G_1 G_2 G_3 N_i + G_2 G_3 n_1 + G_3 n_2 + n_3 \tag{16.36}$$

Substituting n_1, n_2, and n_3 from Eqs. (16.31), (16.33), and (16.35), respectively, into Eq. (16.36) yields

$$N_0 = G N_i + G(F_1 - 1)N_i + G_2 G_3 (F_2 - 1)N_i + G_3 (F_3 - 1)N_i \tag{16.37}$$

where $G = G_1 G_2 G_3$.

The overall F of the system can be expressed from Eq. (16.37):

$$F = \frac{N_0}{G N_i} = \frac{1}{G_1 G_2 G_3} \frac{N_0}{N_i}$$
$$= F_1 + \frac{F_2 - 1}{G_1} + \frac{F_3 - 1}{G_1 G_2} \tag{16.38}$$

or the overall F of the system can be expressed by substituting Eq. (16.36) into Eq. (16.38):

$$F = 1 + \frac{n_1}{G_1 N_i} + \frac{n_2}{G_1 G_2 N_i} + \frac{n_3}{G_1 G_2 G_3 N_i} \tag{16.39}$$

If there is no noise in the system ($F_1 = F_2 = F_3 = 1$), then $F = 1$.

Antenna-temperature noise

Consider that a noise source from the antenna has a temperature T as shown in Fig. 16.8(c). Then, the noise power P_a at the antenna is kTB, or

$$P_a = kT_{ant} B \tag{16.40}$$

If the noise temperature at the antenna T_{ant} is connected to a receiver that has an equivalent noise temperature T_i, then the effective source of temperature is $T = T_{ant} + T_i$.

The output is

$$P_{a0} = kG \left(T_{\mathrm{ant}} + T_i \right) B \qquad (16.41)$$

where G is the antenna gain.

The noise figure F can be expressed as

$$F = 1 + \frac{T_i}{T_{\mathrm{ant}}} \qquad (16.42)$$

The antenna T_{ant} is usually applied to satellite antenna temperature.

16.11   Power-Limited and Bandwidth-Limited Systems

Power-limited systems

A power-limited system is used where the transmitted power is scarce and limited, but system bandwidth is available. In this situation, bandwidth can be increased to compensate for the limited power. Use the formula from Eq. (15.8) to indicate the relationship among the three parameters (C/I, E_b/I_0, and B/R) shown below:

$$\frac{C}{I} = \left(\frac{E_b}{I_0} \right) \!\! \left/ \!\! \left(\frac{B}{R} \right) \right. \qquad (16.43)$$

Let the required C/I be a constant by increasing B, which is the equivalent of a gain in E_b/I_0; or the E_b/I_0 is a constant by increasing B, which is the equivalent of a reduction in C/I. The required C/I reduction means the required transmitted power is reduced.

Bandwidth-limited systems

Bandwidth-limited systems is used where bandwidth is scarce but power is available. Under this situation, we may try to increase transmitted power to compensate for the limited bandwidth. Use the formula of the bandwidth efficiency η:

$$\eta = \frac{R}{B} = \frac{\log_2 M}{BT} \qquad \text{bits/s/Hz} \qquad (16.44)$$

Equation (16.44) indicates the efficiency that a system transmits $\log_2 M$ bits in time T seconds with a bandwidth B in hertz. M is the symbol set size, and R is the transmission rate.

For $B = 1/T$, then

$$\eta = \log_2 M \qquad (16.45)$$

For $M = 4$ (QPSK) and $B = 1/T$, then the bandwidth efficiency is 2 bits/s/Hz. In this case, the efficiency increases with increasing M.

For $B = M/T$, then

$$\eta = \frac{\log_2 M}{M} \tag{16.46}$$

Let $M = 4$ in Eq. (16.46), then $\eta = 1/2$. In this case, the efficiency η decreases with increasing M.

In general, the communication resources are the transmitted power and the channel bandwidth. In the power-limited system, the coding scheme is used to save power at the expense of bandwidth. In bandwidth-limited systems, the spectrally efficient modulation techniques can be used to save bandwidth at the expense of power.

In the mobile radio environment, not only do both power and bandwidth limitations exist, but spectrally efficient modulation cannot be applied because constant envelope modulation is required. The coding scheme cannot be applied efficiently because of the variable duration of fades that causes a nonstationary burst error condition. Therefore, in the mobile radio environment, in addition to using the coding with interlearning and constant-envelop modulation, we use many other techniques such as frequency-reuse schemes, microcells, and smart antennae to overcome both power and bandwidth limitations.

16.12   Mobile and Portable Coverage

The coverage measurement is based on the signal strength received from a carrier signal. In most wireless communication systems, FDD (frequency division duplexing) is used. Every traffic or voice channel consists of a pair of frequencies, one for forward link and one for reverse link. For example, a voice call completed in an AMPS system needs four frequencies: two for setup (forward and reverse) and two for voice (forward and reverse). Now, we must explain how the measurement techniques used to determine coverage for the mobile and the portable are different.

Mobile coverage

Mobile coverage is based on the strength of a carrier signal averaged over time (or over a distance) and received at the mobile unit in motion. The average signal strength is also called the average received power. Fortunately, the same average power is measured from the signal strength received from any frequencies as long as they are within the same radio carrier band. If the transmitting power remains the same, the received average power of the four frequencies for completing an AMPS call are the same. Therefore, we only need to measure the average signal strength of one frequency, *any* one frequency; it does not necessarily need to be one of the four. The average power of all frequencies over the radio carrier band is the same. The mobile coverage is based on the average signal strength of a signal above a given threshold.

Portable coverage

Portable coverage is based on the signal strength at one spot, not the average signal strength. This is because the portable unit usually stands still or moves very slowly. The one-spot signal strengths of four frequencies to complete a call at a certain spot are not the same. This is due to frequency-selective fading. Some signal strengths are strong, and some are weak, as shown in Fig. 16.9. Then, the definition of portable coverage is based on all four frequencies being above a given threshold level S. This is different from the definition of mobile coverage. Portable coverage in an area is measured by the percentage of locations in that area where the call can be completed. In Fig. 16.9, the percentage at the cell boundary is D_1/D, where D_1 is the total length of all frequencies above the level S and D and is therefore the total length of interest. The percentage H of an area for coverage of a single frequency is calculated as shown in Eq. (4.62). Now, since all four frequencies must be received above a level A, the final percentage H' of the coverage in the area becomes [6]

$$H' = P\,(x > A) = H^4 \tag{16.47}$$

If coverage in an area is $H = 90$ percent for a single frequency, then the percentage of the area for the portable coverage is reduced to 65.6, as obtained from Eq. (16.47).

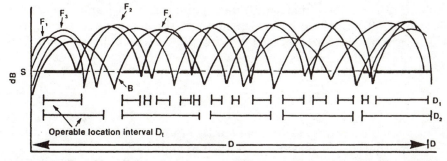

Figure 16.9 Portable cellular—four-frequency selective fading system: F_1 = forward link setup channel, F_2 = reverse link setup channel, F_3 = forward link voice channel, F_4 = reverse link voice channel, D = total length of interest, D_1 = total length that all four frequencies are above level S, and D_2 = total length that only F_4 is above level S.

16.13 Ray-Tracing and Building-Block Approach

Ray-tracing approach

A wave propagation is represented by rays based on geometrical optics [7]. Ray tracing is a technique used to find a signal strength through wave propagation in a simple geographical shape or space such as the waveguide, the fiber cable, or the ionosphere propagation [15]. The rays reflect according to Snell's law. The reflection coefficient is determined based on the wave incident angle and the material property of the reflected wall. The energy loss of the wave propagating through a medium also needs to be known. Then, the energy of the wave dissipated at each location can be calculated by ray-tracing techniques.

In a complex environment, ray-tracing techniques can still be applied, but the accuracy decreases. We cannot match the calculated amplitude and phase of a signal using ray tracing at any point with the measured amplitude and phase of the actual signal. Needless to say, we do not have superior vision to see where the rays travel in a complex structured environment. We can only guess where the ray is traveling. Therefore, ray tracing is a direct method. It is deterministic but inaccurate to use for predicting signal strength at each point in a complex space. However, to average all the predicted signal strength x_{ij} at each point i in a unit area j is

$$\overline{x}_j = \langle x_{ij} \rangle = \frac{\sum_{i=1}^{N} x_{ij}}{N} \tag{16.48}$$

$$\overline{x}_j^2 = \langle x_{ij}^2 \rangle = \frac{\sum_{i=1}^{N} x_{ij}^2}{N} \tag{16.49}$$

where N is the signal-strength data points in area j and the area is about 20–40 wavelengths in size, and $\overline{x}_j$ is the sample average as well as a variable random process following the Gaussian distribution. It is described in Sec. 2.11 as

$$p(x_{ij}) = \frac{1}{\sqrt{2\pi}\sigma_{ij}} \exp\left(-\frac{(x_{ij} - \overline{x}_j)^2}{2\sigma_{ij}^2}\right) \tag{16.50}$$

The variable $\overline{x}_j$ is the sample mean. Its standard deviation σ_j is much smaller than σ_{ij}, as shown in Eqs. (2.106) and (16.51):

$$\sigma_j = (\langle \overline{x}_j^2 \rangle - \langle \overline{x}_j \rangle^2)^{1/2} = \frac{\sigma_{ij}}{\sqrt{N}} \tag{16.51}$$

where

$$\langle \overline{x}_j \rangle = \sum_1^M \frac{\overline{x}_j}{M} \qquad (16.52)$$

$$\langle \overline{x}_j^2 \rangle = \sum_1^M \frac{\overline{x}_j^2}{M} \qquad (16.53)$$

N is the number of samples and M is the number of sample means. Equation (16.51) indicates that the standard deviation σ_j of $\overline{x}_j$ is reduced by a factor $\sqrt{N}$ from σ_{ij} where σ_{ij} is expressed as

$$\sigma_{ij} = (\langle x_{ij}^2 \rangle - \langle x_{ij} \rangle^2)^{1/2}$$

$$= (\overline{x}_j^2 - \overline{x}_j^2)^{1/2} \qquad (16.54)$$

This means that the average of the random process always smoothes the randomness.

Now the comparison of the ray-tracing point x_{ij} to the measured data point y_{ij}:

$$x_{ij} - y_{ij} = \delta_{ij} \qquad (16.55)$$

where δ_{ij} is usually very large. But if we compare the calculated $\overline{x}_j$ with the measured $\overline{y}_j$,

$$\overline{x}_j - \overline{y}_j = \delta_j \qquad (16.56)$$

then δ_j can be small. The reason is that $\overline{x}_j$ is a random variable, as is $\overline{y}_j$; therefore, δ_j is also a variable shown in Eq. (16.56). The error ε between two average values, one from ray-tracing $\overline{x}_j$ and one from the measured $\overline{y}_j$, can be expressed as

$$\varepsilon = \{\langle \delta_j^2 \rangle - \langle \delta_j \rangle^2\}^{1/2} = \frac{\{\langle \delta_{ij}^2 \rangle - \langle \delta_{ij} \rangle^2\}^{1/2}}{\sqrt{N}} \qquad (16.57)$$

The standard deviation ε of δ_j is reduced by a factor $\sqrt{N}$ from the standard deviation of δ_{ij}. Therefore, an eyeball average or curve fit from many individual ray-tracing signal strengths at different locations in a confined area is used to compare with the measured local mean data.

Building-block approach

The building-block approach is used in the microcell propagation prediction described in Sec. 4.7. The advantage of this approach is that it is simple to use. The correlation between the building-block lengths and the additional path loss due to the building blocks is predetermined, although the rays never penetrate through the building. We may say the building-block approach is statistically accurate, since the sum of building-block lengths along the radio path is deterministic. There is no guessing by using this approach. The prediction matches the measurement data quite well, as shown in Fig. 4.20. Besides, the building-block approach is straightforward and simple to use.

16.14   Coding Scheme and Variable Burst-Error Intervals

Most of the forward-error-correction (FEC) codes can only correct random errors. Few of them may correct short burst errors. None of them can correct variable burst errors effectively. In cellular AMPS, the variable duration of fades is due to the variable vehicle speeds that cause the burst errors in variable-length intervals. The random errors of a data stream generated in a Gaussian noise medium are measured by the bit error rate (BER), but the burst errors of a data stream generated in a Rayleigh fading medium must be measured by the frame (word) error rate (FER). The BER does not change with the variable vehicle speed. In a word, the BER can be the same value from different burst-error data streams from different mobile speeds. This is because BER is the first order of statistics that is independent from the variable burst-error intervals due to the variable mobile speeds.

FER is the second order of statistics, and its value depends on the mobile speed. Since the voice quality is affected by the mobile speed, the FER should be used to specify the voice quality in mobile radio communication.

Interleaving is used to randomize the order of bits from an original bit stream before sending. After the bits are received, they are put back into their original order, as shown in Fig. 16.10. The purpose of this process is to change the burst errors of a bit stream generated through the mobile radio medium to the random errors distributed in the original bit stream after receiving and before decoding. Therefore, all the FEC codes have the capability to correct the random errors. The mobile speeds after the interleaving process become ineffective to the received bit stream. This means that FER and BER are the same in any random-error data stream. Interleaving is a good scheme, but there is a time delay before deinterleaving, depending on the interleaving block length L (number of bits). The larger the L, the better the random-error performance but the longer the delay time. However, in reality, the delay time cannot be too long, or interleaving cannot be used very effectively. This is especially true when the mobile speed is very slow (i.e., the duration of fades becomes very long). Therefore, due to the ineffec-

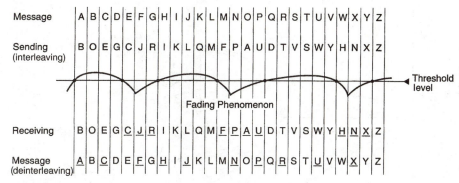

Figure 16.10 Illustration of interleaving and deinterleaving.

tive length of time for interleaving, FER and BER are different in mobile radio communications, even though the interleaving process is used in the bit stream.

16.15 Antenna Downtilt

Sometimes antenna downtilt is used either to reduce interference or fill the coverage holes. We have to be very cautious to use antenna downtilt for cochannel interference reduction. We need to calculate the downtilting antenna pattern and find the interference reduction in dB. If the reduction is less than 2 dB, the effort of antenna downtilting may not be justified. In fact, most of the time, interference reduction has not been seen by downtilting the antenna. We may illustrate this point with the following.

Electronic downtilting

There are two ways of tilting an antenna: electronic tilting and mechanical tilting. Electronic tilting is using an N-element phase-array antenna. Mechanical tilting is positioning by physically tilting an antenna. Both are shown in Fig. 16.11.

Since electrical downtilting is based on a phase-array antenna, there are many sophisticated designs. They all follow a general formula of an N-element antenna array as

$$E_s = \sum_{n=0}^{N-1} A_n e^{j\gamma_n} \tag{16.58}$$

where A_n is the amplitude of the nth element, and γ_n is expressed

$$\gamma_n = \delta_n + \beta d_n \cos \theta \tag{16.59}$$

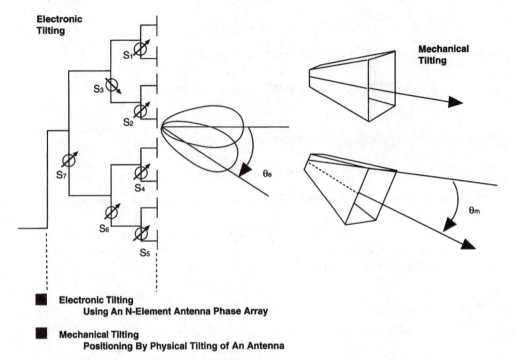

Figure 16.11   Antenna pattern tilting schemes.

All the phase differences δ_n and the space differences d_n are referred to the first element. In Eq. (16.59), many variables, N values of δ_n, and N values of d_n can be played to obtain a desired pattern at a given down-tilting angle θ_e.

For a uniformly spaced array:

$$\delta_n = n\delta$$

$$d_n = nd$$

$$A_n = \frac{1}{n}$$

Therefore, Eq. (16.58) becomes

$$E_s = \frac{1}{n} \sum_{n=0}^{N-1} e^{jn(\delta + \beta d \cos \theta)} \tag{16.60}$$

where δ is the phase difference between the adjacent elements, the electronic downloading angle θ_e is related to δ. By letting

$$\beta d \cos \theta + \delta = 0 \tag{16.61}$$

and specifying the spacing d, then we have a maximum field at a downtilting angle θ_e for adjusting δ in Eq. (16.61) [8]. For example, specifying $d = \lambda/2$ to have a maximum field at $\theta_e = 60°$ requires adjusting $\delta = -(\pi/2)$. The details of Eq. (16.58) appear in Refs. 8, 9, and 10.

Mechanical downtilting

When the center beam is mechanically tilted downward by an angle θ_m (see Fig. 16.11), the off-center beam is tilted down by only an angle $\psi = \psi_1$ observed at a horizontal angle $\phi = \phi_1$. ψ_1 is not the same angle as θ_m, as can be shown in Fig. 16.12.

The calculation of the effective vertical angle ψ after downtilting can be expressed as follows [11]:

$$\psi = \cos^{-1}[1 - \cos^2\phi(1 - \cos\theta_m)] \tag{16.62a}$$

Equation (16.62a) is a close approximation. The exact solution is

$$\tan\psi = \tan\theta_m \cdot \cos\phi \tag{16.62b}$$

Equation (16.62b) is plotted in Fig. 16.13. At angle $\theta_m = 20°$ the vertical angle ψ is different depending on the observed angle ϕ. The following table shows the different ϕ at different ψ:

Horizontal angle	Vertical angle	Horizontal pattern $+ L$ at $\theta = 0°$
$\phi = 0°$	$\psi = \theta_m = 20°$	$L_1 = -15$ dB
$\phi = 45°$	$\psi = 14.°4$	$L_2 = -5$ dB
$\phi = 90°$	$\psi = 0°$	$L_3 = 0$ dB

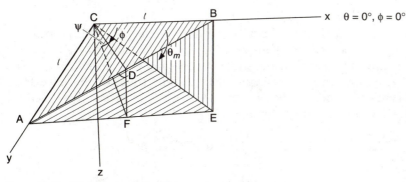

Figure 16.12 Coordinate of the tilting antenna pattern.

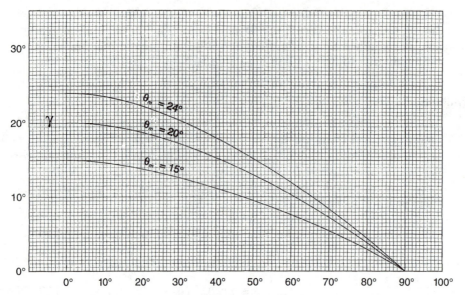

Figure 16.13   Angle ψ versus angle ϕ for different tilting angle θ_m.

Equation (16.62) is applied to any antenna patterns. For example, let us apply Eq. (16.62) to the antenna patterns shown in Fig. 16.14. The vertical pattern is tilted down $\theta_m = 20°$. The horizontal pattern will be plotted following the angle ψ. For $\phi = 45°$, the tilting angle is only $14.°4$, not $20°$. When the vertical pattern is downtilting ($\theta_m = 20°$, as shown in part a of Fig. 16.14), the reduced signal strength at the point ($\phi = 45°$, $\theta = 0°$) is not -15 dB but $L_2 = -5$ dB, and at point ($\phi = 90°$, $\theta = 0°$) is $L_3 = 0$ dB. The horizontal pattern can then be adjusted based on L_1, L_2, and L_3 at their respective points. As a result, the pattern plotted for $\theta_m = 20°$ starts to show a notch as indicated in Fig. 16.14(b). This notch can help in reducing interference [11].

Tilting antenna effect

We may have to point out the tilting antenna effect for a 30-m antenna height and a 3-km cell, as shown in Fig. 16.15. The angle θ_1 is the angle of the beam covered to the serving cell. The angle θ_2 is the angle of the beam reaching to the interfering cell. Assume that the distance from the serving cell antenna to the boundary of the interfering cell is $3.6R$ ($4.6R - R$) and $R = 3$ km. Then

$$\theta_1 \approx \frac{30 \text{ m}}{3 \text{ km}} = 0.573°$$

$$\theta_2 \approx \frac{30 \text{ m}}{10.8 \text{ km}} = 0.159°$$

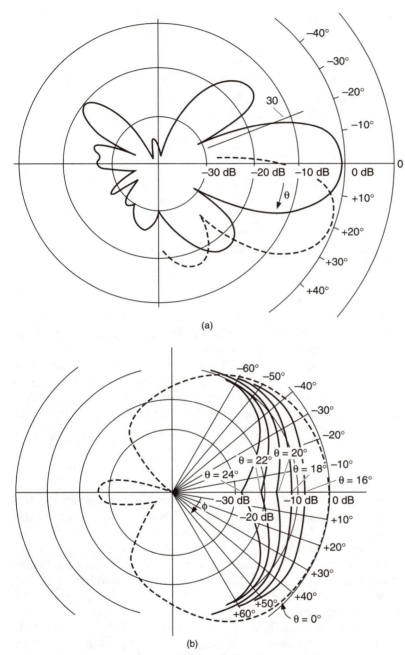

(a)

(b)

Figure 16.14   (a) Vertical antenna pattern of a 120° directional antenna ($\theta_m = 20°$); (b) notch appearing in tilted antenna pattern.

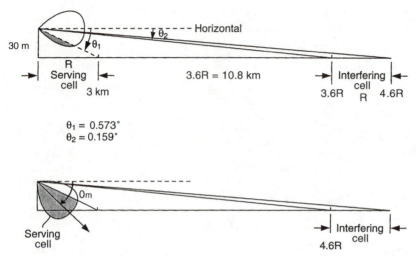

Figure 16.15   Tilting antenna effect.

Now, the gain difference between $\theta_1 = 0.573°$ and $\theta_2 = 0.159°$ can be found from Fig. 16.14(a). It is less than 1 dB. Assume the downtilting angle $\theta_m = 20°$; the gain difference between $\theta_1 + \theta_m$ and $\theta_2 + \theta_m$ seen from Fig. 16.14(a) is still less than 1 dB. This shows that, if θ_1 and θ_2 are very close, downtilting has no effect in reducing the interference. In order to increase the angle separation between θ_1 and θ_2, the antenna height can play a big role. Nevertheless, the downtilting can help in strengthening the weak signal spots in the cell.

16.16   Intermodulation

In cellular systems, when a mobile unit is closing to its base station, the power control reduces the base-station transmitter power so that the LNA (low-noise amplifier) of the mobile receiver will not be overloaded.

The cause of IM interference occurs in a mobile receiver when the mobile unit is close to the foreign system's base station. The strong base-station signal overloads the LNA (low-noise amplifier) of the mobile receiver, and IM components create and fall in the band and cause strong interference. As a result, the call is dropped. The phenomenon occurs in every cellular system, such as AMPS, GSM, TDMA, and so on. However, the IM interference is more prevailing in CDMA receivers. It is because the CDMA band is wider, and the chance to have IM components falling into the CDMA band becomes greater when the CDMA mobile unit is close to the AMPS-only sites. This phenomenon is illustrated in Fig. 16.16. The simulation of IM was done in the laboratory using a multitone generator [13]. The one interfering channel was at 892.66 MHz. The other 27 channels are within a band of 2.34 MHz centered at 882.72 MHz. The 2 B − A components and A + B + C components fall into a CDMA band (1.23 MHz) centered at 872.7 MHz. There are thirteen IM components in the CDMA band, as shown in Fig. 16.17. This scenario is the worst case. The comparison among all the jammer-tone cases is shown in Fig. 16.18. The two to twenty-eight AMPS tones with the same total power that causes IM are plotted in Fig. 16.18. One set of curves is shown with LNA switched in, and another set is shown with LNA switched out. The difference in power residing in the CDMA band between two tones and twenty-eight tones is about 7 dB or less, which can be considered the worst number. Merely a 7-dB increase in jammer power from the two-tone data (a conventional test) to the twenty-eight-tone data is an encouragement in finding possible methods of stopping IM. One way is to switch LNA out when the received signal at the mobile unit is strong. This means that a 15-dB gain will be taken out as shown in Fig. 16.18. Then, from Fig. 16.19, every time the fundamental signal drops in 1 dB, and the created third-order IM power drops in 3 dB. Therefore, the IM interference is reduced.

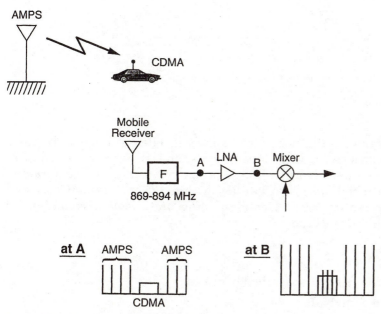

Figure 16.16 The cause of IM in CDMA receivers.

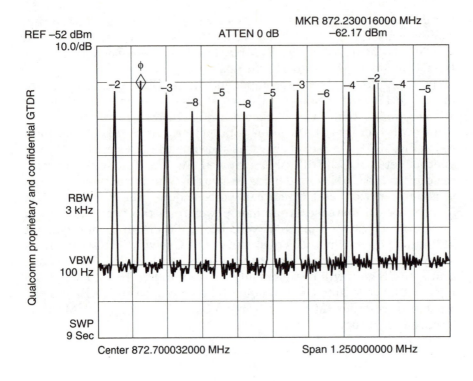

Figure 16.17 Establishing the worst-case scenario (28 tones) and thirteen tones falling in the 1.25-MHz band.

16.17 Mobile Location

Providing the information of locations of mobile terminals or portable units becomes a very important task, as the U.S. FCC mandated E911 service to be provided by all system operators. The accuracy rate of locating a mobile terminal location within 125 meters is 67 percent. There are station-based location-finding and mobile-terminal-based location-finding techniques.

Station-based location finding

1. Using the triangular method based on
   - Angle of arrival
   - Time of arrival
   - Signal strength measurement

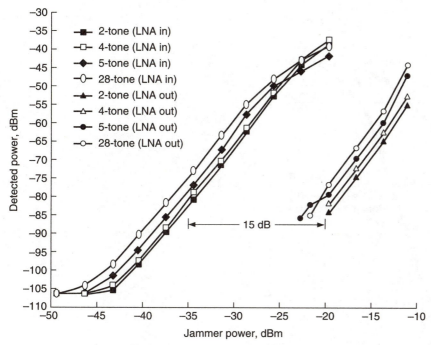

Figure 16.18 Comparison of the IM power due to different AMPS tones and the worst case.

2. Using the transponders along the highways. The mobile unit is sending an ID signal slightly off from a regular QPSK modulated signal. The ID signal is received by the transponders. Both signals can be sent at the same time. The ID signal is weak, but the transmit distance is very short; i.e., from the mobile unit to the transponders. The regular QPSK signal will be strong enough to transmit a long distance, from the mobile unit to the base station (see Chap. 8).

Mobile-unit-based location finding

1. *GPS (Global Positioning Satellite).*   An MEO satellite system (24 satellites). Two GPS satellites provide mobile location, and three satellites provide the location plus the altitude. The accuracy of location obtained from GPS is much higher than all the other methods.

2. *Loren C.*   A system used along U.S. coasts for ships to find their locations. It can be used just about anywhere in the United States, but it is especially effective near the coast. The operating frequency is 455 kHz. The resolution of the location is very poor.

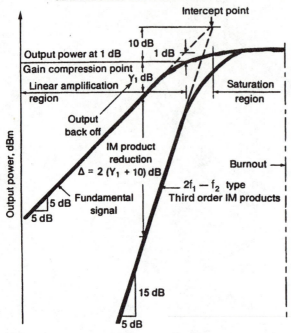

Figure 16.19   Intermodulation (IM) consideration.

3. *Dead-reckoning system.*   Installs a compass and speedometer in each car for determining location.

4. *Sign posts.*   Acts like a beacon to provide location information to the mobile units.

Discussion

Most of the systems operators prefer not to use GPS receivers for the following reasons:

1. The cost of GPS receivers is high.

2. GPS cannot serve inside buildings unless each building installs inexpensive location beacons with the same format as GPS. It is possible to make a sensitive GPS receiver to overcome this building reception issue.

3. The protocol of sending the location information back from the mobile unit needs to be standardized.

The accuracy of location finding from the station-based approach is described in Sec. 16.18.

16.18   Angle Spread with Antenna Height and Its Application

Based on the model shown in Fig. 6.11, we have found that the radius of the local scattering region is only around 69λ as derived in Sec. 6.9.

The spread angle is a function of distance R, expressed as $BW = (2r)/R$, where r is the radius of the local scatters and $2r$ is the scattering diameter. This function is also plotted in Fig. 16.19. The BW of a mobile unit at 2 miles is $1.5°$. The measurement data plotted in Fig. 16.12(c) was obtained from the base-station antenna height at 300 feet, which is high enough to clear the propagation path from the surroundings at the base. From Fig. 16.12(c), we find that the data collected at $R = 3$ miles away matched BW = $0.5°$. Therefore, $r = 69\lambda$ is found from the spread angle function, which does not involve antenna height. In the mobile environment, when the antenna height is lower, obstructions start to get between the mobile unit and the base station, and the angle spread received at the base station becomes wider. This effect cannot be obtained from the function. We may find the relationship between angle spread and the antenna height through the measured data and based on a statistical approach on an extreme but most likely scenario.

Figure 9.4 only shows the empirical curve. The data below the curve are highly probable. Figure 16.20 provides the two empirical curves; the data between the curves are also highly probable. The top curve is the same as the curve shown in Fig. 9.4, denoted as η_{min}. The lower curve is the maximum η from the measurement denoted by η_{max}.

Here we would like to find the angle spread with antenna height by taking the results from Fig. 6.12 and Fig. 9.5. We are taking the broad-

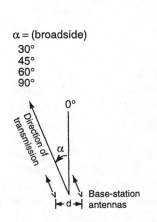
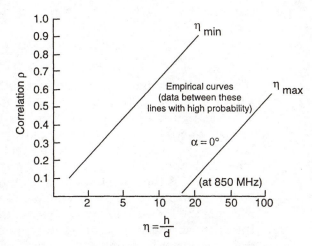

Figure 16.20   Correlation versus antenna height and spacing (broadside case in a suburban area): h, antenna height in wavelength; d, antenna spacing in wavelength.

side case $\alpha = 0°$. From Fig. 6.12, use antenna spacing $d = 10\lambda$ and find the correlations corresponding to the angle spreads. Afterwards, use the values of the correlations from Fig. 6.12 to find the corresponding values of $\eta = h/d$ in Fig. 9.5. In the mathematical derivation, we can obtain the relationship between h_1 and antenna spread BW from the following steps. From Fig. 16.20,

$$\rho = 0.2 + 0.7 \log_{10} \frac{\eta_{min}}{2} \tag{16.63a}$$

$$\rho = -0.47 + 0.57 \log_{10} \frac{\eta_{max}}{2} \tag{16.63b}$$

where $\eta = h_1/d$. From Eq. (6.133), ρ is expressed as

$$\rho = (E[\cos \{\beta d \cos (\phi_i - \alpha)\}])^2 + (E[\sin \{\beta d \cos (\phi_i - \alpha)\}])^2 \tag{16.64}$$

where the distribution of ϕ_i, $f(\phi_i)$ is related to the angle spread BW specified in Eq. (6.134), where BW is the 3-dB beamwidth of $p(\phi_i)$, which is

$$p(\phi_i) = \cos^n \phi_i \tag{16.65}$$

For $n = 3 \times 10^4$, BW $= 0.5°$.

Substituting Eq. (16.64) into Eq. (16.63), the relationship of h' and BW is obtained. The antenna height versus the angle spread is plotted in Fig. 16.21. We may see that the lower the antenna height, the wider the angle spread. Figure 16.21 shows the best-case and worst-case scenarios from the empirical curve of Fig. 9.4.

In theory, the angle BW shown in Fig. 6.11 only varies due to the distance between the base station and the mobile unit R

$$BW = \frac{2r}{R} \tag{16.66}$$

where r is the radius of the local scatters and is obtained from the measurement as 69λ. The measurement data plotted in Fig. 6.12(c) was obtained from the base-station antenna height at 300 ft, which is high enough to clear the propagation path from the surroundings at the base. The distance R is 3 miles, and the BW is found to be 0.5°. From Fig. 16.21, the antenna height for the η_{min} case at BW $= 0.5°$ is $h_1 = 205\lambda$, which is the lowest antenna height for the clear path. From Fig. 16.21, the antenna height for the η_{max} case at BW $= 0.5°$ is $h_1 = 3300\lambda$, which is the lowest antenna height for the clear path. We realize that the angle spread will occur when the angle spread is more than 0.5° from lowering the antenna. At a standard antenna height of 100λ, the angle spread for the η_{max} case is at 5.5°. From BW $= 5.5°$ and $R = 3$ miles

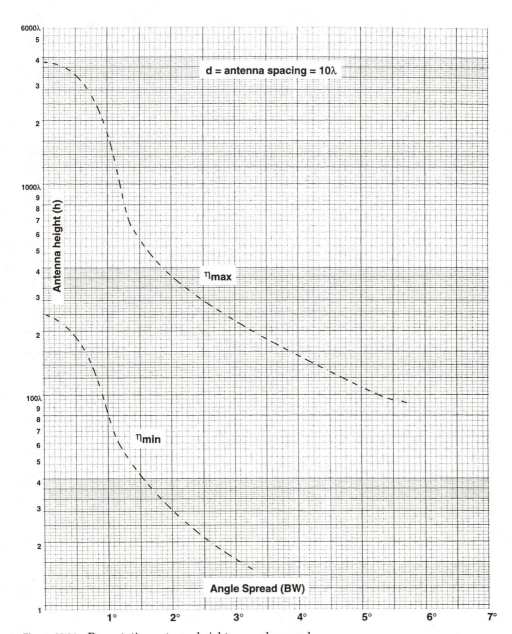

Figure 16.21 Base-station antenna height vs. angle spread.

(1.87 km), we can find the new radius $r_a = 760$ ft. (231.6 m), which is the largest radius of local scattering at a normal height $h_1 = 100$ ft (30.5 m) if the operating frequency is roughly 850 MHz.

The curve of Fig. 16.21 is not applied to the base-station antenna height situated lower than the surroundings at the base.

Location application

1. We have found that the radius of the local scattering region is only around 100λ surrounded at the mobile unit (see Sec. 6.9). If the height of the base station antenna is in the clear path condition, the triangular method would be very accurate, as shown in Fig. 16.22(*a*).
2. If the base-station height is lower, the spread angle is wider, and the location accuracy is degraded as shown in Fig. 16.22(*b*).
3. If the base-station antenna is surrounded by tall buildings, another local scattering region will be formed at the base station and the accuracy of the mobile location is further degraded as shown in Fig. 16.22(*c*).
4. Usually mobile-location finding techniques can increase the accuracy of this method if a frequency in the GHz range is allowed and the propagation path loss at that frequency is not a problem.

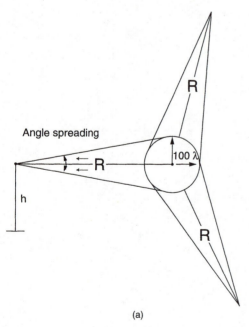

(a)

Figure 16.22   (*a*) Angle spreading.

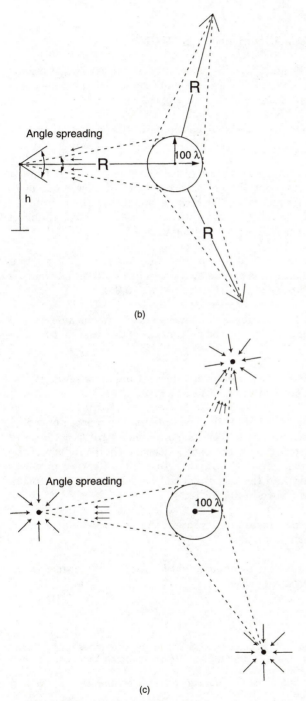

(b)

(c)

Figure 16.22 *(Continued)* (b) The base station, (c) base station surrounded by tall buildings.

Problem Exercises

1. Is the mobile radio-fading environment an AWGN channel?

2. Why are the pdf, CPD, and BER (bit error rate) first-order statistics, and the WER (word error rate), LCR (level crossing rate), and ADF (average duration of fades) second-order statistics?

3. State the difference between the propagation path-loss prediction and the local mean prediction.

4. Two base-station antennae are separated by 5 meters. The height of the lower one is 30 meters. However, the effective height of the lower one is 10 meters. The operational frequency is 850 MHz. Assume that the correlation between the received signals is less than 0.7. How is the power different between the two receptions? Will it meet the condition of Eq. (16.24)?

5. The vertical separation between two mobile antennas is 2λ. Assume that the majority of the signal arrival is at an elevation angle $\theta = 40°$. Find whether this setup meets the conditions of Eq. (16.24) and Eq. (16.25).

6. If the average of four frequencies is at a level –5 dB below the average signal at the boundary of the cell, what is the percentage of the area that the signal is above this level?

7. Plot a horizontal pattern for mechanically downtilting an antenna to 25°. Use the antenna pattern shown in Fig. 16.14 as a given.

8. Why is it that sometimes the downtilt of an antenna beam does not reduce the interference to the cochannel cell? Take the antenna height $h_1 = 150\lambda$, cell radius 1 km, and the distance from the base station to the boundary of the cochannel cell 3.6 km based on $4.6R - R$, then find the difference in power between the cell boundary and the cochannel cell boundary (see Fig. 16.15). Use the antenna patterns shown in Fig. 16.14.

9. Why do we need two parameters to represent the path loss in every city? What is the impact on each parameter?

10. Find the angle spread at an antenna height $h_1 = 50\lambda$. What is the equivalent radius of the local scattering area?

References

1. W. C. Y. Lee, "Theoretical and Experimental Study of the Properties of the Signal from an Energy Density Mobile Radio Antenna," *IEEE Trans. on Vehicular Technology,* vol. VT-16, October 1967, pp. 25–32.
2. W. C. Y. Lee, *Mobile Cellular Telecommunications—Analog and Digital Systems,* McGraw-Hill, 1995, p. 108.
3. W. C. Y. Lee, *Mobile Cellular Communications,* p. 185.

4. W. C. Y. Lee, *Mobile Communications Design Fundamentals,* 2d ed., Wiley & Sons, 1993, pp. 210, 234.
5. W. C. Y. Lee, "Mobile Cellular Telecommunication Systems," *Analog and Digital,* 2d ed., McGraw-Hill, 1995, p. 186.
6. W. C. Y. Lee, "In Cellular Telephone, Complexity Works," *IEEE Circuits & Devices,* vol. 7, no. 1, January 1991, pp. 26–32.
7. Max Born and Emil Wolf, Principles of Optics, 4th ed. Pergamon Press, 1970, p. 109.
8. J. D. Kraus, *Antennas,* McGraw-Hill, 1950, p. 83.
9. K. Fujimoto and J. R. James (ed.), "Mobile Antenna Systems Handbook," Artech House, Inc., 1994, p. 136.
10. R. S. Elliot, *Antenna Theory and Design,* Prentice-Hall, 1981.
11. W. C. Y. Lee, *Mobile Cellular Telecommunications,* p. 208.
12. W. C. Y. Lee, "Cellular Mobile Radiotelephone Systems Using Tilted Antenna Radiation Pattern," U.S. Patent 4,249,181, February 3, 1981.
13. AirTouch internal product "Multi-Channel Generator," according to W. C. Y. Lee's U.S. Patent, Frequency Signal Generator Apparatus and Methods for Simulating Interference in Mobile Communication Systems, U.S. Patent 5,220,680, June 15, 1993.
14. Charles E. Wheatly and J. Maloney, private communication, October 2, 1995.
15. W. C. Y. Lee, "Electromagnetic Scattering by Gyrotropic Cylinders with Axial Magnetic Fields," *NBS/URSI,* vol. 69D, no. 2, February 1965, pp. 227–230.

Chapter 1

New Concepts

17.1 Channel Capacity in a Rayleigh Fading Environment [1]

The channel capacity of Gaussian noise environment was resolved by C. E. Shannon in 1949. It provides an upper bond of maximum transmission rate in a given Gaussian noise environment.

In this section, the channel capacity in a Rayleigh fading environment has been derived. The result shows that the channel capacity in a Rayleigh fading environment is always lower than in a Gaussian noise environment.

In digital transmission operating in a mobile radio environment that has Rayleigh fading statistics, it is very important to know the degradations in channel capacity due to Rayleigh fading. Also, it is very important to know to what degree the diversity schemes can raise the channel capacity in a Rayleigh environment.

Channel capacity in Gaussian noise environments

In 1948, C. E. Shannon's "Mathematical Theory of Communication" [2] perceived that approaching (1) error-free digital communication on a noisy channel, and (2) maximally efficient conversion of analog signal to digital form were dual facets of the same problem. In a Gaussian noise environment (shown in Fig. 17.1), the channel capacity of a white bandlimited Gaussian channel can be expressed as [3, 4]

$$\hat{C} = B \log_2 (1 + \gamma) \text{ bits/sec} \qquad (17.1)$$

where B is the channel bandwidth and γ is the carrier-to-noise ratio, as $\gamma = C/N$; C is the RF carrier power; and N is the noise power within the channel bandwidth. Equation (17.1) is expressed from the Shannon-Hartley theorem and is for a continuous channel. First, it tells us the absolute best that the system can be, given channel parameters C/N and B. Second, with a specified information rate, the power and bandwidth are inversely related to each other. Third, the Shannon-Hartley theorem indicates that a noiseless Gaussian channel has an infinite capacity when C/N approaches infinite. However, the channel capacity does not become infinite when the bandwidth becomes infinite as seen in Eq. (17.1). This is because the noise power increases with the increase of bandwidth. Let $N = N_0 B$ where N_0 is the noise power per Hz. Equation (17.1) then becomes

$$\hat{C} = \lim_{B \to \infty} \hat{C} = \frac{\left(\dfrac{C}{N_0} \right)}{\log_e 2} \qquad (17.2)$$

The upper bounds of the bit rate of a system with an unlimited bandwidth can be derived with the information of Eq. (17.2) [5].

Channel capacity in a Rayleigh fading environment [6]

The channel capacity in Rayleigh fading must calculate in an average sense. The reason for this is that the C/N is a constant over the time in a Gaussian noise environment, but it varies in time due to Rayleigh fading, as shown in Fig. 17.1. The average channel capacity can then indicate the average best over the fading environment. It follows the same concept as obtaining the average bit error rate in the Rayleigh fading environment.

Now we would like to find an equivalent equation to Eq. (17.1) that is suited to a Rayleigh fading environment. The carrier-to-noise ratio will be variable following the Rayleigh fading statistics. A maximum value of the channel capacity in this case can then be obtained in an average sense, as mentioned previously.

The probability density function of a Rayleigh variable is

$$P(\gamma) = \frac{1}{\Gamma} e^{-\gamma/\Gamma} \qquad (17.3)$$

where Γ is the average value of γ, $\Gamma = \langle \gamma \rangle$. We are applying the same technique of obtaining the average bit error rates in a Rayleigh fading environment [7] to find the average channel capacity in the same environment here. The average channel capacity is

$$\langle \hat{C} \rangle = \int_0^\infty B \log_2 (1 + \gamma) \cdot \frac{1}{\Gamma} e^{-\gamma/\Gamma} \, d\gamma$$

$$= \frac{B}{\log_e 2} \cdot e^{-\gamma/\Gamma} \left[-E + \log_e \Gamma + \frac{1}{\Gamma} - \frac{1}{(2 \cdot 2!)\Gamma^2} \right.$$

$$\left. + \frac{1}{(3 \cdot 3!)\Gamma^3} - \frac{1}{(4 \cdot 4!)\Gamma^4} + \ldots \ldots \right] \tag{17.4}$$

where E is the Euler constant ($E = 0.5772157$).

In the case of $\Gamma > 2$, Eq. (17.4) becomes

$$\frac{\langle \hat{C} \rangle}{B} = (\log_2 e) \cdot e^{-1/\Gamma} \left(-E + \log_e \Gamma + \frac{1}{\Gamma} \right) \tag{17.5}$$

Equations (17.1) and (17.5) are plotted in Fig. 17.1. The channel capacity in a Rayleigh fading environment is reduced 32 percent at $\Gamma = 10$ dB and only reduced 11 percent at $\Gamma = 25$ dB, as expected.

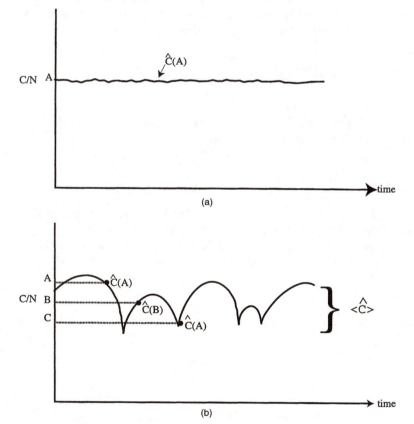

Figure 17.1   Channels capacity: (a) Gaussian noise environment; (b) Rayleigh fading environment.

When the bandwidth of the system becomes unlimited, the average channel capacity is the same as the average of the *C/N* with a constant factor of -1.44 ($= -\log_e 2$) as easily seen from Eq. (17.2).

Channel capacity in a Rayleigh fading environment with *M*-branch diversity scheme

We may use the maximum-ratio combining technique, which is the best combining technique for an *M*-branch diversity signal. The probability diversity function of a carrier-to-noise ratio of a combined signal is expressed as [8, 9]

$$P(\gamma) = \frac{1}{(M-1)!} \cdot \frac{\gamma^{M-1}}{\Gamma^M} \cdot \exp\left(-\frac{\gamma}{\Gamma}\right) \tag{17.6}$$

In Eq. (17.6), we assume the carrier-to-noise ratios of all branches are the same: $\Gamma = \Gamma_M$, where Γ is the carrier-to-noise ratio of a single channel. Equation (17.6) can also be expressed as

$$P(\gamma) = \frac{M^M}{(M-1)!} \cdot \frac{\gamma^{M-1}}{\langle\gamma\rangle^M} \cdot \exp\left(\frac{-M\gamma}{\langle\gamma\rangle}\right) \tag{17.7}$$

where the average value of γ, $\langle\gamma\rangle = M\Gamma$.

Now, the channel capacity in a Rayleigh fading environment with an *M*-branch diversity can be obtained as

$$\langle\hat{C}\rangle = B \int_0^\infty \log_2(1+\gamma) \frac{M^M}{(M-1)!} \cdot \frac{\gamma^{M-1}}{\langle\gamma\rangle^M} \cdot \exp\left(-\frac{M\gamma}{\langle\gamma\rangle}\right) \cdot d\gamma \tag{17.8}$$

Equation (17.8) is calculated numerically for $M \geq 2$ and shown in Fig. 17.2. We realize that the channel capacity in the Rayleigh fading environment for $M = 4$ is greater than for $M < 4$. The $M = 4$ case is very close to channel capacity in the Gaussian noise environment.

The extreme case

The cumulative probability distribution can be found from integration of Eq. (17.7) as

$$P(\gamma \leq \langle\gamma\rangle) = \int_0^{\langle\gamma\rangle} P(\gamma) \, d\gamma \tag{17.9}$$

The extreme case is when $M \to \infty$, then

$$P(\gamma = \langle\gamma\rangle) = 1 \tag{17.10}$$

and

$$P\left(\gamma < \langle\gamma\rangle\right) = 0 \qquad (17.11)$$

which is just the probability density function obtained from Eq. (17.9) as $M \to \infty$.

$$\frac{d}{d\gamma} P\left(\gamma = \langle\gamma\rangle\right)|_{\gamma = \langle\gamma\rangle} = \delta(\gamma) \qquad (17.12)$$

It is an indirect proof that Eq. (17.7) becomes a delta function when $M \to \infty$. Equation (17.8) then becomes:

$$\langle\hat{C}\rangle = B \int_0^\infty \log_2\left(1 + \gamma\right) \cdot \delta\left(\gamma = \langle\gamma\rangle\right) d\gamma$$

$$= B \log_2\left(1 + \gamma\right) \qquad (17.13)$$

Equation (17.13) is the same as Eq. (17.1).

Conclusion

Although the average channel capacity is not an absolute maximum value in the Rayleigh fading environment, it introduces valuable information for the continuous-channel system with finite bandwidth. Comparing the actual transmission rate with the average channel capacity obtained from Fig. 17.2, we can get a feel for how well the system has been designed and how far the actual value will reach to the average channel-capacity value. Several points can be summarized as follows:

1. The channel capacity in a Rayleigh fading environment is measured by an averaging process.
2. The channel capacity in a Rayleigh fading environment is always lower in a Gaussian noise environment.
3. The diversity scheme can raise the channel capacity in a Rayleigh fading environment.
4. When the number of diversity branches M approaches infinite, the channel capacity of an M-branch signal in a Rayleigh fading environment approaches the channel capacity in a Gaussian noise environment.
5. For the same channel capacity, a fading channel needs its C/N 3 dB higher than in the Gaussian channels, as shown in Fig. 17.2.

See Fig. 17.2.

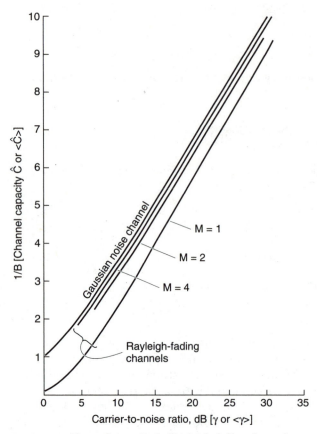

Figure 17.2 Channel capacity in Rayleigh fading channels.

17.2 Real-Time Running Average

In Sec. 3.5, we found the proper length (20λ–40λ) of averaging the instantaneous signal strength data to obtain the local means. In this section, we will discuss the real-time running average.

Why is the real time running average needed? Over fixed radio links, the average level of the signal strength seldom changes in time. However, in the mobile radio environment, the received signal average changes in real time at the mobile radio units. The real-time average changes while the mobile unit is moving due to:

1. The changes in local terrain configuration (at high spot or low spot)

2. The changes in relative distance between base station and the mobile

In this case, the averaging signal levels are different at different spots. Therefore, the conventional averaging formula cannot be applied to this nonstationary case. It follows that there is a different task for obtaining the true real-time average due to the nonstationary nature.

Normally, when the averaging process is applied at a particular time T, the window for the average process should be from $T - (\Delta t)_1$ to $T + (\Delta t)_2$, i.e., $(\Delta t)_1 + (\Delta t)_2$ as shown in Fig. 17.3, where $(\Delta t)_1$ and $(\Delta t)_2$ are on one of two sides of time T, respectively. Usually, $(\Delta t)_1$ and $(\Delta t)_2$ are the same. Nevertheless, when the mobile unit is moving, the real-time average process can only depend on the window $(\Delta t)_1$ at the mobile radio receiver (see Fig. 17.3). This is because we cannot predict the statistical behavior of the future data $(\Delta t)_2$ at time T during the real-time average process.

Therefore, a weight based on the most current data within the window $(\Delta t)_1$ is applied to accommodate the missing part $(\Delta t)_2$ for the running-average value. Let us describe the two processes of average; off-line average and real-time average. Both averages can occur in either the stationary or nonstationary environments.

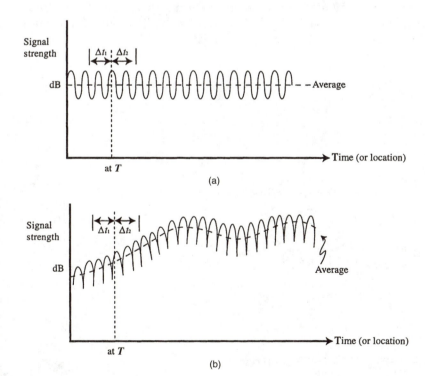

Figure 17.3 Real-time averaging for the (a) stationary case (using the conventional averaging process based on data in $\Delta t_1 + \Delta t_2$); (b) nonstationary case (using the conventional averaging process based on data in $\Delta t_1 + \Delta t_2$).

Off-line average

Average in a stationary environment. In a stationary environment, all the collected stationary samples will fluctuate around their arithmetical average. The arithmetical average is used as the stationary average and, summing all n sample values $a_1, a_2, a_3, \ldots, a_n$ and dividing by the number of sample n_1 is defined as

$$I_n = \frac{\sum_1^n a_i}{n} \tag{17.14}$$

For example, the stationary sample may be the daily temperatures collected at a geographical location. The number of samples n can be as large as many decades. The illustration of daily temperatures is shown in Fig. 17.4(a). The stationary average will not be affected by earlier or later samples. The average device based on Eq. (17.14) is called the *stationary average device*.

Average in a nonstationary environment. On many occasions, the environment is not stationary. To illustrate, take the life span of a human

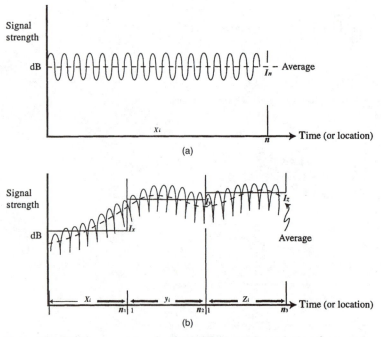

Figure 17.4 Off-line averaging for the (a) stationary case (using the conventional averaging process based on data in all x_i); (b) nonstationary case (using the conventional averaging process based on data in three time intervals).

being—the average life span generally increases over time due to the improvement of hygiene and medicine. Therefore, to determine the average life span over the last century, we cannot sum the age of death samples over a period of 100 years and divide by the age of death samples. Since the average age of death for each year is not a constant, we call this a nonstationary average. This nonstationary average must depend on the group of samples and the number of samples in each group. If obtaining the nonstationary average has no constraint on real time, we can calculate the nonstationary average in an off-line operation after receiving the complete set of data. First, we set up the criterion and calculate the average in each specified time period. In a nonstationary environment, Eq. (17.14) would be redefined as the following:

$$\left.\begin{array}{l} I_x = \dfrac{\sum\limits_{1}^{n_1} x_i}{n_1} \quad \text{(the group } x_i \text{ decreased in 1960--1969)} \\[20pt] I_y = \dfrac{\sum\limits_{1}^{n_2} y_i}{n_2} \quad \text{(the group } y_i \text{ decreased in 1970--1979)} \\[20pt] I_z = \dfrac{\sum\limits_{1}^{n_3} z_i}{n_3} \quad \text{(the group } z_i \text{ decreased in 1980--1990)} \end{array}\right\} \quad (17.15)$$

There are three specified time periods. The total number n of samples is $n = n_1 + n_2 + n_3$. Three sample groups are $x_i, y_i,$ and z_i, where x_i is in the sample group $(1, n_1)$ to generate an average I_x; y_i is in the sample group $(1, n_2)$ to generate an average I_y; and z_i is in the sample group $(1, n_3)$ to generate an average I_z. All three averages are not the same. This is a nonstationary process. An illustration of Eq. (17.15) in a nonstationary environment is shown in Fig. 17.4(b).

Real-time averages

Averages in a stationary environment. Equation (17.14) can also be obtained from a real-time running average device that is called Vincent's device [11]. Because of its stationary nature, Vincent's device will reach a constant as the number of samples n increases. The value of I_{n+1} is different from I_n as n incremented by one, shown in Eq. (17.14) as

$$I_{n+1} = \frac{nI_n + x_{n+1}}{n+1} \qquad (17.16)$$

and

$$I_1 = x_1 \qquad (17.17)$$

It is a straightforward calculation as shown in Fig. 17.5(a).

Averages in nonstationary environments—Lee's device. The true real-time average should use the data x_i in $(\Delta t)_1$ and y_i in $(\Delta t)_2$ to perform the average at time T. As shown in Fig. 17.5(b),

$$I_T = \frac{\sum x_i + \sum y_i}{(\Delta t)_1 + (\Delta t)_2} \qquad (17.18)$$

The running average is shown as a dotted curve in Fig. 17.5(c). However, Eq. (17.18) cannot be realized, because we cannot obtain the samples y_i in the future to calculate the average at T_1. Therefore, a new real-time running device is needed.

Lee's device [12] is used to obtain the real-time average in a nonstationary environment. It is a simple real-time running average device without memory, and expressed by

$$a_n = \frac{(M - 1)a_n + x_{n+1}}{M} \qquad (17.19)$$

and

$$a_0 = x_0 \qquad (17.20)$$

where M is the weight of the nonstationary running average, a fixed number. If the real-time average changes rapidly, then the values of M should be small.

Although the Lee device calculates the sample values up to the sample x_n, which covers the time intervals $(\Delta t)_1$ as shown in Fig. 17.5(b), it predicts the average value at x_n, as shown in Fig. 15.5(c). The merit of Lee's device, expressed in Eq. (17.19), is that it uses more weight on the samples close to x_n such that the average value impact on $(\Delta t)_2$ becomes small. Lee's device uses the entire previous data in the interval $0 \le (\Delta t)_1 \le T$. This is an effective running average device.

Discussion

The value at the time T obtained by Vincent's device is based on the n samples at x_n, which is at time T but represents the average at $T/2$ as shown in Fig. 17.5(c). Vincent's device treats all samples with the same weight; therefore, the conventional average can only be observed at a delay of $T/2$.

In addition, the conventional average obtained from a large averaging interval in a nonstationary environment has no physical meaning. We must demonstrate the difference in obtaining the average values

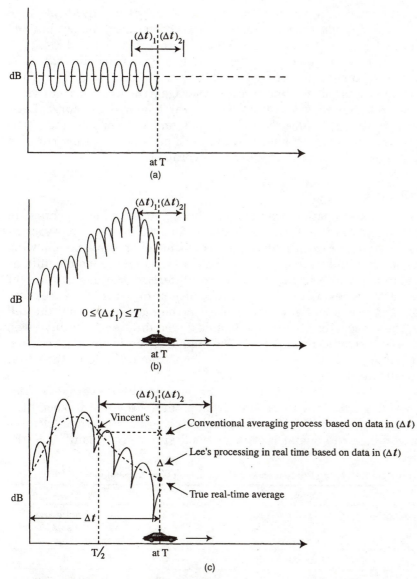

Figure 17.5 Averaging in real time: (*a*) *Stationary case*—the statistical behavior of the future data *can* be predictable using the conventional averaging process based on data within $(\Delta T)_1$; (*b*) *nonstationary case*—the statistical behavior of the future data (ΔT_2) *cannot* be predicted using Lee's invention based on the whole data but weighting them); (*c*) illustration of the differences in process and true average.

from two devices in a nonstationary environment as shown in Table 17.1. An off-time running average curve (true averages) can be obtained by taking $(x_{n+1} + x_n + x_{n+1})/3$, as shown in both Table 17.1 and Fig. 17.6. The mathematical (true) average cannot be generated in real time, because we do not know the value of x_{n+1} at the time of x_n. Therefore, this is known as the off-time running average (true averages) curve.

When the four curves are compared—the true average curve, Lee's curve with $M = 2$, Lee's curve with $M = 3$, and Vincent's average—the off-time running (true) averages and both of Lee's real-time running averages are very close, but Vincent's average is not.

Radio communication environment

In radio communication, we must obtain the nonstationary average in real time and continually make decisions for system operations such as handoffs or power controls based on real-time averages. The following scenario is an illustration. Three cell sites surround the mobile unit, as shown in Fig. 17.7. The mobile unit is at different locations at different times as it moves along the road. Because of distance variances and irregular terrain configurations between the mobile unit and the cell sites, the three sites receive the mobile signal differently at any given time, as illustrated in Fig. 17.8. The signal at the beginning received from cell site A is the strongest, then that from cell site C, then that from cell site B.

Over a period of 40 samples, the conventional average values received at each of the three sites from point Q are almost exactly the same, based on Eq. (17.16). However, as an eyeball average (close to true average), the signal received at cell site B from point Q is the

TABLE 17.1   Illustration of Two Real Averaging Processes and Comparison with True Values

				Sample values					
x_0	x_1	x_2	x_3	x_4	x_5	x_6	x_7	x_8	Lee's sample order
x_1	x_2	x_3	x_4	x_5	x_6	x_7	x_8	x_9	Vincent's sample order
2	5	4	2	6	8	9	7	8	Sample value

		Average values				

Sample value	Average values	From Lee's (a_n) $M = 2$	$M = 3$	From Vincent's (I_n)	True values $(x_{n\pm1} + x_n + x_{n+1})/3$
$x_0 = 2$	a_0, I_1	2	2	2	N/A
$x_1 = 5$	a_1, I_2	3.5	3	3.5	3.66
$x_2 = 4$	a_2, I_3	3.75	3.33	3.66	3.66
$x_3 = 2$	a_3, I_4	2.875	2.889	3.24	4.00
$x_4 = 6$	a_4, I_5	4.44	3.926	3.796	5.33
$x_5 = 8$	a_5, I_6	6.218	5.283	4.496	7.66
$x_6 = 9$	a_6, I_7	7.61	6.52	5.14	8.00
$x_7 = 7$	a_7, I_8	7.30	6.68	5.372	8.00
$x_8 = 8$	a_8, I_9	7.65	7.12	5.664	N/A

x_0	x_1	x_2	x_3	x_4	x_5	x_6	x_7	x_8	Lee's sample order
x_1	x_2	x_3	x_4	x_5	x_6	x_7	x_8	x_9	Vincent's sample order
2	5	4	2	6	8	9	7	8	Sample values (●) in nonstationary case
	3.66	3.66	4.00	5.33	7.66	8.00	8.00		True averages (□) $(x_{n-1} + x_n + x_{n+1})/3$

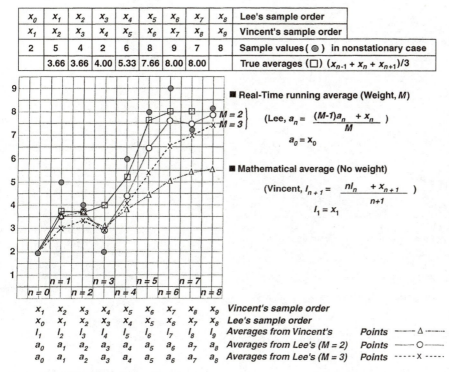

■ Real-Time running average (Weight, M)

$\left.\begin{array}{l} M = 2 \\ M = 3 \end{array}\right\}$ (Lee, $a_n = \dfrac{(M-1)a_n + x_n}{M}$)

$a_0 = x_0$

■ Mathematical average (No weight)

(Vincent, $I_{n+1} = \dfrac{nI_n + x_{n+1}}{n+1}$)

$I_1 = x_1$

x_1	x_2	x_3	x_4	x_5	x_6	x_7	x_8	x_9	Vincent's sample order	
x_0	x_1	x_2	x_3	x_4	x_5	x_6	x_7	x_8	Lee's sample order	
I_1	I_2	I_3	I_4	I_5	I_6	I_7	I_8	I_9	Averages from Vincent's	Points ·····△·····
a_0	a_1	a_2	a_3	a_4	a_5	a_6	a_7	a_8	Averages from Lee's ($M = 2$)	Points ····O····
a_0	a_1	a_2	a_3	a_4	a_5	a_6	a_7	a_8	Averages from Lee's ($M = 3$)	Points ·····x·····

Figure 17.6 Plot from Appendix A.

strongest. Lee's device, Eq. (17.19), can ascertain that cell site B has received a higher signal than the other two cell sites based on the real-time averages. This device is a tool for selecting the proper cell site (the strongest signal) for handoffs. It is also used in selecting the proper zone for intelligent microcells and the proper beam for smart antennas.

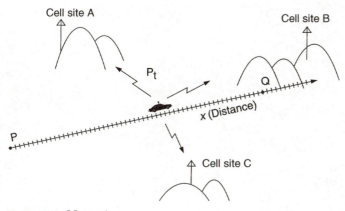

Figure 17.7 Measuring.

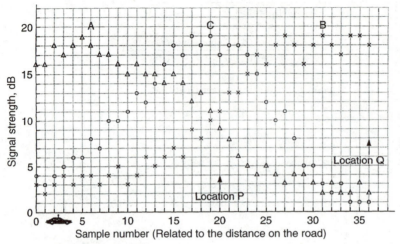

△ Received from cell site A
× Received from cell site B
○ Received from cell site C

Figure 17.8   The mobile signals received from three sites at any given time.

The general form of Lee's device [Eq. (17.19)] is:

$$a_n = \frac{(M-1)\,a_p + x_n}{M} \qquad (17.21)$$

where p is usually set as $p = n - 1$.

$$x_n \ldots\ldots x_5,\, x_4,\, x_3,\, x_2,\, x_1,\, x_0,\ \longrightarrow\ \boxed{\frac{(M-1)\,a_{n-1} + x_n}{M}}\ \longrightarrow a_n \qquad (17.22)$$

$$a_0 = x_0$$

$$a_1 = \frac{(M-1)\,a_0 + x_1}{M} = \frac{x_1}{M} + \frac{M-1}{M}\,x_0$$

$$a_2 = \frac{(M-1)\,a_1 + x_2}{M} = \frac{x_2}{M} + \frac{M-1}{M}\left[\frac{x_1}{M} + \frac{M-1}{M}\,x_0\right]$$

$$= \frac{x_2}{M} + \frac{M-1}{M^2}\,x_1 + \left(\frac{M-1}{M}\right)^2 x_0 = \frac{1}{M}\left[x_2 + \frac{M-1}{M}\,x_1 + M\left(\frac{M-1}{M}\right)^2 x_0\right]$$

$$a_3 = \frac{x_3}{M} + \frac{M-1}{M^2}\,x_2 + \frac{(M-1)^2}{M^3}\,x_1 + \left(\frac{M-1}{M}\right)^3 x_0$$

$$= \frac{1}{M}\left[x_3 + \frac{M-1}{M}\,x_2 + \left(\frac{M-1}{M}\right)^2 x_1 + M\left(\frac{M-1}{M}\right)^3 x_0\right]$$

$$a_n = \frac{1}{M}\left[x_n + \left(\frac{M-1}{M}\right)x_{n-1} + \left(\frac{M-1}{M}\right)^2 x_{n-2} + \cdots + M\left(\frac{M-1}{M}\right)^n x_0 \right] \quad (17.23)$$

From Eq. (17.23), Lee's device can demonstrate the following factors:

1. The samples closer to the time T carry more weight. Insert $M = 5$ in Eq. (17.23); then,

$$a_n = \frac{1}{5}\left[x_n + 0.8x_{n-1} + 0.64x_{n-2} + 0.512x_{n-3} + \ldots\ldots + 5 \cdot \left(\frac{4}{5}\right)^n x_0 \right]$$

$$\uparrow$$

(at T)

2. Equation (17.23) is a general equation that works for both a nonstationary case and a stationary case. The mobile radio signal is received in a nonstationary case. Vincent's formula works only in a stationary case.

3. When set $x_0 = 0$ and $M = n$, where n becomes a large number, then $((n-1)/n)^n \to 1$ and Eq. (17.23) becomes

$$a_n = \frac{1}{n}\left[x_n + x_{n-1} + x_{n-2} + \ldots\ldots x_1 \right] \quad (17.24)$$

Equation (17.24) is the Vincent's formula (conventional); therefore, Vincent's formula is a subset of Eq. (17.23).

17.3 Link Capacities versus Call Drops between GSM and CDMA

The link capacity between GSM and CDMA is always an interesting issue. A. Viterbi's paper [13] analyzed the CDMA link capacity and found that it can be extended due to the soft handoff operation.

A new idea has come from GSM engineers that uses a ping-pong handoff arrangement in one handoff region that can achieve the same selective diversity between two neighboring cells. This section analyzes the performance of this arrangement.

Viterbi's paper points out that the required margin for GSM hard handoff is 10.3 dB, but the required margin for CDMA soft handoff is only 6.2 dB under the same assumed conditions. The difference is 4.1 dB, which is related to the coverage area extension of 1.6 times. The 4-dB advantage is due to selective diversity between two neighboring cells used in the soft handoff area for a CDMA system.

Using rapid ping-pong handoffs instead of a single handoff in a handoff region on a GSM system was suggested to gain the same 4 dB selective diversity as CDMA. When the mobile unit is in a handoff region, the two log-normal fading signals received by two respective neighboring cells (on the reverse link) can be used to select the strongest signal strength in real time. With this information, the rapid ping-pong hand-off operation can take place at the switches. In the meantime, the two neighboring-cell log-normal fading signals received by the mobile unit (on the forward link) can also be used for MAHO (mobile-assisted hand-off) to assist switches in rapid handoffs. Although the idea is very good, the dropped call rate for this ping-pong handoff arrangement should be analyzed.

Dropped call rate for ping-pong handoffs [14]

When a call has gone through n ping-pong handoffs in a handoff region, the call may have k regular handoffs while the mobile unit is passing through $k + 1$ cells. The probability of a dropped call can be expressed as:

$$P = 1 - X^{n + k} \qquad (17.25)$$

$$X = (1 - \delta)(1 - \mu)(1 - \tau\theta)(1 - \beta)^2 \qquad (17.26)$$

where δ = the probability that the signal is below a specified reception threshold (in a noise-limited system)
μ = the probability that the signal is below a specified cochannel interference level (in an interference-limited system)
τ = the probability that no traffic channel is available upon handoff attempt when moving into a new cell
θ = the probability that the cell will return to the original cell
β = the probability of blocking circuits between BSC and MSC during handoff

In a noise-limited system (startup system), there is no frequency reuse configuration. The call drop rate P_A is based on the signal coverage. In an interference-limited system (mature system), the frequency-reuse configuration is applied, and the dropped rate P_B is based on the interference level. The case of P_A and P_B do not occur at the same time. When capacity is based on frequency reuse, the interference level is high, the size of the cell is small, and coverage is not an issue. The call drop rate totally depends on interference.

In a well-designed network, the values of τ, θ, and β are assumed to be very small and can be neglected, Eq. (17.26) becomes

$$X = (1 - \delta)(1 - \mu) \qquad (17.27)$$

Furthermore, in a noise-limited case, $\mu \to 0$ in Eq. (17.27), and Eq. (17.25) becomes

$$P_A = 1 - (1 - \delta)^{n+k} \qquad (17.28)$$

And in an interference-limited system, $\delta \to 0$ in Eq. (17.27), and Eq. (17.25) becomes

$$P_B = 1 - (1 - \mu)^{n+k} \qquad (17.29)$$

Finding the values of δ and μ

The value of δ and μ can be derived for a single cell and in the case of handoff. The single-cell case solution is used for estimating blocked calls. The reason behind this is that the probability of δ and μ in a single case is used for the blocked call rate when setting up calls. Assume that after a call is set up, the call will not be dropped in a cell until the mobile unit travels into the handoff region.

Formula for δ and μ

We must find the value of δ in a single cell by first integrating the cumulative probability function in which the measured level y is greater than a given level A, as shown in Eq. (4.59):

$$P\left(x > \frac{A - \overline{m}}{\sigma}\right) = \int_A^\infty \frac{1}{\sqrt{2\pi}\sigma} \exp\left[\frac{(y' - \overline{m})^2}{2\sigma^2}\right] dy' \qquad (17.30)$$

over the entire cell to find the area Q in which the measured x is greater than $(A(r) - \overline{m})/\sigma$. The mean value $\overline{m}$ is a specified reception level; σ is related to the long-term fading due to the terrain contour; and $A(r)$ is the signal that exceeds $\overline{m}$ at the distance r, which is less or equal to the cell radius R.

$$Q = \int_0^R P\left(x > \frac{A(r) - \overline{m}}{\sigma}\right) \cdot 2\pi r \, dr \qquad (17.31)$$

The probability δ that the signal is below a specified reception threshold $\overline{m}$ in a noise-limited environment system is

$$\delta = \frac{\pi R^2 - Q}{\pi R^2} = 1 - \frac{1}{\pi R^2} \int_0^R \left[1 - P\left(x < \frac{A(r) - \overline{m}}{\sigma}\right)\right] \cdot 2\pi r \, dr \qquad (17.32)$$

The probability μ that the signal is below the specified interference level I in an interference-limited system can also be expressed as:

$$\mu = \frac{\pi R^2 - Q}{\pi R^2} = 1 - \frac{1}{\pi R^2} \int_0^R \left[1 - P\left(x < \frac{A(r) - I}{\sigma}\right)\right] \cdot 2\pi r \, dr \qquad (17.33)$$

We may use the numerical calculation to solve Eq. (17.32) and Eq. (17.33) for dropped calls due to handoffs. The diagram of the dropped calls due to handoffs is shown in Fig. 17.9.

Summary and conclusion

The dropped call rate has been calculated and found to be very high when using the ping-pong handoff. Without the ping-pong handoff arrangement, when a call goes through a handoff region, one handoff occurs. If two regular handoffs occur while crossing two new cells during a call, the dropped call rate under a given condition is 1.76 percent. With the ping-pong handoff arrangement, the call goes through two handoff regions. Assuming 10 ping-pong handoffs take place, the dropped call rate under the same given condition raises to 20 percent. This high drop rate can be even higher in a real environment since more ping-pong handoffs can occur due to the weak signal situation in the handoff region. Therefore, by applying ping-pong handoffs, the 4-dB diversity advantage fades away because of poor performance due to the increased call drop rate.

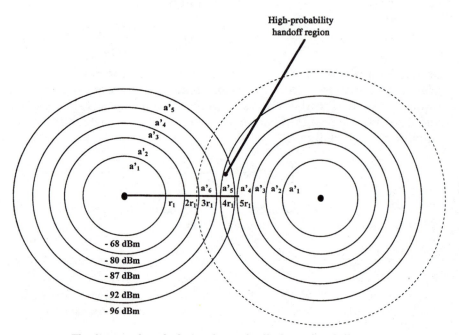

Figure 17.9   The diagram for calculating dropped calls due to handoffs.

17.4   Data Transmission via Cellular Systems

Introduction

Data transmission over the mobile radio is gradually showing its need. However, the question is, are we willing to give up a voice channel to data transmission? If the answer is no, can we use channel hopping for data transmission among the cellular voice channels? Channel hopping occurs when the channel currently used for data is assigned to a voice user, and the data operator must hop the data to another idle voice channel. The merit of applying data transmission over cellular idle channels by using frequency-hopping techniques is studied in this section. On first glance, it seems very logical that data transmission over cellular channels doesn't reduce voice capacity by hopping among idle channels.

In this section, the analytical method is used to study the frequency-hopping scheme. Although the hopping scheme may try to create additional capacity, it creates interference contributed by data transmission. In order to maintain the required performance of voice channels in cellular systems, the load on the voice channels should be reduced in order to reduce the interference generated by the hopping scheme.

Theoretical derivations

When designing a cellular system, two approaches are usually taken:

1. The C/I at the cell boundary is set to 18 dB (some systems use 17 dB), and the blocking probability is 1 or 2 percent.

2. Ninety percent of the cell area will be covered by a C/I equal or greater than 18 dB.

In evaluating these two approaches, we must break the theoretical derivations into four sections.

Find the percentage of a measured C/I above the specified C/I at the cell boundary.   Since the C/I is an average power ratio at the cell boundary, 50 percent of the measured C/I is greater than a specified level, $(C/I)_s$, and 50 percent is lower. Assume that C and I are log-normal variables in dB. The C/I in dB is $[C \text{ (db)} - I \text{ (dB)}]$. The new variable u is

$$u = C - I \qquad (17.34)$$

in dB. Let the measured carrier C be a log-normal with mean C_0 and variance σ_1^2, and let the interference I be a log-normal with mean I_0 and variance σ_2^2. Then u is also a log-normal variable. The joint probability density function of C and I can be found from Eq. E.2.31. The probability density $p(u)$ where $u = C - I$ can be calculated as follows [16]:

$$p(u) = \int_{-\infty}^{\infty} \frac{1}{2\pi(\sigma_1 \sigma_2)} \exp\left\{-\frac{1}{2}\left[\frac{(u - I - C_0)^2}{\sigma_1^2} + \frac{(-I - I_0)^2}{\sigma_2^2}\right]\right\} dI$$

$$= \frac{1}{\sqrt{2\pi(\sigma_1^2 + \sigma_2^2)}}$$

$$\times \exp\left\{-\frac{1}{2}\left[\frac{1}{\sigma_1^2} - \frac{\sigma_2^2}{\sigma_1^2(\sigma_1^2 + \sigma_2^2)}\right](u - C_0)^2 - \frac{I_0}{(\sigma_1^2 + \sigma_2^2)}(u - C_0)\right\}$$

$$\times \exp\left\{\frac{I_0^2}{2\sigma_2^2}\left[\frac{\sigma_1^2}{(\sigma_1^2 + \sigma_2^2)} - 1\right]\right\} \tag{17.35}$$

where C_0 and I_0 are the means of C and I, respectively, and σ_1^2 and σ_2^2 are the variances of C and I, respectively.

The probability that u is greater than a value U can be derived from Eq. (17.35) as:

$$P(u \geq U) = \int_U^{\infty} p(u)\, du$$

$$= \frac{1}{2}\left[1 - \Phi\left(\frac{U - (C_0 + I_0)}{\sqrt{2(\sigma_1^2 + \sigma_2^2)}}\right)\right]$$

$$= \frac{1}{2}\left[1 - \Phi\left(\frac{(U - U_0)}{\sqrt{2}\sigma}\right)\right] \tag{17.36}$$

where $\Phi(x)$ is the probability integration function and U_0 is the average received C/I level.

$$U_0 = C_0 - I_0 \tag{17.37}$$

and

$$\sigma = \sqrt{\sigma_1^2 + \sigma_2^2} \tag{17.38}$$

When a specified C/I level equals U_0, then Eq. (17.36) becomes

$$P(u \geq U_0) = 50\% \tag{17.39}$$

This means that there is a 50 percent chance that a received C/I level u is greater than the specified C/I level U_0. Both of the variance carriers σ_1^2 and the variance of interference σ_2^2 assumes 8 dB in a large cell or 6 dB in a small cell.

The total variance of (C/I), σ^2 is calculated by

$$\sigma = \sqrt{\sum_1^2 \sigma_i^2} = \sqrt{2 \times 64} = 11.3 \text{ dB} \qquad (\text{for } \sigma_i = 8 \text{ dB})$$

$$= \sqrt{2 \times 36} = 8.48 \text{ dB} \qquad (\text{for } \sigma_i = 6 \text{ dB}) \tag{17.40}$$

Equation (17.36) is plotted with a variable U as a specified level of C/I in Fig. 17.10. In general, U_0 is the level at the cell boundary.

When U is 2 dB below U_0, then

$$P\,(u \geq U) = 0.5 - 0.5\Phi\!\left(\frac{-2}{11.3}\right)$$

$$= 60\% \qquad \text{(for } \sigma_i = 8)$$

$$= 0.5 - 0.5\Phi\!\left(\frac{-2}{8.48}\right)$$

$$= 63\% \qquad \text{(for } \sigma_i = 6) \tag{17.41}$$

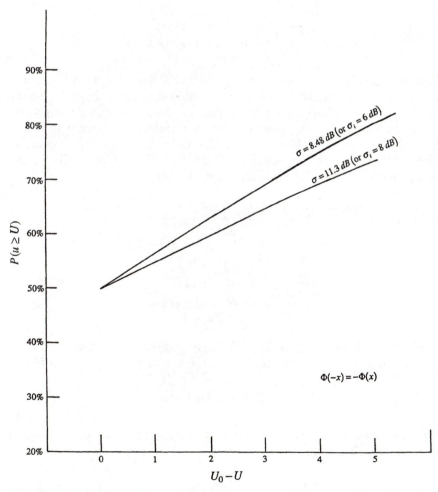

Figure 17.10   Plot of $P(u \geq U)$.

The percentage is increased from 50 to 60 percent for $\sigma_i = 8$, and to 63 percent for $\sigma_i = 6$.

Find the value of U_0 in cellular.   In designing a cellular system, there is a frequency-reuse factor K of 7. The separation cochannel cells D are

$$D = \sqrt{3K} \cdot R = 4.6R$$

where R is the cell radius. Based on the separation D among all the cochannel cells, there are six cochannel cells in the first tier around the cell of interest (see Fig. I.2 of the introduction). Applying the 4th power propagation rule [15],

$$U_0 = \frac{C}{I} = \frac{a \cdot R^{-4}}{a \cdot 6 \times D^{-4}} = \frac{1}{6} \left(\frac{D}{R} \right)^4$$

$$= \frac{1}{6} (4.6)^4 = 74.6 \ (=) \ 18.7 \ \text{dB}$$

The level U_0 at the cell boundary can be as high as 18.7 dB. We can also calculate the interference I in the cell due to six interfering cells shown in Fig. 17.11 and find the I almost constant in a cell with a variance of ± 0.5 dB. This has been observed from a simulation program shown in Sec. 15.10. Therefore, the change of U_0 will be dependent on the carrier level only. If we use the minimum tolerated U level of 17 dB for the acceptable voice quality, then from Eq. (17.36),

$$P \ (u \geq U) = \frac{1}{2} - \frac{1}{2} \ \Phi \left(\frac{1.7}{11.3} \right)$$

$$= 58\% \qquad \text{(for } \sigma = 8 \text{ dB)}$$

$$= \frac{1}{2} - \frac{1}{2} \ \Phi \left(\frac{1.7}{11.3} \right)$$

$$= 61\% \qquad \text{(for } \sigma = 6)$$

Find the area of $u \geq U$.   Take Eq. (17.36) and integrate it in an entire cell to find the area A in which the measured u will be greater than U.

$$A = \int_0^R P(u \geq U) \ 2\pi r \ dr \tag{17.42}$$

Since, as mentioned previously, the U_0 is changed based on the carrier level C_0, which can be converted into the following:

$$C_0 = C_R - \gamma \log \frac{r}{R} \tag{17.43}$$

then Eq. (17.43) becomes

$$C_0 - I = C_R - I - \gamma \log \frac{r}{R} \tag{17.44}$$

or

$$U_0 = U_R - \gamma \log \frac{r}{R} \tag{17.45}$$

where U_R is the C/I measured at the cell boundary, γ is the propagation power rule in dB/dec, and r is the distance within the cell.

Insert Eqs. (17.44) and (17.45) into Eq. (17.36), which is used in Eq. (17.42), and find the percentage of area in a cell Q

$$Q = \frac{A}{\pi R^2} = \frac{2\pi}{\pi R^2} \int_0^R \left\{ \frac{1}{2} - \frac{1}{2} \, \Phi \left[\frac{U - U_R + \gamma \log_{10} \frac{r}{R}}{\sqrt{2}\sigma} \right] \right\} r \, dr$$

$$= \frac{1}{2} + \frac{1}{R^2} \int_0^R \Phi \left[\frac{\Delta U}{\sqrt{2}\sigma} - \frac{\gamma \log_{10} \frac{r}{R}}{\sqrt{2}\sigma} \right] r \, dr \tag{17.46}$$

where $\Delta U = U_R - U$.

Let

$$p = \frac{\Delta U}{\sqrt{2}\sigma}$$

$$q = \frac{\gamma \cdot \log_{10} e}{\sqrt{2}\sigma} \tag{17.47}$$

and

$$v = p - q \log_e \frac{r}{R} \tag{17.48}$$

then

$$r = R \cdot e^{(p/q - v/q)} \tag{17.49}$$

and

$$dr = -\frac{r}{q} \, dv \tag{17.50}$$

Inserting Eq. (17.49) and Eq. (17.50) into Eq. (17.46) yields:

$$Q = \frac{1}{2} + \frac{1}{R^2} \int_{\infty}^{P} \Phi(v) \frac{-r^2}{q} dv$$

$$= \frac{1}{2} + \frac{1}{R^2} \int_{P}^{\infty} \frac{R^2}{q} e^{2(p/q - v/q)} \cdot \Phi(v) \, dv$$

$$= \frac{1}{2} + \frac{1}{q} e^{+2(v/q)} \int_{P}^{\infty} e^{-2(v/q)} \Phi(v) \, dv \qquad (17.51)$$

From the table of integrals of the error functions [17], Eq. (17.51) can be solved and represented as

$$Q = \frac{1}{2} \left\{ 1 + \Phi\left(\frac{\Delta U}{\sqrt{2}\sigma}\right) + \exp\left[\left(\frac{2\Delta U}{\sqrt{2}\sigma}\right) \cdot \left(\frac{\sqrt{2}\sigma}{\gamma b}\right) + \left(\frac{\sqrt{2}\sigma}{\gamma b}\right)^2 \right] \right.$$

$$\left. \times \left[1 - \Phi\left(\frac{\Delta U}{\sqrt{2}\sigma} + \frac{1}{b}\frac{\sqrt{2}\sigma}{\gamma}\right) \right] \right\} \qquad (17.52)$$

where $b = \log_{10} e$ is a constant. Equation (17.52) is plotted in Fig. 17.12. For $U_R = 19$ dB at the cell boundary and $U = 17$ dB as the minimum accepted voice quality, then $\Delta U = 2$ dB. Let $\sigma = 8$ dB and $\gamma = 40$, and we find $Q = 83$ percent from Fig. 17.12.

Find the blocking probability. We have used the blocking probability P_b equals 1 or 2 percent in designing cellular systems. If 54 voice channels are used in an omni-cell, the offered load A is

$$A = 41.5 \text{ Erlangs at } P_b = 1\%$$

$$A = 44 \text{ Erlangs at } P_b = 2\%$$

Since the offered load is less than 54 Erlangs, the interference is also reduced. The reduced interference can be calculated as

$$\Delta I = 10 \log\left(\frac{41.5}{54}\right) = -1.14 \text{ dB at } P_b = 1\%$$

$$\Delta I = 10 \log\left(\frac{44}{54}\right) = -0.89 \text{ dB at } P_b = 2\%$$

Assume that summing all the interference within a cell is almost a constant. Then,

$$U_0 = 19 + 1.14 = 20.14 \text{ dB} \quad \text{at } P_b = 1\%$$

$$\Delta U = 20.14 - 17 = 3.14 \text{ dB}$$

$$Q = 90\% \quad \text{for } \sigma = 6 \text{ dB}$$

$$= 85.5\% \quad \text{for } \sigma = 8 \text{ dB}$$

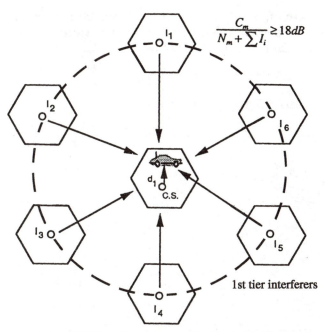

Figure 17.11   Mobile radio environment with six interferers.

1st-tier interferers $= \dfrac{2\pi D}{D} = 6$

2d-tier interferers $= \dfrac{2\pi(2D)}{D} = 12$

3d-tier interferers $= \dfrac{2\pi(3D)}{D} = 18$

No. of interferers at Nth tier $= \dfrac{2\pi(N \cdot D)}{D} = 2\pi \cdot N = 6 \cdot N$

or

$$U_0 = 19 + 0.89 = 19.89 \text{ dB} \quad \text{at } P_b = 2\%$$

$$\Delta U = 19.89 - 17 = 2.89 \text{ dB}$$

$$Q = 89.5\% \quad \text{for } \sigma = 6 \text{ dB}$$

$$= 85.5\% \quad \text{for } \sigma = 8 \text{ dB}$$

Reading from Fig. 17.11, the percentage of area Q increases but cannot match 90 percent of the area.

Analysis

If we use the first approach in designing a cellular system, although the average received $U_0 = C_0/I_0$ is about 3 dB higher than the required

U, the percentage of area *Q* still cannot quite achieve 90 percent. In order to achieve *Q* = 90 percent, we must reduce the traffic further. From Fig. 17.12, for *Q* = 90 percent, ΔU must be:

$$\Delta U = 3.1 \text{ dB} \qquad \sigma = 6 \text{ dB}$$

$$\Delta U = 5 \text{ dB} \qquad \sigma = 8 \text{ dB}$$

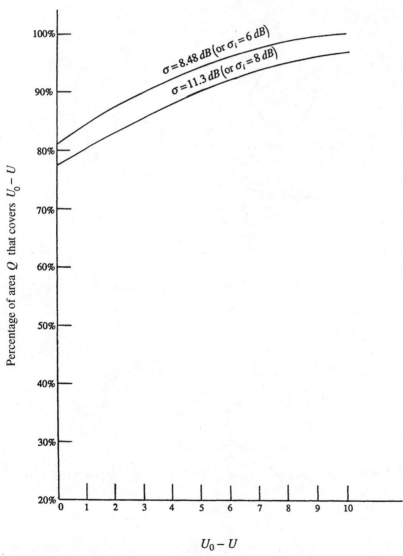

Figure 17.12   Percentage of cell area that covers $(U_0 - U)$.

If so, then the linear ratio of interference for $\sigma = 6$ dB is

$$I = 10^{-3.1/10} = 49\%$$

This means that the offered load will be only 49 percent. The linear ratio of interference for $\sigma = 8$ dB is

$$I = 10^{-5/10} = 31.6\%$$

The offered load in this case will only be 32%.

Summary

We cannot afford to have additional interference in the traffic-loaded interference-limited cellular system. We must reduce traffic in order to meet the requirement of covering 90 percent of the cell with a required C/I, $U = 18$ dB or 17 dB.

Nevertheless, in straightforward thinking, we feel that, as long as the packet data does not block traffic, the additional traffic always increases capacity but degrades voice quality. The packet data transmission on voice channels may be understood by some engineers as being analogous to a jar filled with marbles and sand. Marbles are equivalent to voice channels, and sand is equivalent to data. Sand can be poured into the jar to fill the gaps, even though no additional marbles will fit in the jar. The jar can be totally filled with marbles and still hold sand.

This analogy is applied to the noise-limited environment (i.e., no frequency-reuse situation exists). The wall of the jar in this situation is very thick, as shown in Fig. 17.13(a). However, in cellular, the cochannel interference forms an interference-limited environment. Even if the jar is big, the wall of the jar becomes very thin, as shown in Fig. 17.13(b). A thin-walled jar can only hold so much. Therefore, the total *weight* of the jar's contents is the limit. In this case, we are unable to fill the jar with marbles and/or sand. With a weight (interference) limit, adding a single marble or the equivalent weight of sand would weaken the jar structure to a certain extent. This is what we have not previously considered.

Therefore, the following provides the key points:

1. If the 90 percent coverage required for designing cellular systems is ignored, then the frequency-hopping scheme gains additional capacity, but in the meantime, the voice quality is degraded to a certain degree.

2. If the 90 percent coverage requirement or any other given percentage of coverage requirement for cellular systems is maintained, then

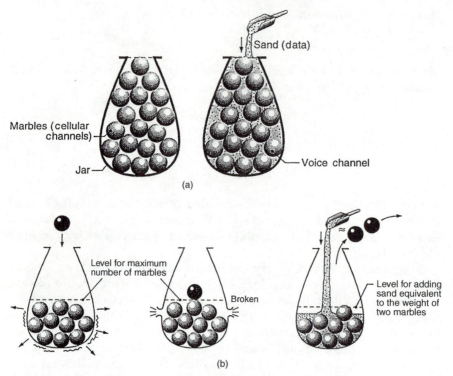

Figure 17.13 An analogy—a jar with marbles and sand: (*a*) the rigid-jar scenario (noise-limited); (*b*) the fragile-jar scenario (interference-limited).

there is no advantage in using the frequency-hopping scheme. Rather, we may just assign a designated channel for data to serve the same purpose.

3. When cellular voice traffic is low, no frequency hopping for data transmission is needed. When cellular voice traffic is high, heavy voice interference is in the radio systems. So, when hopping data over a voice channel in a situation of heavy interference, the data throughput will most likely be undesirably low.

4. Not only is the use of a designated channel for data transmission a better option, but this designated channel can also be assigned to a voice call if the channel is idle. The voice channel will hop out of the data channel when data transmission is needed. Both data and voice transmission should be controlled from a common controller.

5. A thorough analysis has been carried out to describe the conditions needed for data transmission over cellular systems with maximum capacity and minimum interference.

17.5   Multiuser Detection for CDMA

The single-user (conventional) detector that is used to calculate system capacity ignores the presence of multiaccess interference in CDMA. Thus, the multiple access interference (MAI) is usually mitigated by means other than multiuser detection such as:

1. Code waveform design.   Finds a spreading code with good cross-correlation properties; i.e., all the codes were orthogonal, and $\rho_{ij} = 0$ between code i and code j.

2. Power control.   Reduces the near-far interference and the adjacent cell interference.

3. FEC codes/interleaving.   Reduces the error rate at lower carrier-to-interference ratio.

4. Sectored/smart/adaptive antennas.   Uses angle or directed (sectored) beam to isolate the interference. Smart antenna and adaptive antenna can select the directed beam more intelligently.

5. Intelligent microcell system.   Uses small zones (microcells) to isolate the interference and uses intelligence to select those zones.

The single-user detector is usually used at the mobile unit. The multiuser detector is needed at the base station to continue to reduce the MAI effectively. With an optimum multiuser detector [18–20], the performance gains the near-far resistant property and results in lower power consumption (increasing battery life) and higher processing gain requirements (lower bandwidth).

Suppose that K users are sharing the same radio channel, and one of these K users has a traffic channel signal u_k that at the baseband is

$$u_k(t) = A_k\, b_k\, a_k(t) \qquad 0 \le t < T \tag{17.53}$$

where T is the bit interval and A_k, b_k, and $a_k(t)$ are the amplitude, bit sequence, and spreading-chip sequence, respectively. The bit sequence $b_k = \pm 1$ contains the transmitted information. The received signal at the baseband sums up all the users' signals $u_k(t)$ plus the additive noise $n(t)$:

$$r(t) = \sum_{k=1}^{K} u_k(t) + n(t) \tag{17.54}$$

Assume that all user bits are aligned in time (synchronous) with coherent reception. The DS/CDMA detects the chip sequence of user j by a correlator $a_j(t)$. The output of the correlator $a_j(t)$ is the decision statistic over the interval $(0, T)$.

$$y_j = \int_0^T r(t)\, a_j(t)\, dt = \int_0^T a_j(t) \left[\sum_{k=1}^{K} A_k a_k(t)\, b_k + n(t) \right] dt$$

$$= \sum_{k=1}^{K} A_k b_k \int_0^T a_j(t)\, a_k(t)\, dt + \int_0^T a_j(t)\, n(t)\, dt \qquad (17.55)$$

Equation (17.55) can be represented as a vector form by a diagonal matrix Y as follows:

$$Y = RAB + N \qquad (17.56)$$

where N is the noise term after correlation, A is a diagonal matrix of amplitudes, B is the diagonal matrix of bit sequence, and R is the matrix of cross correlation. The element of matrix R is

$$R_{jk} = \int_0^T a_j(t)\, a_k(t)\, dt \qquad (17.57)$$

Equation (15.55) also can be expressed in the following form as

$$y_j = A_j b_j + \sum_{k \neq j}^{K} A_k b_k \int_0^T a_j(t)\, a_k(t)\, dt + \int_0^T a_j(t)\, n(t)\, dt \qquad (17.58)$$

where the correlator $a_j^2(t) = 1$. The decision on the sign of y_j is based on $A_j b_j$ that is the first term in Equation (17.58). The second term is the multiple users' interference, the influence of which we want to remove. The last term is due to noise. Using the knowledge of the chip sequence available at the base station, we may estimate $A_j b_j$ as a whole.

Optimum multiuser interference cancellation

We may obtain the optimum solution by inverting R. Then, applying R^{-1} in Eq. (17.56) yields

$$R^{-1} Y = AB + R^{-1} N \qquad (17.59)$$

In Eq. (17.59), we see that AB is recovered from Eq. (17.58), and the second term in Eq. (15.58) is canceled. The third term is a noise term that is changed to a new noise term and may be enhanced rather than decreased.

Successive interference cancellation

Using a coherent system at the RF stage, the signal received is similar to Eq. (17.54) as

$$r(t) = \sum_{k=1}^{K} A_k b_k\, (t - \tau_k)\, a_k(t - \tau_k) \cos(\omega_c t + \phi_k) + n(t) \qquad (17.60)$$

where ω_c is the carrier frequency, the noise $n(t) = n_I + jn_Q$, τ_k is the time delay, and ϕ_k is the phase of the kth user. τ_k and ϕ_k are assumed to be tracked accurately.

The low-pass filter (LPF) outputs of I-channel and Q-channel are

$$d^I(t) = \text{LPF} \{r(t) \cos (\omega_c t)\}$$

$$= \sum_{k-1}^{K} A_k^I a_k^I (t - \tau_k) b_k^I (t - \tau_k) \frac{\cos (\phi_k)}{2} + \frac{n_I(t)}{2} \qquad (17.61)$$

$$d^Q(t) = \sum A_k^Q a_k^Q (t - \tau_k) b_k^Q (t - \tau_k) \frac{\cos (\phi_k)}{2} + \frac{n_Q(t)}{2} \qquad (17.62)$$

where $n_I(t)$ and $n_Q(t)$ are the in-phase and quadrature components of LPF Gaussian noise. Then after the correlator, the outputs y_j^I and y_j^Q of the jth user are

$$y_j^I = \frac{1}{T} \int_0^T d^I(t) \, a_j^I (t - \tau_j) \cos (\phi_j) \, dt \qquad (17.63)$$

and

$$y_j^Q = \frac{1}{T} \int_0^T d^Q(t) \, a_j^Q (t - \tau_j) \cos (\phi_j) \, dt \qquad (17.64)$$

The decision on the bit of the jth user is based on

$$y_j = y_j^I + y_j^Q$$

In Eqs. (17.63) and (17.64), the improvement of outputs y_j^I and y_j^Q for the jth user's signal depends on $d^I(t)$ and $d^Q(t)$, in which the successive interference reduction can be applied as follows:

Let the LPF outputs $d^I(t)$ and $d^Q(t)$ be used successively with the rest of the $K - 1$ correlators at the base station. We may denote y_i to be the output of the ith correlator

$$y_i = y_i^I + y_i^Q \qquad (17.65)$$

where

$$y_i^I = \frac{1}{T} \int d_{i-1}^I(t) \, a_i(t - \tau_i) \cos (\phi_i) \, dt \qquad (17.66)$$

$$y_i^Q = \frac{1}{T} \int d_{i-1}^Q (t) \, a_i(t - \tau_i) \cos (\phi_i) \, dt \qquad (17.67)$$

For no successive cancellation, $i = 1$ and

$$d_0 (t) = d(t)$$

$$y_1 = y_j$$

The interference in the received $d(t)$ can be reduced by canceling the $K - 1$ interferences successively as from $i = 1, K$ and $i \neq j$.

$$d_i^I(t) = d_{i-1}^I(t) - \underbrace{y_i^I \, a_i(t - \tau_i) \cos{(\phi_i)}}_{\text{the signal through the ith correlator}} \qquad (17.68)$$

$$d_i^Q(t) = d_{i-1}^Q(t) - y_i^Q \, a_i(t - \tau_i) \cos{(\phi_i)} \qquad (17.69)$$

where y_i^I and y_i^Q are found from Eqs. (17.66) and (17.67), respectively.

Therefore, y_{i+1}^I and y_{i+1}^Q are

$$y_{i+1}^I = \frac{1}{T} \int d_i^I(t) \, a_{i+1}(t - \tau_{i+1}) \cos{(\phi_{i+1})} \, dt \qquad (17.70)$$

$$y_{i+1}^Q = \frac{1}{T} \int d_i^Q(t) \, a_{i+1}(t - \tau_{i+1}) \cos{(\phi_{i+1})} \, dt \qquad (17.71)$$

The results of d_{i+1}^I and d_{i+1}^Q are obtained successively from Eqs. (17.68) and (17.69). This process is to reduce the multiuser $K - 1$ interferences, but the noise due to imperfect cancellation in the previous stage increases.

The above process is for a successive interference cancellation (SIC) detector. It takes a serial process to cancel interference. Each stage cancels out one additional direct-sequence user from the received signal so that the remaining users contribute less MAI in the next stage.

The SIC detector can be replaced by a parallel interference cancellation (PIC) detector, which estimates and removes all the MAI for each user in parallel.

From a number of studies [19], SIC is found to be superior in a non-power-controlled fading channel because the canceling scheme is based on the signal strength. On the other hand, PIC is found to be superior in a well-power-controlled channel because the use of the canceling scheme is based on the interfering bits, not the signal strength.

Since we are working in the Rayleigh fading case, the amplitude A is a Rayleigh variable. Let the amplitude A in Eq. (17.53) or Eq. (17.60) be Rayleigh-distributed. The error is introduced as the probability of receiving a weak amplitude increases for the desired signal and a strong amplitude for an interference.

17.6   Spectrum and Technology of a Wireless Local Loop System

Technological advancements and the capabilities of cellular telecommunication since 1980 are growing rapidly worldwide. In the future, subscribers will be allowed virtually unlimited access to information.

The need for the wireless local loop (WLL) system in urban and rural areas and in developed and developing countries is starting to draw a great deal of attention.

Since the January 1996 Telecommunication Reform Act was passed by Congress, everyone can get into everyone else's business. Telephone companies will provide WLL systems to retain their customers. In the meantime, the wireless communication companies will provide WLL systems to take away the wireless telephone companies' customers. Also, there will be many stand-alone WLL companies trying to serve as wireless telephone companies. Those companies will all be competing against each other. They all must make efficient spectrum; that is, high capacity. In this section, we compare the spectrum efficiency of each multiple-access scheme.

The attributes of a WLL system per Fig. 17.14

- *Coverage increases.*   The path loss is based on free space loss.
- *Capacity increases.*   The required $(C/I)_{WLL}$ is much less than the required $(C/I)_{cellular}$.
- *The high-gain directional antenna can be used at both ends.*   Thus, interference decreases and the frequency-reuse distance reduces (i.e., capacity increases).

Concept of deploying wireless systems

The general key drivers of deploying a wireless system.   The general key drivers of deploying a WLL system are capacity, coverage, and quality. However, the three key drivers are interrelated. With an allocated spectral band as a given, then:

$$capacity \propto (quality)^{-1}$$

$$coverage \propto (quality)^{-1}$$

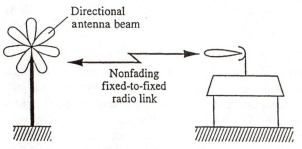

Figure 17.14   The scenario of a WLL system.

The three key drivers related to *C/I*.   The three key drivers (capacity, coverage, and quality) can be expressed as a function of *C/I*.

$$\frac{C}{I} \uparrow = \text{quality is improving}$$

$$\frac{C}{I} \downarrow = \text{capacity and coverage is increasing}$$

Quality *Q* is proportional to *C/I*:

$$Q \propto \frac{C}{I} \qquad (17.72)$$

Radio capacity *m* is inversely proportional to *C/I*:

$$m = \frac{M}{K} \propto \left(\frac{C}{I}\right)^{-1} \qquad \text{channels/cell (for FDMA and TDMA)} \quad (17.73)$$

Coverage *R* is inversely proportional to *C/I*:

$$R \propto D \cdot \left(\frac{C}{I}\right)^{-1} \qquad (17.74)$$

Consideration of deploying a WLL system

Impact from the required *C/I*.   The required *C/I* ratio in each system is determined from the accepted voice quality or corresponds to the specific frame error rate. The required *C/I* of a WLL system under a nonfading fixed-to-fixed condition is always less than the required *C/I* of a cellular system under a mobile radio multipath fading condition.

$$\left(\frac{C}{I}\right)_w < \left(\frac{C}{I}\right)_c \qquad (17.75)$$

It shows that the WLL system can tolerate more Gaussian-like interference than the cellular system can tolerate Rayleigh-like interference. Therefore, the frequency-reuse distance for WLL is supposedly shorter than for that of cellular if the propagation path loss for both systems are the same.

Impact from the propagation path loss

Coverage increases in WLL.   The coverage of a WLL is based on a fixed-to-fixed propagation. The path loss of the fixed-to-fixed propagation in a WLL is based on 20 dB/decade. However, the path loss of mobile radio propagation (fixed-to-mobile) is based on 40 dB/decade, which shows a high excessive loss. Therefore, the same wireless communication system can cover more area for WLL services than for the mobile radio services.

Capacity decreases in WLL if FDMA or TDMA is used. Based on the path loss of 20 dB/decade for WLL and 40 dB/decade for cellular, the formula of C/I and K of both systems can be obtained as follows:

WLL systems

1. Under a condition of six interferers (see Fig. 17.11),

$$\left(\frac{C}{I}\right)_{W_1} = \frac{R_{W_1}^{-2}}{6 \cdot D_{W_1}^{-2}} = \frac{(D/R)_{W_1}^2}{6} \tag{17.76}$$

Substituting Eq. (17.76) into Eq. (15.16) yields

$$K_{W_1} = \frac{1}{3}\left(\frac{D}{R}\right)_{W_2}^{-2} = 2\left(\frac{C}{I}\right)_{W_1} \tag{17.77}$$

2. Under a condition of one interferer (see Fig. 17.15),

$$\left(\frac{C}{I}\right)_{W_2} = \frac{R_{W_2}^{-2}}{D_{W_2}^{-2}} = \left(\frac{D}{R}\right)_{W_2}^2 \tag{17.78}$$

and

$$K_{W_2} = \frac{1}{3}\left(\frac{D}{R}\right)_{W_2}^2 = \frac{1}{3}\left(\frac{C}{I}\right)_{W_2} \tag{17.79}$$

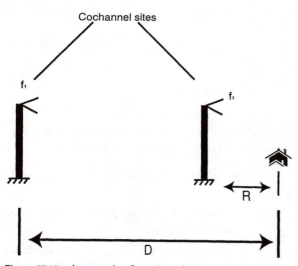

Figure 17.15 A scenario of one interferer.

Cellular systems.   Always under a condition of six interferers (see Fig. 17.11),

$$\left(\frac{C}{I}\right)_c = \frac{R_c^{-4}}{6 \cdot D_c^{-4}} = \frac{(D/R)_c^4}{6} \tag{17.80}$$

Substituting Eq. (17.80) into Eq. (15.16) yields

$$K_c = \frac{1}{3}\left(\frac{D}{R}\right)_c^2 = \sqrt{\frac{2}{3}\left(\frac{C}{I}\right)_c} \tag{17.81}$$

Capacity comparison.   The ratio of K_W/K_c can be used to compare the capacity of two systems, if

$$\frac{K_W}{K_c} < 1 \tag{17.82}$$

The capacity of WLL is greater than that of cellular. The ratio of K_{W_1}/K_c under a condition of six interferers is

$$\frac{K_{W_1}}{K_c} = \frac{2\left(\frac{C}{I}\right)_{W_1}}{\sqrt{\frac{2}{3}\left(\frac{C}{I}\right)_c}} = 2.45 \; \frac{\left(\frac{C}{I}\right)_{W_1}^2}{\sqrt{\left(\frac{C}{I}\right)_c}} \tag{17.83}$$

The ratio of K_{W_2}/K_c under two different conditions, one interferer for WLL and six interferers for cellular, is

$$\frac{K_{W_2}}{K_c} = \frac{1}{6}\frac{K_{W_1}}{K_c} \tag{17.84}$$

Equations (17.83) and (17.84) are plotted in Fig. 17.16 with a variable a:

$$\left(\frac{C}{I}\right)_c = a\left(\frac{C}{I}\right)_W \tag{17.85}$$

where a is always greater than 1, as shown in Eq. (17.85).

Assume that the required $(C/I)_W$ of WLL is 6 dB or greater; then, several observations can be stated from Fig. 17.16 as follows:

1. The region where WLL capacity is greater than cellular capacity, is below the line of $K_W/K_c = 1$.

2. The WLL system under the condition of six interferers cannot have a capacity greater than the capacity of cellular.

3. The WLL system under the condition of one interferer can most likely have a capacity greater than the cellular capacity.

4. When the value a becomes greater, the ratio of K_W/K_c increases.

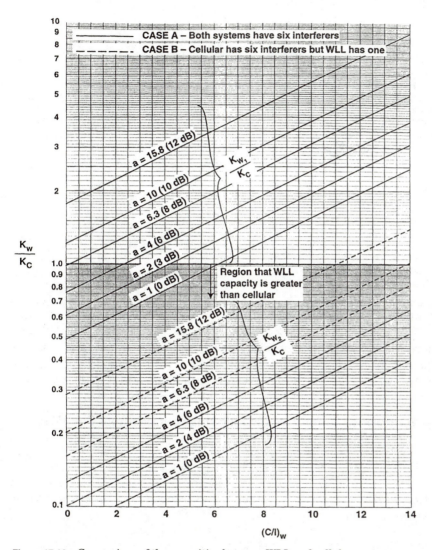

Figure 17.16 Comparison of the capacities between WLL and cellular systems.

Capacity is independent from the propagation path loss if CDMA is used. In a CDMA system, every cell operates the same radio channels; therefore, $D = 2R$ (as stated previously) and K is a constant equal to 1.33. However, M is unknown and is a function of C/I described in Sec. 15.4. The following deviation shows that the propagation path loss does not impact the capacity using the scenario that appears in Fig. 17.17, which is the worst case. The C/I can be expressed as

$$\frac{C}{I} = \frac{C}{I_s + I_a} = \frac{E_b/I_o}{B/R} \tag{17.86}$$

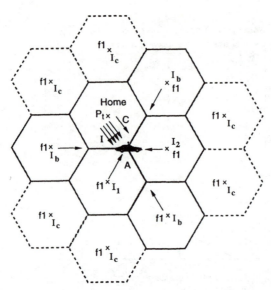

Figure 17.17 A CDMA system and its interference (processing-gain intelligent cell).

where I is the total interference, I_s is the self-interference, and I_a is the adjacent interference, E_b/I_0 is the energy per bit divided by the interference per Hz, B is the allocated spectrum band, and R is the transmit rate per second. In Eq. (17.86) E_b/I_0 and B/R are given.

In WLL systems. Formulate the following equation from Fig. 17.7:

$$\left(\frac{C}{I}\right)_W = \frac{C \cdot R^{-2}}{(M_W - 1) \cdot C \cdot R^{-2} + 2M_W \cdot C \cdot R^{-2} + \Delta_1} \approx \frac{1}{3M_W - 1} \quad (17.87)$$

In Eq. (17.87), we may find the self-interference I_s and the adjacent cell interference I_a as

$$I_s = (M_W - 1) \cdot C \cdot R^{-2}$$

$$I_a = I_1 + I_2 + \Delta_1 = 2M_W \cdot C \cdot R^{-2} + \Delta_1 \approx 2M_W \cdot C \cdot R^{-2}$$

where Δ_1 is the adjacent cell interference coming from three I_b's and six I_c's other than I_1 and I_2. As shown in Fig. 17.17, Δ_1 can be small and neglected. Then, from Eq. (17.87), the total number of traffic channels is found as

$$M_W = \frac{1}{3}\left[\frac{1}{\left(\dfrac{C}{I}\right)_W} + 1\right] \quad (17.88)$$

In cellular systems.   Formulate the following equation from Fig. 17.7:

$$\left(\frac{C}{I}\right)_c = \frac{C \cdot R^{-4}}{(M_c - 1) \cdot C \cdot R^{-4} + 2M_c \cdot C \cdot R^{-4} + \Delta_2} \approx \frac{1}{3M_c - 1} \quad (17.89)$$

In Eq. (17.89), we may find the self-interference I_s and the adjacent cell interference I_a as

$$I_s = (M_c - 1) \cdot C \cdot R^{-4}$$

$$I_a = I_1 + I_2 + \Delta_2 = 2M_c \cdot C \cdot R^{-4} + \Delta_2 \approx 2M_c \cdot C \cdot R^{-4}$$

where Δ_2 is the adjacent cell interference coming from three I_b's and six I_c's other than I_1 and I_2. As shown in Fig. 17.17, Δ_2 can be small and neglected. Then, from Eq. (17.89), the total number of traffic channels is found as

$$M_c = \frac{1}{3}\left[\frac{1}{\left(\dfrac{C}{I}\right)_c} + 1\right] \quad (17.90)$$

Capacity comparison.   Comparing Eq. (17.87) with Eq. (17.89), the two formulas—one based on the path loss of 20 dB/dec and one based on the path loss of 40 dB/dec—are identical. It shows that, in CDMA systems, different propagation-path loss values do not affect the capacity formula. Since from Eq. (17.75), $(C/I)_c > (C/I)_W$ always, then

$$M_W > M_c \quad (17.91)$$

Equation (17.91) is always true in CDMA systems.

Calculated capacity of WLL systems

We may compare the capacity of CDMA with the capacity of FDMA or TDMA. The C/I can be obtained from Eq. (17.86). Let's make some assumptions in the sections that follow.

In CDMA systems
$B_c = 1.23$ MHz
$R = 9.6$ kbps
$s =$ number of sectors $= 3$
$\eta =$ voice activity cycle $= 0.4$
$M'_W = (s/\eta) \cdot M_W =$ the total channels (consider a three-sector cell and 40% voice activity cycle)

Then C/I from Eq. (17.86) becomes

$$\frac{C}{I} = \frac{E_b/I_o}{128} \quad (17.92)$$

Substituting Eq. (17.92) into Eq. (17.88) and then into Eq. (15.15) the total number of channels per cell is

$$m'_W = \frac{M'_W}{K} = \frac{1}{1.33} \cdot \frac{s}{\eta} \left[\frac{1}{3} \left(\frac{128}{E_b/I_0} + 1 \right) \right] = 1.875 \left(\frac{128}{E_b/I_0} + 1 \right) \quad (17.93)$$

In FDMA or TDMA Systems

B_t = total spectrum = 1.23 MHz
B_c = channel bandwidth or equivalent = 25 MHz

then

$$M = \frac{B_t}{B_c} = 49 \text{ channels}$$

Under a condition of six interferers, substituting Eq. (17.77) into Eq. (15.15) yields.

$$m_{W_1} = \frac{M}{K_{W_1}} = \frac{49}{2\left(\frac{C}{I}\right)_{W_1}} = \frac{24.5}{\left(\frac{C}{I}\right)_W} \quad (17.94)$$

Under a condition of one interferer, substituting Eq. (17.79) into Eq. (15.15) yields

$$m_{W_2} = \frac{M}{K_{W_2}} = \frac{49}{\frac{1}{3}\left(\frac{C}{I}\right)_{W_2}} = \frac{147}{\left(\frac{C}{I}\right)_{W_2}} \quad (17.95)$$

Comparing the capacity of CDMA with that of FDMA or TDMA in WLL systems.   The capacity of CDMA (m'_W) shown in Eq. (17.93), the capacity of FDMA or TDMA (m_{W_1}) shown in Eq. (17.94) under a condition of six interferers, and the capacity of FDMA or TDMA (m_{W_2}) shown in Eq. (17.95) under a condition of one interferer are plotted in Fig. 17.18. Assume that the required E_b/I_0 of CDMA is equal to the required $(C/I)_W$ of FDMA or TDMA; then we may conclude that the capacity of CDMA is always larger than that of FDMA or TDMA. Also, the capacity of FDMA or TDMA with one interferer is always greater than that with six interferers.

Comparing the capacity of WLL with the capacity of cellular using CDMA. We have shown that the capacity formulas for both systems are identical in comparing Eq. (17.88) with Eq. (17.90). Therefore, Eq. (17.93) can be used for both systems with a set of given assumptions.

We make a further reasonable assumption that

$$E_b/I_o = 5 \text{ dB for WLL} \quad (17.96)$$

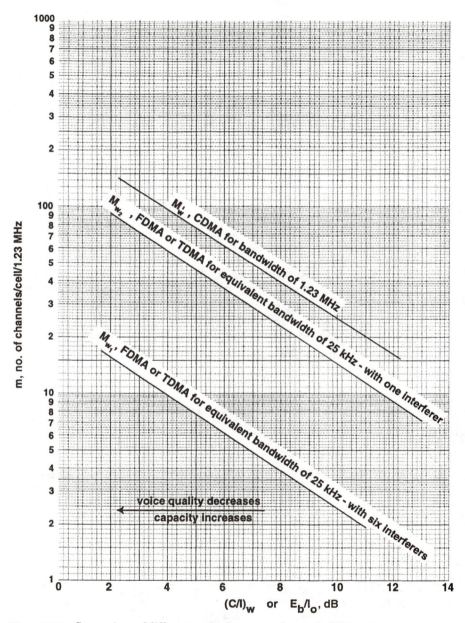

Figure 17.18   Comparison of different multiple access schemes in WLL systems.

then

$$m'_w = 1.875 \left(\frac{128}{E_b/I_0} + 1 \right) = 1.875 \left(\frac{128}{3.16} + 1 \right) = 77.76 \qquad (17.97)$$

and assume that

$$E_b/I_0 = 8 \text{ dB for cellular} \qquad (17.98)$$

then

$$m'_c = 1.875 \left(\frac{128}{6.3} + 1 \right) = 40 \qquad (17.99)$$

Comparing m'_w and m'_c, we may conclude that the capacity of WLL is double the capacity of cellular.

TDD for WLL

The TDD (time division duplexing) system uses one channel to handle both transmitting and receiving in alternating time periods. We may illustrate the channel structure in Fig. 17.19. The guard time τ is the time period that both transmitter and receiver are inactive to avoid the transient. Given the range D to be the maximum distance, then the burst duration Δ has to be

$$\Delta = 2(\delta + T + \tau)$$

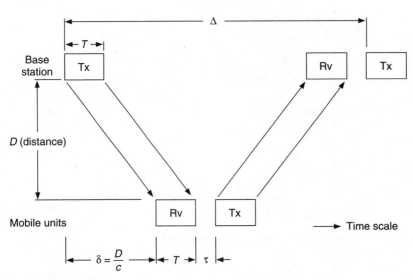

Figure 17.19   A TDD system.

T = burst data in a time slot
τ = guard time
δ = the transit time
Δ = burst duration

where

$$\delta = \frac{D}{c}$$

and T is the time slot for the burst data. The speed of light is c. If the time slot T for the burst data and the transit time δ are not much different (i.e., $T \approx S$) and τ is very small, then the data rate of TDD can be the same as the data rate of FDD.

TDD is not recommended for fast-motion mobile radio. The reason is that the estimation of time of the returning signal from the mobile unit to the base becomes very critical. The power controls for the forward link and the reverse link would be hard to implement in this environment. Thus, reducing both the near-far interference and the multicell interference for the TDD system would be the big challenge in the cellular and PCS systems.

However, in WLL, especially if CDMA is used, the advantages of applying TDD are as follows:

1. No signal fading occurs in WLL. Therefore, no interleaving is needed (see Sec. 16.14), and no decoding delay will be in response.

2. In a fixed-to-fixed system, the power control can be preset. Therefore, no near-far interference is concerned.

3. Directional antennae mounted on both ends can reduce the multisector interference in a cell and the multicell interference in a system.

4. No duplexers are needed. They are a cost item.

5. The transmitters are off when the users are listening. The battery life is longer.

6. Coherent detection can be easily applied on both links, forward and reverse. Thus, reception has better quality.

7. Diversity schemes can be applied at the base station to serve diversity gains at both ends, if needed. This is described in Sec. 16.7.

8. The TDD system is simpler, rugged, and cost-effective.

17.7 Wavelet Representation

A question often asked in signal processing is how best to represent signals. This is related to the applications such as detection, data storage, data compression, signal analysis, signal transmission, and so on.

One wavelet application can be described here. In the 1960s, Bell Labs invented picture phones that used the same twist-wire telephone lines. Normally, to send picture frames, you need a wideband channel for transmitting high-speed data. Here is a clever idea: at first, the picture of a calling party at the sending end is scanned and sent to the called party at the receiving end through a slow data rate. After the entire picture frame is complete at the receiving end, then only the changing (moving) part of a picture frame is sent. It corresponds to a small data stream, and the telephone line can take that kind of load. An entire picture frame in this case is broken down into many wavelets in a wavelet representation. The changing part can be identified by certain wavelets that are the only ones to be sent. This means that all the redundant bits do not need to be sent. To make the wavelet representation more dynamic, the following theory is described.

Wavelet theory deals with the study of the time-scale behavior of the functions, just as the short-time Fourier analysis deals with the time-frequency behavior of functions. The wavelet is a function generated from a *scaling function*. The wavelet function is obtained from an associated transformation that gives a time-scale representation of finite energy functions. The associated transformation is called the *wavelet representation*.

In a wavelet representation, one or both of the time and scale parameters may be discrete or continuous.

If the wavelet is compactly represented, then a finite-power signal can also be localized in time and scale, thus facilitating the study of the time-scale behavior of periodic and nonstationary signals.

The enormous flexibility of choosing the wavelet (the same as choosing the window function in short-time Fourier analysis) permits the use of optimal wavelets for specific applications. Since compactly supported wavelets are determined by a finite sequence of numbers, one can optimize over these to obtain wavelets suited best for specific applications.

The wavelet representations have the desirable property of being localized in both the time (space) and the frequency domains (scale). Before forming a wavelet function, we define a scaling function $\varphi(x)$ such that

$$\varphi(x) = \sum_{k=0}^{2g-1} a(k)\, \varphi\,(2x - k) \tag{17.100}$$

where there are $2g - 1$ different values of $a(k)$ is the finite number $2g - 1$ called the *scaling vector*.

There is an associated set of vectors $b(k)$ that defines the wavelet function associated to the scaling function of the wavelet system. The wavelet function is denoted by

$$\psi(x) = \sum_{k=0}^{2g-1} b(k)\, \varphi\,(2x - k) \tag{17.101}$$

The sequences $a(k)$ and $b(k)$ are called the *scaling vector* and the *wavelet vector*, respectively. From the sampling theorem, it is easy to show that

$$a(k) = \frac{1}{\sqrt{2}} \, \varphi \left(\frac{k}{2}\right) = \sqrt{2} \, \frac{\sin\left(\frac{k}{2}\pi\right)}{k\pi} \qquad (17.102)$$

$$b(k) = \frac{1}{\sqrt{2}} \, \psi \left(\frac{k}{2}\right) = \sqrt{2} \, \varphi \, (k) - \frac{1}{\sqrt{2}} \, \varphi \left(\frac{k}{2}\right) \qquad (17.103)$$

The wavelet system is defined in terms of rescaling by these two functions, $\varphi(x)$ and $\psi(x)$, and these two provide an orthonormal basis. Since space x and time t correspond on a one-to-one basis, $\varphi(x)$ and $\psi(x)$ can be replaced by $\varphi(t)$ and $\psi(t)$. The coefficients $a(k)$ and $b(k)$ form the basis upon which we can compute the approximations that will enable us to scale the wavelet expansion coefficients at any level.

The coefficients of wavelet series representations contain important local information on frequency-time or phase-space. Since there is a localization property in the wavelet, we may send fewer bits in the wavelet representation by removing the redundant bits before sending. The readers who are interested in this topic should obtain Ref. 21.

Problem Exercises

1. Prove that, when the number of branches M becomes very large, the channel capacity of a Rayleigh fading channel converts to that of a Gaussian noise channel.

2. Why can the conventional average formula not apply to the real-time mobile radio environment?

3. Given that the measured C/I is 20 dB, $r/R = 4.6$, $\gamma = 40$ dB/dec, and the minimum tolerated U level is 16 dB, then what is the percentage $P(u \le U)$?

4. Will the propagation-path loss of 20 dB/dec provide higher capacity than that of 40 dB/dec for a system?

5. Prove that the capacity of WLL is independent of path loss.

6. How is the TDD system (time division duplexing) used for WLL?

7. Describe the superiority of applying the successive interference cancellations in a non-power-controlling fading channel.

8. Can a ping-pong handoff arrangement in a handoff region using a selective diversity between two cells increase capacity?

9. How can multiuser detection for CDMA cancel the interference but not single-user detection?

10. How can a multiple-beam directional antenna (m beams) array on a WLL system be able to have more beams than a multiple-beam directional antenna (n beams) array on cellular systems; i.e., $m > n$?

References

1. W. C. Y. Lee, "Estimate of Channel Capacity in Rayleigh Fading Environment," *IEEE Trans. on Vehicular Technology,* vol. VT-39, August 1990, pp. 187–189.
2. C. E. Shannon and W. Weaver, *The Mathematical Theory of Communication,* University of Illinois Press, Urban, 1949.
3. H. Taub and D. L. Schilling, *Principles of Communication Systems,* McGraw-Hill Book Company, 1971, p. 421.
4. A. B. Carlson, *Communication Systems,* McGraw-Hill Book Company, 1975, p. 356.
5. R. S. Kennedy, *Fading Dispersive Communication Channels,* John Wiley & Sons, 1969, p. 109.
6. W. C. Y. Lee, *Mobile Communications Design Fundamentals,* Howard W. Sams and Company, 1986.
7. M. Schwartz, W. Bennett, and S. Stein, *Communication Systems and Techniques,* McGraw-Hill, 1966, p. 407.
8. M. Schwartz, W. Bennett, and S. Stein, ibid., p. 407.
9. W. C. Y. Lee, *Mobile Communications Engineering,* McGraw-Hill Book Company, 1982, p. 310.
10. M. Schwartz, W. Bennett, and S. Stein, *Communication Systems and Techniques,* McGraw-Hill, 1966, p. 444.
11. Allen E. Vincent, Running Average Computer, U.S. patent 4,054,786, October 18, 1997.
12. W. C. Y. Lee, A Real-Time Running Average Device, U.S. patent 08/075,860, June 1, 1994.
13. A. Viterbi et al., "Soft Handoff Extends CDMA Call Coverage and Increases Reserve Link Capacity," *IEEE Journal on Selected Areas in Communications,* October 1994.
14. W. C. Y. Lee, "Link Capacity Between GSM and CDMA," *IEEE International Conference on Communication Proceedings,* Seattle, June 18–22, 1995, pp. 1335–1339.
15. W. C. Y. Lee, *Mobile Cellular Telecommunication Systems,* McGraw-Hill, 1989, p. 185.
16. A. Papolulis, *Random Variables and Stochastic Processes,* McGraw-Hill, 1965, p. 182.
17. "Table of Integrals of the Error Functions," *J. Res. of NBS-B,* vol. 73B, 1969.
18. S. Verdu, *Multiuser Detection Advances in Statistical Signal Processing,* vol. 2, JAI Press, 1993, pp. 369–409.
19. J. M. Holtzman, "DS/CDMA Successive Interference Cancellation," *Proc. IEEE Int'l. Symp. on Spread Spectrum Techniques and Appl.,* Oulu, Finland, July 1994, pp. 69–78.
20. S. Moshavi, "Multiuser Detection for DS-CDMA Communications," *IEEE Communications Magazine,* October 1996, pp. 124–136.
21. Charlie K. Chui, ed., *Wavelets: A Tutorial in Theory and Applications,* San Diego, Academic Press, 1992.

Military Mobile Communications

18.1 Strategies Used in a Jamming Environment

The characteristics of the mobile-radio environment have been described in the preceding chapters, including the phenomena of multipath fading, frequency-selective fading, and the various other transmission losses. In this complex medium, the degradation in performance from unintentional interferences can be calculated. In this chapter, the effect on performance of *intentional* sources of interference, called "jammers," is discussed in terms of its impact on military mobile communications. The various types of jammers can be classified as follows:

1. **Air jammer**—The jamming signal can arrive via true line-of-sight propagation and does not experience any fades, while the reception of the desired signal experiences fading. In this case, the necessary jammer transmitting power is relatively weak.

2. **Ground jammer**—Both the jamming signal and the desired signal experience fading at the receiving end. In this case, the jammer power can be very high.

Jammer strategy

Jammers can be classified as (1) CW jammers operating continuously on one frequency; (2) pulsed jammers consisting of on-off jamming utilizing peak power; (3) wideband pseudo-noise jammers operating at a certain bandwidth; (4) partial-band jammers, which utilize the power of the jamming transmitter to efficiently jam a part of the desired signal band while leaving part of the information intact; and (5) follow-up jammers, which can be used to detect a transmitted signal waveform and then transmit an appropriate jamming signal that will jam the receiving end.

Electronic counter-countermeasure (ECCM) technology

There are two main categories of electronic counter-countermeasure (ECCM) techniques that can be applied. One requires a large degree of spectral spreading, larger than the normal required bandwidth; the second requires prior knowledge of the nature of the jammer to allow the proper non-spectral spreading techniques to be used. Spectral-spreading techniques include spread-spectrum schemes (described in Sec. 8.4) and coding schemes (described in Sec. 13.2). Non-spectral-spreading techniques include any of the following:

1. Site selection and power-control strategy

2. Directivity and antenna-pattern shaping

3. Diversity schemes

4. Adaptive-antenna nulling and adaptive signal canceling

5. Programmable notch filters

In this chapter, four major techniques are discussed: spread-spectrum, coding schemes, diversity schemes, and adaptive-antenna.

Example 18.1 This example is concerned with finding the relationships among processing gain, jammer-to-signal ratio, and signal-to-noise ratio. Although the Shannon channel-capacity formula is applied to a Gaussian noise channel and applied to a Rayleigh-fading channel only when the snr becomes large (as mentioned in Sec. 11.1 of Chap. 11), it is used here to demonstrate the relationships among three parameters: processing gain (PG), jammer-to-signal ratio (J/S), and signal-to-noise ratio (S/N). Suppose that N is replaced by $N + J$ in Eq. (11.1); then

$$C = R_m = B \log_2 \left(1 + \frac{S}{N + J} \right) \tag{E18.1}$$

where R_m is the maximum signaling rate. Equation (E18.1) can be converted into the following expression:

$$10 \log_2 \left(\frac{S}{N} \right) = -10 \log \left[(2^{R_m/B} - 1)^{-1} - \frac{J}{S} \right] \tag{E18.2}$$

where

$$\frac{B}{R_m} = \text{processing gain}$$

$$\frac{J}{S} = \text{jammer-to-signal ratio}$$

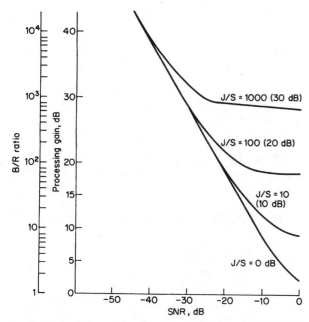

Figure E18.1   Relationships among PG, *J/S,* and *S/N.*

The processing gain can be increased by applying spread-spectrum techniques, and the jammer-to-noise ratio can be suppressed by applying adaptive-antenna nulling. Equation (E18.2) is plotted in Fig. E18.1, which provides a quantitative comparison. To maintain an *S/N* level of 70 dB, PG = 12 dB is required for a *J/S* = 10 dB, and PG = 17.7 dB is required for a *J/S* = 100 dB.

18.2   Spread-Spectrum Scheme—
Frequency Hopping (FH)

In Chap. 8, the use of an FH-DPSK spread-spectrum scheme for providing mobile-radiotelephone services to a large number of customers was described. In this section, the use of the spectrum scheme for antijamming purposes is described. In a mobile-radio environment where jammers are present, a noncoherent frequency-hopping system may be the primary choice for the same reasons given in Chap. 8. The bit-error rate of a noncoherent FSK signal can be obtained from Eq. (8.80). Assume that all the bits received during the time interval of one chip (one hopped frequency) are totally correlated. That means that if one bit is in the fade, most likely all bits are in the fade. On this assumption, the two kinds of bit-error rates passing through the Rayleigh-fading environment are:

Under a nonjamming case:

$$P_{e1} = \int_0^\infty \frac{1}{2} \exp\left(-\frac{\gamma}{2}\right) \frac{1}{\gamma_0} \exp\left(-\frac{\gamma}{\gamma_0}\right) d\gamma = \frac{1}{2 + \gamma_0} \tag{18.1}$$

Under a jamming case, assuming a channel-band jammer:

$$P_{e2} = \int_0^\infty \frac{1}{2} \exp\left(-\frac{\gamma}{2}\right) \frac{1}{\gamma_j} \exp\left(-\frac{\gamma}{\gamma_j}\right) d\gamma = \frac{1}{2 + \gamma_j} \tag{18.2}$$

where γ_0 is the snr and γ_j is the signal/(jammer + noise) ratio, $\gamma_j = S/(J + N)$, and the jammer jams one entire channel (chip) of an FSK signal.

When K jammers are present, the chance that one chip among m chips is jammed is

$$q = \frac{K}{m} \tag{18.3}$$

where m is the maximum number of available chips. While a frequency-hopping channel is taking place, the bit-error rate caused by K jammers, where $K < m$, in a fading environment is

$$P_e = P_{e1}(1 - q) + qP_{e2} \tag{18.4}$$

Substituting Eqs. (18.1) and (18.2) into Eq. (18.4) gives the following:

$$P_e = \frac{m(2 + \gamma_j) + K(\gamma_0 - \gamma_j)}{m(2 + \gamma_0)(2 + \gamma_j)} \tag{18.5}$$

The function for Eq. (18.5) is plotted in Fig. 18.1, for $m = 2000$.

The case of $\gamma_j = 0$ dB occurs when the jammer's power equals the desired signal power at the input to the receiver. The case of $\gamma_j = \gamma_0$ is the nonjamming case. In this analysis, a slow hopping rate of 1000 hops per second or less is assumed. For a case involving infinite jamming power, Eq. (18.5) becomes:

$$P_e = \frac{1}{\gamma_0} + \frac{q}{2} \tag{18.6}$$

providing $\gamma_0 \gg 3$ dB, which is the usual case.

The processing gain (PG) for a frequency-hopping (FH) scheme in a jamming environment is defined:

$$\text{PG}_{\text{FH}} = \text{number of frequency chips } m$$

$$\geq \frac{\text{BW}_{\text{RF}}}{R} \tag{18.7}$$

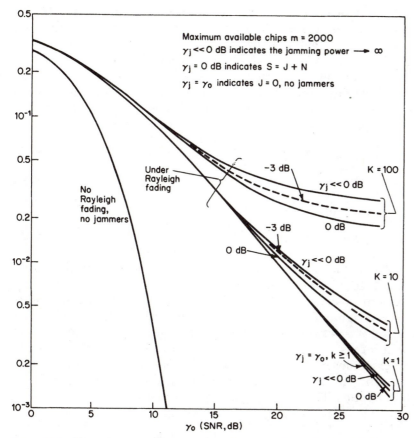

Figure 18.1 Bit-error rate of a noncoherent FSK signal in a Rayleigh-fading and jamming environment ($m = 2000$).

where BW_{RF} is the bandwidth of RF and R is the information rate. For $m = 2000$, PG_{FH} can be expressed:

$$PG_{FH} = 2000 \approx 33 \text{ dB}$$

From Eq. (18.7), the BW_{RF} determines the PG_{FH}. To increase the bandwidth BW_{RF}, it is necessary to operate at a higher frequency range. However, this may cause problems due to other limitations.

Improvement of bit-error rate

To improve the bit-error rate, a scheme for transmitting an identical bit stream from one chip to another over a number of different chips is demonstrated in this section. A similar analysis was described in Chap. 13. When N is an odd number, a majority vote among N repeated bits is necessary to determine a 0 or a 1 value for each information bit. In determining the bit-error rate after a majority vote from N repeated bits, the chance of having more than $(N + 1)/2$ repeated bits in error must be considered. When the frequency separation among chips used for frequency hopping is large and exceeds the coherent bandwidth, then the N repeated bits obtained from N chips for each information bit are mutually uncorrelated. The improved bit-error rate can be obtained by applying the Bernoulli trials method, as follows:

$$P_3 = 1 - (1 - P_e)^2(1 + 2P_e) \qquad N = 3 \tag{18.8}$$

$$P_9 = 1 - (1 - P_e)^2[1 + C_1^5 P_e + C_2^6 P_e^2 + C_3^7 P_e^3 + C_4^8 P_e^4] \qquad N = 9 \tag{18.9}$$

$$P_N = 1 - (1 - P_e)^{(N+1)/2}[1 + C_1^{(N+1)/2} P_e + C_2^{(N+3)/2} P_e^2 + C_3^{(N+5)/2} P_e^3$$
$$+ \cdots + C_{N-1/2}^{N-1} P_e^{N-1/2}] \qquad \text{all } N \tag{18.10}$$

For $m = 2000$ and $\gamma_j = -3$ dB, the improved bit-error rate P_N for values of N from 3 to 9 and of K from 50 to 200 is plotted in Fig. 18.2. An improvement is shown in Fig. 18.2 as N increases. From Eq. (18.10), the required signal levels for a bit-error rate of 10^{-3} are plotted versus the number of repetitions in Fig. 18.3. For an snr of 15 dB, the required repetition rate is 4 for $K = 50$, and 5 for $K = 100$. However, the throughput, which is described in Sec. 13.2 of Chap. 13, decreases accordingly. In order to increase the throughput, effective coding schemes should be applied. These coding schemes are described in the following section.

18.3   Coding Schemes

The two major coding schemes used are the block code and the convolutional code, which were described in Sec. 13.2 of Chap. 13. In the block code, $B(n, k)$, out of n bits transmitted, only k bits contain information. The throughput R_{th} is a fraction of the information-bit rate R, which is expressed:

$$R_{\text{th}} = R \, \frac{k}{n} \tag{18.11}$$

In the convolutional code, $C(n, k)$, out of k message bits fed into the encoder, a block of n coded bits is generated as the encoder output. In this case, the throughput R_{th} becomes:

$$R_{\text{th}} = R \, \frac{k}{n} \tag{18.12}$$

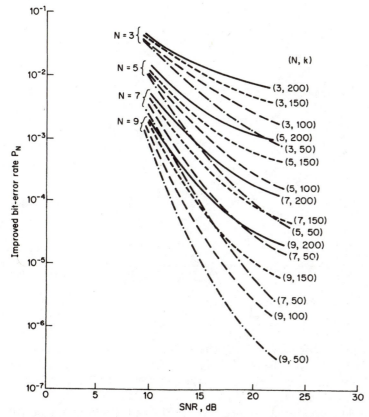

Figure 18.2 Improved bit-error rates for different numbers of repetition ($m = 2000$, $\gamma_j = -3$ dB).

The throughput R_{th}, in both Eqs. (18.11) and (18.12), is linearly proportional to the information-bit rate R. Another important parameter of any coding scheme is the coding gain, which can be expressed:

$$(\text{Coding gain})_{\text{dB}} = \left(\frac{E_b}{\eta_0}\right)_{\text{no coding}} - \left(\frac{E_b}{\eta_0}\right)_{\text{with coding}}$$

where E_b is the average energy per bit and η_0 is the noise-power density. The commonly used notation E_b/η_0 is related to the cnr as follows:

$$\text{cnr} = \frac{P_c}{N_0} = \frac{E_b R}{\eta_0 B} = \frac{E_b}{\eta_0} \frac{R}{B}$$

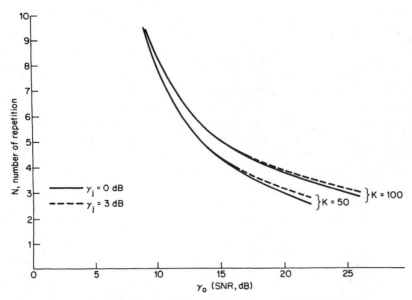

Figure 18.3 Required signal levels for a bit-error rate of 10^{-3} ($m = 2000$).

where P_c is the carrier power, N_0 is the noise power, R is the signaling rate, and B is the bandwidth of the signal. Comparing E_b/η_0 with coding and without coding at a given bit-error rate (ber) reveals a coding gain. Usually a coding gain of 4 to 5 dB is desired.

In the mobile-radio fading environment, the burst errors are generated in a signal-bit stream mixed with Gaussian random errors. Therefore, it is necessary to use burst-error correcting codes, e.g., (1) BCH code, (2) Reed-Solomon code (nonbinary or binary), (3) convolutional code, or (4) interleaved block code. The word-error rates of the first three can be calculated by following the procedure described in Sec. 13.4 for both the slow-fading case and the fast-fading case. The last one depends on the degree of interleaving associated with different types of codes.

Before calculating the word-error rates and the coding gain of a coded signal, the slow- and fast-fading cases have to be defined. Any two adjacent transmitted bits separated in space by approximately 0.75λ or more are uncorrelated in the mobile-radio environment, as mentioned in Sec. 9.4. Take a limiting situation, which is when the maximum vehicle speed is 112.5 km/h (70 mi/h), and find the number of bits S occupying a time frame equal to 0.75λ over which the first bit and last bit of S bits are uncorrelated. The slow- and fast-fading cases, when coding schemes are used, then, can be defined by the code block size C as compared with the number of correlated bits S:

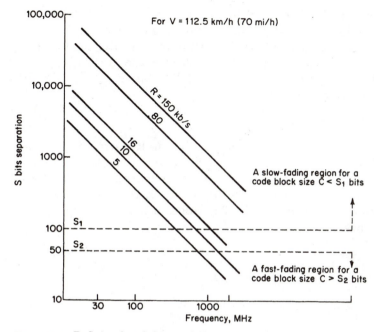

Figure 18.4 Defining fast-fading and slow-fading cases.

$C \ll S$ a slow-fading case: first code block and second code block totally correlated

$C \geq S$ a fast-fading case: first code block and second code block uncorrelated

Figure 18.4 shows the slow- and fast-fading regions for different signaling rates for the equation

$$S = 0.75 \, \frac{R}{V} \, \frac{c}{f} \qquad \text{b}$$

where R is the signaling rate, c is the speed of light, f is the frequency, and $V = 112.5$ km/h.

18.4 Combined Coding, Diversity, and Frequency Hopping in a Jamming Environment

Combining a normally used code such as a BCH code with a diversity scheme in reducing the word-error rate has been demonstrated in Sec. 13.5. In order to show a strong effect on coding in a jamming environment, an outstanding burst-error correcting code called Reed-Solomon code is chosen. An (N, K) Reed-Solomon (R-S) code is used in a noncoherent FSK modulation with frequency hopping (FH). In addition, the advantage of using a space diversity with maximal-ratio combining is included in the analysis. Section 13.2 indicates that the (7, 3) Reed-Solomon code can correct up to two errors. Two kinds of R-S codes are studied separately, as follows.

Binary R-S code

In a binary R-S code, every consecutive 3 bits in the stream identify a code block B_i (where $i = 1, \ldots, 8$), which is one of eight (2^3) possible binary numbers. Then every three consecutive information code blocks are picked and inserted with four parity-check code blocks to form a Reed-Solomon code word (7, 3). The information of a code word is shown as follows:

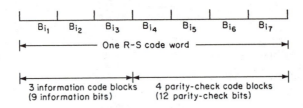

One Reed-Solomon (R-S) code word (W_S) consists of 21 bits, or 7 code blocks ($W_S = 7B_i$), among which 9 bits are information bits and 12 are parity check bits. There are two cases with which we can be concerned, as follows:

1.   Slow-fading case   It has been assumed that in a slow-fading case a word-error rate of all bits in a code word can be treated as equal to a single bit-error rate. Then it is necessary first to find the error probability of having 1 error in every 3-bit interval in a nonfading case and thus to determine the error probability p_{CB} for a code block (3 bits):

$$p_{CB} = 1 - (1-p)^3 \tag{18.13}$$

where p is the bit-error rate of a noncoherent FSK signal obtained from Eq. (8.80). The error rate of a code word which consists of N code blocks and corrects errors up to t code blocks can be shown as:

$$P_{CW}(t \text{ error corrections}) = 1 - \sum_{l=0}^{t} p(N, l)$$

$$= 1 - \sum_{l=0}^{t} \binom{N}{l} (1 - p_{CB})^{N-l} p_{CB}^{l} \tag{18.14}$$

where $p(N, l)$ is the error rate at which exactly l code blocks will be in error in a code word of N code blocks, as shown in Eq. (13.43). Then the average error rate of a code word of N code blocks can be obtained from Eq. (13.44):

$$\langle p(N, l) \rangle = \int_0^\infty p(N, l) p_M(\gamma) \, d\gamma \tag{18.15}$$

where $p_M(\gamma)$ is the M-branch maximal-ratio combiner expressed in Eq. (13.36). The $p(N, l)$ for l errors can be derived as follows:

$$p(N, l) = \binom{N}{l} (1 - p_{CB})^{N-l} p_{CB}^l = \binom{N}{l} (1 - p)^{3(N-l)} [1 - (1 - p)^3]^l$$

$$= \binom{N}{l} (1 - p)^{3(N-l)} \left\{ \sum_{i=0}^l \binom{l}{i} [-(1 - p)^3]^i \right\}$$

$$= \binom{N}{l} \sum_{i=0}^l \binom{l}{i} (-1)^i (1 - p)^{3(N-l+i)}$$

$$= \binom{N}{l} \sum_{i=0}^l \binom{l}{i} (-1)^i \left[\sum_{u=0}^{3(N-l+i)} \binom{3(N-l+i)}{u} (-p)^u \right]$$

$$= \binom{N}{l} \sum_{i=0}^l \sum_{u=0}^{3(N-l+i)} \binom{l}{i} (-1)^{i+u} \binom{3(N-l+i)}{u} p^u \tag{18.16}$$

In Eq. (18.16), only p is a function of γ; hence in inserting Eq. (18.16) into Eq. (18.15) results in the following integration:

$$\int_0^\infty p^u p_M(\gamma) \, d\gamma = \int_0^\infty \left(\frac{1}{2} e^{-\gamma/2} \right)^u \frac{1}{(M-1)!} \frac{\gamma^{M-1}}{\Gamma^M} \exp\left(-\frac{\gamma}{\Gamma} \right) d\gamma$$

$$= \left(\frac{1}{2} \right)^u \frac{1}{\left(1 + \dfrac{u}{2} \Gamma \right)^M}$$

$$= \langle p^u \rangle \tag{18.17}$$

The average error rate of a code word can be derived by integrating Eq. (18.14) over a fading environment with a system of M diversity branches:

$$\langle P_{\text{ew}} \rangle = 1 - \sum_{l=0}^t \langle p(N, l) \rangle$$

$$= 1 - \sum_{l=0}^t \sum_{i=0}^l \sum_{u=0}^{3(N-l+i)} (-1)^{i+u} \binom{N}{l} \binom{l}{i} \binom{3(N-l+i)}{u} \langle p^u \rangle \tag{18.18}$$

where $N = 7$, $t = 2$, M is the number of diversity branches, and Γ is the cnr. Since the 21-bit code word contains 9 information bits, the throughput is $(9/21)R$. Usually, the signaling rate R equals the transmission bandwidth B; hence $\Gamma = E_b/\eta_0$. Equation (18.18) is plotted in Fig. 18.5 versus E_b/η_0.

Now if a word consists of 9 information bits and is not coded, its word-error rate in a slow-fading case should be:

$$P_{\text{ew}} = \int_0^{\infty} [1 - (1 - p)^9] p_M(\gamma)\, d\gamma \tag{18.19}$$

where p is the bit-error rate of a noncoherent FSK signal as shown in Eq. (8.80) and $p_M(\gamma)$ is the pdf of a maximal-ratio combiner as shown in

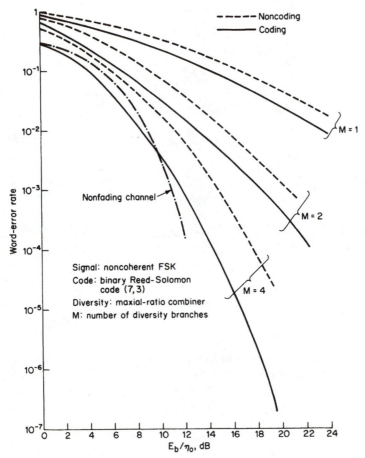

Figure 18.5 Word-error rate improved after coding and diversity in a slow-fading case.

Eq. (13.7). Equation (18.19) is also plotted in Fig. 18.5 for comparison. Comparing a coded signal with a noncoded signal in a nondiversity case ($M = 1$) shows a coding gain of 3 dB for the former at a word-error rate of 10^{-1}. It is also shown that the use of diversity seems more effective than the use of coding.

Now the same technique described in Sec. 18.2 will be applied. Then the bit-error rate of a coded bit stream caused by K jammers, where $K < m$, in a jamming environment is:

$$P_e = P_{e_1}(1 - q) + qP_{e_2} \tag{18.20}$$

where γ_0 replaces γ in Eq. (8.80) for P_{e_1} and γ_j replaces γ in Eq. (8.80) for P_{e_2}. q is defined in Eq. (18.3). P_e of Eq. (18.20) replaces p in Eq. (18.13) in the jamming environment. Then averaging Eq. (18.14) over a fading environment [or Eq. (18.18)] is plotted in Fig. 18.6 as a function of $\gamma_0 = E_b/\eta_0$ with given γ_j. The parameters used in Fig. 18.6 are $K = 50$, $m = 2000$, and $\gamma_j = 0$ dB and -5 dB. In this slow-fading case, the word-error rate changes slightly as the signal/(jammer + noise) ratio γ_j decreases from 0 dB to -5 dB. The limiting conditions due to jamming are shown as E_b/η_0 increases. It seems that the use of diversity is more effective than the use of coding.

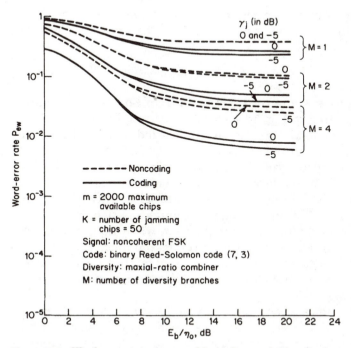

Figure 18.6 Word-error rate improved by coding and diversity in a Rayleigh and jamming environment (slow-fading case).

2. Fast-fading case   The word-error rate for a fast-fading case can be expressed as in Eq. (13.12):

$$P_{ew}(t \text{ error correcting}) = 1 - \sum_{i=0}^{t} p(N, l) \qquad (18.21)$$

where $t = 2$ in this case and

$$p(N, l) = \binom{N}{l} (1 - p_{CB})^{N-l} p_{CB}^{l} \qquad (18.22)$$

where p_{CB} is the error rate of a code block:

$$p_{CB} = 1 - (1 - \langle p \rangle)^3 \qquad (18.23)$$

and $\langle p \rangle$ is the average bit-error rate of a noncoherent FSK signal in a Rayleigh-fading environment. $\langle p \rangle$ can be found from Eq. (18.17) with $u = 1$:

$$\langle p \rangle = \int_{0}^{\infty} p \, p_M(\gamma) \, d\gamma = \frac{1}{2} \frac{1}{\left(1 + \dfrac{\Gamma}{2}\right)^M} \qquad (18.24)$$

Equation (18.21) is plotted in Fig. 18.7 for $N = 7$ and $t = 2$.

For a noncoding word of 9 information bits, the word-error rate can be expressed:

$$P_{ew} = 1 - (1 - \langle p \rangle)^9 \qquad (18.25)$$

in a fast-fading case where $\langle p \rangle$ is obtained from Eq. (18.24). Equation (18.25) is also plotted in Fig. 18.17 for comparison. For a given E_b/η_0, the word-error rate is always less in the fast-fading case than in the slow-fading case (compare Fig. 18.7 with Fig. 18.5). Comparing a coded signal with a noncoded signal in a nondiversity case ($M = 1$) shows a coding gain of 7 dB for the former at a word-error rate of 10^{-1}.

Also in a jamming environment, the two bit-error rates P_{e_1} and P_{e_2}, for a nonjamming condition and for a jamming condition, respectively, are expressed in Eqs. (18.1) and (18.2). Then when Eq. (18.20) is applied, the improved bit-error rate P_e in the jamming environment can be obtained in a fast-fading case. Replacing $\langle p \rangle$ by P_e in Eq. (18.23) and calculating Eq. (18.21) with the new P_e gives an improved word-error rate, plotted in Fig. 18.8.

In the fast-fading case, both diversity and coding effectively reduce the word-error rate, in contrast to the slow-fading case shown in Fig. 18.6. Although the limiting conditions due to the jamming environment still exist, the limiting word-error rate is lower than that shown in Fig. 18.6.

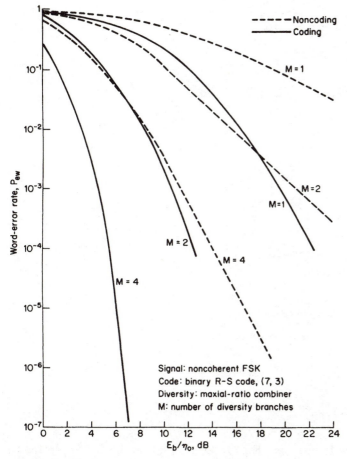

Figure 18.7 Word-error rate improved after coding and diversity in a fast-fading case.

Nonbinary R-S code

The nonbinary R-S code (N, k) consists of N orthogonal components, which can be frequencies or time slots. In the mobile-radio environment, the use of time slots can cause problems when the mobile unit is standing still, as described in Sec. 9.9. Hence, the use of frequencies is desired. For a $(7, 2)$ nonbinary R-S code, seven orthogonal frequencies are needed for distinguishing seven states. The orthogonal frequencies are frequencies each of which is not in the coherent bandwidth of any of the others. Under this condition, the improved word-error rate is:

$$P_{\text{ew}} = 1 - \sum_{l=0}^{2} \binom{7}{l} (1 - \langle p \rangle)^{7-l} (\langle p \rangle)^{l} \qquad (18.26)$$

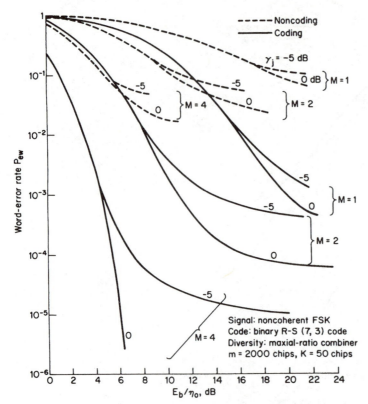

Figure 18.8 Word-error rate improved by coding and diversity in a Rayleigh and jamming environment (a fast-fading case).

where $\langle p \rangle$ can be found from Eq. (18.24). Equation (18.26) is plotted in Fig. 18.9. The word-error rate of a nonbinary R-S coded signal shows a noticeable reduction as either E_b/η_0 or the number of diversity branches increases. The coding scheme does show its effectiveness. However, the use of nonbinary R-S code (7, 2) needs seven orthogonal frequencies. It is, of course, not an efficient way to utilize the frequency spectrum, since seven frequencies are assigned to one particular channel. In a jamming environment, $\langle p \rangle$ shown in Eq. (18.26) should be replaced by a new P_e from Eq. (18.20), where P_{e_1} and P_{e_2} in Eq. (18.20) can be found from Eqs. (18.1) and (18.2). The word-error rate in a jamming environment (for $k = 50$ and $M = 2000$) is also plotted in Fig. 18.9. For $E_b/\eta_0 = 20$ dB and $S/(N + J) = -5$ dB, the word-error rate is 10^{-4} in a nondiversity case. It is thus shown that the nonbinary R-S code is a more effective code than the others.

Example 18.2 If a repetition rate of 3 times is used for each 5-bit code block that is sent, then a two-thirds majority-vote algorithm at the receiving end is applied

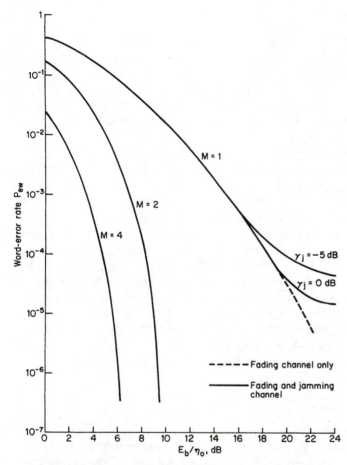

Figure 18.9 Word-error rate improved by a nonbinary code and diversities in a Rayleigh-fading and jamming environment.

to decrease the code-block error rate shown in Eq. (18.13). What will be the new P_{CB}?

solution   The new P_{CB} can be expressed as follows:

$$\text{New } P_{CB} = 1 - (1 - p')^5 \tag{E18.2.1}$$

where p' is the improved bit-error rate after a two-thirds majority-vote algorithm is applied.

$$p' = 1 - \left[\sum_{i=0}^{1} \binom{3}{i} (1 - p)^{3-i} p^i \right] \tag{E18.2.2}$$

where p is found in Eq. (8.80). Equation (E18.2.1) will replace Eq. (18.13) for this new, improved repetition scheme. However, the throughput is dropped by a factor of 3.

18.5   Adaptive Antenna Nulling

The objective of using adaptive antenna nulling is to minimize the power at the direction of jamming and to maximize the power at the direction of desired-signal arrival. The theory of adaptive antenna nulling is very complicated, and it is really a different field which is beyond the scope of this book [1–10]. Here we introduce the possibility of applying the technique to mobile-radio communication when a jamming signal is present.

Basically, there are two algorithms: Widrow [1] and Applebaum [2]. The former one is called the LMS (least mean square error) algorithm and applies an adaptive filter to a phased array; it requires knowledge of the signal waveform. The latter algorithm is based on maximizing the signal-to-noise ratio and requires knowledge of the direction of signal arrival. In military mobile communications, the direction of desired-signal arrival is usually unknown, as base-station antennas have to move from time to time in the battlefield. For this reason, the LMS algorithm may be more suitable to the mobile unit. In general, the mobile units are always considered as moving radios scattered around the base station. Hence the base-station antenna should be omnidirectional. At the mobile unit, since the mobile antenna height is low—usually 2 to 3 m above the ground—the signal will be arriving from more than one direction, and therefore an omnidirectional antenna should be used at the mobile unit also.

The possibility of using the adaptive antenna array

1.   At the mobile unit   As shown in Fig. 18.10, the jamming signal coming from an air jammer usually is in a true line-of-sight condition. Hence adaptive antenna nulling can be effectively applied. However, when a ground jammer is present, jamming may come from more than one direction, as the desired signal does. Therefore adaptive antenna nulling cannot be effectively used against a ground jammer.

2.   At the base station   At the base station, either an air jammer or a ground jammer comes under a true line-of-sight condition, as shown in Fig. 18.11. Hence adaptive antenna nulling can be used against both air and ground jammers.

Consideration

1. The loop convergence of an adaptive algorithm has to be fast enough to overcome the situation in which every time the desired signal received by the mobile gets out of the fades the "self-noise" generated by the searching algorithm degrades the signal performance rather than improving it.

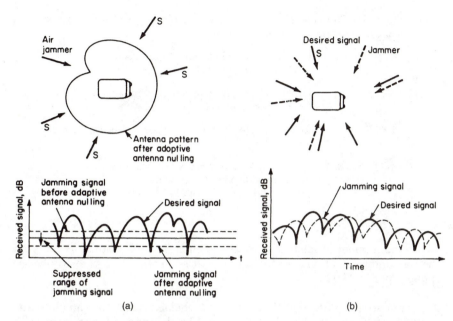

Figure 18.10   The possibility of using adaptive antenna arrays at mobile units: (*a*) air jammer; (*b*) ground jammer.

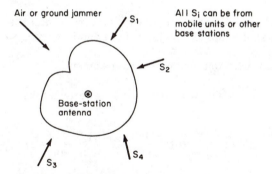

Figure 18.11   The possibility of using adaptive antenna arrays at base stations.

2. The number and direction of jammers also have to be considered in determining whether adaptive antenna nulling should be applied or not.

3. The broadband antenna array design and the integration of adaptive antenna array and modem are also major concerns.

Example 18.3   Find the number of nulls and the gain which can be generated from an N-element antenna array.

solution   From the antenna-array pattern, it is easily shown that $N-1$ nulls can be generated from an N-element array. Hence, $m+1$ elements are needed in order to null m jammers. The gains of different antenna arrays are expressed in terms of the number of elements, as follows:

$G = N$ for a half-wavelength-spaced broadside uniformly illuminated array [11]

$G = N$ for a quarter-wavelength-spaced broadside Chebyshev array [11]

$G = 1.82N$ for a quarter-wavelength-spaced end-fire Chebyshev array [11]

$G = \pi G_x G_y \cos \theta_0$ for an $N \times N$ planar array [12] where the angle θ_0 is determined by the case in which all the phasors align and G_x and G_y are the gains of the two linear arrays

Problem Exercises

1. If the number of maximum available chips is 1000 instead of 2000 but the remaining conditions are the same as those in Sec. 18.2, what will be the changes in Fig. 18.1?

2. If a four-level PSK (QPSK) modulation scheme is used, the signaling rate R equals twice the bandwidth B. What will be the cnr needed to maintain a 10-dB level of E_b/η_0, and what is the difference in cnr compared with a binary PSK modulation scheme?

3. Take a (15, 7) Reed-Solomon code and try to find the improved bit-error rate in a case of combined coding diversity and frequency hopping.

4. Extend Prob. 3 to the jamming environment.

5. Prove that the following expression is derived from a (31, 15) binary R-S coded signal:

$$\langle P_{ew} \rangle = 1 - \left\{ \sum_{u=0}^{155} (-1)^u \frac{115!}{u!} \left\{ A_0 \frac{155 \cdots 116}{(155 - u)!} + \left\{ A_1 \frac{150 \cdots 116}{(150 - u)!} \right. \right. \right.$$

$$+ \left\{ A_2 \frac{145 \cdots 116}{(145 - u)!} + \left\{ A_3 \frac{140 \cdots 116}{(140 - u)!} + \left\{ A_4 \frac{135 \cdots 116}{(135 - u)!} \right. \right. \right.$$

$$+ \left\{ A_5 \frac{130 \cdots 116}{(130 - u)!} + \left\{ A_6 \frac{125 \cdots 116}{(125 - u)!} + \left\{ A_7 \frac{120 \cdots 116}{(120 - u)!} \right. \right. \right.$$

$$+ \left. \frac{A_8}{(115 - u)!} \right\}\right\}\right\}\right\}\right\}\right\}\right\}\right\} \left(\frac{1}{2}\right)^u \frac{1}{(1 + u\Gamma/2)^M} \right\}$$

provided that $1/(c - u)! = 0$ if $c - u < 0$. All A's are constants.

6. Take a Golay code (23, 12) (see Table 13.1) and try to find the improved bit-error rate for the same conditions as given in Prob. 3. Discuss the two results, one from the R-S code and the other from the Golay code.

7. Discuss the difficulty that might occur with an adaptive antenna array when a frequency-hopping scheme is applied.

8. How is an adaptive antenna array applied to a broadband system? (See Ref. 8.)

9. What are the pros and cons of using a repetition-code scheme (redundancy) as opposed to other coding schemes?

10. Design a system at 1 GHz which can null out four jammers with an adaptive antenna gain of 10 dB, a processing gain of 20 dB, and a coding gain of 4 dB.

References

1. B. Widrow et al., "Adaptive Antenna Systems," *Proc. IEEE,* vol. 55, December 1967, pp. 2143–2159.
2. S. P. Applebaum, "Adaptive Arrays," *IEEE Trans. Anten. Prop.,* vol. 24, September 1976, pp. 585–598.
3. L. E. Brennar and I. S. Reed, "Theory of Adaptive Radar," *IEEE Trans. Aerospace and Electronic Sys.,* vol. 9, no. 2, March 1973, pp. 237–252.
4. C. L. Zahm, "Application of Adaptive Arrays to Suppress Strong Jammers in the Presence of Weak Signals," *IEEE Trans. Aerospace and Electronic Sys.,* vol. 9, no. 2, March 1973, pp. 260–271.
5. R. T. Compton, Jr., "The Power-Inversion Adaptive Array: Concept and Performance," *IEEE Trans. Aerospace and Electronic Sys.,* vol. 15, no. 6, November 1979, pp. 803–814.
6. R. L. Riegler and R. T. Compton, Jr., "An Adaptive Array for Interference Rejection," *Proc. IEEE,* vol. 61, no. 6, June 1973, pp. 748–758.
7. L. L. Horowitz, H. Blatt, W. G. Brodsky, and K. D. Senne, "Controlling Adaptive Antenna Arrays with the Sample Matrix Inversion Algorithm," *IEEE Trans. Aerospace and Electronic Sys.,* vol. 15, no. 6, November 1979, pp. 840–847.
8. W. F. Gabriel, "Adaptive Arrays—An Introduction," *Proc. IEEE,* vol. 64, no. 2, February 1976, pp. 239–272.
9. I. S. Reed, J. D. Mallett, and L. E. Brennan, "Rapid Convergence Rate in Adaptive Arrays," *IEEE Trans. Aerospace and Electronic Sys.,* vol. 10, no. 6, November 1974, pp. 853–863.
10. R. T. Compton, Jr., "An Adaptive Array in a Spread-Spectrum Communication System," *Proc. IEEE,* vol. 66, no. 3, March 1978, pp. 289–298.
11. E. A. Wolff, *Antenna Analysis,* Wiley, New York, 1967, pp. 250, 256–257.
12. R. S. Elliott, "Beamwidth and Directivity of Large Scanning Arrays," *Microwave Journal,* vol. 7, January 1964, pp. 74–82.

Index

About the Author

William C. Y. Lee is Vice President and Chief Scientist for Applied Research and Science at AirTouch Communications in Walnut Creek, California. Dr. Lee is one of the foremost educators in the field of mobile communication and is the bestselling author of *Mobile Cellular Telecommunications*, 2d ed., also published by McGraw-Hill.